Modern Probability Theory and Its Applications

Modern Probability Theory

A WILEY PUBLICATION IN MATHEMATICAL STATISTICS

and Its Applications

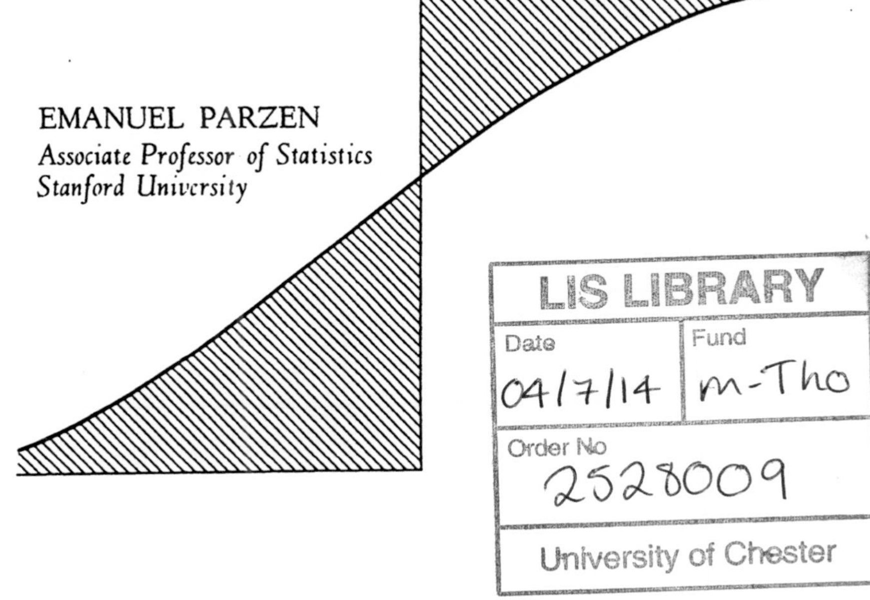

EMANUEL PARZEN
Associate Professor of Statistics
Stanford University

Wiley Classics Library Edition Published 1992

A Wiley-Interscience Publication
JOHN WILEY AND SONS, INC.
New York Chichester Brisbane Toronto Singapore

A NOTE TO THE READER:
This book has been electronically reproduced from digital information stored at John Wiley & Sons, Inc. We are pleased that the use of this new technology will enable us to keep works of enduring scholarly value in print as long as there is a reasonable demand for them. The content of this book is identical to previous printings.

In recognition of the importance of preserving what has been written, it is a policy of John Wiley & Sons, Inc., to have books of enduring value published in the United States printed on acid-free paper, and we exert our best efforts to that end.

Wiley Classics Library Edition Published 1992

ISBN 0-471-57278-0 (pbk)

To the memory

of my mother and father

The conception of chance enters into
the very first steps of scientific activity,
in virtue of the fact that no observation
is absolutely correct. I think chance is a more
fundamental conception than causality; for whether in a
concrete case a cause-effect relation
holds or not can only be judged by applying the laws
of chance to the observations.

MAX BORN
Natural Philosophy
of Cause and Chance

Preface

The notion of probability, and consequently the mathematical theory of probability, has in recent years become of interest to many scientists and engineers. There has been an increasing awareness that not "Will it work?" but "What is the probability that it will work?" is the proper question to ask about an apparatus. Similarly, in investigating the position in space of certain objects, "What is the probability that the object is in a given region?" is a more appropriate question than "Is the object in the given region?" As a result, the feeling is becoming widespread that a basic course in probability theory should be a part of the undergraduate training of all scientists, engineers, mathematicians, statisticians, and mathematics teachers.

A basic course in probability theory should serve two ends.

On the one hand, probability theory is a subject with great charm and intrinsic interest of its own, and an appreciation of the fact should be communicated to the student. Brief explanations of some of the ideas of probability theory are to be found scattered in many books written about many diverse subjects. The theory of probability thus presented sometimes appears confusing because it seems to be a collection of tricks, without an underlying unity. On the contrary, its concepts possess meanings of their own that do not depend on particular applications. Because of this fact, they provide *formal* analogies between real phenomena, which are themselves totally different but which in certain theoretical aspects can be treated similarly. For example, the factors affecting the length of the life of a man of a certain age and the factors affecting the time a light bulb will burn may be quite different, yet similar mathematical ideas may be used to describe both quantities.

On the other hand, a course in probability theory should serve as a background to many courses (such as statistics, statistical physics, industrial engineering, communication engineering, genetics, statistical psychology, and econometrics) in which probabilistic ideas and techniques are employed. Consequently, in the basic course in probability theory one should attempt to provide the student with a confident technique for solving probability problems. To solve these problems, there is no need to employ intuitive witchcraft. In this book it is shown how one may formulate probability problems in a mathematical manner so that they may be systematically attacked by routine methods. The basic step in this procedure is to express any event whose probability of occurrence is being sought as a set of sample descriptions, defined on the sample description space of the random phenomenon under consideration. In a similar spirit, the notion of random variable, together with the sometimes bewildering array of notions that must be introduced simultaneously, is presented in easy stages by first discussing the notion of numerical valued random phenomena.

This book is written as a textbook for a course in probability that can be adapted to the needs of students with diverse interests and backgrounds. In particular, it has been my aim to present the major ideas of modern probability theory without assuming that the reader knows the advanced mathematics necessary for a rigorous discussion.

The first six chapters constitute a one-quarter course in elementary probability theory at the sophomore or junior level. For the study of these chapters, the student need have had only one year of college calculus. Students with more mathematical background would also cover Chapters 7 and 8. The material in the first eight chapters (omitting the last section in each) can be conveniently covered in thirty-nine class hours by students with a good working knowledge of calculus. Many of the sections of the book can be read independently of one another without loss of continuity.

Chapters 9 and 10 are much less elementary in character than the first eight chapters. They constitute an introduction to the limit theorems of probability theory and to the role of characteristic functions in probability theory. These chapters provide careful and rigorous derivations of the law of large numbers and the central limit theorem and contain many new proofs.

In studying probability theory, the reader is exploring a way of thinking that is undoubtedly novel to him. Consequently, it is important that he have available a large number of interesting problems that at once illustrate and test his grasp of the theory. More than 160 examples, 120 theoretical exercises, and 480 exercises are contained in the text. The exercises are divided into two categories and are collected at the end of each section rather than at the end of the book or at the end of each chapter. The theoretical exercises extend the theory; they are stated in the form of assertions that the student is asked to prove. The nontheoretical exercises are numerical problems concerning concrete random phenomena and illustrate the variety of situations to which probability theory may be applied. The answers to odd-numbered exercises are given at the end of the book; the answers to even-numbered exercises are available in a separate booklet.

In choosing the notation I have adopted in this book, it has been my aim to achieve a symbolism that is self-explanatory and that can be read as if it were English. Thus the symbol $F_X(x)$ is defined as "the distribution function of the random variable X evaluated at the real number x." The terminology adopted agrees, I believe, with that used by most recent writers on probability theory.

The author of a textbook is indebted to almost everyone who has touched the field. I especially desire to express my intellectual indebtedness to the authors whose works are cited in the brief literature survey given in section 8 of Chapter 1.

To my colleagues at Stanford, and especially to Professors A. Bowker and S. Karlin, I owe a great personal debt for the constant encouragement they have given me and for the stimulating atmosphere they have provided. All have contributed much to my understanding of probability theory and statistics.

I am very grateful for the interest and encouragement accorded me by various friends and colleagues. I particularly desire to thank Marvin Zelen for his valuable suggestions.

To my students at Stanford who have contributed to this book by their comments, I offer my thanks. Particularly valuable assistance has been rendered by E. Dalton and D. Ylvisaker and also by M. Boswell and P. Williams.

To the cheerful, hard-working staff of the Applied Mathematics and Statistics Laboratory at Stanford, I wish to express my gratitude for their encouragement. Great thanks are due also to Mrs. Mary Alice McComb and Mrs. Isolde Field for their excellent typing and to Mrs. Betty Jo Prine for her excellent drawings.

EMANUEL PARZEN

Stanford, California
January 1960

Contents

List of Important Tables

CHAPTER 1

Probability Theory as the Study of Mathematical Models of Random Phenomena

The purpose of this chapter is to discuss the nature of probability theory. In section 1 we point out the existence of a certain body of phenomena that may be called *random*. In section 2 we state the view, which is adopted in this book, that probability theory is the study of mathematical models of random phenomena. The language and notions that are used to formulate mathematical models are discussed in sections 3 to 7.

1. PROBABILITY THEORY AS THE STUDY OF RANDOM PHENOMENA

One of the most striking features of the present day is the steadily increasing use of the ideas of probability theory in a wide variety of scientific fields, involving matters as remote and different as the prediction by geneticists of the relative frequency with which various characteristics occur in groups of individuals, the calculation by telephone engineers of the density of telephone traffic, the maintenance by industrial engineers of manufactured products at a certain standard of quality, the transmission (by engineers concerned with the design of communications and automatic-control systems) of signals in the presence of noise, and the study by physicists of thermal noise in electric circuits and the Brownian motion of particles immersed in a liquid or gas. What is it that is studied in probability theory that enables it to have such diverse applications? In order to answer this question, we must first define the property that is possessed in common by phenomena such as the number of individuals possessing a certain genetical characteristic, the number of telephone calls made in a given city between given hours of the day, the standard of quality of the items manufactured by a certain process, the number of automobile accidents each day on a given highway, and so on. Each of these phenomena may often be considered a *random phenomenon* in the sense of the following definition.

A *random* (*or chance*) *phenomenon* is an empirical phenomenon characterized by the property that its observation under a given set of circumstances does not always lead to the same observed outcome (so that there is no deterministic regularity) but rather to different outcomes in such a way that there is *statistical regularity*. By this is meant that numbers exist between 0 and 1 that represent the relative frequency with which the different possible outcomes may be observed in a series of observations of independent occurrences of the phenomenon.

Closely related to the notion of a random phenomenon are the notions of a random event and of the probability of a random event. A *random event* is one whose relative frequency of occurrence, in a very long sequence of observations of randomly selected situations in which the event may occur, approaches a stable limit value as the number of observations is increased to infinity; the limit value of the relative frequency is called *the probability of the random event*.

In order to bring out in more detail what is meant by a random phenomenon, let us consider a typical random event; namely, an automobile accident. It is evident that just where, when, and how a particular accident takes place depends on an enormous number of factors, a slight change in any one of which could greatly alter the character of the accident or even avoid it altogether. For example, in a collision of two cars, if one of the motorists had started out ten seconds earlier or ten seconds later, if he had stopped to buy cigarettes, slowed down to avoid a cat that happened to cross the road, or altered his course for any one of an unlimited number of similar reasons, this particular accident would never have happened; whereas even a slightly different turn of the steering wheel might have prevented the accident altogether or changed its character completely, either for the better or for the worse. For any motorist starting out on a given highway it cannot be predicted that he will or will not be involved in an automobile accident. Nevertheless, if we observe all (or merely some very large number of) the motorists starting out on this highway on a given day, we may determine the proportion that will have automobile accidents. If this proportion remains the same from day to day, then we may adopt the belief that what happens to a motorist driving on this highway is a random phenomenon and that the event of his having an automobile accident is a random event.

Another typical random phenomenon arises when we consider the experiment of drawing a ball from an urn. In particular, let us examine an urn (or a bowl) containing six balls, of which four are white, and two are red. Except for color, the balls are identical in every detail. Let a ball be drawn and its color noted. We might be tempted to ask "what will be the color of a ball drawn from the urn?" However, it is clear that there is no answer to this question. If one actually performs the experiment of drawing a ball from an urn, such as the one described, the color of the ball one draws will sometimes be white and sometimes red. Thus the outcome of the experiment of drawing a ball is unpredictable.

Yet there are things that are predictable about this experiment. In Table 1A the results of 600 independent trials are given (that is, we have

TABLE 1A

The number of white balls drawn in 600 trials of the experiment of drawing a ball from an urn containing four white balls and two red balls.

In Trials Numbered	Number of White Balls Drawn	In Trials Numbered	Proportion of White Balls Drawn
1–100	69	1–100	0.690
101–200	70	1–200	0.695
201–300	59	1–300	0.660
301–400	63	1–400	0.653
401–500	76	1–500	0.674
501–600	64	1–600	0.668

taken an urn containing four white balls and two red balls, mixed the balls well, drawn a ball, and noted its color, after which the ball drawn was returned to the urn; these operations were repeated 600 times). It is seen that in each block of 100 trials (as well as in the entire set of 600 trials) the proportion of experiments in which a white ball is drawn is approximately equal to $\frac{2}{3}$. Consequently, one may be tempted to assert that the proportion $\frac{2}{3}$ has some real significance for this experiment and that in a reasonably long series of trials of the experiment $\frac{2}{3}$ of the balls drawn will be colored white. If one succumbs to this temptation, then one has asserted that the outcome of the experiment (of drawing a ball from an urn containing six balls, of which four are white and two are red) is a random phenomenon.

More generally, if one believes that the experiment of drawing a ball from an urn will, in a long series of trials, yield a white ball in some definite proportion (which one may not know) of the trials of the experiment, then one has asserted (i) that the drawing of a ball from such an urn is a random phenomenon and (ii) that the drawing of a white ball is a random event.

Let us give an illustration of the way in which one may use the knowledge (or belief) that a phenomenon is random. Consider a group of 300 persons who are candidates for admission to a certain school at which there are facilities for only 200 students. In the interest of fairness it is decided to use a random mechanism to choose the students from among the candidates. In one possible random method the 300 candidates are assembled in a room. Each candidate draws a ball from an urn containing six balls, of which four are white; those who draw white balls are admitted as students. Given an individual student, it cannot be foretold whether or not he will be admitted by this method of selection. Yet, if we believe that the outcome of the experiment of drawing a ball possesses the property of statistical regularity, then on the basis of the experiment represented by Table 1A, which indicates that the probability of drawing a white ball is $\frac{2}{3}$, we believe that the number of candidates who will draw white balls, and consequently be admitted as students, will be approximately equal to 200 (note that 200 represents the product of (i) the number of trials of the experiment and (ii) the probability of the event that the experiment will yield a white ball). By a more careful analysis, one can show that the probability is quite high that the number of candidates who will draw white balls is between 186 and 214.

One of the aims of this book is to show how by means of probability theory the same mathematical procedure can be used to solve quite different problems. To illustrate this point, we consider a variation of the foregoing problem which is of great practical interest. Many colleges find that only a certain proportion of the students they admit as students actually enroll. Consequently a college must decide how many students to admit in order to be sure that enough students will enroll. Suppose that a college finds that only two-thirds of the students it admits enroll; one may then say that the probability is $\frac{2}{3}$ that a student will enroll. If the college desires to ensure that about 200 students will enroll, it should admit 300 students.

EXERCISES

1.1. Give an example of a random phenomenon that would be studied by (i) a physicist, (ii) a geneticist, (iii) a traffic engineer, (iv) a quality-control engineer, (v) a communications engineer, (vi) an economist, (vii) a psychologist, (viii) a sociologist, (ix) an epidemiologist, (x) a medical researcher, (xi) an educator, (xii) an executive of a television broadcasting company.

1.2. *The Statistical Abstract of the United States* (1957 edition, p. 57) reports that among the several million babies born in the United States the number of boys born per 1000 girls was as follows for the years listed:

Year	Male Births per 1000 Female Births
1935	1053
1940	1054
1945	1055
1950	1054
1951	1052
1952	1051
1953	1053
1954	1051
1955	1051

Would you say the event that a newborn baby is a boy is a random event? If so, what is the probability of this random event? Explain your reasoning.

1.3. **A discussion question.** Describe how you would explain to a layman the meaning of the following statement: An insurance company is not gambling with its clients because it knows with sufficient accuracy what will happen to every thousand or ten thousand or a million people even when the company cannot tell what will happen to any individual among them.

2. PROBABILITY THEORY AS THE STUDY OF MATHEMATICAL MODELS OF RANDOM PHENOMENA

One view that one may take about the nature of probability theory is that it is part of the study of nature in the same way that physics, chemistry, and biology are. Physics, chemistry, and biology may each be defined as the study of certain observable phenomena, which we may call, respectively, the physical, chemical, and biological phenomena. Similarly, one might be tempted to define probability theory as the study of certain observable phenomena, namely the random phenomena. However, a random phenomenon is generally also a phenomenon of some other type; it is a random physical phenomenon, or a random chemical phenomenon, and so on. Consequently, it would seem overly ambitious for researchers in probability theory to take as their province of research all random phenomena. In this book we take the view that probability theory is not directly concerned with the study of random phenomena but rather with the study of the methods of thinking that can be used in the study of random phenomena. More precisely, we make the following definition.

The theory of probability is concerned with the study of those methods of analysis that are common to the study of random phenomena in all the fields in which they arise. Probability theory is thus the study of the study of random phenomena, in the sense that it is concerned with those properties of random phenomena that depend essentially on the notion of randomness and not on any other aspects of the phenomenon considered. More fundamentally, the notions of randomness, of a random phenomenon, of statistical regularity, and of "probability" cannot be said to be obvious or intuitive. Consequently, one of the main aims of a study of the theory of probability is to clarify the meaning of these notions and to provide us with an understanding of them, in much the same way that the study of arithmetic enables us to count concrete objects and the study of electromagnetic wave theory enables us to transmit messages by wireless.

We regard probability theory as a part of mathematics. As is the case with all parts of mathematics, probability theory is constructed by means of the axiomatic method. One begins with certain undefined concepts. One then makes certain statements about the properties possessed by, and the relations between, these concepts. These statements are called the axioms of the theory. Then, by means of logical deduction, without any appeal to experience, various propositions (called *theorems*) are obtained from the axioms. Although the propositions do not refer directly to the real world, but are merely logical consequences of the axioms, they do represent conclusions about real phenomena, namely those real phenomena one is willing to assume possess the properties postulated in the axioms.

We are thus led to the notion of a *mathematical model of a real phenomenon.* A mathematical theory constructed by the axiomatic method is said to be a *model* of a real phenomenon, if one gives a rule for translating propositions of the mathematical theory into propositions about the real phenomenon. This definition is vague, for it does not state the character of the rules of translation one must employ. However, the foregoing definition is not meant to be a precise one but only to give the reader an intuitive understanding of the notion of a mathematical model. Generally speaking, to use a mathematical theory as a model for a real phenomenon, one needs only to give a rule for identifying the abstract objects about which the axioms of the mathematical theory speak with aspects of the real phenomenon. It is then expected that the theorems of the theory will depict the phenomenon to the same extent that the axioms do, for the theorems are merely logical consequences of the axioms.

As an example of the problem of building models for real phenomena, let us consider the problem of constructing a mathematical theory (or explanation) of the experience recorded in Table 1A, which led us to believe that a long series of trials (of the experiment of drawing a ball from an urn containing six balls, of which four are white and two red) would yield a white ball in approximately $\frac{2}{3}$ of the trials. In the remainder of this chapter we shall construct a mathematical theory of this phenomenon, which we believe to be a satisfactory model of certain features of it. It may clarify the ideas involved, however, if we consider here an explanation of this phenomenon, which we shall then criticize.

We imagine that we are permitted to label the six balls in the urn with numbers 1 to 6, labeling the four white balls with numbers 1 to 4. When a ball is drawn from the urn, there are six possible outcomes that can be recorded; namely, that ball number 1 was drawn, that ball number 2 was drawn, etc. Now four of these outcomes correspond to the outcome that a white ball is drawn. Therefore the ratio of the number of outcomes of the experiment favorable to a white ball being drawn to the number of all possible outcomes is equal to $\frac{2}{3}$. Consequently, in order to "explain" why the observed relative frequency of the drawing of a white ball from the urn is equal to $\frac{2}{3}$, one need only adopt this assumption (stated rather informally): the probability of an event (by which is meant the relative frequency with which an event, such as the drawing of a white ball, is observed to occur in a long series of trials of some experiment) is equal to the ratio of the number of outcomes of the experiment in which the event may be observed to the number of all possible outcomes of the experiment.

There are several grounds on which one may criticize the foregoing explanation. First, one may state that it is *not* mathematical, since it does not possess a structure of axioms and theorems. This defect may perhaps be remedied by using the tools that we develop in the remainder of this chapter; consequently, we shall not press this criticism. However, there is a second defect in the explanation that cannot be repaired. *The assumption stated, that the probability of an event is equal to a certain ratio, does not lead to an explanation of the observed phenomenon because by counting in different ways one can obtain different values for the ratio.* We have already obtained a value of $\frac{2}{3}$ for the ratio; we next obtain a value of $\frac{1}{2}$. If one argues that there are merely two outcomes (either a white ball or a nonwhite ball is drawn), then exactly one of these outcomes is favorable to a white ball being drawn. Therefore, the ratio of the number of outcomes favorable to a white ball being drawn to the number of possible outcomes is $\frac{1}{2}$.

We now proceed to develop the mathematical tools we require to construct satisfactory models of random phenomena.

3. THE SAMPLE DESCRIPTION SPACE OF A RANDOM PHENOMENON

It has been stated that probability theory is the study of mathematical models of random phenomena; in other words, probability theory is concerned with the statements one can make about a random phenomenon about which one has postulated certain properties. The question immediately arises: how does one formulate postulates concerning a random phenomenon? This is done by introducing the sample description space of the random phenomenon.

The *sample description* space of a random phenomenon, usually denoted by the letter S, is the space of descriptions of all possible outcomes of the phenomenon.

To be more specific, suppose that one is performing an experiment or observing a phenomenon. For example, one may be tossing a coin, or two coins, or 100 coins; or one may be measuring the height of people, or both their height and weight, or their height, weight, waist size, and chest size; or one may be measuring and recording the voltage across a circuit at one point of time, or at two points of time, or for a whole interval of time (by photographing the effect of the voltage upon an oscilloscope). In all these cases one can imagine a space that consists of all possible descriptions of the outcome of the experiment or observation. We call it the *sample description space,* since the outcome of an experiment or observation is usually called a *sample.* Thus a sample is something that has been observed; a sample description is the name of something that is observable.

A remark may be in order on the use of the word "space." The reader should not confuse the notion of space as used in this book with the use of the word space to denote certain parts of the world we live in, such as the region between planets. A notion of great importance in modern mathematics, since it is the starting point of all mathematical theories, is the notion of a *set*. A set is a collection of objects (either concrete objects, such as books, cities, and people, or abstract objects, such as numbers, letters, and words). A set that is in some sense complete, so that only those objects in the set are to be considered, is called a *space*. In developing any mathematical theory, one has first to define the class of things with which the theory will deal; such a class of things, which represents the universe of discourse, is called a space. A space has neither dimension nor volume; rather, a space is a complete collection of objects.

Techniques for the construction of the sample description space of a random phenomenon are systematically discussed in Chapter 2. For the present, to give the reader some idea of what sample description spaces look like, we consider a few simple examples.

Suppose one is drawing a ball from an urn containing six balls, of which four are white and two are red. The possible outcomes of the draw may be denoted by W and R, and we write W or R accordingly, as the ball drawn is white or red. In symbols, we write $S = \{W, R\}$. On the other hand, we may regard the balls as numbered 1 to 6; then we write $S = \{1, 2, 3, 4, 5, 6\}$ to indicate that the possible outcome of a draw is a number, 1 to 6.

Next, let us suppose that one draws two balls from an urn containing six balls, numbered 1 to 6. We shall need a notation for recording the outcome of the two draws. Suppose that the first ball drawn bears number 5 and the second ball drawn bears number 3; we write that the outcome of the two draws is (5, 3). The object (5, 3) is called a 2-tuple. We assume that the balls are drawn one at a time and that the order in which the balls are drawn matters. Then (3, 5) represents the outcome that first ball 3 and then ball 5 were drawn. Further, (3, 5) and (5, 3) represent different possible outcomes. In terms of this notation, the sample description space of the experiment of drawing two balls from an urn containing balls numbered 1 to 6 (assuming that the balls are drawn in order and that the ball drawn on the first draw is not returned to the urn before the second draw is made) has 30 members:

$$
\begin{aligned}
(3.1) \qquad S = \{&(1, 2), && (1, 3), && (1, 4), && (1, 5), && (1, 6) \\
&(2, 1), && (2, 3), && (2, 4), && (2, 5), && (2, 6) \\
&(3, 1), && (3, 2), && (3, 4), && (3, 5), && (3, 6) \\
&(4, 1), && (4, 2), && (4, 3), && (4, 5), && (4, 6) \\
&(5, 1), && (5, 2), && (5, 3), && (5, 4), && (5, 6) \\
&(6, 1), && (6, 2), && (6, 3), && (6, 4), && (6, 5)\}
\end{aligned}
$$

We next consider an example that involves the measurement of numerical quantities. Suppose one is observing the ages (in years) of couples who apply for marriage licenses in a certain city. We adopt the following notation to record the outcome of the observation. Suppose one has observed a man and a woman (applying for a marriage license) whose ages are 24 and 22, respectively; we record this observation by writing the 2-tuple (24, 22). Similarly, (18, 80) represents the age of a couple in which the man's age is 18 and the woman's age is 80. Now let us suppose that the age (in years) at which a man or a woman may get married is any number, 1 to 200. It is clear that the number of possible outcomes of the observation of the ages of a marrying couple is too many to be conveniently listed; indeed, there are $(200)(200) = 40{,}000$ possible outcomes! One thus sees that it is often more convenient to *describe*, rather than to list, the sample descriptions that constitute the sample description space S. To describe S in the example at hand, we write

$$S = \{\text{2-tuples } (x, y)\text{: } x \text{ is any integer, 1 to 200, } y \text{ is any integer, 1 to 200}\}. \tag{3.2}$$

We have the following notation for forming sets. We draw two braces to indicate that a set is being defined. Next, we can define the set either by *listing* its members (for example, $S = \{W, R\}$ and $S = \{1, 2, 3, 4, 5, 6\}$) or by *describing* its members, as in (3.2). When the latter method is used, a *colon* will always appear between the braces. On the left side of the colon, one will describe objects of some general kind; on the right side of the colon, one will specify a property that these objects must have in order to belong to the set being defined.

All of the sample description spaces so far considered have been of finite size.* However, there is no logical necessity for a sample description space to be finite. Indeed, there are many important problems that require sample description spaces of infinite size. We briefly mention two examples. Suppose that we are observing a Geiger counter set up to record cosmic-ray

* Given any set A of objects of any kind, the *size* of A is defined as the number of members of A. Sets are said to be of finite size if their size is one of the *finite* numbers $\{1, 2, 3, \ldots\}$. Examples of sets of finite size are the following: the set of all the continents in the world, which has size 7; the set of all the planets in the universe, which has size 9; the set $\{1, 2, 3, 5, 7, 11, 13\}$ of all prime numbers from 1 to 15, which has size 7; the set $\{(1, 4), (2, 3), (3, 2), (4, 1)\}$ of 2-tuples of whole numbers between 1 and 6 whose sum is 5, which has size 4.

However, there are also sets of *infinite* (that is, nonfinite) size. Examples are the set of all prime numbers $\{1, 2, 3, 5, 7, 11, 13, 17, \ldots\}$ and the set of all points on the real line between the numbers 0 and 1, called the interval between 0 and 1. If a set A has as many members as there are integers $1, 2, 3, 4, \ldots$ (by which is meant that a one-to-one correspondence may be set up between the members of A and the members of the set $\{1, 2, 3, \ldots\}$ of all integers) then A is said to be *countably infinite*. The set of even integers $\{2, 4, 6, 8 \ldots\}$ contains a countable infinity of members, as does the set of odd integers $\{1, 3, 5, \ldots\}$ and the set of primes. A set that is neither finite nor countably infinite is said to be *noncountably infinite*. An interval on the real line, say the interval between 0 and 1, contains a noncountable infinity of members.

counts. The number of counts recorded may be any integer. Consequently, as the sample description space S we would adopt the set $\{1, 2, 3, \ldots\}$ of all positive integers. Next, suppose we were measuring the time (in microseconds) between two neighboring peaks on an electrocardiogram or some other wiggly record; then we might take the set $S = \{\text{real numbers } x\colon\ 0 < x < \infty\}$ of all positive real numbers as our sample description space.

It should be pointed out that the sample description space of a random phenomenon is capable of being defined in more than one way. Observers with different conceptions of what could possibly be observed will arrive at different sample description spaces. For example, suppose one is tossing a single coin. The sample description space might consist of two members, which we denote by H (for heads) and T (for tails). In symbols, $S = \{H, T\}$. However, the sample description space might consist of three members, if we desired to include the possibility that the coin might stand on its edge or rim. Then $S = \{H, T, R\}$, in which the description R represents the possibility of the coin standing on its rim. There is yet a fourth possibility; the coin might be lost by being tossed out of sight or by rolling away when it lands. The sample description space would then be $S = \{H, T, R, L\}$, in which the description L denotes the possibility of loss.

Insofar as probability theory is the study of mathematical models of random phenomena, it cannot give rules for the construction of sample description spaces. Rather the sample description space of a random phenomenon is one of the undefined concepts with which the mathematical theory begins. The considerations by which one chooses the correct sample description space to describe a random phenomenon are a part of the art of applying the mathematical theory of probability to the study of the real world.

4. EVENTS

The notion of the sample description space of a random phenomenon derives its importance from the fact that it provides a means to define the notion of an event.

Let us first consider what is intuitively meant by an event. Let us consider an urn containing six balls, of which two are white. Let the balls be numbered 1 to 6, the white balls being numbered 1 to 2. Let two balls be drawn from the urn, one after the other; the first ball drawn is *not* returned to the urn before the second ball is drawn. The sample description space S of this experiment is given by (3.1). Now some possible events are (i) the event that the ball drawn on the first draw is white, (ii) the event that the ball drawn on the second draw is white, (iii) the event that both balls drawn are white, (iv) the event that the sum of the numbers on the balls drawn is 7, (v) the event that the sum of the numbers on the balls drawn is less than or equal to 4.

The mathematical formulation that we shall give of the notion of an event depends on the following fact. *For each of the events just described there is a set of descriptions such that the event occurs if and only if the observed outcome of the two draws has a description that lies in the set.* For example, the event that the ball drawn on the first draw is white can be reformulated as the event that the description of the outcome of the experiment belongs to the set $\{(1, 2), (1, 3), (1, 4), (1, 5), (1, 6), (2,1), (2, 3), (2, 4), (2, 5), (2, 6)\}$. Similarly, events (ii) to (v) described above may be reformulated as the events that the description of the outcome of the experiment belongs to the set (ii) $\{(2, 1), (3, 1), (4, 1), (5, 1), (6, 1), (1, 2), (3, 2), (4, 2), (5, 2), (6, 2)\}$, (iii) $\{(1, 2), (2, 1)\}$, (iv) $\{(1, 6), (2, 5), (3, 4), (4, 3), (5, 2), (6, 1)\}$, (v) $\{(1, 2), (2, 1), (1, 3), (3, 1)\}$.

Consequently, we *define an event as a set of descriptions. To say that an event E has occurred is to say that the outcome of the random situation under consideration has a description that is a member of E.* Note that there are two notions being defined here, the notion of "an event" and the notion of "the occurrence of an event." The first notion represents a basic tool for the construction of mathematical models of random phenomena; the second notion is the basis of all translations of statements made in the mathematical model into statements about the real phenomenon.

An alternate way in which the definition of an event may be phrased is in terms of the notion of *subset.* Consider two sets, E and F, of objects of any kind. We say that E is a subset of F, denoted $E \subset F$, if every member of the set E is also a member of the set F. We now *define an event as any subset of the sample description space S.* In particular, the sample description space S is a subset of itself and is thus an event. We call the sample description space S the *certain* event, since by the method of construction of S it will always occur.

It is to be emphasized that in studying a random phenomenon our interest is in the events that can occur (or more precisely, in the probabilities with which they can occur). The sample description space is of interest not for the sake of its members, which are the descriptions, but for the sake of its subsets, which are the events!

We next consider the relations that can exist among events and the operations that can be performed on events. One can perform on events algebraic operations similar to those of addition and multiplication that one can perform on ordinary numbers. The concepts to be presented in the remainder of this section may be called the *algebra of events.* If one speaks of sets rather than of events, then the concepts of this section constitute what is called *set theory*.

Given any event E, it is as natural to ask for the probability that E will *not* occur as it is to ask for the probability that E will occur. Thus, to any event E, there is an event denoted by E^c and called the complement of E (or E complement). The event E^c is the event that E does not occur and consists of all descriptions in S which are not in E.

Let us next consider two events, E and F. We may ask whether E and F both occurred or whether at least one of them (and possibly both) occurred. Thus we are led to define the events EF and $E \cup F$, called, respectively, the *intersection* and *union* of the events E and F.

The intersection EF is defined as consisting of the descriptions that belong to both E and F; consequently, the event EF is said to occur if and only if both E and F occur, which is to say that the observed outcome has a description that is a member of both E and F.

The union $E \cup F$ is defined as consisting of the descriptions that belong to at least one of the events E and F; consequently, the event $E \cup F$ is said to occur if and only if either E or F occurs, which is to say that the observed outcome has a description that is a member of either E or F (or of both).

It should be noted that many writers denote the intersection of two events by $E \cap F$ rather than by EF.

We may give a symbolic representation of these operations in a diagram called a Venn diagram (Figs. 4A to 4C). Let the sample description space S be represented by the interior of a rectangle in the plane; let the event E be represented by the interior of a circle that lies within the rectangle; and let the event F be represented by the interior of a square also lying within the rectangle (but not necessarily overlapping the circle, although in Fig. 4B it is drawn that way). Then E^c, the complement of E, is represented in Fig. 4A by the points within the rectangle outside the circle; EF, the intersection of E and F, is represented in Fig. 4B by the points within the circle and the square; $E \cup F$, the union of E and F, is represented in Fig. 4C by the points lying within the circle or the square.

As another illustration of the notions of the complement, union, and intersection of events, let us consider the experiment of drawing a ball from an urn containing twelve balls, numbered 1 to 12. Then $S = \{1, 2, \ldots, 12\}$. Consider events $E = \{1, 2, 3, 4, 5, 6\}$ and $F = \{4, 5, 6, 7, 8, 9\}$. Then $E^c = \{7, 8, 9, 10, 11, 12\}$, $EF = \{4, 5, 6\}$ and $E \cup F = \{1, 2, 3, 4, 5, 6, 7, 8, 9\}$.

One of the main problems of the calculus of events is to establish the equality of two events defined in two different ways. Two events E and F are said to be *equal*, written $E = F$, if every description in one event belongs to the other. The definition of equality of two events may also be phrased in terms of the notion of *subevent*. An event E is said to be a *subevent* of an event F, written $E \subset F$, if the occurrence of E necessarily implies the occurrence of F. In order for this to be true, every description in E must belong also to F, so E is a subevent of F if and only if E is a subset of F.

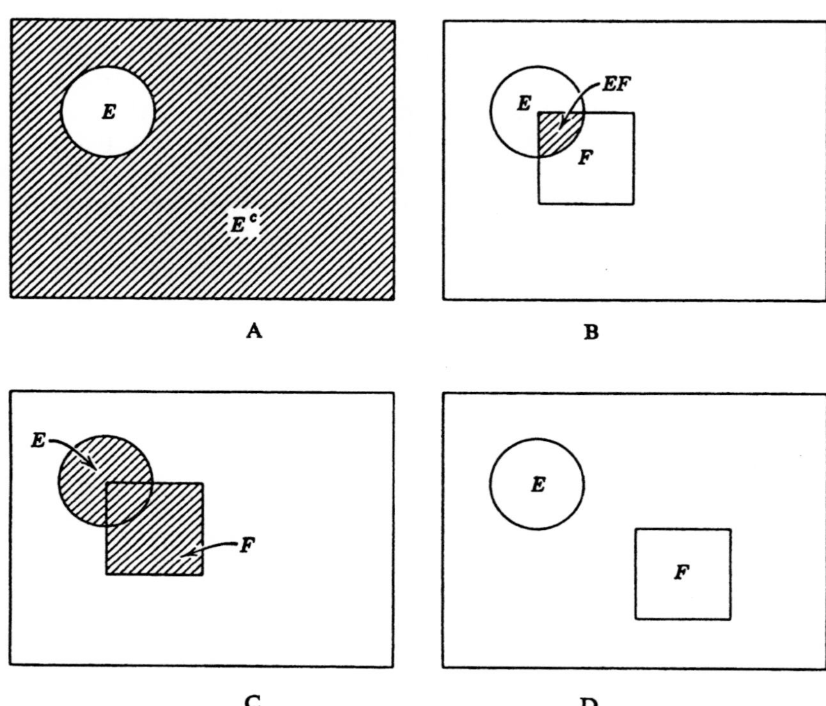

Fig. 4A. A Venn diagram. The shaded area represents E^c.

Fig. 4B. A Venn diagram. The shaded area represents EF.

Fig. 4C. A Venn diagram. The shaded area represents $E \cup F$.

Fig. 4D. A Venn diagram. The shaded area (or rather the lack of a shaded area) represents the impossible event $\emptyset$, which is the intersection of the two mutually exclusive events E and F.

We then have the basic principle that E equals F if and only if E is a subevent of F and F is a subevent of E. In symbols,

$$(4.1) \qquad E = F \quad \text{if and only if } E \subset F \quad \text{and} \quad F \subset E.$$

The interesting question arises whether the operations of event union and event intersection may be applied to an arbitrary pair of events E and F. In particular, consider two events, E and F, that contain no descriptions in common; for example, suppose $S = \{1, 2, 3, 4, 5, 6\}$, $E = \{1, 2\}$, $F = \{3, 4\}$. The union $E \cup F = \{1, 2, 3, 4\}$ is defined. However, what meaning is to be assigned to the intersection EF? To meet this need, we introduce the notion of the *impossible event*, denoted by $\emptyset$. *The impossible event* $\emptyset$ *is defined as the event that contains no descriptions and therefore cannot occur.* In set theory the impossible event is called the empty set. One important property of the impossible event is that it is the complement of the certain event S; clearly $S^c = \emptyset$, for it is impossible for S not to occur. A second important property of the impossible event is that it is equal to the intersection of any event E and its complement E^c; clearly, $EE^c = \emptyset$, for it is impossible for both an event and its complement to occur simultaneously.

Any two events, E and F, that cannot occur simultaneously, so that their intersection EF is the impossible event, are said to be *mutually exclusive* (or disjoint). Thus, two events, E and F, are mutually exclusive if and only if $EF = \emptyset$.

Two mutually exclusive events may be represented on a Venn diagram by the interiors of two geometrical figures that do not overlap, as in Fig. 4D. The impossible event may be represented by the shaded area on a Venn diagram, in which there is no shading, as in Fig. 4D.

Events may be defined verbally, and it is important to be able to express them in terms of the event operations. For example, let us consider two events, E and F. The event that exactly one of the events, E and F, will occur is equal to $EF^c \cup E^cF$; the event that exactly none of the events, E and F, will occur is equal to E^cF^c. The event that at least one (that is, one or more) of the events, E or F, will occur is equal to $E \cup F$. The event that at most one (that is, one or less) of the events will occur is equal to $(EF)^c = E^c \cup F^c$.

The operations of event union and event intersection have many of the algebraic properties of ordinary addition and multiplication of numbers (although they are conceptually quite distinct from the latter operations). Among the important algebraic properties of the operations $E \cup F$ and EF are the following relations, which hold for any events E, F, and G:

Commutative law	$E \cup F = F \cup E$	$EF = FE$
Associative law	$E \cup (F \cup G) = (E \cup F) \cup G$	$E(FG) = (EF)G$
Distributive law	$E(F \cup G) = EF \cup EG$	$E \cup (FG) = (E \cup F)(E \cup G)$
Idempotency law	$E \cup E = E$	$EE = E$

Because the operations of union and intersection are commutative and associative, there is no difficulty in defining the union and intersection of an arbitrary number of events, $E_1, E_2, \ldots, E_n, \ldots$. *The union, written* $E_1 \cup E_2 \cup \ldots E_n \cup \ldots$, *is defined as the event consisting of all descriptions that belong to at least one of the events. The intersection, written*

$E_1E_2 \ldots E_n \ldots$, *is defined as the event consisting of all descriptions that belong to all the events.*

An unusual property of the event operations, which is used very frequently, is given by *de Morgan's laws,* which state, for any two events, E and F,

$$(4.2) \qquad (E \cup F)^c = E^cF^c, \qquad (EF)^c = E^c \cup F^c,$$

and for n events, $E_1, E_2, \ldots, E_n$,

$$(4.3) \qquad (E_1 \cup E_2 \cup \cdots \cup E_n)^c = E_1{}^cE_2{}^c \cdots E_n{}^c,$$
$$(E_1E_2 \cdots E_n)^c = E_1{}^c \cup E_2{}^c \cup \cdots \cup E_n{}^c.$$

An intuitive justification for (4.2) and (4.3) may be obtained by considering Venn diagrams.

In section 5 we require the following formulas for the equality of certain events. Let E and F be two events defined on the same sample description space S. Then

$$(4.4) \qquad E\emptyset = \emptyset, \qquad E \cup \emptyset = E.$$

$$(4.5) \qquad F = FE \cup FE^c, \qquad E \cup F = F \cup EF^c = E \cup FE^c.$$

$$(4.6) \qquad F \subset E \text{ implies } EF = F \qquad E \cup F = E.$$

In order to verify these identities, one can establish in each case that the left-hand side of the identity is a subevent of the right-hand side and that the right-hand side is a subevent of the left-hand side.

EXERCISES

4.1. An experiment consists of drawing 3 radio tubes from a lot and testing them for some characteristic of interest. If a tube is defective, assign the letter D to it. If a tube is good, assign the letter G to it. A drawing is then described by a 3-tuple, each of whose components is either D or G. For example, (D, G, G) denotes the outcome that the first tube drawn was defective and the remaining 2 were good. Let A_1 denote the event that the first tube drawn was defective, A_2 denote the event that the second tube drawn was defective, and A_3 denote the event that the third tube drawn was defective. Write down the sample description space of the experiment and list all sample descriptions in the events A_1, A_2, A_3, $A_1 \cup A_2$, $A_1 \cup A_3$, $A_2 \cup A_3$, $A_1 \cup A_2 \cup A_3$, A_1A_2, A_1A_3, A_2A_3, $A_1A_2A_3$.

4.2. For each of the following 16 events draw a Venn diagram similar to Figure 4*A* or 4*B* and on it shade the area corresponding to the event. Only 7 diagrams will be required to illustrate the 16 events, since some of the events described are equivalent. (i) AB^c, (ii) $AB^c \cup A^cB$, (iii) $(A \cup B)^c$, (iv) A^cB^c, (v) $(AB)^c$, (vi) $A^c \cup B^c$, (vii) the event that exactly 0 of the events, A and B, occurs, (viii) the event that exactly 1 of the events, A and B, occurs, (ix) the event that exactly 2 of the events, A and B, occur, (x) the event that at least 0 of the events A and B, occurs, (xi) the event that at least 1 of the events, A and B, occurs, (xii) the event that at least 2 of the events, A and B, occur, (xiii) the event that no more than 0 of the events, A and B, occurs, (xiv) the event that no more than 1 of the events, A and B, occurs, (xv) the event that no more than 2 of the events, A and B, occur, (xvi) the event that A occurs and B does not occur. Remark: By "at least 1" we mean "1 or more," by "no more than 1" we mean "1 or less," and so on.

4.3. Let $S = \{1, 2, 3, 4, 5, 6, 7, 8, 9, 10, 11, 12\}$, $A = \{1, 2, 3, 4, 5, 6\}$, and $B = \{4, 5, 6, 7, 8, 9\}$. For each of the events described in exercise 4.2, write out the numbers that are members of the event.

4.4. For each of the following 12 events draw a Venn diagram and on it shade the area corresponding to the event: the event that of the events A, B, C, there occur (i) exactly 0, (ii) exactly 1, (iii) exactly 2, (iv) exactly 3, (v) at least 0, (vi) at least 1, (vii) at least 2, (viii) at least 3, (ix) no more than 0, (x) no more than 1, (xi) no more than 2, (xii) no more than 3.

4.5. Let S, A, B be as in exercise 4.3, and let $C = \{7, 8, 9\}$. For each of the events described in exercise 4.4, write out the numbers that are members of the event.

4.6. Prove (4.4). Note that (4.4) states that the impossible event behaves under the operations of intersection and union in a manner similar to the way in which the number 0 behaves under the operations of multiplication and addition.

4.7. Prove (4.5). Show further that the events F and EF^c are mutually exclusive.

5. THE DEFINITION OF PROBABILITY AS A FUNCTION OF EVENTS ON A SAMPLE DESCRIPTION SPACE

The mathematical notions are now at hand with which one may state the postulates of a mathematical model of a random phenomenon. Let us recall that in our heuristic discussion of the notion of a random phenomenon in section 1 we accepted the so-called "frequency" interpretation of probability, according to which the probability of an event E is a number (which we denote by $P[E]$). This number can be known to us only by experience as the result of a very long series of observations of independent trials of the event E. (By a trial of E is meant an occurrence of the phenomenon on which E is defined.) Having observed a long series of trials, the probability of E represents the fraction of trials whose outcome has a description that is a member of E. In view of the frequency interpretation of $P[E]$, it follows that a mathematical definition of the probability of an event cannot tell us the value of $P[E]$ for any particular event E. Rather a mathematical theory of probability must be concerned with the properties of the probability of an event considered as a function defined on all events. With these considerations in mind, we now give the following definition of probability.

The definition of probability as a function of events on the subsets of a sample description space of a random phenomenon:

Given a random situation, which is described by a sample description space S, probability is a function* $P[\cdot]$ that to every event E assigns a nonnegative real number, denoted by $P[E]$ and called the probability of the event E. The probability function must satisfy three axioms:

AXIOM 1. $P[E] \geq 0$ for every event E,

AXIOM 2. $P[S] = 1$ for the certain event S,

AXIOM 3. $P[E \cup F] = P[E] + P[F]$, if $EF = \emptyset$, or in words, the probability of the union of two mutually exclusive events is the sum of their probabilities.

It should be clear that the properties stated by the foregoing axioms do constitute a formal statement of some of the properties of the numbers $P[E]$ and $P[F]$, interpreted to represent the relative frequency of occurrence of the events E and F in a large number N of occurrences of the random phenomenon on which they are defined. For any event, E, let N_E be the number of occurrences of E in the N occurrences of the phenomenon. Then, by the frequency interpretation of probability, $P[E] = N_E/N$. Clearly, $P[E] \geq 0$. Next, $N_S = N$, since, by the construction of S, it occurs on every occurrence of the random phenomenon. Therefore, $P[S] = 1$. Finally, for two mutually exclusive events, E and F, $N_{(E \cup F)} = N_E + N_F$. Thus axiom 3 is satisfied.

It therefore follows that any property of probabilities that can be shown to be logical consequences of axioms 1 to 3 will hold for probabilities interpreted as relative frequencies. We shall see that for many purposes axioms 1 to 3 constitute a sufficient basis from which to derive the properties of probabilities. In advanced studies of probability theory, in which more delicate questions concerning probability are investigated, it is found necessary to strengthen the axioms somewhat. At the end of this section we indicate briefly the two most important modifications required.

We now show how one can derive from axioms 1 to 3 some of the important properties that probability possesses. In particular, we show how axiom 3 suffices to enable us to compute the probabilities of events constructed by means of complementations and unions of other events in terms of the probabilities of these other events.

* Definition: A *function* is a rule that assigns a real number to each element of a set of objects (called the domain of the function). Here the domain of the probability function $P[\cdot]$ is the set of all events on S.

In order to be able to state briefly the hypotheses of the theorems subsequently proved, we need some terminology. It is to be emphasized that one can speak of the probability of an event only if the event is a subset of a definite sample description space S, on whose subsets a probability function has been defined. Consequently, the hypothesis of a theorem concerning events should begin, "Let S be a sample description space on the subsets of which a probability function $P[\cdot]$ has been defined. Let E and F be any two events on S." For the sake of brevity, we write instead "Let E and F be any two events on a probability space"; by a *probability space* we mean a sample description space on which a probability function (satisfying axioms 1, 2, and 3) has been defined.

FORMULA FOR THE PROBABILITY OF THE IMPOSSIBLE EVENT $\emptyset$.

$$P[\emptyset] = 0. \tag{5.1}$$

Proof: By (4.4) it follows that the certain event S and the impossible event are mutually exclusive; further, their union $S \cup \emptyset = S$. Consequently, $P[S] = P[S \cup \emptyset] = P[S] + P[\emptyset]$, from which it follows that $P[\emptyset] = 0$.

FORMULA FOR THE PROBABILITY OF A DIFFERENCE FE^c OF TWO EVENTS E AND F. For any two events, E and F, on a probability space

$$P[FE^c] = P[F] - P[EF]. \tag{5.2}$$

Proof: The events FE and FE^c are mutually exclusive, and their union is F [compare (4.5)]. Then, by axiom 3, $P[F] = P[EF] + P[FE^c]$, from which (5.2) follows immediately.

FORMULA FOR THE PROBABILITY OF THE COMPLEMENT OF AN EVENT. For any event E on a probability space

$$P[E^c] = 1 - P[E]. \tag{5.3}$$

Proof: Let $F = S$ in (5.2). Since $SE^c = E^c$, $SE = E$, and $P[S] = 1$, we have obtained (5.3).

FORMULA FOR THE PROBABILITY OF A UNION $E \cup F$ OF TWO EVENTS E AND F. For any two events, E and F, on a probability space

$$P[E \cup F] = P[E] + P[F] - P[EF]. \tag{5.4}$$

Proof: We use the fact that the event $E \cup F$ may be written as the union of the two mutually exclusive events, E and FE^c. Then, by axiom 3, $P[E \cup F] = P[E] + P[FE^c]$. By evaluating $P[FE^c]$ by (5.2), one obtains (5.4).

Note that (5.4) extends axiom 3 to the case in which the events whose union is being formed are not necessarily mutually exclusive.

We next obtain a basic property of the probability function, namely, that if an event F is a subevent of another event E, then the probability that F will occur is less than or equal to the probability that E will occur.

INEQUALITY FOR THE PROBABILITY OF A SUBEVENT. Let E and F be events on a probability space S such that $F \subset E$ (that is, F is a subevent of E). Then

$$(5.5) \qquad P[EF^c] = P[E] - P[F] \qquad \text{if } F \subset E,$$

$$(5.6) \qquad P[F] \leq P[E], \qquad \text{if } F \subset E.$$

Proof: By (5.2), $P[E] - P[EF] = P[EF^c]$. Now, since $F \subset E$, it follows that, as in (4.6), $EF = F$. Therefore, $P[E] - P[F] = P[EF^c]$, which proves (5.5). Next, $P[EF^c] \geq 0$, by axiom 1. Therefore, $P[E] - P[F] \geq 0$, from which it follows that $P[F] \leq P[E]$, which proves (5.6).

From the preceding inequality we may derive the basic fact that probabilities are numbers between 0 and 1:

$$(5.7) \qquad \text{for any event } E \qquad 0 \leq P[E] \leq 1.$$

This is proved as follows. By axiom 1, $0 \leq P[E]$. Next, any event E is a subevent of the certain event. Therefore, by (5.6), $P[E] \leq P[S]$. However, by axiom 2, $P[S] = 1$, and the proof of the assertion is completed.

FORMULA FOR THE PROBABILITY OF THE UNION OF A FINITE NUMBER OF MUTUALLY EXCLUSIVE EVENTS. For any positive integer n the probability of the union of n mutually exclusive events $E_1, E_2, \ldots, E_n$ is equal to the sum of the probabilities of the events; in symbols,

$$(5.8) \qquad P[E_1 \cup E_2 \cup \cdots \cup E_n] = P[E_1] + P[E_2] + \cdots + P[E_n],$$

if, for every two integers i and j which are not equal and which are between 1 and n, inclusive, $E_iE_j = \emptyset$.

Proof: To prove (5.8), we make use of the *principle of mathematical induction*, which states that a proposition $p(n)$, which depends on an integer n, is true for $n = 1, 2, \ldots$, if one shows that (i) it is true for $n = 1$, and (ii) it satisfies the implication: $p(n)$ implies $p(n + 1)$. Now, for any positive integer n let $p(n)$ be the proposition that for any set of n mutually exclusive events, $E_1, \ldots, E_n$, (5.8) holds. That $p(1)$ is true is obvious, since in the case that $n = 1$ (5.8) states that $P[E_1] = P[E_1]$. Next, let n be a definite integer, and let us assume that $p(n)$ is true. Let us show that from the assumption that $p(n)$ is true it follows that $p(n + 1)$ is true. Let $E_1, E_2, \ldots, E_n, E_{n+1}$ be $n + 1$ mutually exclusive events. Since the events $E_1 \cup E_2 \cup \ldots \cup E_n$ and E_{n+1} are then mutually exclusive, it follows, by axiom 3, that

$$(5.9) \qquad P[E_1 \cup E_2 \cup \cdots \cup E_{n+1}] = P[E_1 \cup E_2 \cup \cdots \cup E_n] + P[E_{n+1}]$$

From (5.9), and the assumption that $p(n)$ is true, it follows that $P[E_1 \cup \ldots \cup E_{n+1}] = P[E_1] + \ldots + P[E_{n+1}]$. We have thus shown that $p(n)$ implies $p(n + 1)$. By the principle of mathematical induction, it holds that the proposition $p(n)$ applies to any positive integer n. The proof of (5.8) is now complete.

The foregoing axioms are completely adequate for the study of random phenomena whose sample description spaces are finite. For the study of infinite sample description spaces, however, it is necessary to modify axiom 3. We may wish to consider an infinite sequence of mutually exclusive events, $E_1, E_2, \ldots, E_n, \ldots$. That the probability of the union of an infinite number of mutually exclusive events is equal to the sum of the probabilities of the events cannot be proved by axiom 3 but must be postulated separately. Consequently, in advanced studies of probability theory, instead of axiom 3, the following axiom is adopted.

Axiom 3′. For any infinite sequence of mutually exclusive events, $E_1, E_2, \ldots, E_n, \ldots,$

$$(5.10) \qquad P[E_1 \cup E_2 \cup \cdots \cup E_n \cup \cdots] = P[E_1] + P[E_2] + \cdots + P[E_n] + \cdots.$$

A somewhat more esoteric modification in the foregoing axioms becomes necessary when we consider a random phenomenon whose sample description space S is noncountably infinite. It may then turn out that there are subsets of S that are nonprobabilizable, in the sense that it is not possible to assign a probability to these sets in a manner consistent with the axioms. If such is the case, then only probabilizable subsets of S are defined as events. Since it may be proved that the union, intersection, and complements of events are events, this restriction of the notion of event causes no difficulty in application and renders the mathematical theory rigorous.

EXERCISES

5.1. **Boole's inequality.** For a finite set of events, $A_1, A_2, \ldots, A_n$,

$$(5.11) \qquad P[A_1 \cup A_2 \cup \cdots \cup A_n] \leq P[A_1] + P[A_2] + \cdots + P[A_n].$$

Prove this assertion by means of the principle of mathematical induction.

5.2. **Formula for the probability that exactly 1 of 2 events will occur.** Show that for any 2 events, A and B, on a probability space

$$P[AB^c \cup BA^c] = P[A] + P[B] - 2P[AB]. \tag{5.12}$$

The event $AB^c \cup BA^c$ is the event that exactly 1 of the events, A and B, will occur. Contrast (5.12) with (5.4), which could be called the formula for the probability that at least 1 of 2 events will occur.

5.3. Show that for any 3 events, A, B, and C, defined on a probability space, the probability of the event that at least 1 of the events will occur is given by

$$P[A \cup B \cup C] = P[A] + P[B] + P[C] - P[AB] - P[AC] - P[BC] + P[ABC].$$

5.4. Let A and B be 2 events on a probability space. Show that

$$P[AB] \leq P[A] \leq P[A \cup B] \leq P[A] + P[B].$$

5.5. Let A and B be 2 events on a probability space. In terms of $P[A]$, $P[B]$, and $P[AB]$, express (i) for $k = 0, 1, 2$, P[exactly k of the events, A and B, occur], (ii) for $k = 0, 1, 2$, P[at least k of the events, A and B, occur], (iii) for $k = 0, 1, 2$, P[at most k of the events, A and B, occur], (iv) P[A occurs and B does not occur].

5.6. Let A, B, and C be 3 events on a probability space. In terms of $P[A]$, $P[B]$, $P[C]$, $P[AB]$, $P[AC]$, $P[BC]$, and $P[ABC]$ express for $k = 0, 1, 2, 3$ (i) P[exactly k of the events, A, B, C, occur], (ii) P[at least k of the events, A, B, C, occur], (iii) P[at most k of the events, A, B, C, occur].

5.7. Evaluate the probabilities asked for in exercise 5.5 in the case that (i) $P[A] = P[B] = \frac{1}{3}$, $P[AB] = \frac{1}{6}$, (ii) $P[A] = P[B] = \frac{1}{3}$, $P[AB] = \frac{1}{9}$, (iii) $P[A] = P[B] = \frac{1}{3}$, $P[AB] = 0$.

5.8. Evaluate the probabilities asked for in exercise 5.6 in the case that (i) $P[A] = P[B] = P[C] = \frac{1}{3}$, $P[AB] = P[AC] = P[BC] = \frac{1}{9}$, $P[ABC] = \frac{1}{27}$, (ii) $P[A] = P[B] = P[C] = \frac{1}{3}$, $P[AB] = P[AC] = P[BC] = P[ABC] = 0$.

The size of sets: The various formulas that have been developed for probabilities continue to hold true if one replaces P by N and for any set A define $N[A]$ as the number of elements in, or the size of, the set A. Further, replace 1 by $N[S]$.

5.9. Suppose that a study of 900 college graduates 25 years after graduation revealed that 300 were "successes," 300 had studied probability theory in college, and 100 were both "successes" and students of probability theory. Find, for $k = 0, 1, 2$, the number of persons in the group who had done of these two things: (i) exactly k, (ii) at least k, (iii) at most k.

5.10. In a very hotly fought battle in a small war 270 men fought. Of these, 90 lost an eye, 90 lost an arm, and 90 lost a leg: 30 lost both an eye and an arm, 30 lost both an arm and a leg, and 30 lost both a leg and an eye; 10 lost all three. Find, for $k = 0, 1, 2, 3$, the number of men who suffered of these injuries: (i) exactly k, (ii) at least k, (iii) no more than k.

5.11. Certain data obtained from a study of a group of 1000 subscribers to a certain magazine relating to their sex, marital status, and education were reported as follows: 312 males, 470 married, 525 college graduates, 42 male college graduates, 147 married college graduates, 86 married males, and 25 married male college graduates. Show that the numbers reported in the various groups are not consistent.

6. FINITE SAMPLE DESCRIPTION SPACES

To gain some insight into the amount of freedom we have in defining probability functions, it is useful to consider finite sample description spaces. The sample description space S of a random observation or experiment is defined as finite if it is of finite size, which is to say that the random observation or experiment under consideration possesses only a finite number of possible outcomes.

Consider now a finite sample description space S, of size N. We may then list the descriptions in S. If we denote the descriptions in S by D_1, $D_2, \ldots, D_N$, then we may write $S = \{D_1, D_2, \ldots, D_N\}$. For example, let S be the sample description space of the random experiment of tossing two coins; if we define $D_1 = (H, H)$, $D_2 = (H, T)$, $D_3 = (T, H)$, $D_4 = (T, T)$, then $S = \{D_1, D_2, D_3; D_4\}$.

It is shown in section 1 of Chapter 2 that 2^N possible events may be defined on a sample description space of finite size N. For example, if $S = \{D_1, D_2, D_3, D_4\}$, then there are sixteen possible events that may be defined; namely, $S, \emptyset, \{D_1\}, \{D_2\}, \{D_3\}, \{D_4\}, \{D_1, D_2\}, \{D_1, D_3\}, \{D_1, D_4\}, \{D_2, D_3\}, \{D_2, D_4\}, \{D_3, D_4\}, \{D_1, D_2, D_3\}, \{D_1, D_2, D_4\}, \{D_1, D_3, D_4\}, \{D_2, D_3, D_4\}$.

Consequently, to define a probability function $P[\cdot]$ on the subsets of S, one needs to specify the 2^N values that $P[A]$ assumes as A varies over the events on S. However, the values of the probability function cannot be specified arbitrarily but must be such that axioms 1 to 3 are satisfied.

There are certain events of particularly simple structure, called the *single-member* events, on which it will suffice to specify the probability function $P[\cdot]$ in order that it be specified for all events. *A single-member event is an event that contains exactly one description.* If an event E has as its only member the description D_i, this fact may be expressed in symbols by writing $E = \{D_i\}$. Thus $\{D_i\}$ is the event that occurs if and only if the random situation being observed has description D_i. The reader should note the distinction between D_i and $\{D_i\}$; the former is a description, the latter is an event (which because of its simple structure is called a single-member event).

▶ **Example 6A. The distinction between a single-member event and a sample description.** Suppose that we are drawing a ball from an urn containing six balls, numbered 1 to 6 (or, alternately, we may be observing the outcome of the toss of a die, bearing numbers 1 to 6 on its sides). As sample description space S, we take $S = \{1, 2, 3, 4, 5, 6\}$. The event, denoted by $\{2\}$, that the outcome of the experiment is a 2 is a single-member event. The event, denoted by $\{2, 4, 6\}$, that the outcome of the experiment is an even number is not a single-member event. Note that 2 is a description, whereas $\{2\}$ is an event. ◀

A probability function $P[\cdot]$ defined on S can be specified by giving its value $P[\{D_i\}]$ on the single-member events $\{D_i\}$ which correspond to the members of S. Its value $P[E]$ on any event E may then be computed by the following formula:

FORMULA FOR CALCULATING THE PROBABILITIES OF EVENTS WHEN THE SAMPLE DESCRIPTION SPACE IS FINITE. Let E be any event on a finite sample description space $S = \{D_1, D_2, \ldots, D_N\}$. Then the probability $P[E]$ of the event E is the sum, over all descriptions D_i that are members of E, of the probabilities $P[\{D_i\}]$; we express this symbolically by writing that if $E = \{D_{i_1}, D_{i_2}, \ldots, D_{i_k}\}$ then

$$P[E] = P[\{D_{i_1}\}] + P[\{D_{i_2}\}] + \cdots + P[\{D_{i_k}\}]. \tag{6.1}$$

To prove (6.1), one need note only that if E consists of the descriptions $D_{i_1}, D_{i_2}, \ldots, D_{i_k}$ then E can be written as the union of the mutually exclusive single-member events $\{D_{i_1}\}, \{D_{i_2}\}, \ldots, \{D_{i_k}\}$. Equation (6.1) follows immediately from (5.8).

▶ **Example 6B. Illustrating the use of (6.1).** Suppose one is drawing a sample of size 2 from an urn containing white and red balls. Suppose that as the sample description space of the experiment one takes $S = \{(W, W), (W, R), (R, W), (R, R)\}$. To specify a probability function $P[\cdot]$ on S, one may specify the values of $P[\cdot]$ on the single-member events by a table:

x	(W, W)	(W, R)	(R, W)	(R, R)
$P[\{x\}]$	$\frac{6}{15}$	$\frac{4}{15}$	$\frac{4}{15}$	$\frac{1}{15}$.

Let E be the event that the ball drawn on the first draw is white. The event E may be represented as a set of descriptions by $E = \{(W, W), (W, R)\}$. Then, by (6.1), $P[E] = P[\{(W, W)\}] + P[\{(W, R)\}] = \frac{2}{3}$. ◀

7. FINITE SAMPLE DESCRIPTION SPACES WITH EQUALLY LIKELY DESCRIPTIONS

In many probability situations in which finite sample description spaces arise it may be assumed that all descriptions are equally likely; that is, all descriptions in S have equal probability of occurring. More precisely, we define *the sample description space* $S = \{D_1, D_2, \ldots, D_N\}$ *as having equally likely descriptions if all the single-member events on* S *have equal probabilities*, so that

$$P[\{D_1\}] = P[\{D_2\}] = \cdots = P[\{D_N\}] = \frac{1}{N}. \tag{7.1}$$

It should be clear that each of the single-member events $\{D_i\}$ has probability $(1/N)$, since there are N such events, each of which has equal probability, and the sum of their probabilities must equal 1, the probability of the certain event.

The computation of the probability of an event, defined on a sample description space with equally likely descriptions, can be reduced to the computation of the size of the event. By (6.1), the probability of E is equal to $(1/N)$, multiplied by the number of descriptions in E. In other words, *the probability of E is equal to the ratio of the size of E to the size of S.* If, for a set E of finite size, we let $N[E]$ denote the size of E (the number of members of E), then the foregoing conclusions can be summed up in a basic formula:

FORMULA FOR CALCULATING THE PROBABILITIES OF EVENTS WHEN THE SAMPLE DESCRIPTION SPACE S IS FINITE AND ALL DESCRIPTIONS ARE EQUALLY LIKELY: For any event E on S

$$P[E] = \frac{N[E]}{N[S]} = \frac{\text{size of } E}{\text{size of } S}. \tag{7.2}$$

This formula can be stated in words. If an event is defined as a subset of a finite sample description space, whose descriptions are all equally likely, then the probability of the event is the ratio of the number of descriptions belonging to it to the total number of descriptions. This statement may be regarded as a precise formulation of the classical "equal-likelihood" definition of the probability of an event, first explicitly formulated by Laplace in 1812.

THE LAPLACEAN "EQUAL-LIKELIHOOD" DEFINITION OF THE PROBABILITY OF A RANDOM EVENT. The probability of a random event is the ratio of the number of cases favoring it to the number of all possible cases, when nothing leads us to believe that one of these cases ought to occur rather than the others. This renders them, for us, equally possible.

In view of (7.2), one sees that in adopting the axiomatic definition of probability given in section 5 one does not thereby reject the Laplacean definition of probability. Rather, the Laplacean definition is a special case of the axiomatic definition, corresponding to the case in which the sample description space is finite and the probability distribution on the sample description space is a uniform one. This is an alternate way of saying that all descriptions are equally likely.

We may now state a mathematical model for the experiment of drawing a ball from an urn containing six balls, numbered 1 to 6, of which balls one to four are colored white and the remaining two balls are nonwhite. For the sample description space S of the experiment we take $S = \{1, 2, 3, 4, 5, 6\}$. The event A that the ball drawn is white is then given as a subset of S by $A = \{1, 2, 3, 4\}$. To compute the probability of A, we must adopt a probability function $P[\cdot]$ on S. If we assume that the descriptions in S are equally likely, then $P[\cdot]$ is determined by (7.2), and $P[A] = \frac{2}{3}$. On the other hand, we may specify a different probability function $P[\cdot]$, specified on the single-member events of S:

$$P[\{1\}] = P[\{2\}] = P[\{3\}] = P[\{4\}] = \tfrac{1}{8}, \qquad P[\{5\}] = P[\{6\}] = \tfrac{1}{4}.$$

Then the function $P[\cdot]$ is determined by (6.1), and $P[A] = \frac{1}{2}$.

We have thus stated two different mathematical models for the experiment of drawing a ball from an urn. Only the results of actual experiments can decide which of the two models is realistic. However, as we study the properties of various models in the course of this book, theoretical grounds will appear for preferring some kinds of models over others.

▶ **Example 7A.** Find the probability that the thirteenth day of a randomly chosen month is a Friday.

Solution: The sample description space of the experiment of observing the day of the week upon which the thirteenth day of a randomly chosen month will fall is clearly $S = \{$Sunday, Monday, Tuesday, Wednesday, Thursday, Friday, Saturday$\}$. We are seeking $P[\{\text{Friday}\}]$. If we assume equally likely descriptions, then $P[\{\text{Friday}\}] = \frac{1}{7}$. However, would one believe this conclusion in the face of the following alternative mathematical model? To define a probability function on S, note that our calendar has a period of 400 years, since every fourth year is a leap year, except for years such as 1700, 1800, and 1900, at which a new century begins (or an old century ends) but which are not multiples of 400. In 400 years there are 97 leap years and exactly 20,871 weeks. For each of the 4800 dates between 1600 and 2000 that is the thirteenth day of some month one may determine the day of the week on which it falls. For any given day x of the week let us define $P[\{x\}]$ as the relative frequency of occurrence of x in the list of

4800 days of the week which arise as the thirteenth day of some month. It may be shown by a direct but tedious enumeration [see *American Mathematical Monthly*, Vol. 40 (1933), p. 607] that

(7.3)

x	Sunday	Monday	Tuesday	Wednesday	Thursday	Friday	Saturday
$P[\{x\}]$	$\frac{687}{4800}$	$\frac{685}{4800}$	$\frac{685}{4800}$	$\frac{687}{4800}$	$\frac{684}{4800}$	$\frac{688}{4800}$	$\frac{684}{4800}$

Note that the probability model given by (7.3) leads to the conclusion that the thirteenth of the month is more likely to be a Friday than any other day of the week! ◀

▶ **Example 7B.** Consider a state (such as Illinois) in which the license plates of automobiles are numbered serially, beginning with 1. Assuming that there are 3,000,000 automobiles registered in the state, what is the probability that the first digit on the license plate of an automobile selected at random will be the digit 1?

Solution: As the first digit on the license of a car, one may observe any integer in the set $\{1, 2, 3, 4, 5, 6, 7, 8, 9\}$. Consequently, one may be tempted to adopt this set as the sample description space. If one assumes that all sample descriptions in this space are equally likely, then one would arrive at the conclusion that the probability is $\frac{1}{9}$ that the digit 1 will be the first digit on the license plate of an automobile randomly selected from the automobiles registered in Illinois. However, would one believe this conclusion in the face of the following alternative model? As a result of observing the number on a license plate, one may observe any number in the set S consisting of all integers 1 to 3,000,000. The event A that one observes a license plate whose first digit is 1 consists of the integers enumerated in Table 7A. The set A has size $N[A] = 1{,}111{,}111$. If the

TABLE 7A

LICENSE PLATES WITH FIRST DIGIT 1

All License Plates in the Following Intervals Have First Digit 1	Number of Integers in this Interval
1	1
10–19	10
100–199	100
1000–1999	1000
10,000–19,999	10,000
100,000–199,999	100,000
1,000,000–1,999,999	1,000,000

set S is adopted as the sample description space and all descriptions in S are assumed to be equally likely, then

$$P[A] = \frac{N[A]}{N[S]} = \frac{1,111,111}{3,000,000} = 0.37037. \blacktriangleleft$$

EXERCISES

7.1. Suppose that a die (with faces marked 1 to 6) is loaded in such a manner that, for $k = 1, \ldots, 6$, the probability of the face marked k turning up when the die is tossed is proportional to k. Find the probability of the event that the outcome of a toss of the die will be an even number.

7.2. What is the probability that the thirteenth of the month will be (i) a Friday or a Saturday, (ii) a Saturday, Sunday, or Monday?

7.3. Let a number be chosen from the integers 1 to 100 in such a way that each of these numbers is equally likely to be chosen. What is the probability that the number chosen will be (i) a multiple of 7, (ii) a multiple of 14?

7.4. Consider a state in which the license plates of automobiles are numbered serially, beginning with 1. What is the probability that the first digit on the license plate of an automobile selected at random will be the digit 1, assuming that the number of automobiles registered in the state is equal to (i) 999,999, (ii) 1,000,000, (iii) 1,500,000, (iv) 2,000,000, (v) 6,000,000?

7.5. What is the probability that a ball, drawn from an urn containing 3 red balls, 4 white balls, and 5 blue balls, will be white? State carefully any assumptions that you make.

7.6. **A research problem.** Using the same assumptions as those with which the table in (7.3) was derived, find the probability that Christmas (December 25) is a Monday. Indeed, show that the probability that Christmas will fall on a given day of the week is supplied by the following table:

x	Sunday	Monday	Tuesday	Wednesday	Thursday	Friday	Saturday
$P[\{x\}]$	$\frac{58}{400}$	$\frac{56}{400}$	$\frac{58}{400}$	$\frac{57}{400}$	$\frac{57}{400}$	$\frac{58}{400}$	$\frac{56}{400}$

8. NOTES ON THE LITERATURE OF PROBABILITY THEORY

The first book on probability theory, *De Ratiociniis in Ludo Aleae*, a treatise on problems of games of chance, was published by Huyghens in 1657. There were no published writings on this subject before 1657, although evidence exists that a number of fifteenth- and sixteenth-century Italian mathematicians worked out the solutions to various probability problems concerning games of chance. General methods of attack on such problems seem first to have been given by Pascal and Fermat in a celebrated correspondence, beginning in 1654. It is a fascinating cultural puzzle that the calculus of probability did not emerge until the seventeenth century, although random phenomena, such as those arising in games of chance, have always been present in man's environment. For some enlightening remarks on this puzzle see M. G. Kendall, *Biometrika*, Vol. 43 (1956), pp. 9–12. A complete history of the development of probability theory during the period 1575 to 1825 is given by I. Todhunter, *A History of the Mathematical Theory of Probability from the Time of Pascal to Laplace*, originally published in 1865 and reprinted in 1949 by Chelsea, New York.

The work of Laplace marks a natural division in the history of probability, since in his great treatise *Théorie Analytique des Probabilités*, first published in 1812, he summed up his own extensive work and that of his predecessors. Laplace also wrote a popular exposition for the educated general public, which is available in English translation as *A Philosophical Essay on Probabilities* (with an introduction by E. T. Bell, Dover, New York, 1951).

The breadth of probability theory is today too immense for any one man to be able to sum it up. One can list only the main references in English of which the student should be aware.* The literature of probability theory divides into three broad categories: (i) the nature (or foundations) of probability, (ii) mathematical probability theory, and (iii) applied probability theory.

The nature of probability theory is a subject about which competent men differ. There are at least two main classes of concepts that historically have passed under the name of "probability." It has been suggested that one distinguish between these two concepts by calling one probability$_1$ and the other probability$_2$ (this terminology is suggested by R. Carnap, *Logical Foundations of Probability*, University of Chicago Press, 1950). The theory of probability$_1$ is concerned with the problem of inductive inference, with the nature of scientific proof, with the credibility of propositions given empirical evidence, and in general with ways of reasoning from empirical data to conclusions about future experiences. The theory of probability$_2$ is concerned with the study of repetitive events that appear to possess the property that their relative frequency of occurrence in a large number of trials has a stable limit value. Enlightening discussions of the theories of probability$_1$ and probability$_2$ are given, respectively, by Sir Harold Jeffreys, *Scientific Inference*, Second Edition, Cambridge

* Important contributions to probability theory have been made by men of all nationalities. In this section are mentioned only books available in the English language. However, the reader should be aware that important works on probability theory have been written in all the major languages of the world.

University Press, Cambridge, 1957, and Richard von Mises, *Probability, Statistics, and Truth*, Second Edition, Macmillan, New York, 1957. The viewpoint of professional philosophers in regard to the nature of probability theory is debated in "A Symposium on Probability," *Philosophy and Phenomenological Research*, Vol. 5 (1945), pp. 449–532, Vol. 6 (1946), pp. 11–86 and pp. 590–622. The philosophical implications of the use of probability theory in scientific explanation are examined from the point of view of the physicist in two books written for the educated layman: Max Born, *Natural Philosophy of Cause and Chance*, Oxford University Press, 1949, and David Bohm, *Causality and Chance in Modern Physics*, London, Routledge and Kegan Paul, 1957.

The mathematical theory of probability may be defined as consisting of those writings in which the viewpoint is the axiomatic one formulated in this chapter. This viewpoint developed in the twentieth century at the hands of such great probabilists as E. Borel, H. Steinhaus, P. Lévy, and A. Kolmogorov.* The first systematic presentation of probability theory on an axiomatic basis was made in 1933 by Kolmogorov in a monograph available in English translation as *Foundations of the Theory of Probability*, Chelsea, New York, 1950. Several comprehensive treatises, in which are summarized the development of mathematical probability theory up to, say, 1950, are available: J. L. Doob, *Stochastic Processes*, Wiley, New York, 1953; B. V. Gnedenko and A. N. Kolmogorov, *Limit Distributions for Sums of Independent Random Variables* (translated by K. L. Chung), Addison-Wesley, Cambridge, 1954; and M. Loève, *Probability Theory: Foundations, Random Sequences*, Van Nostrand, New York, 1955. A number of monographs covering the developments of the last twenty years are in process of preparation by various authors. The reader may gain some idea of the scope of recent work in the mathematical theory of probability by consulting the section "Probability" in the monthly publication *Mathematical Reviews*, which abstracts all published material on probability theory.

Applied probability theory may be defined as consisting of those writings in which probability theory enters as a tool in a scientific or scholarly investigation. There are so many fields of engineering and the physical, natural, and social sciences to which probability theory has been applied that it is not possible to cite a short list of representative references. A number of references are given in this book in the examples in which we

* For exact references, see page 259 of the excellent book by Mark Kac, entitled *Probability and Related Topics in Physical Sciences*, Interscience, New York, 1959, and also Paul Lévy, "Random Functions: General Theory with Special Reference to Laplacian Random Functions," *University of California Publications in Statistics*, Vol. 1 (1953), p. 340.

discuss various applications of probability theory. Some idea of the diverse applications of probability theory can be gained by consulting M. S. Bartlett, *Stochastic Processes*, Cambridge University Press, 1955, or the book by Feller cited below. The role of probability theory in mathematical statistics is discussed in H. Cramér, *Mathematical Methods of Statistics*, Princeton University Press, 1946.

The following books are classic introductions to probability theory that the reader can consult for alternate treatments of some of the topics discussed in this book: W. Feller, *An Introduction to Probability Theory and its Applications*, Second Edition, Wiley, New York, 1957; T. C. Fry, *Probability and its Engineering Uses*, Van Nostrand, New York, 1928; J. V. Uspensky, *Introduction to Mathematical Probability*, McGraw-Hill, New York, 1937. Feller's inimitable book is especially recommended, since it is simultaneously an introductory textbook and a treatise on mathematical and applied probability theory.

CHAPTER 2

Basic Probability Theory

Many of the basic concepts of probability theory, as well as a large number of important problems of applied probability theory, may be considered in the context of finite sample description spaces and thus can be studied with a minimum of mathematical technique. In Chapters 2 and 3 only finite sample description spaces are considered. In this chapter we further restrict ourselves to finite sample description spaces with equally likely descriptions.

1. SAMPLES AND n-TUPLES

A basic tool for the construction of sample description spaces of random phenomena is provided by the notion of an n-tuple. *An n-tuple* $(z_1, z_2, \ldots, z_n)$ *is an array of n symbols,* $z_1, z_2, \ldots, z_n$, *which are called, respectively, the first component, the second component, and so on, up to the nth component, of the n-tuple.* The order in which the components of an n-tuple are written is of importance (and consequently one sometimes speaks of ordered n-tuples). Two n-tuples $(z_1, z_2, \ldots, z_n)$ and $(z_1', z_2', \ldots, z_n')$ are said to be identical, or indistinguishable, if and only if they consist of the same components written in the same order; symbolically, $z_k = z_k'$ for $k = 1, 2, \ldots, n$. The usefulness of n-tuples derives from the fact that they are convenient devices for reporting the results of a drawing of a sample of size n.

A basic random phenomenon with whose analysis we are concerned in probability theory is that of *sampling*. Suppose we have an urn containing M balls, which are numbered 1 to M. Suppose we draw balls from the urn one at a time, until n balls have been drawn; for brevity, we say we have drawn a *sample* (or an ordered sample) of size n. Of course, we must also specify whether the sample has been drawn with replacement or without replacement.

The drawing is said to be done *with replacement*, and the sample is said to be drawn with replacement, if after each draw the number of the ball drawn is recorded, but the ball itself is returned to the urn. The drawing is said to be done *without replacement*, and the sample is said to be drawn without replacement, if the ball drawn is not returned to the urn after each draw, so that the number of balls available in the urn for the kth draw is $M - k + 1$. Consequently, if the drawing is done without replacement, then the size n of the sample drawn must be less than or equal to M, the original number of balls in the urn. On the other hand, if the drawing is done with replacement, then n may be any number.

To report the result of drawing a sample of size n, an n-tuple $(z_1, z_2, \ldots, z_n)$ is used, in which z_1 represents the number of the ball drawn on the first draw, z_2 represents the number of the ball drawn on the second draw, and so on, up to z_n, which represents the number of the ball drawn on the nth draw.

▶ **Example 1A. All possible samples of size 3 from an urn containing four balls.** Let us consider an urn which contains four balls, numbered 1 to 4, and let a sample of size 3 be drawn. If the sampling is done without replacement, then the possible samples that can be drawn are

$$\begin{array}{llll}
(1, 2, 3), & (2, 3, 1), & (3, 4, 1), & (4, 1, 2) \\
(1, 2, 4), & (2, 3, 4), & (3, 4, 2), & (4, 1, 3) \\
(1, 3, 2), & (2, 4, 1), & (3, 1, 2), & (4, 2, 3) \\
(1, 3, 4), & (2, 4, 3), & (3, 1, 4), & (4, 2, 1) \\
(1, 4, 2), & (2, 1, 3), & (3, 2, 1), & (4, 3, 1) \\
(1, 4, 3), & (2, 1, 4), & (3, 2, 4), & (4, 3, 2)
\end{array}$$

If the sampling is done with replacement, then the possible samples that can be drawn are

$$\begin{array}{llll}
(1, 1, 1), & (2, 1, 1), & (3, 1, 1), & (4, 1, 1) \\
(1, 1, 2), & (2, 1, 2), & (3, 1, 2), & (4, 1, 2) \\
(1, 1, 3), & (2, 1, 3), & (3, 1, 3), & (4, 1, 3) \\
(1, 1, 4), & (2, 1, 4), & (3, 1, 4), & (4, 1, 4) \\
(1, 2, 1), & (2, 2, 1), & (3, 2, 1), & (4, 2, 1) \\
(1, 2, 2), & (2, 2, 2), & (3, 2, 2), & (4, 2, 2)
\end{array}$$

(1, 2, 3),	(2, 2, 3),	(3, 2, 3),	(4, 2, 3)
(1, 2, 4),	(2, 2, 4),	(3, 2, 4),	(4, 2, 4)
(1, 3, 1),	(2, 3, 1),	(3, 3, 1),	(4, 3, 1)
(1, 3, 2),	(2, 3, 2),	(3, 3, 2),	(4, 3, 2)
(1, 3, 3),	(2, 3, 3),	(3, 3, 3),	(4, 3, 3)
(1, 3, 4),	(2, 3, 4),	(3, 3, 4),	(4, 3, 4)
(1, 4, 1),	(2, 4, 1),	(3, 4, 1),	(4, 4, 1)
(1, 4, 2),	(2, 4, 2),	(3, 4, 2),	(4, 4, 2)
(1, 4, 3),	(2, 4, 3),	(3, 4, 3),	(4, 4, 3)
(1, 4, 4),	(2, 4, 4),	(3, 4, 4),	(4, 4, 4) ◀

As indicated in section 7 of Chapter 1, many probability problems defined on finite sample description spaces may be reduced to problems of counting. Consequently, it is useful to know the basic principles of combinatorial analysis by which the size of sets of n-tuples, which arise in various ways, may be counted. We now state a formula that is basic to the theory of counting sets of n-tuples and that may be called the *basic principle of combinatorial analysis.*

Suppose there is a set A whose members are ordered n-tuples of objects of some sort. In order to compute the size of A, first determine the number N_1 of objects that may be used as the first component of an n-tuple in A. Next determine (*if it exists**) the number N_2 of objects that may be second components of an n-tuple, of which the first component is known. Then determine (if it exists) the number N_3 of objects that may be third components of an n-tuple, of which the first two components are known. Continue in this manner until the number N_n (if it exists) of objects that may be the nth component of an n-tuple, of which the first $(n - 1)$ components are known, has been determined. *The size of the set A of n-tuples is then given by the product of the numbers* $N_1, N_2, \ldots, N_n$; *in symbols.*

$$N[A] = N_1 N_2 \cdots N_n. \tag{1.1}$$

As a first application of this basic principle, suppose that we have n different kinds of objects. Suppose that we have N_1 objects $a_1^{(1)}, \ldots, a_{N_1}^{(1)}$ of the first kind, N_2 objects $a_1^{(2)}, \ldots, a_{N_2}^{(2)}$ of the second kind, and so on, up to N_n objects $a_1^{(n)}, \ldots, a_{N_n}^{(n)}$ of the nth kind. *We may then form* $N_1 N_2 \ldots N_n$ *ordered n-tuples* $(a_{j_1}^{(1)}, a_{j_2}^{(2)}, \ldots, a_{j_n}^{(n)})$ *containing one element of each kind.*

▶ **Example 1B.** A man has five suits, three pairs of shoes, and two hats. How many different combinations of attire can he wear?

* The number N_2 exists if the number of possible second components that may occur in an n-tuple, of which the first component is known, does not depend on which first component has occurred.

Solution: A combination of attire is a 3-tuple $(a^{(1)}, a^{(2)}, a^{(3)})$, in which $a^{(1)}, a^{(2)}, a^{(3)}$ denote, respectively, the suit, shoes, and hat worn. By the basic principle of combinatorial analysis there are $5 \cdot 3 \cdot 2 = 30$ combinations of attire. ◀

We next apply the basic principle of combinatorial analysis to determine the number of samples of size n that can be drawn with or without replacement from an urn containing M distinguishable balls.

The number of ways in which one can draw a sample of n balls from an urn containing M distinguishable balls is $M(M-1)\cdots(M-n+1)$, if the sampling is done without replacement, and M^n, if the sampling is done with replacement.

To show the first of these statements, note that there are M possible choices of numbers for the first ball drawn, $(M-1)$ choices of numbers for the second ball drawn, and finally $M-n+1 = M-(n-1)$ choices of numbers for the nth ball drawn. The second statement follows by a similar argument, since for each of the n balls in the sample there are M choices.

Various notations have been adopted to denote the product $M(M-1)\ldots(M-n+1)$. We adopt the notation $(M)_n$. We thus define, for any positive integer $M = 1, 2, \ldots$, and for any integer $n = 1, 2, \ldots, M$,

$$(M)_n = M(M-1)\cdots(M-n+1). \tag{1.2}$$

Another notation with which the reader should be familiar is that of the factorial. Given any positive integer M, we define $M!$ (read, M factorial) as the product of all the integers, 1 to M. Thus

$$M! = 1 \cdot 2 \cdots (M-1)M. \tag{1.3}$$

We can write $(M)_n$ in terms of the factorial notation by

$$(M)_n = \frac{M!}{(M-n)!} \qquad n = 0, 1, 2, \cdots, M. \tag{1.4}$$

In order that (1.4) may hold for $n = M$, we define

$$0! = 1. \tag{1.5}$$

In order that (1.4) may hold for $n = 0$, we define

$$(M)_0 = 1. \tag{1.6}$$

▶ **Example 1C.** $(4)_0 = 1$, $(4)_1 = 4$, $(4)_2 = 12$, $(4)_3 = 24$, $(4)_4 = 4! = 24$. Note that $(4)_5$ is undefined at present. It is later defined as having value 0. ◀

An important application of the foregoing relations is to the problem of finding *the number of subsets of a set*. Consider the set $S = \{1, 2, \ldots, N\}$, which consists of all integers, 1 to N. How many possible subsets of S can be formed? In order to solve this problem, we first find for $k = 1, 2, \ldots, N$ the number of subsets of S of size k that can be formed. Let x_k be the number of subsets of S of size k. We shall prove that x_k satisfies the relationship $x_k \cdot k! = (N)_k$, so that

$$x_k = \frac{(N)_k}{k!}. \tag{1.7}$$

To see this, regard each subset of S of size k as an urn (containing k distinguishable balls) from which samples of size k are being drawn without replacement; the number of samples that can be drawn in this manner is $k!$. On the other hand, the number of samples of size k, drawn without replacement, that can be drawn from the set S, regarded as an urn containing N distinguishable balls, is $(N)_k$. A little reflection will convince the reader that all the samples without replacement of size k that can be drawn from S can be obtained by first choosing a subset of S of size k from which one then draws all possible samples without replacement of size k. Consequently, $x_k \cdot k! = (N)_k$, or, in words, *the number of subsets of S of size k, multiplied by the number of samples that can be drawn without replacement from a subset of size k, is equal to the number of samples of size k that can be drawn without replacement from S itself.*

We now introduce some notation. We define, for any integer $N = 1, 2, \ldots$, and integer $k = 0, 1, \ldots, N$, the symbol $\binom{N}{k}$ by

$$\binom{N}{k} = \frac{(N)_k}{k!} = \frac{N(N-1)\cdots(N-k+1)}{1 \cdot 2 \cdots k} = \frac{N!}{k!(N-k)!} \tag{1.8}$$

Equation (1.7) may be restated as follows: *the number of subsets of size k that may be formed from the members of a set of size N is* $\binom{N}{k}$.

▶ **Example 1D. The subsets of size 3 of a set of size 4.** Consider the set $\{1, 2, 3, 4\}$. There are $\binom{4}{3} = 4$ subsets of size 3 that can be formed, namely, $\{1, 2, 3\}$, $\{1, 2, 4\}$, $\{1, 3, 4\}$, $\{2, 3, 4\}$. Notice that from each of these subsets one may draw without replacement six possible samples, so that there are twenty-four possible samples of size 3 to be drawn without replacement from an urn containing four balls. ◀

The quantities $\binom{N}{k}$ are generally called *binomial coefficients* because of the role they play in the binomial theorem, which states that for any two real numbers a and b and any positive integer N

$$(1.9) \qquad (a+b)^N = \sum_{k=0}^{N} \binom{N}{k} a^{N-k} b^k$$

$$= \binom{N}{0} a^N + \binom{N}{1} a^{N-1} b + \binom{N}{2} a^{N-2} b^2$$

$$+ \cdots + \binom{N}{k} a^{N-k} b^k + \cdots + \binom{N}{N-2} a^2 b^{N-2}$$

$$+ \binom{N}{N-1} a b^{N-1} + \binom{N}{N} b^N.$$

It is convenient to extend the definitions of $\binom{N}{k}$ and $(N)_k$ to any positive or negative integer k. We define, for $N = 1, 2, \ldots,$

$$(1.10) \qquad (N)_0 = \binom{N}{0} = 1, \qquad (N)_k = \binom{N}{k} = 0,$$

if either $k < 0$ or $k > N$.

We next note the extremely useful relation, holding for $N = 1, 2, \ldots,$ and $k = 0, \pm 1, \pm 2, \ldots,$

$$(1.11) \qquad \binom{N}{k-1} + \binom{N}{k} = \binom{N+1}{k}.$$

This relation may be verified directly from the definition of binomial coefficients. An intuitive justification of (1.11) can be obtained. Given a set S, with $N + 1$ members, choose an element t in S. The number of subsets of S of size k in which t is not present is equal to $\binom{N}{k}$, whereas the number of subsets of S of size k in which t is present is $\binom{N}{k-1}$; the sum of these two quantities is equal to $\binom{N+1}{k}$, the total number of subsets of S of size k.

Equation (1.11) is the algebraic expression of a fact represented in tabular form by *Pascal's triangle:*

$$\binom{1}{0} = 1 \quad \binom{1}{1} = 1$$

$$\binom{2}{0} = 1 \quad \binom{2}{1} = 2 \quad \binom{2}{2} = 1$$

$$\binom{3}{0} = 1 \quad \binom{3}{1} = 3 \quad \binom{3}{2} = 3 \quad \binom{3}{3} = 1$$

$$\binom{4}{0} = 1 \quad \binom{4}{1} = 4 \quad \binom{4}{2} = 6 \quad \binom{4}{3} = 4 \quad \binom{4}{4} = 1$$

$$\binom{5}{0} = 1 \quad \binom{5}{1} = 5 \quad \binom{5}{2} = 10 \quad \binom{5}{3} = 10 \quad \binom{5}{4} = 5 \quad \binom{5}{5} = 1$$

. . . .

and so on. Equation (1.11) expresses the fact that each term in Pascal's triangle is the sum of the two terms above it.

One also notices in Pascal's triangle that the entries on each line are symmetric about the middle entry (or entries). More precisely, the binomial coefficients have the property that for any positive integer N and $k = 0, 1, 2, \ldots, N$

$$\binom{N}{N-k} = \binom{N}{k}. \tag{1.12}$$

To prove (1.12) one need note only that each side of the equation is equal to $N!/k!(N-k)!$.

It should be noted that with (1.11) and the aid of the principle of mathematical induction one may prove the binomial theorem.

The mathematical facts are now at hand to determine how many subsets of a set of size N one may form. From the binomial theorem (1.9), with $a = b = 1$, it follows that

$$1 + \binom{N}{1} + \binom{N}{2} + \cdots + \binom{N}{N-1} + \binom{N}{N} = 2^N. \tag{1.13}$$

From (1.13) it follows that *the number of events (including the impossible event) that can be formed on a sample description space of size N is* 2^N. For there is one impossible event, $\binom{N}{1}$ events of size 1, $\binom{N}{2}$ events of size 2, . . . , $\binom{N}{k}$ events of size k, . . . , $\binom{N}{N-1}$ events of size $N - 1$, and $\binom{N}{N}$ events of size N. There is an alternate way of showing that if S has

N members then it has 2^N subsets. Let the members of S be numbered 1 to N. To describe a subset A of S, we may write an N-tuple $(t_1, t_2, \ldots, t_N)$, whose jth component t_j is equal to 1 or 0, depending on whether the jth member of S does or does not belong to the subset A. Since one can form 2^N N-tuples, it follows that S possesses 2^N subsets.

Another counting problem whose solution we shall need is that of finding the *number of partitions of a set of size N* and, in particular, of the set $S = \{1, 2, \ldots, N\}$. Let r be a positive integer and let $k_1, k_2, \ldots, k_r$ be positive integers such that $k_1 + k_2 + \ldots + k_r = N$. By a partition of S, with respect to r and $k_1, k_2, \ldots, k_r$, we mean a division of S into r subsets (ordered so that one may speak of a first subset, a second subset, etc.) such that the first subset has size k_1, the second subset has size k_2, and so on, up to the rth subset, which has size k_r.

▶ **Example 1E. Partitions of a set of size 4.** The possible partitions of the set $\{1, 2, 3, 4\}$ into three subsets, the first subset of size 1, the second subset of size 2, and the third subset of size 1, may be listed as follows:

$$\begin{array}{llllll}
(\{1\}, & \{2, 3\}, & \{4\}), & (\{2\}, & \{1, 3\}, & \{4\}), \\
(\{1\}, & \{2, 4\}, & \{3\}), & (\{2\}, & \{1, 4\}, & \{3\}), \\
(\{1\}, & \{3, 4\}, & \{2\}), & (\{2\}, & \{3, 4\}, & \{1\}), \\
(\{3\}, & \{1, 2\}, & \{4\}), & (\{4\}, & \{1, 2\}, & \{3\}) \\
(\{3\}, & \{1, 4\}, & \{2\}), & (\{4\}, & \{1, 3\}, & \{2\}) \\
(\{3\}, & \{2, 4\}, & \{1\}), & (\{4\}, & \{2, 3\}, & \{1\})
\end{array}$$

◀

We now prove that *the number of ways in which one can partition a set of size N into r ordered subsets so that the first subset has size k_1, the second subset has size k_2, and so on, where $k_1 + k_2 + \ldots + k_r = N$, is the product*

$$(1.14) \qquad \binom{N}{k_1}\binom{N-k_1}{k_2}\binom{N-k_1-k_2}{k_3}\cdots\binom{N-k_1-k_2-\cdots-k_{r-1}}{k_r}.$$

To prove (1.14) we proceed as follows. For the first subset of k_1 items there are N items available, so that there are $\binom{N}{k_1}$ ways in which the subset of k_1 items can be selected. There are $N - k_1$ items available from which to select the k_2 items that go into the second subset; consequently, the second subset, containing k_2 items, can be selected in $\binom{N-k_1}{k_2}$ ways. Continuing in this manner, we determine that the rth subset, containing k_r items, can be selected in $\binom{N-k_1-\cdots-k_{r-1}}{k_r}$ ways. By multiplying these expressions, we obtain the number of ways in which a set of size N can be partitioned in the manner described.

The expression (1.14) may be written in a more convenient form. It is clear by use of the definition of $\binom{N}{k}$ that

$$\binom{N}{k_1}\binom{N-k_1}{k_2} = \frac{N!}{k_1!k_2!(N-k_1-k_2)!}. \tag{1.15}$$

Next, one obtains

$$\binom{N}{k_1}\binom{N-k_1}{k_2}\binom{N-k_1-k_2}{k_3} = \frac{N!}{k_1!k_2!k_3!(N-k_1-k_2-k_3)!}.$$

Continuing in this manner, one finds that (1.14) is equal to

$$\frac{N!}{k_1!k_2!\cdots k_r!}. \tag{1.16}$$

Quantities of the form of (1.16) arise frequently, and a special notation is introduced to denote them. For any integer N, and r nonnegative integers $k_1, k_2, \ldots, k_r$ whose sum is N, we define the *multinomial* coefficient:

$$\binom{N}{k_1 k_2 \cdots k_r} = \frac{N!}{k_1!k_2!\cdots k_r!}. \tag{1.17}$$

The multinomial coefficients derive their name from the fact that they are the coefficients in the expansion of the Nth power of the multinomial form $a_1 + a_2 + \ldots + a_r$ in terms of powers of $a_1, a_2, \ldots, a_r$:

$$\begin{aligned}(a_1 + a_2 + \cdots + a_r)^N \\ = \sum_{\substack{k_1=0\\k_1+k_2+\cdots+k_r=N}}^{N} \sum_{k_2=0}^{N} \cdots \sum_{k_r=0}^{N} \binom{N}{k_1 k_2 \cdots k_r} a_1^{k_1} a_2^{k_2} \cdots a_r^{k_r}.\end{aligned} \tag{1.18}$$

It should be noted that the summation in (1.18) is over all nonnegative integers $k_1, k_2, \ldots, k_r$ which sum to N.

▶ **Example 1F. Bridge hands.** The number of different hands a player in a bridge game can obtain is

$$\binom{52}{13} = 635,013,559,600 \doteq (6.35)\,10^{11}, \tag{1.19}$$

since a bridge hand constitutes a set of thirteen cards selected from a set of

52. The number of ways in which a bridge deck may be dealt into four hands (labeled, as is usual, North, West, South, and East) is

$$(1.20)\quad \binom{52}{13}\binom{39}{13}\binom{26}{13}\binom{13}{13} = \binom{52}{13\ 13\ 13\ 13} = \frac{(52)!}{(13)!^4} \doteq (5.36)\ 10^{28}. \blacktriangleleft$$

The symbol $\doteq$ is used in this book to denote approximate equality.

It should be noted that tables of factorials and logarithms of factorials are available and may be used to evaluate expressions such as those in (1.20).

EXERCISES

1.1. A restaurant menu lists 3 soups, 10 meat dishes, 5 desserts, and 3 beverages. In how many ways can a meal (consisting of soup, meat dish, dessert, and beverage) be ordered?

1.2. Find the value of (i) $(5)_3$, (ii) $(5)^3$, (iii) $5!$ (iv) $\binom{5}{3}$.

1.3. How many subsets of size 3 does a set of size 5 possess? How many subsets does a set of size 5 possess?

1.4. In how many ways can a bridge deck be partitioned into 4 hands, each of size 13?

1.5. Five politicians meet at a party. How many handshakes are exchanged if each politician shakes hands with every other politician once and only once?

1.6. Consider a college professor who every year tells exactly 3 jokes in his course. If it is his policy never to tell the same 3 jokes in any year that he has told in any other year, what is the minimum number of jokes he will tell in 35 years? If it is his policy never to tell the same joke twice, what is the minimum number of jokes he will tell in 35 years?

1.7. In how many ways can a student answer an 8-question, true-false examination if (i) he marks half the questions true and half the questions false, (ii) he marks no two consecutive answers the same?

1.8. State, by inspection, the value of

$$3^4 + 4 \cdot 3^3 + \frac{4 \cdot 3}{1 \cdot 2} \cdot 3^2 + \frac{4 \cdot 3 \cdot 2}{1 \cdot 2 \cdot 3} \cdot 3 + 1.$$

1.9. If $\binom{n}{11} = \binom{n}{7}$, find n. If $\binom{18}{r} = \binom{18}{r-2}$, find r.

1.10. Find the value of (i) $\binom{5}{2\ 1\ 2}$, (ii) $\binom{5}{2\ 2\ 1}$, (iii) $\binom{5}{5\ 0\ 0}$, (iv) $\binom{5}{3\ 2\ 0}$.

Explain why $\binom{5}{3\ 2\ 0} = \binom{5}{3}$.

1.11. Evaluate the following sums:

$$\sum_{k=1}^{8} k^2, \qquad \sum_{i=1}^{3}\sum_{j=2}^{4} ij, \qquad \sum_{i=1}^{3}\sum_{j=i+1}^{4} i^j, \qquad \sum_{i=1}^{3}\sum_{j=i+1}^{4} j^i.$$

1.12. Given an alphabet of n symbols, in how many ways can one form words consisting of exactly k symbols? Consequently, find the number of possible 3 letter words that can be formed in the English language.

1.13. Find the number of 3-letter words that can be formed in the English language whose first and third letters are consonants and whose middle letter is a vowel.

1.14. Use (1.11) and the principle of mathematical induction to prove the binomial theorem, which is stated by (1.9).

2. POSING PROBABILITY PROBLEMS MATHEMATICALLY

The principle that lies at the foundation of the mathematical theory of probability is the following: to speak of the probability of a random event A, a probability space on which the event is defined must first be set up. In this section we show how several problems, which arise frequently in applied probability theory, may be formulated so as to be mathematically well posed. The examples discussed also illustrate the use of combinatorial analysis to solve probability problems that are posed in the context of finite sample description spaces with equally likely descriptions.

▶ **Example 2A. An urn problem.** Two balls are drawn with replacement (without replacement) from an urn containing six balls, of which four are white and two are red. Find the probability that (i) both balls will be white, (ii) both balls will be the same color, (iii) at least one of the balls will be white.

Solution: To set up a mathematical model for the experiment described, assume that the balls in the urn are distinguishable; in particular, assume that they are numbered 1 to 6. Let the white balls bear numbers 1 to 4, and let the red balls be numbered 5 and 6.

Let us first consider that the balls are drawn without replacement. The sample description space S of the experiment is then given by (3.1) of Chapter 1; more compactly we write

$$S = \{(z_1, z_2)\colon \quad \text{for } i = 1, 2, z_i = 1, \cdots, 6, \text{ but } z_1 \neq z_2\}. \tag{2.1}$$

In words, one may read (2.1) as follows: S is the set of all 2-tuples (z_1, z_2) whose components are any numbers, 1 to 6, subject to the restriction that no two components of a 2-tuple are equal. The jth component z_j of a description represents the number of the ball drawn on the jth draw. Now let A be the event that both balls drawn are white, let B be the event that both balls drawn are red, and let C be the event that at least one of the balls drawn is white. The problem at hand can then be stated as one of finding (i) $P[A]$, (ii) $P[A \cup B]$, (iii) $P[C]$. It should be noted that $C = B^c$, so that $P[C] = 1 - P[B]$. Further, A and B are mutually exclusive, so that $P[A \cup B] = P[A] + P[B]$. Now

$$(2.2) \quad A = \{(1, 2), (1, 3), (1, 4), (2, 1), (2, 3), (2, 4), (3, 1), (3, 2), (3, 4), (4, 1), (4, 2), (4, 3)\}$$

$$= \{(z_1, z_2): \quad \text{for } i = 1, 2, z_i = 1, \cdots, 4, \text{ but } z_1 \neq z_2\},$$

whereas $B = \{(5, 6), (6, 5)\}$. Let us assume that all descriptions in S are equally likely. Then

$$(2.3) \qquad P[A] = \frac{N[A]}{N[S]} = \frac{4 \cdot 3}{6 \cdot 5} = 0.4, \qquad P[B] = \frac{2 \cdot 1}{6 \cdot 5} = 0.066.$$

The answers to the questions posed in example 2A are given, in the case of sampling without replacement, by (i) $P[A] = 0.4$, (ii) $P[A \cup B] = 0.466$, (iii) $P[C] = 0.933$. These probabilities have been obtained under the assumption that the balls in the urn may be regarded as numbered (distinguishable) and that all descriptions in the sample description space S given in (2.1) are equally likely. In the case of sampling with replacement, a similar analysis may be carried out; one obtains the answers

$$(2.4) \qquad P[A] = \frac{4 \cdot 4}{6 \cdot 6} = 0.444, \qquad P[B] = \frac{2 \cdot 2}{6 \cdot 6} = 0.11,$$

$$P[A \cup B] = 0.555, \qquad P[C] = 0.888.$$

It is interesting to compare the values obtained by the foregoing model with values obtained by two other possible models. One might adopt as a sample description space $S = \{(W, W), (W, R), (R, W), (R, R)\}$. This space corresponds to recording the outcome of each draw as W or R, depending on whether the outcome of the draw is white or red. If one were to assume that all descriptions in S were equally likely, then $P[A] = \frac{1}{4}$, $P[A \cup B] = \frac{1}{2}$, $P[C] = \frac{3}{4}$. Note that the answers given by this model do not depend on whether the sampling is done with or without replacement. One arrives at a similar conclusion if one lets $S = \{0, 1, 2\}$, in which 0 signifies that no white balls were drawn, 1 signifies that exactly 1 white ball was drawn, and 2 signifies that exactly two white balls were drawn.

Under the assumption that all descriptions in S are equally likely, one would conclude that $P[A] = \frac{1}{3}$, $P[A \cup B] = \frac{2}{3}$, $P[C] = \frac{2}{3}$. ◀

The next example illustrates the treatment of problems concerning urns of arbitrary composition. It also leads to a conclusion that the reader may find startling if he considers the following formulation of it. Suppose that at a certain time the milk section of a self-service market is known to contain 150 quart bottles, of which 100 are fresh. If one assumes that each bottle is equally likely to be drawn, then the probability is $\frac{2}{3}$ that a bottle drawn from the section will be fresh. However, suppose that one selects one bottle after each of fifty other persons have selected a bottle. Is one's probability of drawing a fresh bottle changed from what it would have been had one been the first to draw? By the reasoning employed in example 2B it can be shown that the probability that the fifty-first bottle drawn will be fresh is the same as the probability that the first bottle drawn will be fresh.

▶ **Example 2B. An urn of arbitrary composition.** An urn contains M balls, of which M_W are white and M_R are red. A sample of size 2 is drawn with replacement (without replacement). What is the probability that (i) the first ball drawn will be white, (ii) the second ball drawn will be white, (iii) both balls drawn will be white?

Solution: Let A denote the event that the first ball drawn is white, B denote the event that the second ball drawn is white, and C denote the event that both balls drawn are white. It should be noted that $C = AB$. Let the balls in the urn be numbered 1 to M, the white balls bearing numbers 1 to M_W, and the red balls bearing numbers $M_W + 1$ to M.

We consider first the case of *sampling with replacement.* The sample description space S of the experiment consists of ordered 2-tuples (z_1, z_2), in which z_1 is the number of the ball drawn on the first draw and z_2 is the number of the ball drawn on the second draw. Clearly, $N[S] = M^2$. To compute $N[A]$, we use the fact that a description is in A if and only if its first component is a number 1 to M_W (meaning a white ball was drawn on the first draw) and its second component is a number 1 to M (due to the sampling with replacement the color of the ball drawn on the second draw is not affected by the fact that the first ball drawn was white). Thus there are M_W possibilities for the first component, and for each of these M possibilities for the second component of a description in A. Consequently, by (1.1), the size of A is $M_W M$. Similarly, $N[B] = MM_W$, since there are M possibilities for the first component and M_W possibilities for the second component of a description in B. The reader may verify by a similar argument that the event AB, (a white ball is drawn on both draws), has size $N[AB] = M_W M_W$. Thus in the case of sampling with replacement one obtains the result, if all descriptions are equally likely, that

$$P[A] = P[B] = \frac{M_W}{M}, \qquad P[AB] = \left(\frac{M_W}{M}\right)^2. \tag{2.5}$$

We next consider the case of *sampling without replacement*. The sample description space of the experiment again consists of ordered 2-tuples (z_1, z_2), in which z_j (for $j = 1, 2$) denotes the number of the ball drawn on the jth draw. As in the case of sampling with replacement, each z_j is a number 1 to M. However, in sampling without replacement a description (z_1, z_2) must satisfy the requirement that its components are *not* the same. Clearly, $N[S] = (M)_2 = M(M-1)$. Next, $N[A] = M_W(M-1)$, since there are M_W possibilities for the first component of a description in A and $M - 1$ possibilities for the second component of a description in A; the urn from which the second ball is drawn contains only $(M-1)$ balls. To compute $N[B]$, we first concentrate our attention on the second component of a description in B. Since B is the event that the ball drawn on the second draw is white, there are M_W possibilities for the second component of a description in B. To each of these possibilities, there are only $M - 1$ possibilities for the first component, since the ball which is to be drawn on the second draw is known to us and cannot be drawn on the first draw. Thus $N[B] = (M-1)M_W$ by (1.1). The reader may verify that the event AB has size $N[AB] = M_W(M_W - 1)$. Consequently, in sampling without replacement one obtains the result, if all descriptions are equally likely, that

$$P[A] = P[B] = \frac{M_W}{M}, \qquad P[AB] = \frac{M_W(M_W - 1)}{M(M-1)}. \tag{2.6}$$

Another way of computing $P[B]$, which the reader may find more convincing on first acquaintance with the theory of probability, is as follows. Let B_1 denote the event that the first ball drawn is white and the second ball drawn is white. Let B_2 denote the event that the first ball drawn is red and the second ball drawn is white. Clearly, $N[B_1] = M_W(M_W - 1)$, $N[B_2] = (M - M_W)M_W$. Since $P[B] = P[B_1] + P[B_2]$, we have

$$P[B] = \frac{M_W(M_W - 1)}{M(M-1)} + \frac{(M - M_W)M_W}{M(M-1)} = \frac{M_W}{M}.$$

To illustrate the use of (2.5) and (2.6), let us consider an urn containing $M = 6$ balls, of which $M_W = 4$ are white. Then $P[A] = P[B] = \frac{2}{3}$ and $P[AB] = \frac{4}{9}$ in sampling with replacement, whereas $P[A] = P[B] = \frac{2}{3}$ and $P[AB] = \frac{2}{5}$ in sampling without replacement.

The reader may find (2.6) startling. It is natural, in the case of sampling with replacement, in which $P[A] = P[B]$, that the probability of drawing a white ball is the same on the second draw as it is on the first draw, since the composition of the urn is the same in both draws. However, it seems very unnatural, if not unbelievable, that in sampling without replacement $P[A] = P[B]$. The following remarks may clarify the meaning of (2.6).

Suppose that one desired to regard the event that a white ball is drawn on the second draw as an event defined on the sample description space, denoted by S', which consists of all possible outcomes of the second draw. To begin with, one might write $S' = \{1, 2, \ldots, M\}$. However, how is a probability function to be defined on the subsets of S' in the case in which the sample is drawn without replacement. If one knows nothing about the outcome of the first draw, perhaps one might regard all descriptions in S' as being equally likely; then, $P[B] = M_W/M$. However, suppose one knows that a white ball was drawn on the first draw. Then the descriptions in S' are no longer equally likely; rather, it seems plausible to assign probability 0 to the description corresponding to the (white) ball, which is not available on the second draw, and assume the remaining descriptions to be equally likely. One then computes that the probability of the event B (that a white ball will be drawn on the second draw), given that the event A (that a white ball was drawn on the first draw) has occurred, is equal to $(M_W - 1)/(M - 1)$. Thus $(M_W - 1)/(M - 1)$ represents a conditional probability of the event B (and, in particular, the conditional probability of B, given that the event A has occurred), whereas M_W/M represents the unconditional probability of the event B. The distinction between unconditional and conditional probability is made precise in section 4. ◀

The next example we shall consider is a generalization of the celebrated problem of *repeated birthdays*. Suppose that one is present in a room in which there are n people. What is the probability that no two persons in the room have the same birthday? Let it be assumed that each person in the room can have as his birthday any one of the 365 days in the year (ignoring the existence of leap years) and that each day of the year is equally likely to be the person's birthday. Then selecting a birthday for each person is the same as selecting a number randomly from an urn containing $M = 365$ balls, numbered 1 to 365. It is shown in example 2C that the probability that no two persons in a room containing n persons will have the same birthday is given by

$$(2.7) \qquad \frac{(365)_n}{(365)^n} = \left(1 - \frac{1}{365}\right)\left(1 - \frac{2}{365}\right)\cdots\left(1 - \frac{n-1}{365}\right).$$

The value of (2.7) for various values of n appears in Table 2A.

TABLE 2A

In a room containing n persons let P_n be the probability that there *are not* two or more persons in the room with the same birthday and let Q_n be the probability that there *are* two or more persons with the same birthday.

n	P_n	Q_n
4	0.984	0.016
8	0.926	0.074
12	0.833	0.167
16	0.716	0.284
20	0.589	0.411
22	0.524	0.476
23	0.493	0.507
24	0.462	0.538
28	0.346	0.654
32	0.247	0.753
40	0.109	0.891
48	0.039	0.961
56	0.012	0.988
64	0.003	0.997

From Table 2A one determines a fact that many students find startling and completely contrary to intuition. How many people must there be in a room in order for the probability to be greater than 0.5 that at least two of them will have the same birthday? Students who have been asked this question have given answers as high as 100, 150, 365, and 730. In fact, the answer is 23!

▶ **Example 2C. The probability of a repetition in a sample drawn with replacement.** Let a sample of size n be drawn with replacement from an urn containing M balls, numbered 1 to M. Let P denote the probability that there are no repetitions in the sample (that is, that all the numbers in the sample occur just once). Let us show that

$$(2.8) \qquad P = \frac{(M)_n}{M^n} = \left(1 - \frac{1}{M}\right)\left(1 - \frac{2}{M}\right)\cdots\left(1 - \frac{n-1}{M}\right).$$

The sample description space S of the experiment of drawing with replacement a sample of size n from an urn containing M balls, numbered 1 to M, is

$$(2.9)\quad S = \{(z_1, z_2, \cdots, z_n)\colon \text{ for } i = 1, \cdots, n, z_i = 1, \cdots, M\}.$$

The jth component z_j of a description represents the number of the ball drawn on the jth draw. The event A that there are no repetitions in the sample is the set of all n-tuples in S, none of whose components are equal. The size of A is given by $N[A] = (M)_n$, since for any description in A there are M possibilities for its first component, $(M - 1)$ possibilities for its second component, and so on. The size of S is $N[S] = M^n$. If we assume that all descriptions in S are equally likely, then (2.8) follows. ◀

▶ **Example 2D. Repeated random digits.** Another application of (2.8) is to the problem of *repeated random digits*. Consider the following experiment. Take any telephone directory and open it to any page. Choose 100 telephone numbers from the page. Count the numbers whose last four digits are all different. If it is assumed that each of the last four digits is chosen (independently) from the numbers 0 to 9 with equal probability, then the probability that the last four digits of a randomly chosen telephone number will be different is given by (2.8), with $n = 4$ and $M = 10$. The probability is $(10)_4/10^4 = 0.504$. ◀

The next example is concerned with a celebrated problem, which we call here the problem of *matches*. Suppose you are one of M persons, each of whom has put his hat in a box. Each person then chooses a hat randomly from the box. What is the probability that you will choose your own hat? It seems reasonable that the probability of choosing one's own hat should be $1/M$, since one could have chosen any one of M hats. However, one might prefer to adopt a more detailed model that takes account of the fact that other persons may already have selected hats. A suitable mathematical model is given in example 2E. In section 6 the model given in example 2E is used to find the probability that at least one person will choose his own hat. But whether the number of hats involved is 8, 80, or 8,000,000, the rather startling result obtained is that the probability is approximately equal to $e^{-1} \doteq 0.368$ that no man will choose his own hat and approximately equal to $1 - e^{-1} \doteq 0.632$ that at least one man will choose his own hat.

▶ **Example 2E. Matches (rencontres).** Suppose that we have M urns, numbered 1 to M, and M balls, numbered 1 to M. Let one ball be inserted in each urn. If a ball is put into the urn bearing the same number as the ball, a match is said to have occurred. In section 6 formulas are given (for each integer $n = 0, 1, \ldots, M$) for the probability that exactly

n matches will occur. Here we consider only the problem of obtaining, for $k = 1, 2, \ldots, M$ the probability of the event A_k that a match will occur in the kth urn. The probability $P[A_k]$ corresponds, in the case of the M persons selecting their hats randomly from a box, to the probability that the kth person will select his own hat.

To write the sample description space S of the experiment of distributing M balls in M urns, let z_j represent the number of the ball inserted in the jth urn (for $j = 1, \ldots, M$). Then S is the set of M-tuples $(z_1, z_2, \ldots, z_M)$, in which each component z_j is a number 1 to M, but no two components are equal. The event A_k is the set of descriptions $(z_1, \ldots, z_M)$ in S such that $z_k = k$; in symbols, $A_k = \{(z_1, z_2, \ldots, z_M)\colon\ z_k = k\}$. It is clear that $N[A_k] = (M - 1)!$ and $N[S] = M!$. If it is assumed that all descriptions in S are equally likely, then $P[A_k] = 1/M$. Thus we have proved that the probability of a person's choosing his own hat does not depend on whether he is the first, second, or even the last person to choose a hat. ◀

Sample description spaces in which the descriptions are subsets and partitions rather than n-tuples are systematically discussed in section 5. The following example illustrates the ideas.

▶ **Example 2F. How to tell a prediction from a guess.** In order to verify the contention of the existence of extrasensory perception, the following experiment is sometimes performed. Eight cards, four red and four black, are shuffled, and then each is looked at successively by the experimenter. In another room the subject of study attempts to guess whether the card looked at by the experimenter is red or black. He is required to say "black" four times and "red" four times. If the subject of the study has no extrasensory perception, what is the probability that the subject will "guess" correctly the colors of exactly six of eight cards? Notice that the problem is unchanged if the subject claimed the gift of "prophecy" and, before the cards were dealt, stated the order in which he expected the cards to appear.

Solution: Let us call the first card looked at by the experimenter card 1; similarly, for $k = 2, 3, \ldots, 8$, let the kth card looked at by the experimenter be called card k. To describe the subject's response during the course of the experiment, we write the subset $\{z_1, z_2, z_3, z_4\}$ of the numbers $\{1, 2, 3, 4, 5, 6, 7, 8\}$, which consists of the numbers of all the cards the subject said were red. The sample description space S then consists of all subsets of size 4 of the set $\{1, 2, 3, 4, 5, 6, 7, 8\}$. Therefore, $N[S] = \binom{8}{4}$. The event A that the subject made exactly six correct guesses may be represented as the set of those subsets $\{z_1, z_2, z_3, z_4\}$, exactly three of whose members are equal to the numbers of cards that were, in fact, red. To compute the size of A, we notice that the three numbers in a description in A, corresponding to a correct guess, may be chosen in $\binom{4}{3}$ ways, whereas the one number in a description in A, corresponding to an incorrect guess, may be chosen in $\binom{4}{1}$ ways. Consequently, $N[A] = \binom{4}{3}\binom{4}{1}$, and

$$P[A] = \frac{\binom{4}{3}\binom{4}{1}}{\binom{8}{4}} = \frac{8}{35}. \qquad \blacktriangleleft$$

EXERCISES

In solving the following problems, state carefully any assumptions made. In particular, describe the probability space on which the events, whose probabilities are being found, are defined.

2.1. Two balls are drawn with replacement (without replacement) from an urn containing 8 balls, of which 5 are white and 3 are black. Find the probability that (i) both balls will be white, (ii) both balls will be the same color, (iii) at least 1 of the balls will be white.

2.2. An urn contains 3 red balls, 4 white balls, and 5 blue balls. Another urn contains 5 red balls, 6 white balls, and 7 blue balls. One ball is selected from each urn. What is the probability that (i) both will be white, (ii) both will be the same color?

2.3. An urn contains 6 balls, numbered 1 to 6. Find the probability that 2 balls drawn from the urn with replacement (without replacement), (i) will have a sum equal to 7, (ii) will have a sum equal to k, for each integer k from 2 to 12.

2.4. Two fair dice are tossed. What is the probability that the sum of the dice will be (i) equal to 7, (ii) equal to k, for each integer k from 2 to 12?

2.5. An urn contains 10 balls, bearing numbers 0 to 9. A sample of size 3 is drawn with replacement (without replacement). By placing the numbers in a row in the order in which they are drawn, an integer 0 to 999 is formed. What is the probability that the number thus formed is divisible by 39? *Note:* regard 0 as being divisible by 39.

2.6. Four probabilists arrange to meet at the Grand Hotel in Paris. It happens that there are 4 hotels with that name in the city. What is the probability that all the probabilists will choose different hotels?

2.7. What is the probability that among the 32 persons who were President of the United States in the period 1789–1952 at least 2 were born on the same day of the year.

2.8. Given a group of 4 people, find the probability that at least 2 among them have (i) the same birthday, (ii) the same birth month.

2.9. Suppose that among engineers there are 12 fields of specialization and that there is an equal number of engineers in each field. Given a group of 6 engineers, what is the probability that no 2 among them will have the same field of specialization?

2.10. Two telephone numbers are chosen randomly from a telephone book. What is the probability that the last digits of each are (i) the same, (ii) different?

2.11. Two friends, Irwin and Danny, are members of a group of 6 persons who have placed their hats on a table. Each person selects a hat randomly from the hats on the table. What is the probability that (i) Irwin will get his own hat, (ii) both Irwin and Danny will get their own hats, (iii) at least one, either Irwin or Danny, will get his own hat?

2.12. Two equivalent decks of 52 different cards are put into random order (shuffled) and matched against each other by successively turning over one card from each deck simultaneously. What is the probability that (i) the first, (ii) the 52nd card turned over from each deck will coincide? What is the probability that both the first and 52nd cards turned over from each deck will coincide?

2.13. In example 2F what is the probability that the subject will guess correctly the colors of (i) exactly 5 of the 8 cards, (ii) 4 of the 8 cards?

2.14. In his paper "Probability Preferences in Gambling," *American Journal of Psychology*, Vol. 66 (1953), pp. 349–364, W. Edwards tells of a farmer who came to the psychological laboratory of the University of Washington. The farmer brought a carved whalebone with which he claimed that he could locate hidden sources of water. The following experiment was conducted to test the farmer's claim. He was taken into a room in which there were 10 covered cans. He was told that 5 of the 10 cans contained water and 5 were empty. The farmer's task was to divide the cans into 2 equal groups, 1 group containing all the cans with water, the other containing those without water. What is the probability that the farmer correctly put at least 3 cans into the water group just by chance?

3. THE NUMBER OF "SUCCESSES" IN A SAMPLE

A basic problem of the theory of sampling is the following. An urn contains M balls, of which M_W are white (where $M_W < M$) and $M_R = M - M_W$ are red. A sample of size n is drawn either without replacement (in which case $n \leq M$), or with replacement. Let k be an integer between 0 and n (that is, $k = 0, 1, 2, \ldots,$ or n). What is the probability that the sample will contain exactly k white balls?

This problem is a prototype of many problems, which, as stated, do not involve the drawing of balls from an urn.

▶ **Example 3A. Acceptance sampling of a manufactured product.** Consider the problem of *acceptance sampling of a manufactured product.* Suppose we are to inspect a lot of size M of manufactured articles of some kind, such as light bulbs, screws, resistors, or anything else that is manufactured to meet certain standards. An article that is below standard is said to be *defective.* Let a sample of size n be drawn without replacement from the lot. A basic role in the theory of statistical quality control is played by the following problem. Let k and M_D be integers such that $k \leq n$ and $M_D \leq M$. What is the probability that the sample will contain k defective articles if the lot contains M_D defective articles? This is the same problem as that stated above, with defective articles playing the role of white balls. ◀

▶ **Example 3B. A sample-minded game warden.** Consider a fisherman who has caught 10 fish, 2 of which were smaller than the law permits to be caught. A game warden inspects the catch by examining two that he selects randomly from among the fish. What is the probability that he will not select either of the undersized fish? This problem is an example of those previously stated, involving sampling without replacement, with undersized fish playing the role of white balls, and $M = 10$, $M_W = 2$, $n = 2$, $k = 0$. By (3.1), the required probability is given by $\binom{2}{0}(2)_0(8)_2/(10)_2 = 28/45$. ◀

▶ **Example 3C. A sample-minded die.** Another problem, which may be viewed in the same context but which involves sampling with replacement, is the following. Let a fair die be tossed four times. What is the probability that one will obtain the number 3 exactly twice in the four tosses? This problem can be stated as one involving the drawing (with replacement) of balls from an urn containing balls numbered 1 to 6, among which ball number 3 is white and the other balls, red (or, more strictly, nonwhite). In the notation of the problem introduced at the beginning of the section this problem corresponds to the case $M = 6$, $M_W = 1$, $n = 4$, $k = 2$. By (3.2), the required probability is given by $\binom{4}{2}(1)^2(5)^2/(6)^4 = 25/216$. ◀

To emphasize the wide variety of problems, of which that stated at the beginning of the section is a prototype, it may be desirable to avoid references to white balls in the statement of the solution of the problem (although not in the statement of the problem itself) and to speak instead of scoring "successes." Let us say that we score a success whenever we draw a white ball. Then the problem can be stated as that of finding, for $k = 0, 1, \ldots, n$, the probability of the event A_k that one will score exactly k successes when one draws a sample of size n from an urn containing M balls, of which M_W are white. We now show that in the case of *sampling without replacement*

$$(3.1)\qquad P[A_k] = \binom{n}{k} \frac{(M_W)_k (M - M_W)_{n-k}}{(M)_n}, \qquad k = 0, 1, \ldots, n.$$

whereas in the case of *sampling with replacement*

$$(3.2)\qquad P[A_k] = \binom{n}{k} \frac{(M_W)^k (M - M_W)^{n-k}}{M^n}, \qquad k = 0, 1, \ldots, n.$$

It should be noted that in sampling without replacement if the number M_W of white balls in the urn is less than the size n of the sample drawn then clearly $P[A_k] = 0$ for $k = M_W + 1, \ldots, n$. Equation (3.1) embodies this fact, in view of (1.10).

Before indicating the proofs of (3.1) and (3.2), let us state some useful alternative ways of writing these formulas. For many purposes it is useful to express (3.1) and (3.2) in terms of

$$(3.3)\qquad p = \frac{M_W}{M},$$

the proportion of white balls in the urn. The formula for $P[A_k]$ can then be compactly written, in the case of *sampling with replacement*,

$$(3.4)\qquad P[A_k] = \binom{n}{k} p^k (1 - p)^{n-k}.$$

Equation (3.4) is a special case of a very general result, called the binomial law, which is discussed in detail in section 3 of Chapter 3. The expression given by (3.1) for the probability of k successes in a sample of size n *drawn without replacement* may be expressed in terms of p by

$$(3.5)\qquad P[A_k] = \binom{n}{k} p^k (1 - p)^{n-k}$$

$$\times \frac{\left[\left(1 - \frac{1}{M_W}\right)\left(1 - \frac{2}{M_W}\right) \cdots \left(1 - \frac{k-1}{M_W}\right)\left(1 - \frac{1}{M - M_W}\right) \times \left(1 - \frac{2}{M - M_W}\right) \cdots \left(1 - \frac{n-k-1}{M - M_W}\right)\right]}{\left(1 - \frac{1}{M}\right)\left(1 - \frac{2}{M}\right) \cdots \left(1 - \frac{n-1}{M}\right)}$$

Consequently, one sees that in the case in which k/M_W, $(n-k)/(M-M_W)$, and n/M are small (say, less than 0.1) then the probability of the event A_k is approximately the same in sampling without replacement as it is in sampling with replacement.

Another way of writing (3.1) is in the computationally simpler form

$$P[A_k] = \frac{\binom{M_W}{k}\binom{M-M_W}{n-k}}{\binom{M}{n}}. \tag{3.6}$$

It may be verified algebraically that (3.1) and (3.6) agree. In section 5 we discuss the intuitive meaning of (3.6).

We turn now to the proof of (3.1). Let the balls in the urn be numbered 1 to M, the white balls bearing numbers 1 to M_W. The sample description space S then consists of n-tuples $(z_1, z_2, \ldots, z_n)$, in which, for $i = 1, \ldots, n$, z_i is a number 1 to M, subject to the condition that no two components of an n-tuple may be the same. The size of S is given by $N[S] = (M)_n$. The event A_k consists of all sample descriptions in S, exactly k components of which are numbers 1 to M_W. To compute the size of A_k, we first compute the size of events B of the following form. Let $J = \{j_1, j_2, \ldots, j_k\}$ be a subset of size k of the set of integers $\{1, 2, \ldots, n\}$. Define B_J as the event that white balls are drawn in and only in those draws whose draw numbers are in J; that is, B_J is the set of descriptions $(z_1, z_2, \ldots, z_n)$ whose j_1st, j_2nd, $\ldots, j_k$th components are numbers 1 to M_W and whose remaining components are numbers $M_W + 1$ to M. The size of B_J may be obtained immediately by means of the basic principle of combinatorial analysis. We obtain $N[B_J] = (M_W)_k(M - M_W)_{n-k}$, since there are $(M_W)_k$ ways in which white balls may be assigned to the k components of a description in B_J in which white balls occur and $(M - M_W)_{n-k}$ ways in which nonwhite balls may be assigned to the remaining $(n - k)$ components. Now, by (1.8), there are $\binom{n}{k}$ subsets of size k of the integers $\{1, 2, \ldots, n\}$. For any two such subsets J and J' the corresponding events B_J and $B_{J'}$ are mutually exclusive. Further, the event A may be regarded as the union, over such subsets J, of the events B_J. Consequently, the size of A is given by $N[A] = \binom{n}{k}(M_W)_k(M - M_W)_{n-k}$. If we assume that all the descriptions in S are equally likely, we obtain (3.1). To prove (3.2), we use a similar argument.

▶ **Example 3D. The difference between k successes and successes on k specified draws.** Let a sample of size 3 be drawn without replacement from an urn containing six balls, of which four are white. The probability that the first and second balls drawn will be white and the third ball black is equal to $(4)_2(2)_1/(6)_3$. However, the probability that the sample will contain exactly two white balls is equal to $\binom{3}{2}(4)_2(2)_1/(6)_3$. If the sample is drawn with replacement, then the probability of white balls on the first and second draws and a black ball on the third is equal to $(4)^2(2)^1/(6)^3$, whereas the probability of exactly two white balls in the sample is equal to $\binom{3}{2}(4)^2(2)^1/(6)^3$. ◀

▶ **Example 3E. Acceptance sampling.** Suppose that we wish to inspect a certain product by means of a sample drawn from a lot. Probability theory cannot tell us how to constitute a lot or how to inspect the sample or even how large a sample to draw. Rather, probability theory can tell us the consequences of certain actions, given that certain assumptions are true. Suppose we decide to inspect the product by forming lots of size 1000, from which we will draw a sample of size 100. Each of the items in the sample is classified as defective or nondefective. It is unreasonable to demand that the lot be perfect. Consequently, we may decide to accept the lot if the sample contains one or fewer defectives and to reject the lot if two or more of the items inspected are defective. The question naturally arises as to whether this acceptance scheme is too lax or too stringent; perhaps we ought to demand that the sample contain no defectives, or perhaps we ought to permit the sample to contain two or fewer defectives. In order to decide whether or not a given acceptance scheme is suitable, we must determine the probability P that a randomly chosen lot will be accepted. However, we do not possess sufficient information to compute P. In order to compute the probability P of acceptance of a lot, using a given acceptance sampling plan, we must know the proportion p of defectives in a lot. Thus P is a function of p, and we write $P(p)$ to denote the probability of acceptance of a lot in which the proportion of defectives is p. Now for the acceptance sampling plan, which consists in drawing a sample of size 100 from a lot of size 1000 and accepting it if the lot contains one or fewer defectives, $P(p)$ is given by

$$P(p) = \frac{(1000q)_{100}}{(1000)_{100}} + 100\,\frac{1000p(1000q)_{99}}{(1000)_{100}}, \tag{3.7}$$

where we have let $q = 1 - p$. The graph of $P(p)$ as a function of p is called the operating characteristic curve, or OC curve, of the acceptance sampling plan. In Fig. 3A we have plotted the OC curve for the sampling scheme described. We see that the probability of accepting a lot is 0.95 if it contains 0.4% defective items, whereas the probability of accepting a lot is only 0.50 if it contains 1.7% defective items. ◀

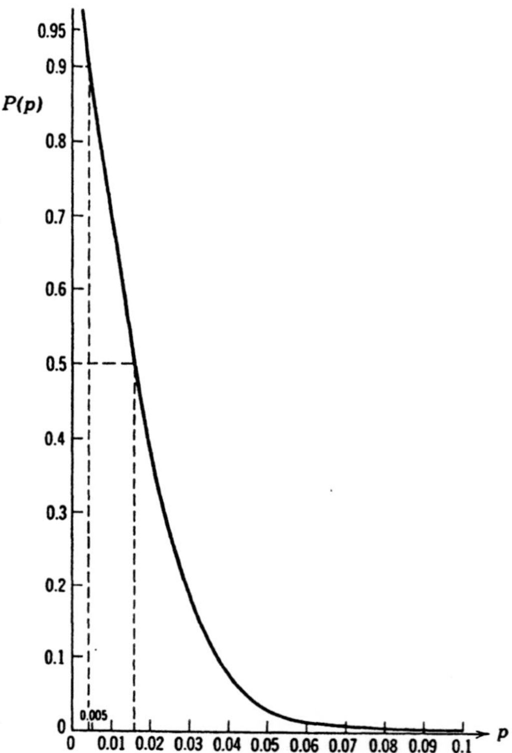

Fig. 3A. An operating characteristic, or OC, curve. $P(p)$ is the probability of accepting a lot containing proportion defective p for sample size $n = 100$ and acceptance number 1.

▶ **Example 3F. Winning a prize in a lottery.** Consider a lottery that sells n^2 tickets and awards n prizes. If one buys n tickets, what is the probability of winning a prize?

Solution: The probability P_1 of winning a prize is related to the probability P_0 of not winning a prize by $P_1 = 1 - P_0$. Now P_0 is the probability that a sample of size n drawn without replacement from an urn containing n^2 tickets will not contain any of n specified tickets. Consequently,

$$P_0 = \frac{(n^2 - n)_n}{(n^2)_n}. \tag{3.8}$$

In the case that $n = 10$, $P_0 = (90)_{10}/(100)_{10} = 0.330$, so that $P_1 = 0.670$. In the case that n is large it may be shown approximately, that,

$$P_0 = \frac{1}{e} = (2.718)^{-1} = 0.368, \qquad P_1 = 1 - e^{-1} = 0.632. \tag{3.9}$$ ◀

In the foregoing we have considered the problem of drawing a sample from an urn containing balls of only two colors. However, one may desire to consider urns containing balls of more than two colors. In theoretical exercises 3.1 to 3.3 we obtain formulas for this case. The following example illustrates the ideas involved.

▶ **Example 3G. Sampling from three plumbers.** Consider a town in which there are three plumbers, whom we call A, B, and C. On a certain day six residents of the town telephone for a plumber. If each resident selects a plumber at random from the telephone directory, what is the probability that three residents will call A, two residents will call B, and one resident will call C?

Solution: For $j = 1, 2, \ldots, 6$ let $z_j = A$, B, or C, depending on whether the plumber called by the jth resident is A, B, or C. The sample description space S of the observation is then a space of 6-tuples, $S = \{(z_1, z_2, \ldots, z_6)\colon \text{ for } j = 1, \ldots, 6, z_j = A, B, \text{ or } C\}$. Clearly, $N[S] = 3^6$. Next, the event E that three residents call A, two call B, and one calls C has size

$$N[E] = \binom{6}{3\ 2\ 1} = \frac{6!}{3!\ \ 2!\ \ 1!} = 60, \tag{3.10}$$

so that $P[E] = 60/3^6 = 0.082$. To prove (3.10), we note that the number of samples of size 6, which contain three calls for A, two calls for B, and one call for C, is the number of ways one can partition the set $\{1, 2, 3, 4, 5, 6\}$ into three ordered subsets of sizes 3, 2, and 1, respectively. ◀

THEORETICAL EXERCISES

3.1. Consider an urn containing M balls of r different colors. Let M_1, $M_2, \ldots, M_r$ denote, respectively, the number of balls of color 1, color $2, \ldots,$ color r. Show the probability that a sample of size n will contain k_1 balls of color 1, k_2 balls of color $2, \ldots, k_r$ balls of color r, where $k_1 + k_2 + \cdots + k_r = n$, in the case of sampling with replacement is given by

$$\binom{n}{k_1 k_2 \cdots k_r} \frac{(M_1)^{k_1}(M_2)^{k_2} \cdots (M_r)^{k_r}}{(M)^n}. \tag{3.11}$$

and, in the case of sampling without replacement, is given by

$$(3.12) \qquad \binom{n}{k_1 k_2 \cdots k_r} \frac{(M_1)_{k_1}(M_2)_{k_2} \cdots (M_r)_{k_r}}{(M)_n}.$$

Hint: The number of samples of size n that contain k_1 balls of color 1, k_2 balls of color 2, ..., k_r balls of color r is equal to the number of ways one can partition a set of size n into r ordered subsets of sizes $k_1, k_2, \ldots, k_r$, respectively.

3.2. Show that, in terms of the proportions

$$(3.13) \qquad p_1 = \frac{M_1}{M}, \qquad p_2 = \frac{M_2}{M}, \cdots, p_r = \frac{M_r}{M},$$

one may express (3.11) by

$$(3.14) \qquad \binom{n}{k_1 k_2 \cdots k_r} p_1^{k_1} p_2^{k_2} \cdots p_r^{k_r}.$$

3.3. Consider an urn containing n balls, each of a different color. Let r be any integer. Show the probability that a sample of size r drawn with replacement will contain r_1 balls of color 1, r_2 balls of color 2, ..., r_n balls of color n, where $r_1 + r_2 + \cdots + r_n = r$ is given by

$$\frac{1}{n^r}\binom{r}{r_1 r_2 \cdots r_n}.$$

3.4. An urn contains M balls, numbered 1 to M. Let N numbers be designated "lucky," where $N \leq M$. Let a sample of size n be drawn either without replacement (in which case $n \leq M$), or with replacement. Show that the probability that the sample will contain exactly k balls with "lucky" numbers is given by (3.1) and (3.2), respectively, with M_W replaced by N.

EXERCISES

3.1. An urn contains 52 balls, numbered 1 to 52. Suppose that numbers 1 through 13 are considered "lucky." A sample of size 2 is drawn from the urn with replacement (without replacement). What is the probability that (i) both balls drawn will be "lucky," (ii) neither ball drawn will be "lucky," (iii) at least 1 of the balls drawn will be "lucky," (iv) exactly 1 of the balls drawn will be "lucky"?

3.2. An urn contains 52 balls, numbered 1 to 52. Suppose that the numbers 1, 14, 27, and 40 are considered "lucky." A sample of size 13 is drawn from the urn with replacement (without replacement). What is the probability that the sample will contain (i) exactly 1 "lucky" number, (ii) at least 1 lucky number, (iii) exactly 4 "lucky" numbers?

3.3. A man tosses a fair coin 10 times. Find the probability that he will have (i) heads on the first 5 tosses, tails on the second 5 tosses, (ii) heads on tosses 1, 3, 5, 7, 9, tails on tosses 2, 4, 6, 8, 10, (iii) 5 heads and 5 tails, (iv) at least 5 heads, (v) no more than 5 heads.

3.4. A group of n men toss fair coins simultaneously. Find the probability that the n coins (i) are all heads, (ii) are all tails, (iii) contain exactly 1 head, (iv) contain exactly 1 tail, (v) are all alike. Evaluate these probabilities for $n = 2, 3, 4, 5$.

3.5. Consider 3 urns; urn I contains 2 white and 4 red balls, urn II contains 8 white and 4 red balls, urn III contains 1 white and 3 red balls. One ball is selected from each urn. Find the probability that the sample drawn will contain exactly 2 white balls.

3.6. A box contains 24 bulbs, 4 of which are known to be defective and the remainder of which is known to be nondefective. What is the probability that 4 bulbs selected at random from the box will be nondefective?

3.7. A box contains 50 razor blades, 5 of which are known to be used, the remainder unused. What is the probability that 5 razor blades selected from the box will be unused?

3.8. A fisherman caught 10 fish, 3 of which were smaller than the law permits to be caught. A game warden inspects the catch by examining 2, which he selects at random among the fish. What is the probability that he will not select any undersized fish?

3.9. A professional magician named Sebastian claimed to be able to "read minds." In order to test his claims, an experiment is conducted with 5 cards, numbered 1 to 5. A person concentrates on the numbers of 2 of the cards, and Sebastian attempts to "read his mind" and to name the 2 cards. What is the probability that Sebastian will correctly name the 2 cards, under the assumption that he is merely guessing?

3.10. Find approximately the probability that a sample of 100 items drawn from a lot of 1000 items contains 1 or fewer defective items if the proportion of the lot that is defective is (i) 0.01, (ii) 0.02, (iii) 0.05.

3.11. The contract between a manufacturer of electrical equipment (such as resistors or condensors) and a purchaser provides that out of each lot of 100 items 2 will be selected at random and subjected to a test. In negotiations for the contract the following two acceptance sampling plans are considered. Plan (*a*): reject the lot if both items tested are defective; otherwise accept the lot. Plan (*b*): accept the lot if both items tested are good; otherwise reject the lot. Obtain the operating characteristic curves of each of these plans. Which plan is more satisfactory to (i) the purchaser, (ii) the manufacturer? If you were the purchaser, would you consider either of the plans acceptable?

3.12. Consider a lottery that sells 25 tickets, and offers (i) 3 prizes, (ii) 5 prizes. If one buys 5 tickets, what is the probability of winning a prize?

3.13. Consider an electric fixture (such as Christmas tree lights) containing 5 electric light bulbs which are connected so that none will operate if any one of them is defective. If the light bulbs in the fixture are selected randomly from a batch of 1000 bulbs, 100 of which are known to be defective, find the probability that all the bulbs in the electric fixture will operate.

3.14. An urn contains 52 balls, numbered 1 to 52. Find the probability that a sample of 13 balls drawn without replacement will contain (i) each of the numbers 1 to 13, (ii) each of the numbers 1 to 7.

3.15. An urn contains balls of 4 different colors, each color being represented by the same number of balls. Four balls are drawn, with replacement. What is the probability that at least 3 different colors are represented in the sample?

3.16. From a committee of 3 Romans, 4 Babylonians, and 5 Philistines a subcommittee of 4 is selected by lot. Find the probability that the committee will consist of (i) 2 Romans and 2 Babylonians, (ii) 1 Roman, 1 Babylonian, and 2 Philistines; (iii) 4 Philistines.

3.17. Consider a town in which there are 3 plumbers; on a certain day 4 residents telephone for a plumber. If each resident selects a plumber at random from the telephone directory, what is the probability that (i) all plumbers will be telephoned, (ii) exactly 1 plumber will be telephoned?

3.18. Six persons, among whom are A and B, are arranged at random (i) in a row, (ii) in a ring. What is the probability that (*a*) A and B will stand next to each other, (*b*) A and B will be separated by one and only one person?

4. CONDITIONAL PROBABILITY

In section 3 we have been concerned with problems of the following type. Suppose one has a box containing 100 light bulbs, of which five are defective. What is the probability that a bulb selected from the box will be defective? A natural extension of this problem is the following. Suppose a light bulb (chosen from a box containing 100 light bulbs, of which five are defective) is found to be defective; what is the probability that a second light bulb drawn from the box (now containing 99 bulbs, of which four are defective) will be defective? A mathematical model for the statement and solution of this problem is provided by the notion of conditional probability.

Given two events, A and B, by the conditional probability of the event B, given the event A, denoted by $P[B \mid A]$, we mean intuitively the probability that B will occur, under the assumption that A has occurred. In other words, $P[B \mid A]$ represents our re-evalaution of the probability of B in the light of the information that A has occurred.

To motivate the formal definition of $P[B \mid A]$, which we shall give, let us consider the meaning of $P[B \mid A]$ from the point of view of the frequency interpretation of probability (since it is our desire to give to $P[B \mid A]$ a mathematical meaning that corresponds to its meaning as a relative frequency). Suppose one observes a large number N of occurrences of a random phenomenon in which the events A and B are defined. Let N_A denote the number of occurrences of the event A in the N occurrences of the random phenomenon. Similarly, let N_B denote the number of occurrences of B. Next, let N_{AB} denote the number of occurrences of the random phenomenon in which both the events A and B occur.

▶ **Example 4A. Thirty observed samples of size 2.** Consider the following results of thirty repetitions of the experiment of drawing, without replacement, a sample of size 2 from an urn containing six balls, numbered 1 to 6:

(1, 6),	(4, 5),	(1, 4),	(5, 3),	(3, 2),	(4, 3)
(3, 1),	(5, 1),	(2, 1),	(2, 3),	(4, 5),	(5, 6)
(5, 4),	(3, 1),	(6, 3),	(5, 6),	(2, 5),	(6, 4)
(1, 3),	(6, 2),	(4, 1),	(1, 5),	(4, 6),	(6, 3)
(2, 3),	(5, 2),	(3, 6),	(6, 4),	(6, 4),	(1, 2)

If the balls numbered 1 to 4 are colored white, and the balls numbered 5 and 6 are colored red, then the outcome of the thirty trials can be recorded as follows:

(W, R),	(W, R),	(W, W),	(R, W),	(W, W),	(W, W)
(W, W),	(R, W),	(W, W),	(W, W),	(W, R),	(R, R)
(R, W),	(W, W),	(R, W),	(R, R),	(W, R),	(R, W)
(W, W),	(R, W),	(W, W),	(W, R),	(W, R),	(R, W)
(W, W),	(R, W),	(W, R),	(R, W),	(R, W),	(W, W)

Let N_A denote the number of experiments in which a white ball appeared on the first trial. Let N_B denote the number of experiments in which a white ball appeared on the second trial, and let N_{AB} denote the number of experiments in which white balls appeared at both trials. By direct enumeration, one obtains that $N_A = 18$, $N_B = 21$, and $N_{AB} = 11$. ◀

In terms of the frequency definition, the unconditional probabilities of the events A, B, and AB are given by

$$P[A] = \frac{N_A}{N}, \qquad P[B] = \frac{N_B}{N}, \qquad P[AB] = \frac{N_{AB}}{N} \tag{4.1}$$

On the other hand, the conditional probability $P[B \mid A]$ of the event B, given the event A, represents the fraction of experiments in which A occurred that B also occurred; in symbols,

$$P[B \mid A] = \frac{N_{AB}}{N_A} \tag{4.2}$$

It should be noted that (4.2) makes sense only if N_A is not zero. If N_A is zero, we must regard $P[B \mid A]$ as being undefined.

Equation (4.2) represents the meaning of the notion of conditional probability from the frequency point of view. Now, (4.2) may be written in a manner that will indicate a formal definition of $P[B \mid A]$, which will embody the properties of conditional probability as it is intuitively conceived. We rewrite (4.2) (in the case that N_A is not zero):

$$P[B \mid A] = \frac{(N_{AB}/N)}{(N_A/N)} = \frac{P[AB]}{P[A]}. \tag{4.3}$$

In analogy with (4.3) we now give the following formal definition of $P[B \mid A]$:

FORMAL DEFINITION OF CONDITIONAL PROBABILITY. Let A and B be two events on a sample description space S, on the subsets of which is defined a probability function $P[\cdot]$. The conditional probability of the event B, given the event A, denoted by $P[B \mid A]$, is defined by

$$P[B \mid A] = \frac{P[AB]}{P[A]} \qquad \text{if } P[A] > 0, \tag{4.4}$$

and if $P[A] = 0$, then $P[B \mid A]$ is undefined.

▶ **Example 4B. Computing a conditional probability.** Consider the problem of drawing, without replacement, a sample of size 2 from an urn containing four white and two red balls. Let A denote the event that the first ball drawn is white, and B, the event that the second ball drawn is white. Let us compute $P[B \mid A]$. By (2.6), it follows that $P[AB] = (4 \cdot 3)/(6 \cdot 5) = \frac{12}{30}$, whereas $P[A] = \frac{4}{6} = \frac{20}{30}$. Therefore, $P[B \mid A] = \frac{12}{20} = 0.6$, which accords with our intuitive ideas, since the second ball is drawn from an urn containing five balls, of which three are white. Compare these theoretically computed probabilities with the observed relative frequencies in example 4A. We have $N_{AB}/N = \frac{11}{30}$, $N_A/N = \frac{18}{30}$, $N_{AB}/N_A = \frac{11}{18} = 0.611$. ◀

We next give a formula that may help to clarify the difference between the unconditional and the conditional probability of an event B. We have, for any events B and A such that $0 < P[A] < 1$,

$$P[B] = P[B \mid A]P[A] + P[B \mid A^c]P[A^c]. \tag{4.5}$$

Equation (4.5) is proved as follows. From the definition of conditional probability given by (4.4) one has the *basic formula*

$$P[AB] = P[A]\, P[B \mid A]. \tag{4.6}$$

Similarly, one has $P[A^cB] = P[A^c]P[B \mid A^c]$. Now, the events AB and A^cB are mutually exclusive, and their union is B. Consequently, $P[B] = P[AB] + P[A^cB]$. The desired conclusion may now be inferred.

▶ **Example 4C. A numerical verification of (4.5).** Consider again the problem in example 4B. One has $P[A] = \frac{2}{3}$. Therefore, $P[A^c] = \frac{1}{3}$. Next, one has $P[B \mid A] = \frac{3}{5}$. However, from this it does not follow that $P[B \mid A^c] = \frac{2}{5}$. Rather, by use of definition (4.4), $P[B \mid A^c] = \frac{4}{5}$; one may also obtain this result by intuitive reasoning (which is made rigorous in section 4 of Chapter 3), for if a white ball were not drawn on the first draw, there would be four white balls among the five balls in the urn from which the second draw would be made. Then, by (4.5), $P[B] = (\frac{3}{5})(\frac{2}{3}) + (\frac{4}{5})(\frac{1}{3}) = \frac{10}{15} = \frac{2}{3}$. ◀

Example 4D yields conclusions which students, on first acquaintance, often think startling and contrary to intuition.

▶ **Example 4D.** Consider a family with two children. Assume that each child is as likely to be a boy as it is to be a girl. What is the conditional probability that both children are boys, given that (i) the older child is a boy, (ii) at least one of the children is a boy?

Solution: Let A be the event that the older child is a boy, and let B be the event that the younger child is a boy. Then $A \cup B$ is the event that at least one of the children is a boy, and AB is the event that both children are boys. The probability that both children are boys, given that the older is a boy, is equal to

$$P[AB \mid A] = \frac{P[AB]}{P[A]} = \frac{1/4}{1/2} = \frac{1}{2}. \tag{4.7}$$

The probability that both children are boys, given that at least one of them is a boy, is equal to since $(AB)(A \cup B) = AB$

$$P[AB \mid A \cup B] = \frac{P[AB]}{P[A \cup B]} = \frac{1/4}{3/4} = \frac{1}{3}. \tag{4.8}$$ ◀

▶ **Example 4E. The outcome of a draw, given the outcome of a sample.** Let a sample of size 4 be drawn with replacement (without replacement), from an urn containing twelve balls, of which eight are white. Find the conditional probability that the ball drawn on the third draw was white, given that the sample contains three white balls.

Solution: Let A be the event that the sample contains exactly three white balls, and let B be the event that the ball drawn on the third draw was white. The problem at hand is to find $P[B \mid A]$. In the case of sampling with replacement

$$(4.9)\quad P[A] = \frac{\binom{4}{3}8^3 4}{(12)^4}, \qquad P[AB] = \frac{\binom{3}{2}8^3 4}{(12)^4}, \qquad P[B \mid A] = \frac{\binom{3}{2}}{\binom{4}{3}} = \frac{3}{4}.$$

In the case of sampling without replacement

$$(4.10)\quad P[A] = \frac{\binom{4}{3}(8)_3 4}{(12)_4}, \qquad P[AB] = \frac{\binom{3}{2}(8)_3 4}{(12)_4}, \qquad P[B \mid A] = \frac{\binom{3}{2}}{\binom{4}{3}} = \frac{3}{4}.$$

More generally, it may be proved (see theoretical exercise 4.4) that if a sample of size n contains k white balls then the probability is k/n that on any specified draw a white ball was drawn. Note that this result is the same, no matter what the composition of the urn and irrespective of whether the sample was drawn with or without replacement. In a sense, one may express the results just stated by the statement that on any given draw all balls in the sample are equally likely to occur. Many students attempt to solve the problem given here by reasoning that on the third draw any one of the four balls in the sample could have occurred and of these three are white, so that the (conditional) probability of a white ball on the third draw is $\frac{3}{4}$, in agreement with the foregoing equations. However, this line of reasoning consists in making assumptions *in addition* to those made in our derivation of these equations. It is desirable to prove that these new assumptions are a *consequence of the model postulated in* deriving (4.9) and (4.10). ◀

THEORETICAL EXERCISES

4.1. Prove the following statements, for any events A, B, and C, such that $P[C] > 0$. These relations illustrate the fact that all general theorems on probabilities are also valid for conditional probabilities with respect to any particular event C.

(i) $P[S \mid C] = 1$ where S is the certain event.

(ii) $P[A \mid C] = 1$ if C is a subevent of A.

(iii) $P[A \mid C] = 0$ if $P[A] = 0$.

(iv) $P[A \cup B \mid C] = P[A \mid C] + P[B \mid C] - P[AB \mid C]$.

(v) $P[A^c \mid C] = 1 - P[A \mid C]$.

4.2. Let B be an event of positive probability. Show that for any event A,

(i) $A \subset B$ implies $P[A \mid B] = P[A]/P[B]$,

(ii) $B \subset A$ implies $P[A \mid B] = 1$.

4.3. Let A and B be two events, each with positive probability. Show that statement (i) is true, whereas statements (ii) and (iii) are, in general, false:

(i) $$P[A \mid B] + P[A^c \mid B] = 1.$$

(ii) $$P[A \mid B] + P[A \mid B^c] = 1.$$

(iii) $$P[A \mid B] + P[A^c \mid B^c] = 1.$$

4.4. An urn contains M balls, of which M_W are white (where $M_W \leq M$). Let a sample of size n be drawn from the urn either with replacement or without replacement. For $j = 1, 2, \ldots, n$ let B_j be the event that the ball drawn on the jth draw is white. For $k = 1, 2, \ldots, n$ let A_k be the event that the sample (of size n) contains exactly k white balls. Show that $P[B_j \mid A_k] = k/n$. Express this fact in words.

4.5. An urn contains M balls, of which M_W are white. n balls are drawn and laid aside (not replaced in the urn), their color unnoted. Another ball is drawn (it is assumed that n is less than M). What is the probability that it will be white? *Hint:* Compare example 2B.

EXERCISES

4.1. A man tosses 2 fair coins. What is the conditional probability that he has tossed 2 heads, given that he has tossed at least 1 head?

4.2. An urn contains 12 balls, of which 4 are white. Five balls are drawn and laid aside (not replaced in the urn), their color unnoted.

(i) Another ball is drawn. What is the probability that it will be white?

(ii) A sample of size 2 is drawn. What is the probability that it will contain exactly one white ball?

(iii) What is the conditional probability that it will contain exactly 2 white balls, given that it contains at least 1 white ball.

4.3. In the milk section of a self-service market there are 150 quarts, 100 of which are fresh, and 50 of which are a day old.

(i) If 2 quarts are selected, what is the probability that both will be fresh?

(ii) Suppose that the 2 quarts are selected after 50 quarts have been removed from the section. What is the probability that both will be fresh?

(iii) What is the conditional probability that both will be fresh, given that at least 1 of them is fresh?

4.4. The student body of a certain college is composed of 60% men and 40% women. The following proportions of the students smoke cigarettes: 40% of the men and 60% of the women. What is the probability that a student who is a cigarette smoker is a man? A woman?

4.5. Consider two events A and B such that $P[A] = \frac{1}{4}$, $P[B \mid A] = \frac{1}{2}$, $P[A \mid B] = \frac{1}{4}$. For each of the following 4 statements, state whether it is true or false: (i) The events A and B are mutually exclusive, (ii) A is a subevent of B, (iii) $P[A^c \mid B^c] = \frac{3}{4}$; (iv) $P[A \mid B] + P[A \mid B^c] = 1$.

4.6. Consider an urn containing 12 balls, of which 8 are white. Let a sample of size 4 be drawn with replacement (without replacement). What is the conditional probability that the first ball drawn will be white, given that the sample contained exactly (i) 2 white balls, (ii) 3 white balls?

4.7. Consider an urn containing 6 balls, of which 4 are white. Let a sample of size 3 be drawn with replacement (without replacement). Let A denote the event that the sample contains exactly 2 white balls, and let B denote the event that the ball drawn on the third draw is white. Verify numerically that (4.5) holds in this case.

4.8. Consider an urn containing 12 balls, of which 8 are white. Let a sample of size 4 be drawn with replacement (without replacement). What is the conditional probability that the second and third balls drawn will be white, given that the sample contains exactly three white balls?

4.9. Consider 3 urns; urn I contains 2 white and 4 red balls, urn II contains 8 white and 4 red balls, urn III contains 1 white and 3 red balls. One ball is selected from each urn. What is the probability that the ball selected from urn II will be white, given that the sample drawn contains exactly 2 white balls?

4.10. Consider an urn in which 4 balls have been placed by the following scheme. A fair coin is tossed; if the coin falls heads, a white ball is placed in the urn, and if the coin falls tails, a red ball is placed in the urn.
(i) What is the probability that the urn will contain exactly 3 white balls?
(ii) What is the probability that the urn will contain exactly 3 white balls, given that the first ball placed in the urn was white?

4.11. A man tosses 2 fair dice. What is the (conditional) probability that the sum of the 2 dice will be 7, given that (i) the sum is odd, (ii) the sum is greater than 6, (iii) the outcome of the first die was odd, (iv) the outcome of the second die was even, (v) the outcome of at least 1 of the dice was odd, (vi) the 2 dice had the same outcomes, (vii) the 2 dice had different outcomes, (viii) the sum of the 2 dice was 13?

4.12. A man draws a sample of 3 cards one at a time (without replacement) from a pile of 8 cards, consisting of the 4 aces and the 4 kings in a bridge deck. What is the (conditional) probability that the sample will contain at least 2 aces, given that it contains (i) the ace of spades, (ii) at least one ace? Explain why the answers to (i) and (ii) need not be equal.

4.13. Consider 4 cards, on each of which is marked off a side 1 and side 2. On card 1, both side 1 and side 2 are colored red. On card 2, both side 1 and side 2 are colored black. On card 3, side 1 is colored red and side 2 is colored black. On card 4, side 1 is colored black and side 2 is colored red. A card is chosen at random. What is the (conditional) probability that if one side of the card selected is red the other side of the card will be black? What is the (conditional) probability that if side 1 of the card selected is examined and found to be red side 2 of the card will be black? *Hint:* Compare example 4D.

4.14. A die is loaded in such a way that the probability of a given number turning up is proportional to that number (for instance, a 4 is twice as probable as a 2).

(i) What is the probability of rolling a 5, given that an odd number turns up.

(ii) What is the probability of rolling an even number, given that a number less than 5 turns up.

5. UNORDERED AND PARTITIONED SAMPLES—OCCUPANCY PROBLEMS

We have insisted in the foregoing that the experiment of drawing a sample from an urn should always be performed in such a manner that one may speak of the first ball drawn, the second ball drawn, etc. Now it is clear that sampling need not be done in this way. Especially if one is sampling without replacement, the balls in the sample may be extracted from the urn not one at a time but all at once. For example, as in a bridge game, one may extract 13 cards from a deck of cards and examine them after all have been received and rearranged. If n balls are extracted all at once from an urn containing M balls, numbered 1 to M, the outcome of the experiment is a *subset* $\{z_1, z_2, \ldots, z_n\}$ of the numbers 1 to M, rather than an n-tuple $(z_1, z_2, \ldots, z_n)$ whose components are numbers 1 to M.

We are thus led to define the notions of *ordered* and *unordered* samples. A sample is said to be ordered if attention is paid to the order in which the numbers (on the balls in the sample) appear. A sample is said to be unordered if attention is paid only to the numbers that appear in the sample but not to the order in which they appear. The sample description space of the random experiment of drawing (with or without replacement) an ordered sample of size n from an urn containing M balls numbered 1 to M consists of n-tuples $(z_1, z_2, \ldots, z_n)$, in which each component z_j is a number 1 to M. The sample description space S of the random experiment of drawing (with or without) replacement an unordered sample of size n from an urn containing M balls numbered 1 to M consists of sets $\{z_1, z_2, \ldots, z_n\}$ of size n, in which each member z_j is a number 1 to M.

▶ **Example 5A. All possible unordered samples of size 3 from an urn containing four balls.** In example 1A we listed all possible sample descriptions in the case of the random experiment of drawing, with or without replacement, an ordered sample of size 3 from an urn containing four balls. We now list all possible unordered samples. If the sampling is done without replacement, then the possible unordered samples of size 3 that can be drawn are

$$\{1, 2, 3\}, \qquad \{1, 2, 4\}, \qquad \{1, 3, 4\}, \qquad \{2, 3, 4\}$$

If the sampling is done with replacement, then the possible unordered samples of size 3 that can be drawn are

$$\begin{array}{llll}
\{1, 1, 1\}, & \{2, 2, 2\}, & \{3, 3, 3\}, & \{4, 4, 4\} \\
\{1, 1, 2\}, & \{2, 2, 3\}, & \{3, 3, 4\}, & \\
\{1, 1, 3\}, & \{2, 2, 4\}, & \{3, 4, 4\}, & \\
\{1, 1, 4\}, & \{2, 3, 3\}, & & \\
\{1, 2, 2\}, & \{2, 3, 4\}, & & \\
\{1, 2, 3\}, & \{2, 4, 4\}, & & \\
\{1, 2, 4\}, & & & \\
\{1, 3, 3\}, & & & \\
\{1, 3, 4\}, & & & \\
\{1, 4, 4\}, & & &
\end{array}$$

◀

We next compute the size of S. In the case of unordered samples, drawn without replacement, it is clear that $N[S] = \binom{M}{n}$, since the number of unordered samples of size n is the same as the number of subsets of size n of the set $\{1, 2, \ldots, M\}$. In the case of unordered samples drawn with replacement, one may show (see theoretical exercise 5.2) that $N[S] = \binom{M + n - 1}{n}$.

In section 3 the problem of the number of successes in a sample was considered under the assumption that the sample was ordered. Suppose now that an *unordered sample* of size n is drawn from an urn containing M balls, of which M_W are white. Let us find, for $k = 0, 1, \ldots, n$, the probability of the event A_k that the sample will contain exactly k white balls. We consider first the case of sampling without replacement. Then $N[S] = \binom{M}{n}$. Next, $N[A_k] = \binom{M_W}{k}\binom{M - M_W}{n - k}$, since any description $\{z_1, z_2, \ldots, z_n\}$ in A_k contains k white balls, which can be chosen in $\binom{M_W}{k}$ ways, and $(n - k)$ nonwhite balls, which can be chosen in $\binom{M - M_W}{n - k}$ ways. Consequently, in the case of unordered samples drawn without replacement

$$P[A_k] = \frac{\binom{M_W}{k}\binom{M - M_W}{n - k}}{\binom{M}{n}}. \tag{5.1}$$

It is readily verified that the value of $P[A_k]$, given by the model of unordered samples, agrees with the value of $P[A_k]$, given by the model of ordered samples, in the case of sampling without replacement. However, in the case of sampling with replacement the probability that an unordered sample of size n, drawn from an urn containing M balls, of which M_W are white, will contain exactly k white balls is equal to

$$P[A_k] = \frac{\binom{M_W + k - 1}{k}\binom{M - M_W + n - k - 1}{n - k}}{\binom{M + n - 1}{n}}, \tag{5.2}$$

which does not agree with the value of $P[A_k]$, given by the model of ordered samples.

▶ **Example 5B. Distributing balls among urns (the occupancy problem).** Suppose that we are given M urns, numbered 1 to M, among which we are to distribute n balls, where $n < M$. What is the probability that each of the urns numbered 1 to n will contain exactly 1 ball?

Solution: Let A be the event that each of the urns numbered 1 to n will contain exactly 1 ball. In order to determine the probability space on which the event A is defined, we must first make assumptions regarding (i) the distinguishability of the balls and (ii) the manner in which the distribution of balls is to be carried out.

If the balls are regarded as being distinguishable (by being labeled with the numbers 1 to n), then to describe the results of distributing n balls among the N urns one may write an n-tuple $(z_1, z_2, \ldots, z_n)$, whose jth component z_j designates the number of the urn in which ball j was deposited. If the balls are regarded as being all alike, and therefore indistinguishable, then to describe the results of distributing n balls among the N urns one may write a set $\{z_1, z_2, \ldots, z_n\}$ of size n, in which each member z_j represents the number of an urn into which a ball has been deposited. Thus *ordered and unordered samples correspond in the occupancy problem to distributing distinguishable and indistinguishable balls, respectively.*

Next, in distributing the balls, *one may or may not impose an exclusion rule* to the effect that in distributing the balls one ball at most may be put into any urn. It is clear that imposing an exclusion rule is equivalent to choosing the urn numbers (sampling) without replacement, since an urn may be chosen once at most. If an exclusion rule is not imposed, so that in any urn one may deposit as many balls as one pleases, then one is choosing the urn numbers (sampling) with replacement.

Let us now return to the problem of computing $P[A]$. The size of the

TABLE 5A

<table>
<tr><td colspan="3">The number of ways in which n balls may be distributed into M distinguishable urns</td><td></td></tr>
<tr><td>Balls distributed</td><td>Distinguishable balls</td><td>Indistinguishable balls</td><td></td></tr>
<tr><td>Without exclusion</td><td>M^n
Maxwell-Boltzmann statistics</td><td>$\binom{M+n-1}{n}$
Bose-Einstein statistics</td><td>With replacement</td></tr>
<tr><td>With exclusion</td><td>$(M)_n$</td><td>$\binom{M}{n}$
Fermi-Dirac statistics</td><td>Without replacement</td></tr>
<tr><td></td><td>Ordered samples</td><td>Unordered samples</td><td>Samples drawn</td></tr>
<tr><td></td><td colspan="3">The number of ways in which samples of size n may be drawn from an urn containing M distinguishable balls</td></tr>
</table>

sample description space is given in Table 5A for each of the various possible cases. Next, let us determine the size of A. Whether or not an exclusion rule is imposed, we obtain $N[A] = n!$ if the balls are distinguishable and $N[A] = 1$ if the balls are indistinguishable. Consequently, if the balls are distinguishable and distributed without exclusion,

$$P[A] = \frac{n!}{M^n}; \tag{5.3}$$

if the balls are indistinguishable and distributed without exclusion,

$$P[A] = \frac{1}{\binom{M+n-1}{n}}; \tag{5.4}$$

if the balls are distributed with exclusion, it makes no difference whether the balls are considered distinguishable or indistinguishable, since

$$P[A] = \frac{n!}{(M)_n} = \frac{1}{\binom{M}{n}}. \tag{5.5}$$

Each of the different probability models for occupancy problems, described in the foregoing, find application in statistical physics. Suppose one seeks to determine the equilibrium state of a physical system composed of a very large number n of "particles" of the same nature: electrons, protons, photons, mesons, neutrons, etc. For simplicity, assume that there are M microscopic states in which each of the particles can be (for example, there are M energy levels that a particle can occupy). To describe the macroscopic state of the system, suppose that it suffices to state the M-tuple $(n_1, n_2, \ldots, n_M)$ whose jth component n_j is the number of "particles" in the jth microscopic state. The equilibrium state of the system of particles is defined as that macroscopic state $(n_1, n_2, \ldots, n_M)$ with the highest probability of occurring. To compute the probability of any given macroscopic state, an assumption must be made as to whether or not the particles obey the Pauli exclusion principle (which states that there cannot be more than one particle in any of the microscopic states). If the indistinguishable particles are assumed to obey the exclusion principle, then they are said to possess Fermi-Dirac statistics. If the indistinguishable particles are not required to obey the exclusion principle, then they are said to possess Bose-Einstein statistics. If the particles are assumed to be distinguishable and do not obey the exclusion principle, then they are said to possess Maxwell-Boltzmann statistics. Although physical particles cannot be considered distinguishable, Maxwell-Boltzmann statistics are correct as approximations in certain circumstances to Bose-Einstein and Fermi-Dirac statistics. ◀

The probability of various events defined on the general occupancy and sampling problems are summarized in Table 6A on p. 84.

Partitioned Samples. If we examine certain card games, we may notice still another type of sampling. We may extract n distinguishable balls (or cards) from an urn (or deck of cards), which can then be divided into a number of subsets (in a bridge game, into four hands). More precisely, we may specify a positive integer r and nonnegative integers $k_1, k_2, \ldots, k_r$, such that $k_1 + k_2 + \ldots + k_r = n$. We then divide the sample of size n into r subsets; a first subset of size k_1, a second subset of size $k_2, \ldots$, an rth subset of size k_r. For example, in the game of bridge there are four hands (subsets), each of size 13, called East, North, West, and South (instead of first, second, third, and fourth subsets). The outcome of a sample taken in this way is an *r-tuple of subsets,*

$$(5.6)\quad (\{z_1, \cdots, z_{k_1}\}, \{z_{k_1+1}, \cdots, z_{k_1+k_2}\}, \cdots, \{z_{k_1+\cdots+k_{r-1}+1}, \cdots, z_{k_1+\cdots+k_r}\})$$

whose first component is the first subset, second component is the second subset, . . . , rth component is the rth subset. We call a sample of the form of (5.6) a *partitioned sample*, with *partitioning scheme* $(r;\ k_1, k_2, \ldots, k_r)$.

▶ **Example 5C. An example of partitioned samples.** Consider again the experiment of drawing a sample of size 3 from an urn containing four balls, numbered 1 to 4. If the sampling is done without replacement, and the sample is partitioned, with partitioning scheme (2; 1, 2), then the possible samples that could have been drawn are

$$\begin{array}{llll}
(\{1\}, \{2, 3\}), & (\{2\}, \{1, 3\}), & (\{3\}, \{1, 2\}), & (\{4\}, \{1, 2\}) \\
(\{1\}, \{2, 4\}), & (\{2\}, \{1, 4\}), & (\{3\}, \{1, 4\}), & (\{4\}, \{1, 3\}) \\
(\{1\}, \{3, 4\}), & (\{2\}, \{3, 4\}), & (\{3\}, \{2, 4\}), & (\{4\}, \{2, 3\})
\end{array}$$

If the sampling is done with replacement, and the sample is partitioned, with partitioning scheme (2; 1, 2), then the possible samples that could have been drawn are

$$\begin{array}{llll}
(\{1\}, \{1, 1\}), & (\{2\}, \{1, 1\}), & (\{3\}, \{1, 1\}), & (\{4\}, \{1, 1\}) \\
(\{1\}, \}1, 2\}), & (\{2\}, \{1, 2\}), & (\{3\}, \{1, 2\}), & (\{4\}, \{1, 2\}) \\
(\{1\}, \{1, 3\}), & (\{2\}, \{1, 3\}), & (\{3\}, \{1, 3\}), & (\{4\}, \{1, 3\}) \\
(\{1\}, \{1, 4\}), & (\{2\}, \{1, 4\}), & (\{3\}, \{1, 4\}), & (\{4\}, \{1, 4\}) \\
(\{1\}, \{2, 2\}), & (\{2\}, \{2, 2\}), & (\{3\}, \{2, 2\}), & (\{4\}, \{2, 2\}) \\
(\{1\}, \{2, 3\}), & (\{2\}, \{2, 3\}), & (\{3\}, \{2, 3\}), & (\{4\}, \{2, 3\}) \\
(\{1\}, \{2, 4\}), & (\{2\}, \{2, 4\}), & (\{3\}, \{2, 4\}), & (\{4\}, \{2, 4\}) \\
(\{1\}, \{3, 3\}), & (\{2\}, \{3, 3\}), & (\{3\}, \{3, 3\}), & (\{4\}, \{3, 3\}) \\
(\{1\}, \{3, 4\}), & (\{2\}, \{3, 4\}), & (\{3\}, \{3, 4\}), & (\{4\}, \{3, 4\}) \\
(\{1\}, \{4, 4\}), & (\{2\}, \{4, 4\}), & (\{3\}, \{4, 4\}), & (\{4\}, \{4, 4\})
\end{array}$$ ◀

We next derive formulas for the number of ways in which partitioned samples may be drawn.

In the case of sampling without replacement from an urn containing M balls, numbered 1 to M, the number of possible partitioned samples of size n, with partitioning scheme $(r;\ k_1, k_2, \ldots, k_r)$, is equal to

$$(5.7)\quad \binom{M}{k_1}\binom{M-k_1}{k_2}\cdots\binom{M-k_1-\cdots-k_{r-1}}{k_r} = \binom{M}{k_1\,k_2 \cdots k_r\; M-n}.$$

Since there are $\binom{M}{k_1}$ possible subsets of k_1 balls, $\binom{M-k_1}{k_2}$ possible subsets of k_2 balls (there are $M - k_1$ balls available from which to select the k_2 balls to go into the second subset), it follows that there are $\binom{M-k_1-\cdots-k_{r-1}}{k_r}$ ways in which to select the rth subset.

In the case of sampling with replacement from an urn containing M balls, numbered 1 to M, the number ot possible partitioned samples of size n, with partitioning scheme $(r;\ k_1, k_2, \cdots, k_r)$, is equal to

$$\binom{M+k_1-1}{k_1}\binom{M+k_2-1}{k_2}\cdots\binom{M+k_r-1}{k_r}. \tag{5.8}$$

The next example illustrates the theory of partitioned samples and provides a technique whereby card games such as bridge may be analyzed.

▶ **Example 5D.** An urn contains fifty-two balls, numbered 1 to 52. Let the balls be drawn one at a time and divided among four players in the following manner: for $j = 1, 2, 3, 4$, balls drawn on trials numbered $j + 4k$ (for $k = 0, 1, \ldots, 12$) are given to player j. Thus player 1 gets the balls drawn on the first, fifth, . . . , forty-ninth draws, player 2 gets the balls drawn on the second, sixth, . . . , fiftieth draws, and so on. Suppose that the balls numbered 1, 11, 31, and 41 are considered "lucky." What is the probability that each player will have a "lucky" ball?

Solution: Dividing the fifty-two balls drawn among four players in the manner described is exactly the same process as drawing, without replacement, a partitioned sample of size 52, with partitioning scheme (4; 13, 13, 13, 13). The sample description space S of the experiment being performed here consists of 4-tuples of mutually exclusive subsets, of size 13, of the numbers 1 to 52, in which (for $j = 1, 2, \ldots, 4$) the jth subset represents the balls held by the jth player. The size of the sample description space is the number of ways in which a sample of fifty-two balls, partitioned in the way we have described, may be drawn from an urn containing fifty-two distinguishable balls. Thus

$$N[S] = \binom{52}{13}\binom{39}{13}\binom{26}{13}\binom{13}{13} = \frac{52!}{(13!)^4}. \tag{5.9}$$

We next calculate the size of the event A that each of the four players will have exactly one "lucky" ball. First, consider a description in A that has the following properties: player 1 has ball number 11, player 2 has ball number 41, player 3 has ball number 1, and player 4 has ball number 31. Each description has forty-eight members about which nothing has been specified; consequently there are $(48!)(12!)^{-4}$ descriptions, for in this many ways can the remaining forty-eight balls be distributed among the members of the description. Now the four "lucky" balls can be distributed among the four hands in 4! ways. Consequently,

$$N[A] = 4!\,\frac{(48!)}{(12!)^4} \tag{5.10}$$

and the probability that each player will possess exactly one "lucky" ball is given by the quotient of (5.10) and (5.9). ◀

The interested reader may desire to consider for himself the theory of partitions that are unordered, rather than ordered, arrays of subsets.

THEORETICAL EXERCISES

5.1. An urn contains M balls, numbered 1 to M. A sample of size n is drawn without replacement, and the numbers on the balls are arranged in increasing order of their numbers: $x_1 < x_2 < \ldots < x_n$. Let K be a number 1 to M, and k, a number 1 to n. Show the probability that $x_k = K$ is

$$\binom{K-1}{k-1}\binom{M-K}{n-k}\bigg/\binom{M}{n} \tag{5.11}$$

5.2. The number of unordered samples with replacement. Let $U(M, n)$ denote the number of unordered samples of size n that one may draw, by sampling with replacement, from an urn containing M distinguishable balls. Show that $U(M, n) = \binom{M+n-1}{n}$.

Hint. To prove the assertion, make use of the principle of mathematical induction. Let $P(n)$ be the proposition that, whatever M, $U(M, n) = \binom{M+n-1}{n}$. $P(1)$ is clearly true, since there are M unordered samples of size 1. To complete the proof, we must show that $P(n)$ implies $P(n + 1)$. The following formula is immediately obtained: for any $M = 1, 2, \ldots$, and $n = 1, 2, \ldots$:

$$U(M, n+1) = U(M, n) + U(M-1, n) + \cdots + U(1, n)$$

To obtain this formula, let the balls be numbered 1 to M. Let each unordered sample be arranged so that the numbers of the balls in the sample are in nondecreasing order (as in the example in the text involving unordered samples of size 3 from an urn containing 4 balls). Then there are $U(M, n)$ samples of size $(n + 1)$ whose first entry is 1, $U(M - 1, n)$ samples of size $(n + 1)$ whose first entry is 2, and so on, until there are $U(1, n)$ whose first entry is M. Now, by the induction hypothesis, $U(k, n) = \binom{k+n-1}{n}$. Consequently, $U(k, n) = \binom{k+n}{n+1} - \binom{k+n-1}{n+1}$. We thus determine that $U(M, n + 1) = \binom{M+n}{n+1}$, so that $P(n + 1)$ is proved, and the asserted formula for $U(M, n)$ is proved by mathematical induction.

5.3. Show that the number of ways in which n indistinguishable objects may be arranged in M distinguishable cells is $\binom{M+n-1}{n} = \binom{M+n-1}{M-1}$.

5.4. Let $n > M$. Show that the number of ways in which n indistinguishable objects may be arranged in M distinguishable cells so that no cell will be empty is $\binom{n-1}{n-M} = \binom{n-1}{M-1}$. *Hint.* It suffices to find the number of ways in which $(n - M)$ indistinguishable objects may be arranged in M distinguishable cells, since after placing 1 object in each cell the remaining objects may be arranged without restriction.

EXERCISES

5.1. On an examination the following question was posed: From a point on the base of a certain mountain there are 5 paths leading to the top of the mountain. In how many ways can one make a round trip (from the base to the top and back again)? Explain why each of the following 4 answers was graded as being correct: (i) $(5)_2 = 20$, (ii) $5^2 = 25$, (iii) $\binom{5}{2} = 10$, (iv) $\binom{6}{2} = 15$.

5.2. A certain young woman has 3 men friends. She is told by a fortune teller that she will be married twice and that both her husbands will come from this group of 3 men. How many possible marital histories can this woman have? Consider 4 cases. (May she marry the same man twice? Does the order in which she marries matter?)

5.3. The legitimate theater in New York gives both afternoon and evening performances on Saturdays. A man comes to New York one Saturday to attend 2 performances (1 in the afternoon and 1 in the evening) of the living theater. There are 6 shows that he might consider attending. In how many ways can he choose 2 shows? Consider 4 cases.

5.4. An urn contains 52 balls, numbered 1 to 52. Let the balls be drawn 1 at a time and divided among 4 people. Suppose that the balls numbered 1, 11, 31, and 41 are considered "lucky." What is the probability that (i) each person will have a "lucky" ball, (ii) 1 person will have all 4 "lucky" balls?

5.5. A bridge player announces that his hand (of 13 cards) contains (i) an ace (that is, at least 1 ace), (ii) the ace of hearts. What is the probability that it will contain another one?

5.6. What is the probability that in a division of a deck of cards into 4 bridge hands, 1 of the hands will contain (i) 13 cards of the same suit, (ii) 4 aces and 4 kings, (iii) 3 aces and 3 kings?

5.7. Prove that the probability of South's receiving exactly k aces when a bridge deck is divided into 4 hands is the same as the probability that a hand of 13 cards drawn from a bridge deck will contain exactly k aces.

5.8. An urn contains 8 balls numbered 1 to 8. Four balls are drawn without replacement; suppose x is the second smallest of the 4 numbers drawn. What is the probability that $x = 3$?

5.9. A red card is removed from a bridge deck of 52 cards; 13 cards are then drawn and found to be the same color. Show that the (conditional) probability that all will be black is equal to $\frac{2}{3}$.

5.10. A room contains 10 people who are wearing badges numbered 1 to 10. What is the probability that if 3 persons are selected at random (i) the largest (ii) the smallest badge number chosen will be 5?

5.11. From a pack of 52 cards an even number of cards is drawn. Show that the probability that half of these cards will be red and half will be black is

$$\left(\frac{52!}{(26!)^2} - 1\right) \div (2^{51} - 1).$$

Hint. Show, and then use (with $n = 52$), the facts that for any integer n

$$\binom{n}{0} + \binom{n}{2} + \binom{n}{4} + \cdots = \binom{n}{1} + \binom{n}{3} + \binom{n}{5} + \cdots \tag{5.12}$$

$$= (\tfrac{1}{2}) \sum_{k=0}^{n} \binom{n}{k} - (\tfrac{1}{2}) \sum_{k=0}^{n} (-1)^k \binom{n}{k} = 2^{n-1}.$$

$$\binom{n}{0}^2 + \binom{n}{1}^2 + \cdots + \binom{n}{n}^2 = \binom{2n}{n} = \frac{(2n)!}{(n!)^2}. \tag{5.13}$$

6. THE PROBABILITY OF OCCURRENCE OF A GIVEN NUMBER OF EVENTS

Consider M events $A_1, A_2, \ldots, A_M$ defined on a probability space. In this section we shall develop formulas for the probabilities of various events, defined in terms of the events $A_1, \ldots, A_M$, especially for $m = 0, 1, \ldots, M$, that (i) exactly m of them, (ii) at least m of them, (iii) no more than m of them will occur. With the aid of these formulas, a variety of questions connected with sampling and occupancy problems may be answered.

THEOREM. Let $A_1, A_2, \ldots, A_M$ be M events defined on a probability space. Let the quantities $S_0, S_1, \ldots, S_M$ be defined as follows:

$$\begin{aligned}
S_0 &= 1 \\
S_1 &= \sum_{k=1}^{M} P[A_k] \\
S_2 &= \sum_{k_1=1}^{M} \sum_{k_2=k_1+1}^{M} P[A_{k_1} A_{k_2}] \\
&\;\;\vdots \\
S_r &= \sum_{k_1=1}^{M} \sum_{k_2=k_1+1}^{M} \cdots \sum_{k_r=k_{r-1}+1}^{M} P[A_{k_1} A_{k_2} \cdots A_{k_r}] \\
&\;\;\vdots \\
S_M &= P[A_1 A_2 \cdots A_M]
\end{aligned} \tag{6.1}$$

The definition of S_r is usually written

$$(6.1') \qquad S_r = \sum_{\{k_1, \cdots, k_r\}} P[A_{k_1} A_{k_2} \cdots A_{k_r}],$$

in which the summation in (6.1') is over the $\binom{M}{r}$ possible subsets $\{k_1, \ldots, k_r\}$ of size r of the set $\{1, 2, \ldots, M\}$.

Then, for any integer $m = 0, 1, \ldots, M$, the probability of the event B_m that exactly m of the M events $A_1, \ldots, A_M$ will occur simultaneously is given by

$$(6.2) \quad P[B_m] = \sum_{r=m}^{M} (-1)^{r-m} \binom{r}{m} S_r$$

$$= S_m - \binom{m+1}{m} S_{m+1} + \binom{m+2}{m} S_{m+2} - \cdots \pm \binom{M}{m} S_M.$$

In particular, for $m = 0$,

$$(6.3) \qquad P[B_0] = 1 - S_1 + S_2 - S_3 + S_4 - \cdots \pm (S_M).$$

The probability that at least m of the M events $A_1, \ldots, A_M$ will occur is given by (for $m \geq 1$)

$$(6.4) \quad P[B_m] + P[B_{m+1}] + \cdots + P[B_M] = \sum_{r=m}^{M} (-1)^{r-m} \binom{r-1}{m-1} S_r$$

Before giving the proof of this theorem, we shall discuss various applications of it.

▶ **Example 6A. The matching problem (case of sampling without replacement).** Suppose that we have M urns, numbered 1 to M, and M balls, numbered 1 to M. Let the balls be inserted randomly in the urns, with one ball in each urn. If a ball is put into the urn bearing the same number as the ball, a *match* is said to have occurred. Show that the probability that (i) at least one match will occur is

$$(6.5) \qquad 1 - \frac{1}{2!} + \frac{1}{3!} - \cdots \pm \frac{1}{M!} \doteq 1 - e^{-1} = 0.63212,$$

(ii) exactly m matches will occur, for $m = 0, 1, \ldots, M$, is

$$(6.6) \quad \frac{1}{m!}\left\{1 - 1 + \frac{1}{2!} - \frac{1}{3!} + \cdots \pm \frac{1}{(M-m)!}\right\} = \frac{1}{m!} \sum_{k=0}^{M-m} (-1)^k \frac{1}{k!}$$

$$\doteq \frac{1}{m!} e^{-1} \qquad \text{for } M - m \text{ large.}$$

The matching problem may be formulated in a variety of ways. *First variation:* if M married gentlemen and their wives (in a monogamous society) draw lots for a dance in such a way that each gentleman is equally likely to dance with any of the M wives, what is the probability that exactly m gentlemen will dance with their own wives? *Second variation:* if M soldiers who sleep in the same barracks arrive home one evening so drunk that each soldier chooses at random a bed in which to sleep, what is the probability that exactly m soldiers will sleep in their own beds? *Third variation:* if M letters and M corresponding envelopes are typed by a tipsy typist and the letters are put into the envelopes in such a way that each envelope contains just one letter that is equally likely to be any one of the M letters, what is the probability that exactly m letters will be inserted into their corresponding envelopes? *Fourth variation:* If two similar decks of M cards (numbered 1 to M) are shuffled and dealt simultaneously, one card from each deck at a time, what is the probability that on just m occasions the two cards dealt will bear the same number?

There is a considerable literature on the matching problem that has particularly interested psychologists. The reader may consult papers by D. E. Barton, *Journal of the Royal Statistical Society*, Vol. 20 (1958), pp. 73–92, and P. E. Vernon, *Psychological Bulletin*, Vol. 33 (1936), pp. 149–77, which give many references. Other references may be found in an editorial note in the *American Mathematical Monthly*, Vol. 53 (1946), p. 107. The matching problem was stated and solved by the earliest writers on probability theory. It may be of value to reproduce here the statement of the matching problem given by De Moivre (*Doctrine of Chances*, 1714, Problem 35): "Any number of letters a, b, c, d, e, f, etc., all of them different, being taken promiscuously as it happens; to find the Probability that some of them shall be found in their places according to the rank they obtain in the alphabet and that others of them shall at the same time be displaced."

Solution: To describe the distribution of the balls among the urns, write an n-tuple $(z_1, z_2, \ldots, z_n)$ whose jth component z_j represents the number of the ball inserted in the jth urn. For $k = 1, 2, \ldots, M$ the event A_k that a match will occur in the kth urn may be written $A_k = \{(z_1, z_2, \ldots, z_n): z_k = k\}$. It is clear that for any integer $r = 1, 2, \ldots, M$ and any r unequal integers $k_1, k_2, \ldots, k_r$, 1 to M,

$$P[A_{k_1}A_{k_2}\cdots A_{k_r}] = \frac{(M-r)!}{M!}. \tag{6.7}$$

It then follows that the sum S_r, defined by (6.1), is given by

$$S_r = \binom{M}{r}\frac{(M-r)!}{M!} = \frac{1}{r!}. \tag{6.8}$$

Equations (6.5) and (6.6) now follow immediately from (6.8), (6.3), and (6.2). ◀

▶ **Example 6B. Coupon collecting (case of sampling with replacement).** Suppose that a manufacturer gives away in packages of his product certain items (which we take to be coupons, each bearing one of the integers 1 to M) in such a way that each of the M items available is equally likely to be found in any package purchased. If n packages are bought, show that the probability that exactly m of the M integers, 1 to M, will not be obtained (or, equivalently, that exactly $M - m$ of the integers, 1 to M, will be obtained) is equal to

$$\binom{M}{m} \frac{\Delta^{M-m}(0^n)}{M^n} \tag{6.9}$$

where we define, for any integer n, and $r = 0, 1, \ldots, n$,

$$\Delta^r(0^n) = \sum_{k=0}^{r} (-1)^k \binom{r}{k} (r - k)^n \tag{6.10}$$

The symbol Δ is used with the meaning assigned to it in the calculus of finite differences as an operator defined by $\Delta f(x) = f(x + 1) - f(x)$. We write $\Delta^r(0^n)$ to mean the value at $x = 0$ of $\Delta^r(x^n)$.

A table of $\Delta^r(0^n)/r!$ for $n = 2(1)25$ and $r = 2(1)n$ is to be found in *Statistical Tables for Agricultural, Biological, and Medical Research* (1953), Table XXII.

The problem of coupon collecting has many variations and practical applications. *First variation* (the occupancy problem): if n distinguishable balls are distributed among M urns, numbered 1 to M, what is the probability that there will be exactly m urns in which no ball was placed (that is, exactly m urns remain empty after the n balls have been distributed)? *Second variation* (measuring the intensity of cosmic radiation): if M counters are exposed to a cosmic ray shower and are hit by n rays, what is the probability that precisely $M - m$ counters will go off? *Third variation* (genetics): if each mouse in a litter of n mice can be classified as belonging to any one of M genotypes, what is the probability that $M - m$ genotypes will be represented among the n mice?

Solution: To describe the coupons found in the n packages purchased, we write an n-tuple $(z_1, z_2, \ldots, z_n)$, whose jth component z_j represents the number of the coupon found in the jth package purchased. We now define events $A_1, A_2, \ldots, A_M$. For $k = 1, 2, \ldots, M$, A_k is the event that the number k will not appear in the sample; in symbols,

$$A_k = \{(z_1, z_2, \cdots, z_n)\colon \text{ for } j = 1, 2, \cdots, n, z_j \neq k\}. \tag{6.11}$$

It is easy to obtain the probability of the intersection of any number of the events $A_1, \ldots, A_M$. We have

$$P[A_k] = \left(\frac{M-1}{M}\right)^n = \left(1 - \frac{1}{M}\right)^n, \qquad k = 1, \cdots, M,$$

$$(6.12) \qquad P[A_{k_1}A_{k_2}] = \left(1 - \frac{2}{M}\right)^n, \qquad k_1 = 1, \cdots, n, \quad k_2 = k_1 + 1, \cdots, n,$$

$$P[A_{k_1}A_{k_2} \cdots A_{k_r}] = \left(1 - \frac{r}{M}\right)^n, \qquad k_1 = 1, \cdots, n,$$

$$k_2 = k_1 + 1, \cdots, n, \cdots, \qquad k_r = k_{r-1} + 1, \cdots, n.$$

The quantities S_r, defined by (6.1), are then given by

$$(6.13) \qquad S_r = \binom{M}{r}\left(1 - \frac{r}{M}\right)^n, \qquad r = 0, 1, \cdots, M.$$

Let B_m be the event that exactly m of the integers 1 to M will not be found in the sample. Clearly B_m is the event that exactly m of the events A_1, $A_2, \cdots, A_M$ will occur. By (6.2) and (6.13),

$$(6.14) \qquad P[B_m] = \sum_{r=m}^{M} (-1)^{r-m}\binom{r}{m}\binom{M}{r}\left(1 - \frac{r}{M}\right)^n$$

$$= \binom{M}{m}\sum_{k=0}^{M-m}(-1)^k\binom{M-m}{k}\left(1 - \frac{m+k}{M}\right)^n,$$

which coincides with (6.9). ◀

Other applications of the theorem stated at the beginning of this section may be found in a paper by J. O. Irwin, "A Unified Derivation of Some Well-Known Frequency Distributions of Interest in Biometry and Statistics," *Journal of the Royal Statistical Society*, Series A, Vol. 118 (1955), pp. 389–404 (including discussion).

The remainder of this section* is concerned with the proof of the theorem stated at the beginning of the section. Our proof is based on *the method of indicator functions* and is the work of M. Loève. Our proof has the advantage of being constructive in the sense that it is not merely a verification that (6.2) is correct but rather obtains (6.2) from first principles.

The method of indicator functions proceeds by interpreting operations

* The remainder of this section may be omitted in a first reading of the book.

on events in terms of arithmetic operations. Given an event A, on a sample description space S, we define its indicator function, denoted by $I(A)$, as a function defined on S, with value at any description s, denoted by $I(A; s)$, equal to 1 or 0, depending on whether the description s does or does not belong to A.

The two basic properties of indicator functions, which enable us to operate with them, are the following.

First, *a product of indicator functions* can always be replaced by a single indicator function; more precisely, for any events $A_1, A_2, \ldots, A_n$,

$$I(A_1)I(A_2) \cdots I(A_n) = I(A_1A_2 \cdots A_n), \tag{6.15}$$

so that the product of the indicator functions of the sets $A_1, A_2, \ldots, A_n$ is equal to the indicator function of the intersection $A_1, A_2, \ldots, A_n$. Equation (6.15) is an equation involving functions; strictly speaking, it is a brief method of expressing the following family of equations: for every description s

$$I(A_1; s)I(A_2; s) \cdots I(A_n; s) = I(A_1A_2 \cdots A_n; s).$$

To prove (6.15), one need note only that $I(A_1A_2 \ldots A_n; s) = 0$ if and only if s does not belong to $A_1A_2 \ldots A_n$. This is so if and only if, for some $j = 1, \ldots, n$, s does not belong to A_j, which is equivalent to, for some $j = 1, \ldots, n$, $I(A_j; s) = 0$, which is equivalent to the product $I(A_1; s) \cdots I(A_n; s) = 0$.

Second, *a sum of indicator functions* can in certain circumstances be replaced by a single indicator function; more precisely, if the events $A_1, A_2, \ldots, A_n$ are mutually exclusive, then

$$I(A_1) + I(A_2) + \cdots + I(A_n) = I(A_1 \cup A_2 \cup \cdots \cup A_n). \tag{6.16}$$

The proof of (6.16) is left to the reader. One case, in which $n = 2$ and $A_2 = A_1^c$, is of especial importance. Then $A_1 \cup A_2 = S$, and $I(A_1 \cup A_2)$ is identically equal to 1. Consequently, we have, for any event A,

$$I(A) + I(A^c) = 1, \qquad I(A^c) = 1 - I(A). \tag{6.17}$$

From (6.15) to (6.17) we may derive expressions for the indicator functions of various events. For example, let A and B be any two events. Then

$$\begin{aligned} I(A \cup B) &= 1 - I(A^cB^c) \\ &= 1 - I(A^c)I(B^c) \\ &= 1 - (1 - I(A))(1 - I(B)) \\ &= I(A) + I(B) - I(AB) \end{aligned} \tag{6.18}$$

Our ability to write expressions in the manner of (6.18) for the indicator functions of compound events, in terms of the indicator functions of the events of which they are composed, derives its importance from the following fact: *an equation involving only sums and differences* (but not products) *of indicator functions leads immediately to a corresponding equation involving probabilities; this relation is obtained by replacing* $I(\cdot)$ *by* $P[\cdot]$. For example, if one makes this replacement in (6.18), one obtains the well-known formula $P[A \cup B] = P[A] + P[B] - P[AB]$.

The principle just enunciated is a special case of the additivity property of the operation of taking the expected value of a function defined on a probability space. This is discussed in Chapter 8, but we shall sketch a proof here of the principle stated. We prove a somewhat more general assertion. Let $f(\cdot)$ be a function defined on a sample description space. Suppose that the possible values of f are integers from $-N(f)$ to $N(f)$ for some integer $N(f)$ that will depend on $f(\cdot)$. We may then represent $f(\cdot)$ as a linear combination of indicator functions:

$$f(\cdot) = \sum_{k=-N(f)}^{N(f)} kI[D_k(f)], \tag{6.19}$$

in which $D_k(f) = \{s\colon\ f(s) = k\}$ is the set of descriptions at which $f(\cdot)$ takes the value k. Define the expected value of $f(\cdot)$, denoted by $E[f(\cdot)]$:

$$E[f(\cdot)] = \sum_{k=-N(f)}^{N(f)} kP[D_k(f)]. \tag{6.20}$$

In words, $E[f(\cdot)]$ is equal to the sum, over all possible values k of the function $f(\cdot)$, of the product of the value k and the probability that $f(\cdot)$ will assume the value k. In particular, if $f(\cdot)$ is an indicator function, so that $f(\cdot) = I(A)$ for some set A, then $E[f(\cdot)] = P[A]$. Consider now another function $g(\cdot)$, which may be written

$$g(\cdot) = \sum_{j=-N(g)}^{N(g)} jI[D_j(g)] \tag{6.21}$$

We now prove the basic additivity theorem that

$$E[f(\cdot) + g(\cdot)] = E[f(\cdot)] + E[g(\cdot)]. \tag{6.22}$$

The sum $f(\cdot) + g(\cdot)$ of the two functions is a function whose possible values are numbers, $-N$ to N, in which $N = N(f) + N(g)$. However, we may represent the function $f(\cdot) + g(\cdot)$ in terms of the indicator functions $I[D_k(f)]$ and $I[D_j(g)]$:

$$f(\cdot) + g(\cdot) = \sum_{k=-N(f)}^{N(f)} \sum_{j=-N(g)}^{N(g)} (k + j)I[D_k(f)D_j(g)].$$

Therefore,

$$E[f(\cdot)+g(\cdot)] = \sum_{k=-N(f)}^{N(f)} \sum_{j=-N(g)}^{N(g)} (k+j)P[D_k(f)D_j(g)]$$
$$= \sum_{k=-N(f)}^{N(f)} k \sum_{j=-N(g)}^{N(g)} P[D_k(f)D_j(g)] + \sum_{j=-N(g)}^{N(g)} j \sum_{k=-N(f)}^{N(f)} P[D_k(f)D_j(g)]$$
$$= \sum_{k=-N(f)}^{N(f)} kP[D_k(f)] + \sum_{j=-N(g)}^{N(g)} jP[D_j(g)]$$
$$= E[f(\cdot)] + E[g(\cdot)],$$

and the proof of (6.22) is complete. By mathematical induction we determine from (6.22) that, for any n functions, $f_1(\cdot), f_2(\cdot), \ldots, f_n(\cdot)$,

(6.23) $$E[f_1(\cdot) + \cdots + f_n(\cdot)] = E[f_1(\cdot)] + \cdots + E[f_n(\cdot)].$$

Finally, from (6.23), and the fact that $E[I(A)] = P[A]$, we obtain the principle we set out to prove; namely, that, for any events $A, A_1, \ldots, A_n$, if

(6.24) $$I(A) = c_1I(A_1) + c_2I(A_2) + \cdots + c_nI(A_n),$$

in which the c_i are either $+1$ or -1, then

(6.25) $$P[A] = c_1P[A_1] + c_2P[A_2] + \cdots + c_nP[A_n].$$

We now apply the foregoing considerations to derive (6.2). The event B_m, that exactly m of the M events $A_1, A_2, \ldots, A_M$ occur, may be expressed as the union, over all subsets $J_m = \{i_1, \ldots, i_m\}$ of size m of the integers $\{1, 2, \ldots, M\}$, of the events $A_{i_1} \ldots A_{i_m}A^c_{i_{m+1}} \ldots A^c_{i_M}$; there are $\binom{M}{m}$ such events, and they are mutually exclusive. Consequently, by (6.15)–(6.17),

(6.26) $$I(B_m) = \sum_{J_m} I(A_{i_1}) \cdots I(A_{i_m})[1 - I(A_{i_{m+1}})] \cdots (1 - I(A_{i_M})]$$

Now each term in (6.26) may be written

(6.27) $$I(A_{i_1}) \cdots I(A_{i_m})\{1 - H_1(J_m) + \cdots + (-1)^k H_k(J_m) + \cdots \pm H_{M-m}(J_m)\},$$

where we define $H_k(J_m) = \Sigma I(A_{j_1} \ldots A_{j_k})$, in which the summation is over all subsets of size k of the set of integers $\{i_{m+1}, \ldots, i_M\}$. Now because of symmetry and (6.15), one sees that

(6.28) $$\sum_{J_m} I(A_{i_1}) \cdots I(A_{i_m})H_k(J_m) = \binom{m+k}{m} H_{k+m}.$$

TABLE 6A THE PROBABILITIES OF VARIOUS EVENTS DEFINED ON THE GENERAL OCCUPANCY AND SAMPLING PROBLEMS

	The probability that		Distributing n balls into M distinguishable urns		
			Without exclusion		With exclusion
	Occupancy problem	Sampling problem	Distinguishable balls	Indistinguishable balls	Either distinguishable or indistinguishable balls
I	A specified urn will contain k balls, where $k \le n$	A specified ball will appear k times in the sample, where $k \le n$	$\frac{\binom{n}{k}}{M^n}(M-1)^{n-k}$	$\frac{\binom{M+n-k-2}{n-k}}{\binom{M+n-1}{n}}$	$\frac{\binom{M-1}{n-k}}{\binom{M}{n}}$ if $k = 0, 1$
II	First urn contains k_1 balls; second urn contains k_2 balls; . . . ; the Mth urn contains k_M balls, where $k_1 + k_2 + \cdots + k_M = n$	In the sample, the first ball appears k_1 times; the second ball appears k_2 times; . . . ; the Mth ball appears k_M times, where $k_1 + k_2 + \cdots + k_M = n$	$\frac{\binom{n}{k_1 k_2 \cdots k_M}}{M^n}$	$\frac{1}{\binom{M+n-1}{n}}$	$\frac{1}{\binom{M}{n}}$ if $k_j \le 1$ for $j = 1, \cdots, n$
III	Each of N specified urns will be occupied, where $N \le M$	Each of N specified balls is contained in the sample, where $N \le M$	$\sum_{k=0}^{N} (-1)^k \binom{N}{k}\left(1-\frac{k}{M}\right)^n$	$\frac{\binom{M-N+n-1}{n-N}}{\binom{M+n-1}{n}}$	$\sum_{k=0}^{N} (-1)^k \binom{N}{k}\frac{(M-k)_n}{(M)_n} = \binom{M-N}{n-N} \div \binom{M}{n}$
IV	Exactly m of N specified urns will be empty where $N \le M$, $m = 0, 1, \ldots, N$.	Exactly m of N specified balls are not contained in the sample where $N \le M$, $m = 0, 1, \ldots, N$.	$\sum_{k=m}^{M} (-1)^{k-m}\binom{N}{k}\binom{k}{m} \times \left(1-\frac{k}{M}\right)^n = \binom{N}{m}\sum_{k=0}^{M-m}(-1)^k\binom{N-m}{k} \times \left(1-\frac{m+k}{M}\right)^n$	$\frac{\binom{N}{m}\binom{M-m+n-(N-m)-1}{n-(N-m)}}{\binom{M+n-1}{n}} = \frac{\binom{N}{m}\binom{M-N+n-1}{M-m-1}}{\binom{M+n-1}{n}}$	$\sum_{k=m}^{M} (-1)^k \binom{N}{k}\binom{k}{m}\frac{(M-k)_n}{(M)_n} = \binom{N}{m}\binom{M-N}{n-N+m} \div \binom{M}{n}$
			Ordered samples	Unordered samples	Either unordered or ordered samples
			With replacement		Without replacement
			Drawing samples of size n from an urn containing M distinguishable balls		

$H_{k+m} = \Sigma I(A_{j_1} \ldots A_{j_{k+m}})$, in which the summation is over all subsets of size $k + m$ of the set of integers $\{1, 2, \ldots, M\}$. To see (6.28), note that there are $\binom{M}{m}$ terms in J_m, $\binom{M-m}{k}$ terms in $H_k(J_m)$, and $\binom{M}{m+k}$ terms in H_{k+m} and use the fact that $\binom{M}{m}\binom{M-m}{k} \div \binom{M}{m+k} = \binom{m+k}{m}$.

Finally, from (6.24) to (6.28) and (6.1) we obtain

$$P[B_m] = \sum_{k=0}^{M-m} (-1)^k \binom{m+k}{m} S_{k+m}, \tag{6.29}$$

which is the same as (6.2). Equation (6.4) follows immediately from (6.2) by induction.

THEORETICAL EXERCISES

6.1. Matching problem (case of sampling without replacement). Show that for $j = 1, \ldots, M$ the conditional probability of a match in the jth urn, given that there are m matches, is m/M. Show, for any 2 unequal integers j and k, 1 to M, that the conditional probability that the ball number j was placed in urn number k, given that there are m matches, is equal to $(M - m)(M - m - 1)/M(M - 1)$.

6.2. Matching (case of sampling with replacement). Consider the matching problem under the assumption that, for $j = 1, 2, \ldots, M$, the ball inserted in the jth urn was chosen randomly from all the M balls available (and then made available as a candidate for insertion in the $(j + 1)$st urn). Show that the probability of at least 1 match is $1 - [1 - (1/M)]^M \doteq 1 - e^{-1} = 0.63212$. Find the probability of exactly m matches.

6.3. A man addresses n envelopes and writes n checks in payment of n bills.

(i) If the n bills are placed at random in the n envelopes, show that the probability that each bill will be placed in the wrong envelope is

$$\sum_{k=2}^{n} (-1)^k (1/k!).$$

(ii) If the n bills and n checks are placed at random in the n envelopes, 1 in each envelope, show that the probability that in no instance will the enclosures be completely correct is $\sum_{k=0}^{n} (-1)^k (n-k)!/(n!k!)$.

(iii) In part (ii) the probability that each bill and each check will be in a wrong envelope is equal to the square of the answer to part (i).

6.4. A sampling (or coupon collecting) problem. Consider an urn that contains rM balls, for given integers r and M. Suppose that for each integer j, 1 to M, exactly r balls bear the integer j. Find the probability that in a sample of size n (in which $n \geq M$), drawn without replacement from the urn, exactly m of the integers 1 to M will be missing.

Hint: $$S_j = \binom{M}{j}[r(M-j)]_n/(rM)_n.$$

6.5. Verify the formulas in row I of Table 6A.

6.6. Verify the formulas in row II of Table 6A.

6.7. Verify the formulas in row III of Table 6A.

6.8. Verify the formulas in row IV of Table 6A.

EXERCISES

6.1. If 10 indistinguishable balls are distributed among 7 urns in such a way that all arrangements are equally likely, what is the probability that (i) a specified urn will contain 3 balls, (ii) all urns will be occupied, (iii) exactly 5 urns will be empty?

6.2. If 7 indistinguishable balls are distributed among 10 urns in such a way that not more than 1 ball may be put in any urn and all such arrangements are equally likely, what is the probability that (i) a specified urn will contain 1 ball (ii) exactly 3 of the first 4 urns will be empty?

6.3. If 10 distinguishable balls are distributed among 4 urns in such a way that all arrangements are equally likely, what is the probability that (i) a specified urn will contain 6 balls, (ii) the first urn will contain 4 balls, the second urn will contain 3 balls, the third urn will contain 2 balls, and the fourth urn will contain 1 ball, (iii) all urns will be occupied?

6.4. Consider 5 families, each consisting of 4 persons. If it is reported that 6 of the 20 individuals in these families have a contagious disease, what is the probability that (i) exactly 2, (ii) at least 3 of the families will be quarantined?

6.5. Write out (6.2) and (6.4) for (i) $M = 2$ and $m = 0, 1, 2$, (ii) $M = 3$ and $m = 0, 1, 2, 3$, (iii) $M = 4$ and $m = 0, 1, 2, 3, 4$.

CHAPTER 3

Independence and Dependence

In this chapter we show how to treat probability problems involving finite sample description spaces, in which the descriptions are not necessarily equally likely, by using the notions of independent and dependent events and trials.

1. INDEPENDENT EVENTS AND FAMILIES OF EVENTS

The notions of independent and dependent events play a central role in probability theory. Certain relations, which recur again and again in probability problems, may be given a general formulation in terms of these notions. If the events A and B have the property that the conditional probability of B, given A, is equal to the unconditional probability of B, one intuitively feels that the event B is statistically independent of A, in the sense that the probability of B having occurred is not affected by the knowledge that A has occurred. We are thus led to the following formal definition.

DEFINITION OF AN EVENT B BEING INDEPENDENT OF AN EVENT A WHICH HAS POSITIVE PROBABILITY. Let A and B be events defined on the same probability space S. Assume $P[A] > 0$, so that $P[B \mid A]$ is well defined.

The event B is said to be independent (or statistically independent) of the event A if the conditional probability of B, given A, is equal to the unconditional probability of B; in symbols, B is independent of A if

$$P[B \mid A] = P[B]. \tag{1.1}$$

Now suppose that both A and B have positive probability. Then both $P[A \mid B]$ and $P[B \mid A]$ are well defined, and from (4.6) of Chapter 2 it follows that

$$P[AB] = P[B \mid A]P[A] = P[A \mid B]P[B]. \tag{1.2}$$

If B is independent of A, it then follows that A is independent of B, since from (1.1) and (1.2) it follows that $P[A \mid B] = P[A]$. It further follows from (1.1) and (1.2) that

$$P[AB] = P[A]P[B]. \tag{1.3}$$

By means of (1.3), a definition may be given of two events being independent, in which the two events play a symmetrical role.

DEFINITION OF INDEPENDENT EVENTS. Let A and B be events defined on the same probability space. The events A and B are said to be independent if (1.3) holds.

► **Example 1A.** Consider the problem of drawing with replacement a sample of size 2 from an urn containing four white and two red balls. Let A denote the event that the first ball drawn is white and B, the event that the second ball drawn is white. By (2.5), in Chapter 2, $P[AB] = (\frac{4}{6})^2$, whereas $P[A] = P[B] = \frac{4}{6}$. In view of (1.3), the events A and B are independent. ◄

Two events that do not satisfy (1.3) are said to be *dependent* (although a more precise terminology would be *nonindependent*). Clearly, to say that two events are dependent is not very informative, for two events, A and B, are dependent if and only if $P[AB] \neq P[A]P[B]$. However, it is possible to classify dependent events to a certain extent, and this is done later. (See section 5.)

It should be noted that two mutually exclusive events, A and B, are independent if and only if $P[A]P[B] = 0$, which is so if and only if either A or B has probability zero.

► **Example 1B. Mutually exclusive events.** Let a sample of size 2 be drawn from an urn containing six balls, of which four are white. Let C denote the event that exactly one of the balls drawn is white, and let D denote the event that both balls drawn are white. The events C and D are mutually exclusive and are not independent, whether the sample is drawn with or without replacement. ◄

▶ **Example 1C. A paradox?** Choose a summer day at random on which both the Dodgers and the Giants are playing baseball games. Let A be the event that the Dodgers win, and let B be the event that the Giants win. If the Dodgers and the Giants are not playing each other, then we may consider the events A and B as independent but not mutually exclusive. If the Giants and the Dodgers are playing each other, then we may consider the events A and B as mutually exclusive but not independent. To resolve this paradox, one need note only that the probability space on which the events A and B are defined is not the same in the two cases. (See example 2B.) ◀

The notions of independent events and of conditional probability may be extended to more than two events. Suppose one has three events A, B, and C defined on a probability space. What are we to mean by the conditional probability of the event C, given that the events A and B have occurred, denoted by $P[C \mid A, B]$? From the point of view of the frequency interpretation of probability, by $P[C \mid A, B]$ we mean the fraction of occurrences of both A and B on which C also occurs. Consequently, we make the formal definition that

$$P[C \mid A, B] = P[C \mid AB] = \frac{P[CAB]}{P[AB]} \tag{1.4}$$

if $P[AB] > 0$; $P[C \mid A, B]$ is undefined if $P[AB] = 0$.

Next, what do we mean by the statement that the event C is independent of the events A and B? It would seem that we should mean that the conditional probability of C, given either A or B or the intersection AB, is equal to the unconditional probability of C. We therefore make the following formal definition.

The events A, B, and C, defined on the same probability space, are said to be independent (or statistically independent) if

$$P[AB] = P[A]P[B], \qquad P[AC] = P[A]P[C], \qquad P[BC] = P[B]P[C], \tag{1.5}$$

$$P[ABC] = P[A]P[B]P[C]. \tag{1.6}$$

If (1.5) and (1.6) hold, it then follows that (assuming that the events A, B, C, AB, AC, BC have positive probability, so that the conditional probabilities written below are well defined)

$$\begin{aligned} P[A \mid B, C] &= P[A \mid B] = P[A \mid C] = P[A] \\ P[B \mid A, C] &= P[B \mid A] = P[B \mid C] = P[B] \\ P[C \mid A, B] &= P[C \mid A] = P[C \mid B] = P[C] \end{aligned} \tag{1.7}$$

Conversely, if all the relations in (1.7) hold, then all the relations in (1.5) and (1.6) hold.

It is to be emphasized that (1.5) does not imply (1.6), so that three events, A, B, and C, which are pairwise independent [in the sense that (1.5) holds], are not necessarily independent. To see this, consider the following example.

▶ **Example 1D. Pairwise independent events that are not independent.** Let a ball be drawn from an urn containing four balls, numbered 1 to 4. Assume that $S = \{1, 2, 3, 4\}$ possesses equally likely descriptions. The events $A = \{1, 2\}$, $B = \{1, 3\}$, and $C = \{1, 4\}$ satisfy (1.5) but do not satisfy (1.6). Indeed, $P[C \mid A, B] = 1 \neq \frac{1}{2} = P[C] = P[C \mid A] = P[C \mid B]$. The reader may find it illuminating to explain in words why $P[C \mid A, B] = 1$. ◀

▶ **Example 1E. The joint credibility of witnesses.** Consider an automobile accident on a city street in which car I stops suddenly and is hit from behind by car II. Suppose that three persons, whom we call A', B', and C', witness the accident. Suppose the probability that each witness has correctly observed that car I stopped suddenly is estimated by having the witnesses observe a number of contrived incidents about which each is then questioned. Assume that it is found that A' has probability 0.9 of stating that car I stopped suddenly, B' has probability 0.8 of stating that car I stopped suddenly, and C' has probability 0.7 of stating that car I stopped suddenly. Let A, B, and C denote, respectively, the events that persons A', B', and C' will state that car I stopped suddenly. Assuming that A, B, and C are independent events, what is the probability that (i) A', B', and C' will state that car I stopped suddenly, (ii) exactly two of them will state that car I stopped suddenly?

Solution: By independence, the probability $P[ABC]$ that all three witnesses will state that car I stopped suddenly is given by $P[ABC] = P[A]P[B]P[C] = (0.9)(0.8)(0.7) = 0.504$. It is subsequently shown that if A, B, and C are independent events then A, B, and C^c are independent events. Consequently, the probability that *exactly* two of the witnesses will state that car I stopped suddenly is given by

$$\begin{aligned} P[ABC^c \cup AB^cC \cup A^cBC] & \\ &= P[A]P[B]P[C^c] + P[A]P[B^c]P[C] + P[A^c]P[B]P[C] \\ &= (0.9)(0.8)(0.3) + (0.9)(0.2)0.7) + (0.1)(0.8)(0.7) \\ &= 0.398. \end{aligned}$$

The probability that at least two of the witnesses will state that car I stopped suddenly is $0.504 + 0.398 = 0.902$. It should be noted that the sample description space S on which the events A, B, and C are defined is the space of 3-tuples (z_1, z_2, z_3) in which z_1 is equal to "yes" or "no," depending on whether person A' says that car I did or did not stop suddenly; components z_2 and z_3 are defined similarly with respect to persons B' and C'. ◀

We next define the notions of independence and of conditional probability for n events $A_1, A_2, \ldots, A_n$.

We define the conditional probability of A_n, given that the events $A_1, A_2, \ldots, A_{n-1}$ have occurred, denoted by $P[A_n \mid A_1, A_2, \ldots, A_{n-1}]$;

$$(1.8) \qquad P[A_n \mid A_1, A_2, \cdots, A_{n-1}] = P[A_n \mid A_1A_2 \cdots A_{n-1}] = \frac{P[A_1A_2 \cdots A_n]}{P[A_1A_2 \cdots A_{n-1}]}$$

if $P[A_1A_2 \cdots A_{n-1}] > 0$.

We define the events $A_1, A_2, \ldots, A_n$ as independent (or statistically independent) if for every choice of k integers $i_1 < i_2 < \ldots < i_k$ from 1 to n

$$(1.9) \qquad P[A_{i_1}A_{i_2} \cdots A_{i_k}] = P[A_{i_1}]P[A_{i_2}] \cdots P[A_{i_k}].$$

Equation (1.9) implies that for any choice of integers $i_1 < i_2 < \ldots < i_k$ from 1 to n (for which the following conditional probability is defined) and for any integer j from 1 to n not equal to $i_1, i_2, \ldots, i_k$ one has

$$(1.10) \qquad P[A_j \mid A_{i_1}, A_{i_2}, \cdots, A_{i_k}] = P[A_j].$$

We next consider *families of independent events*, for independent events never occur alone. Let $\mathscr{A}$ and $\mathscr{B}$ be two families of events; that is, $\mathscr{A}$ and $\mathscr{B}$ are sets whose members are events on some sample description space S. *Two families of events $\mathscr{A}$ and $\mathscr{B}$ are said to be independent if any two events A and B, selected from $\mathscr{A}$ and $\mathscr{B}$, respectively, are independent.* More generally, *n families of events $(\mathscr{A}_1, \mathscr{A}_2, \ldots, \mathscr{A}_n)$ are said to be independent if any set of n events $A_1, A_2, \cdots, A_n$ (where A_1 is selected from $\mathscr{A}_1$, A_2 is selected from $\mathscr{A}_2$, and so on, until A_n is selected from $\mathscr{A}_n$) is independent,* in the sense that it satisfies the relation

$$(1.11) \qquad P[A_1A_2 \cdots A_n] = P[A_1]P[A_2] \cdots P[A_n].$$

As an illustration of the fact that independent events occur in families, let us consider two independent events, A and B, which are defined on a sample description space S. Define the families $\mathscr{A}$ and $\mathscr{B}$ by

$$(1.12) \qquad \mathscr{A} = \{A, A^c, S, \emptyset\}, \qquad \mathscr{B} = \{B, B^c, S, \emptyset\},$$

so that $\mathscr{A}$ consists of A, its complement A^c, the certain event S, and the impossible event $\emptyset$, and, similarly, $\mathscr{B}$ consists of B, B^c, S, and $\emptyset$.

We now show that *if the events A and B are independent then the families of events $\mathscr{A}$ and $\mathscr{B}$ defined by* (1.12) *are independent*. In order to prove this assertion, we must verify the validity of (1.11) with $n = 2$ for each pair of events, one from each family, that may be chosen. Since each family has four members, there are sixteen such pairs. We verify (1.11) for only four of these pairs, namely (A, B), (A, B^c), (A, S), and $(A, \emptyset)$, and leave to the reader the verification of (1.11) for the remaining twelve pairs. We have that A and B satisfy (1.11) by hypothesis. Next, we show that A and B^c satisfy (1.11). By (5.2) of Chapter 1, $P[AB^c] = P[A] - P[AB]$. Since, by hypothesis, $P[AB] = P[A]P[B]$, it follows that

$$P[AB^c] = P[A](1 - P[B]) = P[A]P[B^c],$$

for by (5.3) of Chapter 1 $P[B^c] = 1 - P[B]$. Next, A and S satisfy (1.11), since $AS = A$ and $P[S] = 1$, so that $P[AS] = P[A] = P[A]P[S]$. Next, A and $\emptyset$ satisfy (1.11), since $A\emptyset = \emptyset$ and $P[\emptyset] = 0$, so that $P[A\emptyset] = P[\emptyset] = P[A]P[\emptyset] = 0$.

More generally, by the same considerations, we may prove the following important theorem, which expresses (1.9) in a very concise form.

THEOREM. Let $A_1, A_2, \ldots, A_n$ be n events on a probability space. The events $A_1, A_2, \ldots, A_n$ are independent if and only if the families of events

$$\mathscr{A}_1 = \{A_1, A_1^c, S, \emptyset\}, \qquad \mathscr{A}_2 = \{A_2, A_2^c, S, \emptyset\}, \cdots, \mathscr{A}_n = \{A_n, A_n^c, S, \emptyset\}$$

are independent.

THEORETICAL EXERCISES

1.1. Consider n independent events $A_1, A_2, \ldots, A_n$. Show that

$$P[A_1 \cup A_2 \cup \cdots \cup A_n] = 1 - P[A_1^c]P[A_2^c] \cdots P[A_n^c].$$

Consequently, obtain the probability that in 6 independent tosses of a fair die the number 3 will appear at least once. *Answer:* $1 - (5/6)^6$.

1.2. Let the events $A_1, A_2, \ldots, A_n$ be independent and $P[A_i] = p_i$ for $i = 1, \ldots, n$. Let P_0 be the probability that none of the events will occur. Show that $P_0 = (1 - p_1)(1 - p_2) \cdots (1 - p_n)$.

1.3. Let the events $A_1, A_2, \ldots, A_n$ be independent and have equal probability $P[A_i] = p$. Show that the probability that exactly k of the events will occur is (for $k = 0, 1, \ldots, n$)

$$\binom{n}{k} p^k q^{n-k}. \tag{1.13}$$

Hint: $P[A_1 \cdots A_k A_{k+1}^c \cdots A_n^c] = p^k q^{n-k}$.

1.4. **The multiplicative rule for the probability of the intersection of n events $A_1, A_2, \ldots, A_n$.** Show that, for n events for which $P[A_1 A_2 \ldots A_{n-1}] > 0$,

$$P[A_1A_2A_3 \cdots A_n] = P[A_1]P[A_2 \mid A_1]P[A_3 \mid A_1, A_2] \cdots P[A_n \mid A_1, A_2, \cdots, A_{n-1}].$$

1.5. Let A and B be independent events. In terms of $P[A]$ and $P[B]$, express, for $k = 0, 1, 2$, (i) P[exactly k of the events A and B will occur], (ii) P[at least k of the events A and B will occur], (iii) P[at most k of the events A and B will occur].

1.6. Let A, B, and C be independent events. In terms of $P[A]$, $P[B]$, and $P[C]$, express, for $k = 0, 1, 2, 3$, (i) P[exactly k of the events A, B, C will occur], (ii) P[at least k of the events A, B, C will occur], (iii) P[at most k of the events A, B, C will occur].

EXERCISES

1.1. Let a sample of size 4 be drawn with replacement (without replacement) from an urn containing 6 balls, of which 4 are white. Let A denote the event that the ball drawn on the first draw is white, and let B denote the event that the ball drawn on the fourth draw is white. Are A and B independent? Prove your answers.

1.2. Let a sample of size 4 be drawn with replacement (without replacement) from an urn containing 6 balls, of which 4 are white. Let A denote the event that exactly 1 of the balls drawn on the first 2 draws is white. Let B be the event that the ball drawn on the fourth draw is white. Are A and B independent? Prove your answers.

1.3. (Continuation of 1.2). Let A and B be as defined in exercise 1.2. Let C be the event that exactly 2 white balls are drawn in the 4 draws. Are A, B, and C independent? Are B and C independent? Prove your answers.

1.4. Consider example 1E. Find the probability that (i) both A' and B' will state that car I stopped suddenly, (ii) neither A' nor C' will state that car I stopped suddenly, (iii) at least 1 of A', B', and C' will state that car I stopped suddenly.

1.5. A manufacturer of sports cars enters 3 drivers in a race. Let A_1 be the event that driver 1 "shows" (that is, he is among the first 3 drivers in the race to cross the finish line), let A_2 be the event that driver 2 shows, and let A_3 be the event that driver 3 shows. Assume that the events A_1, A_2, A_3 are independent and that $P[A_1] = P[A_2] = P[A_3] = 0.1$. Compute the probability that (i) none of the drivers will show, (ii) at least 1 will show, (iii) at least 2 will show, (iv) all of them will show.

1.6. Compute the probabilities asked for in exercise 1.5 under the assumption that $P[A_1] = 0.1$, $P[A_2] = 0.2$, $P[A_3] = 0.3$.

1.7. A manufacturer of sports cars enters n drivers in a race. For $i = 1, \ldots, n$ let A_i be the event that the ith driver shows (see exercise 1.5). Assume that the events $A_1, \ldots, A_n$ are independent and have equal probability $P[A_i] = p$. Show that the probability that exactly k of the drivers will show is $\binom{n}{k} p^k q^{n-k}$ for $k = 0, 1, \ldots, n$.

1.8. Suppose you have to choose a team of 3 persons to enter a race. The rules of the race are that a team must consist of 3 people whose respective probabilities p_1, p_2, p_3 of showing must add up to $\frac{1}{2}$; that is, $p_1 + p_2 + p_3 = \frac{1}{2}$. What probabilities of showing would you desire the members of your team to have in order to maximize the probability that at least 1 member of your team will show? (Assume independence.)

1.9. Let A and B be 2 independent events such that the probability is $\frac{1}{6}$ that they will occur simultaneously and $\frac{1}{3}$ that neither of them will occur. Find $P[A]$ and $P[B]$; are $P[A]$ and $P[B]$ uniquely determined?

1.10. Let A and B be 2 independent events such that the probability is $\frac{1}{6}$ that they will occur simultaneously and $\frac{1}{3}$ that A will occur and B will not occur. Find $P[A]$ and $P[B]$; are $P[A]$ and $P[B]$ uniquely determined?

2. INDEPENDENT TRIALS

The notion of independent families of events leads us next to the notion of independent trials. Let S be a sample description space of a random observation or experiment on which is defined a probability function $P[\cdot]$. *Suppose further that each description in S is an n-tuple. Then the random phenomenon which S describes is defined as consisting of n trials.* For example, suppose one is drawing a sample of size n from an urn containing M balls. The sample description space of such an experiment consists of n-tuples. It is also useful to regard this experiment as a series of trials, in each of which a ball is drawn from the urn. Mathematically, the fact that in drawing a sample of size n one is performing n trials is expressed by the fact that the sample description space S consists of n-tuples $(z_1, z_2, \ldots, z_n)$; the first component z_1 represents the outcome of the first trial, the second component z_2 represents the outcome of the second trial, and so on, until z_n represents the outcome of the nth trial.

We next define the important notion of *event depending on a trial.* Let S be a sample description space consisting of n trials, and let A be an event on S. Let k be an integer, 1 to n. We say that *A depends on* the kth trial if the occurrence or nonoccurrence of A depends only on the outcome of the kth trial. In other words, in order to determine whether or not A has occurred, one must have a knowledge only of the outcome of the kth trial. From a more abstract point of view, an event A is said to depend on the kth trial if the decision as to whether a given description in S belongs to the event A depends only on the kth component of the description. It should be especially noted that the certain event S and the impossible event $\emptyset$ may be said to depend on every trial, since the occurrence or nonoccurrence of these events can be determined without knowing the outcome of any trial.

▶ **Example 2A.** Suppose one is drawing a sample of size 2 from an urn containing white and black balls. The event A that the first ball drawn is white depends on the first trial. Similarly, the event B that the second ball drawn is white depends on the second trial. However, the event C that exactly one of the balls drawn is white does *not* depend on any one trial. Note that one may express C in terms of A and B by $C = AB^c \cup A^cB$. ◀

▶ **Example 2B.** Choose a summer day at random on which both the Dodgers and the Giants are playing baseball games, but not with one another. Let $z_1 = 1$ or 0, depending on whether the Dodgers win or lose their game, and, similarly, let $z_2 = 1$ or 0, depending on whether the Giants win or lose their game. The event A that the Dodgers win depends on the first trial of the sample description space $S = \{(z_1, z_2)\colon \; z_1 = 1 \text{ or } 0, z_2 = 1 \text{ or } 0\}$. ◀

We next define the very important notion of *independent trials.* Consider a sample description space S consisting of n trials. For $k = 1, 2, \ldots, n$ let $\mathscr{A}_k$ be the family of events on S that depends on the kth trial. *We define the n trials as independent (and we say that S consists of n independent trials) if the families of events* $\mathscr{A}_1, \mathscr{A}_2, \ldots, \mathscr{A}_n$ *are independent.* Otherwise, the n trials are said to be dependent or nonindependent. More explicitly, the n trials are said to be independent if (1.11) holds for every set of events $A_1, A_2, \ldots, A_n$, such that, for $k = 1, 2, \ldots, n$, A_k depends only on the kth trial.

If the reader traces through the various definitions that have been made in this chapter, it should become clear to him that the mathematical definition of the notion of independent trials embodies the intuitive meaning of the notion, which is that two trials (of the same or different experiments) are independent if *the outcome of one does not affect the outcome of the other* and are otherwise dependent.

In the foregoing definition of independent trials it was assumed that the probability function $P[\cdot]$ was already defined on the sample description space S, which consists of n-tuples. If this were the case, it is clear that to establish that S consists of n independent trials requires the verification of a large number of relations of the form of (1.11). However, in practice, one does not start with a probability function $P[\cdot]$ on S and then proceed to verify all of the relations of the form of (1.11) in order to show that S consists of n independent trials. *Rather, the notion of independent trials derives its importance from the fact that it provides an often-used method for setting up a probability function on a sample description space.* This is done in the following way.*

* The remainder of this section may be omitted in a first reading of the book if the reader is willing to accept intuitively the ideas made precise here.

Let $Z_1, Z_2, \ldots, Z_n$ be n sample description spaces (which may be alike) on whose subsets, respectively, are defined probability functions $P_1, P_2, \ldots, P_n$. For example, suppose we are drawing, with replacement, a sample of size n from an urn containing N balls, numbered 1 to N. We define (for $k = 1, 2, \ldots, n$) Z_k as the sample description space of the outcome of the kth draw; consequently, $Z_k = \{1, 2, \ldots, N\}$. If the descriptions in Z_k are assumed to be equally likely, then the probability function P_k is defined on the events C_k of Z_k by $P_k[C_k] = N[C_k]/N[Z_k]$.

Now suppose we perform in succession the n random experiments whose sample description spaces are $Z_1, Z_2, \ldots, Z_n$, respectively. The sample description space S of this series of n random experiments consists of n-tuples $(z_1, z_2, \ldots, z_n)$, which may be formed by taking for the first component z_1 any member of Z_1, by taking for the second component z_2 any member of Z_2, and so on, until for the nth component z_n we take any member of Z_n. We introduce a notation to express these facts; we write $S = Z_1 \otimes Z_2 \otimes \ldots \otimes Z_n$, which we read "$S$ is the combinatorial product of the spaces $Z_1, Z_2, \ldots, Z_n$." More generally, we define the notion of a *combinatorial product event* on S. For any events C_1 on Z_1, C_2 on Z_2, and C_n on Z_n we define the *combinatorial product event* $C = C_1 \otimes C_2 \otimes \ldots \otimes C_n$ as the set of all n-tuples $(z_1, z_2, \ldots, z_n)$, which can be formed by taking for the first component z_1 any member of C_1, for the second component z_2 any member of C_2, and so on, until for the nth component z_n we take any member of C_n.

We now define a probability function $P[\cdot]$ on the subsets of S. For every event C on S that is a combinatorial product event, so that $C = C_1 \otimes C_2 \otimes \ldots \otimes C_n$ for some events $C_1, C_2, \ldots, C_n$, which belong, respectively, to $Z_1, Z_2, \ldots, Z_n$, we define

$$P[C] = P_1[C_1]P_2[C_2] \cdots P_n[C_n]. \tag{2.1}$$

Not every event in S is a combinatorial product event. However, it can be shown that it is possible to define a unique probability function $P[\cdot]$ on the events of S in such a way that (2.1) holds for combinatorial product events.

It may help to clarify the meaning of the foregoing ideas if we consider the special (but, nevertheless, important) case, in which each sample description space $Z_1, Z_2, \ldots, Z_n$ is finite, of sizes $N_1, N_2, \ldots, N_n$, respectively. As in section 6 of Chapter 1, we list the descriptions in $Z_1, Z_2, \ldots, Z_n$: for $j = 1, \ldots, n$.

$$Z_j = \{D_1^{(j)}, D_2^{(j)}, \cdots, D_{N_j}^{(j)}\}.$$

Now let $S = Z_1 \otimes Z_2 \otimes \ldots \otimes Z_n$ be the sample description space of the random experiment, which consists in performing in succession the n

random experiments whose sample description spaces are $Z_1, Z_2, \ldots, Z_n$, respectively. A typical description in S can be written $(D_{i_1}^{(1)}, D_{i_2}^{(2)}, \ldots, D_{i_n}^{(n)})$ where, for $j = 1, \ldots, n$, $D_{i_j}^{(j)}$ represents a description in Z_j and i_j is some integer, 1 to N_j. To determine a probability function $P[\cdot]$ on the subsets of S, it suffices to specify it on the single-member events of S. Given probability functions $P_1[\cdot], P_2[\cdot], \ldots, P_n[\cdot]$ defined on $Z_1, Z_2, \ldots, Z_n$, respectively, we define $P[\cdot]$ on the subsets of S by defining

$$(2.2)\quad P[\{(D_{i_1}^{(1)}, D_{i_2}^{(2)}, \cdots, D_{i_n}^{(n)})\}] = P_1[\{D_{i_1}^{(1)}\}]P_2[\{D_{i_2}^{(2)}\}] \cdots P_n[\{D_{i_n}^{(n)}\}].$$

Equation (2.2) is a special case of (2.1), since a single-member event on S can be written as a combinatorial product event; indeed,

$$(2.3)\quad \{(D_{i_1}^{(1)}, D_{i_2}^{(2)}, \cdots, D_{i_n}^{(n)})\} = \{D_{i_1}^{(1)}\} \otimes \{D_{i_2}^{(2)}\} \otimes \cdots \otimes \{D_{i_n}^{(n)}\}.$$

▶ **Example 2C.** Let $Z_1 = \{H, T\}$ be the sample description space of the experiment of tossing a coin, and let $Z_2 = \{1, 2, \ldots, 6\}$ be the sample description space of the experiment of throwing a fair die. Let S be the sample description space of the experiment, which consists of first tossing a coin and then throwing a die. What is the probability that in the jointly performed experiment one will obtain heads on the coin toss and a 5 on the die toss? The assumption made by (2.2) is that it is equal to the product of (i) the probability that the outcome of the coin toss will be heads and (ii) the probability that the outcome of the die throw will be a 5. ◀

We now desire to show that the probability space, consisting of the sample description space $S = Z_1 \otimes Z_2 \otimes \ldots \otimes Z_n$, *on whose subsets a probability function* $P[\cdot]$ *is defined by means of* (2.1), *consists of n independent trials.*

We first note that an event A_k in S, which depends only on the kth trial, is necessarily a combinatorial product event; indeed, for some event C_k in Z_k

$$(2.4)\quad A_k = Z_1 \otimes \cdots \otimes Z_{k-1} \otimes C_k \otimes Z_{k+1} \otimes \cdots \otimes Z_n.$$

Equation (2.4) follows from the fact that an event A_k depends on the kth trial if and only if the decision as to whether or not a description $(z_1, z_2, \ldots, z_n)$ belongs to A_k depends only on the kth component z_k of the description. Next, let $A_1, A_2, \ldots, A_n$ be events depending, respectively, on the first, second, $\ldots$, nth trial. For each A_k we have a representation of the form of (2.4). We next assert that the intersection may be written as a combinatorial product event:

$$(2.5)\quad A_1A_2 \cdots A_n = C_1 \otimes C_2 \otimes \cdots \otimes C_n.$$

We leave the verification of (2.5), which requires only a little thought, to the reader. Now, from (2.1) and (2.5)

$$P[A_1A_2 \cdots A_n] = P_1[C_1]P_2[C_2] \cdots P_n[C_n], \tag{2.6}$$

whereas from (2.1) and (2.4)

$$\begin{aligned} P[A_k] &= P_1[Z_1] \cdots P_{k-1}[Z_{k-1}]P_k[C_k]P_{k+1}[Z_{k+1}] \cdots P_n[Z_n] \\ &= P_k[C_k] \end{aligned} \tag{2.7}$$

From (2.6) and (2.7) it is seen that (1.11) is satisfied, so that S consists of n independent trials.

The foregoing considerations are not only sufficient to define a probability space that consists of independent trials but are also necessary in the sense of the following theorem, which we state without proof. *Let the sample description space S be a combinatorial product of n sample description spaces $Z_1, Z_2, \ldots, Z_n$. Let $P[\cdot]$ be a probability function defined on the subsets of S. The probability space S consists of n independent trials if and only if there exist probability functions $P_1[\cdot], P_2[\cdot], \ldots, P_n[\cdot]$, defined, respectively, on the subsets of the sample description spaces $Z_1, Z_2, \ldots, Z_n$, with respect to which $P[\cdot]$ satisfies* (2.6) *for every set of n events $A_1, A_2, \ldots, A_n$ on S such that, for $k = 1, \ldots, n$, A_k depends only on the kth trial* (*and then C_k is defined by* (2.4)).

To illustrate the foregoing considerations, we consider the following example.

▶ **Example 2D.** A man tosses two fair coins independently. Let C_1 be the event that the first coin tossed is a head, let C_2 be the event that the second coin tossed is a head, and let C be the event that both coins tossed are heads. Consider sample description spaces: $S = \{(H, H), (H, T), (T, H), (T, T)\}$, $Z_1 = Z_2 = \{H, T\}$. Clearly S is the sample description space of the outcome of the two tosses, whereas Z_1 and Z_2 are the sample description spaces of the outcome of the first and second tosses, respectively. We assume that each of these sample description spaces has equally likely descriptions.

The event C_1 may be defined on either S or Z_1. If defined on Z_1, $C_1 = \{H\}$. If defined on S, $C_1 = \{(H, H), (H, T)\}$. The event C_2 may in a similar manner be defined on either Z_2 or S. However, the event C can be defined only on S; $C = \{(H, H)\}$.

The spaces on which C_1 and C_2 are defined determines the relation that exists between C_1, C_2, and C. If both C_1 and C_2 are defined on S, then $C = C_1C_2$. If C_1 and C_2 are defined on Z_1 and Z_2, respectively, then $C = C_1 \otimes C_2$.

In order to speak of the independence of C_1 and C_2, we must regard them as being defined on the same sample description space. That C_1 and C_2 are independent events is intuitively clear, since S consists of two independent trials and C_1 depends on the first trial, whereas C_2 depends on the second trial. Events can be independent without depending on independent trials. For example, consider the event $D = \{(H, H), (T, T)\}$ that the two tosses have the same outcome. One may verify that D and C_1 are independent and also that D and C_2 are independent. On the other hand, the events D, C_1, and C_2 are not independent. ◀

EXERCISES

2.1. Consider a man who has made 2 tosses of a die. State whether each of the following six statements is true or false.

Let A_1 be the event that the outcome of the first throw is a 1 or a 2.
Statement 1: A_1 depends on the first throw.

Let A_2 be the event that the outcome of the second throw is a 1 or a 2.
Statement 2: A_1 and A_2 are mutually exclusive events.

Let B_1 be the event that the sum of the outcomes is 7.
Statement 3: B_1 depends on the first throw.

Let B_2 be the event that the sum of the outcomes is 3.
Statement 4: B_1 and B_2 are mutually exclusive events.

Let C be the event that one of the outcomes is a 1 and the other is a 2.
Statement 5: $A_1 \cup A_2$ is a subevent of C.
Statement 6: C is a subevent of B_2.

2.2. Consider a man who has made 2 tosses of a coin. He assumes that the possible outcomes of the experiment, together with their probability, are given by the following table:

Sample Descriptions D	(H, H)	(H, T)	(T, H)	(T, T)
$P[\{D\}]$	$\frac{1}{3}$	$\frac{1}{6}$	$\frac{1}{6}$	$\frac{1}{3}$

Show that this probability space does not consist of 2 independent trials. Is there a unique probability function that must be assigned on the subsets of the foregoing sample description space in order that it consist of 2 independent trials?

2.3. Consider 3 urns; urn I contains 1 white and 2 black balls, urn II contains 3 white and 2 black balls, and urn III contains 2 white and 3 black balls. One ball is drawn from each urn. What is the probability that among the balls drawn there will be (i) 1 white and 2 black balls, (ii) at least 2 black balls, (iii) more black than white balls?

2.4. If you had to construct a mathematical model for events A and B, as described below, would it be appropriate to assume that A and B are independent? Explain the reasons for your opinion.

(i) A is the event that a subscriber to a certain magazine owns a car, and B is the event that the same subscriber is listed in the telephone directory.

(ii) A is the event that a married man has blue eyes, and B is the event that his wife has blue eyes.

(iii) A is the event that a man aged 21 is more than 6 feet tall, and B is the event that the same man weighs less than 150 pounds.

(iv) A is the event that a man lives in the Northern Hemisphere, and B is the event that he lives in the Western Hemisphere.

(v) A is the event that it will rain tomorrow, and B is the event that it will rain within the next week.

2.5. Explain the meaning of the following statements:

(i) A random phenomenon consists of n trials.

(ii) In drawing a sample of size n, one is performing n trials.

(iii) An event A depends on the third trial.

(iv) The event that the third ball drawn is white depends on the third trial.

(v) In drawing with replacement a sample of size 6, one is performing 6 independent trials of an experiment.

(vi) If S is the sample description space of the experiment of drawing with replacement a sample of size 6 from an urn containing balls, numbered 1 to 10, then $S = Z_1 \otimes Z_2 \otimes \ldots \otimes Z_6$, in which $Z_j = \{1, 2, \ldots, 10\}$ for $j = 1, \ldots, 6$.

(vii) If, in (vi), balls numbered 1 to 7 are white and if A is the event that all balls drawn are white, then $A = C_1 \otimes C_2 \otimes \ldots \otimes C_6$, in which $C_j = \{1, 2, \ldots, 7\}$ for $j = 1, \ldots, 6$.

3. INDEPENDENT BERNOULLI TRIALS

Many problems in probability theory involve independent repeated trials of an experiment whose outcomes have been classified in two categories, called "successes" and "failures" and represented by the letters s and f, respectively. Such an experiment, which has only two possible outcomes, is called a *Bernoulli trial.* The probability of the outcome s is usually denoted by p, and the probability of the outcome f is usually denoted by q, where

$$p \geq 0, \qquad q \geq 0, \qquad p + q = 1. \tag{3.1}$$

In symbols, the sample description space of a Bernoulli trial is $Z = \{s, f\}$, on whose subsets is given a probability function $P_Z[\cdot]$, satisfying $P_Z[\{s\}] = p$, $P_Z[\{f\}] = q$.

Consider now *n independent repeated Bernoulli trials*, in which the word "repeated" is meant to indicate that the probabilities of success and failure remain the same throughout the trials. The sample description space S of n independent repeated Bernoulli trials contains 2^n descriptions, each an

n-tuple $(z_1, z_2, \ldots, z_n)$, in which each z_i is either an s or an f. The sample description space S is finite. However, to specify a probability function $P[\cdot]$ on the subsets of S, we shall not assume that all descriptions in S are equally likely. Rather, we shall use the ideas in section 2.

In order to specify a probability function $P[\cdot]$ on the subsets of S, it suffices to specify it on the single-member events $\{(z_1, \ldots, z_n)\}$. However, a single-member event may be written as a combinatorial product event; indeed, $\{(z_1, \ldots, z_n)\} = \{z_1\} \otimes \ldots \otimes \{z_n\}$. Since it has been assumed that $P_Z[\{s\}] = p$ and $P_Z[\{f\}] = q$, we obtain the following basic rule.*

If a probability space consists of n independent repeated Bernoulli trials, then the probability $P[\{(z_1, \ldots, z_n)\}]$ *of any single-member event is equal to* $p^k q^{n-k}$, *in which k is the number of successes s among the components of the description* $(z_1, \ldots, z_n)$.

▶ **Example 3A.** Suppose that a man tosses ten times a possibly unfair coin, whose probability of falling heads is p, which may be any number between 0 and 1, inclusive, depending on the construction of the coin. On each trial a success s is said to have occurred if the coin falls heads. Let us find the probability of the event A that the coin will fall heads on the first four tosses and tails on the last six tosses, assuming that the tosses are independent. It is equal to p^4q^6, since the event A is the same as the single-member event $\{(s, s, s, s, f, f, f, f, f, f)\}$. ◀

One usually encounters Bernoulli trials by considering a random event E, whose probability of occurrence is p. In each trial one is interested only in the occurrence or nonoccurrence of E. A success s corresponds to an occurrence of the event E, and a failure f corresponds to a nonoccurrence of E. Thus, for example, one may be tossing darts at a target, and E may be the event that the target is hit; or one may be tossing a pair of dice, and E may represent the event that the sum of the dice is 7 (for fair dice, $p = \frac{1}{6}$); or 3 men may be tossing coins simultaneously, and E may be the event that all of the coins fall heads (for fair coins, $p = \frac{1}{8}$); or a woman may be pregnant, and E is the event that her child is a boy; or a man may be celebrating his 21st birthday, and E may be the event that he will live to be 22 years old.

The Probability of k Successes in n Independent Repeated Bernoulli Trials. Frequently, the only fact about the outcome of a succession of n Bernoulli trials in which we are interested is the *number of successes*. We now compute the probability that the number of successes will be k, for any integer k from $0, 1, 2, \ldots, n$. The event "k successes in n trials" can

* A reader who has omitted the preceding section may take this rule as the definition of n independent repeated Bernoulli trials.

happen in as many ways as k letters s may be distributed among n places; this is the same as the number of subsets of size k that may be formed from a set containing n members. Consequently, there are $\binom{n}{k}$ descriptions containing exactly k successes and $n - k$ failures. Each such description has probability $p^k q^{n-k}$. Thus we have obtained a basic formula.

The Binomial Law. The probability, denoted by $b(k; n, p)$, that n independent repeated Bernoulli trials, with probabilities p for success, and $q = 1 - p$ for failure, will result in k successes and $n - k$ failures (in which $k = 0, 1, \ldots, n$) is given by

$$b(k; n, p) = \binom{n}{k} p^k q^{n-k}. \tag{3.2}$$

The law expressed by (3.2) is called the binomial law because of the role the quantities in (3.2) play in the binomial theorem, which states that

$$\sum_{k=0}^{n} \binom{n}{k} p^k q^{n-k} = (p + q)^n = 1, \tag{3.3}$$

since $p + q = 1$.

The reader should note that (3.2) is very similar to (3.4) of Chapter 2. However, (3.2) represents the solution to a probability problem that does not involve equally likely descriptions. The importance of this fact is illustrated by the following example. Suppose one is throwing darts at a target. It is difficult to see how one could compute the probability of the event E that one will hit the target by setting up some appropriate sample description space with equally likely descriptions. Rather, p may have to be estimated approximately by means of the frequency definition of probability. Nevertheless, even though p cannot be computed, once one has *assumed* a value for p one can compute by the methods of this section the probability of any event A that can be expressed in terms of independent trials of the event E.

The reader should also note that (3.2) is very similar to (1.13). By means of the considerations of section 2, it can be seen that (3.2) and (1.13) are equivalent formulations of the same law.

The binomial law, and consequently the quantity $b(k; n, p)$, occurs frequently in applications of probability theory. The quantities $b(k; n, p)$, $k = 0, 1, \ldots, n$, are tabulated for $p = 0.01$ (0.01) 0.50 and $n = 2(1)$ 49 (that is, for all values of p and n in the ranges $p = 0.01, 0.02, 0.03, \ldots, 0.50$ and $n = 2, 3, 4, \ldots, 49$) in "Tables of the Binomial Probability Distribution," National Bureau of Standards, Applied Mathematics Series 6, Washington, 1950. A short table of $b(k; n, p)$ for various values of p between 0.01 and 0.5 and for $n = 2, 3, \ldots, 10$ is given in Table II on

p. 442. It should be noted that values of $b(k; n, p)$ for $p > 0.5$ can be obtained from Table II by means of the formula

$$b(k; n, p) = b(n - k; n, 1 - p). \tag{3.4}$$

▶ **Example 3B.** By a series of tests of a certain type of electrical relay, it has been determined that in approximately 5% of the trials the relay will fail to operate under certain specified conditions. What is the probability that in ten trials made under these conditions the relay will fail to operate one or more times?

Solution: To describe the results of the ten trials, we write a 10-tuple $(z_1, z_2, \ldots, z_{10})$ whose kth component $z_k = s$ or f, depending on whether the relay did or did not operate on the kth trial. We next assume that the ten trials constitute ten independent repeated Bernoulli trials, with probability of success $p = 0.95$ at each trial. The probability of no failures in the ten trials is $b(10; 10, 0.95) = (0.95)^{10} = b(0; 10, 0.05)$. Consequently, the probability of one or more failures in the ten trials is equal to

$$1 - (0.95)^{10} = 1 - b(0; 10, 0.05) = 1 - 0.5987 = 0.4013. \quad ◀$$

▶ **Example 3C. How to tell skill from luck.** A rather famous personage in statistical circles is the tea-tasting lady whose claims have been discussed by such outstanding scholars as R. A. Fisher and J. Neyman; see J. Neyman, *First Course in Probability and Statistics*, Henry Holt, New York, 1950, pp. 272–289. "A Lady declares that by tasting a cup of tea made with milk she can discriminate whether the milk or the tea infusion was first added to the cup." Specifically, the lady's claim is "not that she could draw the distinction with invariable certainty, but that, though sometimes mistaken, she would be right more often than not." To test the lady's claim, she will be subjected to an experiment. She will be required to taste and classify n pairs of cups of tea, each pair containing one cup of tea made by each of the two methods under consideration. Let p be the probability that the lady will correctly classify a pair of cups. Assuming that the n pairs of cups are classified under independent and identical conditions, the probability that the lady will correctly classify k of the n pairs is $\binom{n}{k} p^k q^{n-k}$. Suppose that it is decided to grant the lady's claims if she correctly classifies at least eight of ten pairs of cups. Let $P(p)$ be the probability of granting the lady's claims, given that her true probability of classifying a pair of cups is p. Then $P(p) = \binom{10}{8} p^8 q^2 + \binom{10}{9} p^9 q + p^{10}$, since $P(p)$ is equal to the probability that the lady will correctly classify at least eight of ten pairs. In particular, the probability that the lady will establish her claim, given that she is skillful (say, $p = 0.85$) is given by

$P(0.85) = 0.820$, whereas the probability that the lady will establish her claim, given that she is merely lucky (that is, $p = 0.50$) is given by $P(0.50) = 0.055$. ◀

▶ **Example 3D. The game of "odd man out".** Let N distinguishable coins be tossed simultaneously and independently, where $N \geq 3$. Suppose that each coin has probability p of falling heads. What is the probability that either exactly one of the coins will fall heads or that exactly one of the coins will fall tails?

Application: In a game, which we shall call "odd man out," N persons toss coins to determine one person who will buy refreshments for the group. If there is a person in the group whose outcome (be it heads or tails) is not the same as that of any other member of the group, then that person is called an odd man and must buy refreshment for each member of the group. The probability asked for in this example is the probability that in any play of the game there will be an odd man. The next example is concerned with how many plays of the game will be required to determine an odd man.

Solution: To describe the results of the N tosses, we write an N-tuple $(z_1, z_2, \ldots, z_N)$ whose kth component is s or f, depending on whether the kth coin tossed fell heads or tails. We are then considering N independent repeated Bernoulli trials, with probability p of success at each trial. The probability of exactly one success is $\binom{N}{1} pq^{N-1}$, whereas the probability of exactly one failure is $\binom{N}{N-1} p^{N-1}q$. Consequently, the probability that either exactly one of the coins will fall heads or exactly one of the coins will fall tails is equal to $N(p^{N-1}q + p^{N-1})$. If the coins are fair, so that $p = \frac{1}{2}$, then the probability is $N/2^{N-1}$. Thus, if five persons play the game of "odd man out" with fair coins, the probability that in any play of the game there will be a loser is $\frac{5}{16}$. ◀

▶ **Example 3E. The duration of the game of "odd man out".** Let N persons play the game of "odd man out" with fair coins. What is the probability for $n = 1, 2, \ldots$ that n plays will be required to conclude the game (that is, the nth play is the first play in which one of the players will have an outcome on his coin toss different from those of all the other players)?

Solution: Let us rephrase the problem. (See theoretical exercise 3.3.) Suppose that n independent plays are made of the game of "odd man out." What is the probability that on the nth play, but not on any preceding play, there will be an odd man? Let P be the probability that on any play there will be an odd man. In example 3D it was shown that $P = N/2^{N-1}$ if N persons are tossing fair coins. Let $Q = 1 - P$. To describe the results of

n plays, we write an n-tuple $(z_1, z_2, \ldots, z_n)$ whose kth component is s or f, depending on whether the kth play does or does not result in an odd man. Assuming that the plays are independent, the n plays thus constitute repeated independent Bernoulli trials with probability $P = N/2^{N-1}$ of success at each trial. Consequently, the event $\{(f, f, \ldots, f, s)\}$ of failure at all trials but the nth has probability $Q^{n-1}P$. Thus, if five persons toss fair coins, the probability that four tosses will be required to produce an odd man is $(11/16)^3(5/16)$. ◀

Various approximations that exist for computing the binomial probabilities are discussed in section 2 of Chapter 6. We now briefly indicate the nature of one of these approximations, namely, that of the binomial probability law by the Poisson probability law.

The Poisson Law. A random phenomenon whose sample description space S consists of all the integers from 0 onward, so that $S = \{0, 1, 2, \ldots\}$, and on whose subsets a probability function $P[\cdot]$ is defined in terms of a parameter $\lambda > 0$ by

$$P[\{k\}] = e^{-\lambda}\frac{\lambda^k}{k!}, \qquad k = 0, 1, 2, \cdots \tag{3.5}$$

is said to obey the Poisson probability law with parameter λ. Examples of random phenomena that obey the Poisson probability law are given in section 3 of Chapter 6. For the present, let us show that under certain circumstances the number of successes in n independent repeated Bernoulli trials, with probability of success p at each trial, approximately obeys the Poisson probability law with parameter $\lambda = np$.

More precisely, we show that for any fixed $k = 0, 1, 2, \ldots$, and $\lambda > 0$

$$\lim_{n\to\infty}\binom{n}{k}\left(\frac{\lambda}{n}\right)^k\left(1 - \frac{\lambda}{n}\right)^{n-k} = e^{-\lambda}\frac{\lambda^k}{k!}. \tag{3.6}$$

To prove (3.6), we need only rewrite its left-hand side:

$$\frac{1}{k!}\lambda^k\left(1 - \frac{\lambda}{n}\right)^{n-k}\frac{n(n-1)\cdots(n-k+1)}{n^k}.$$

Since $\lim_{n\to\infty}[1 - (\lambda/n)]^n = e^{-\lambda}$, we obtain (3.6).

Since (3.6) holds in the limit, we may write that it is approximately true for large values of n that

$$\binom{n}{k}p^k(1-p)^{n-k} = e^{-np}\frac{(np)^k}{k!}. \tag{3.7}$$

We shall not consider here the remainder terms for the determination of the accuracy of the approximation formula (3.7). In practice, the approximation represented by (3.7) is used if $p \leq 0.1$. A short table of the Poisson probabilities defined in (3.5) is given in Table III (see p. 444).

▶ **Example 3F.** It is known that the probability that an item produced by a certain machine will be defective is 0.1. Let us find the probability that a sample of ten items, selected at random from the output of the machine, will contain no more than one defective item. The required probability, based on the binomial law, is $\binom{10}{0}(0.1)^0(0.9)^{10} + \binom{10}{1}(0.1)^1(0.9)^9 = 0.7361$, whereas the Poisson approximation given by (3.7) yields the value $e^{-1} + e^{-1} = 0.7358$. ◀

▶ **Example 3G. Safety testing vaccine.** Suppose that at a certain stage in the production process of a vaccine the vaccine contains, on the average, m live viruses per cubic centimeter and the constant m is known to us. Consequently, let it be assumed that in a large vat containing V cubic centimeters of vaccine there are $n = mV$ viruses. Let a sample of vaccine be drawn from the vat; the sample's volume is v cubic centimeters. Let us find for $k = 0, 1, \ldots, n$ the probability that the sample will contain k viruses. Let us write an n-tuple $(z_1, z_2, \ldots, z_n)$ to describe the location of the n viruses in the vat, the jth component z_j being equal to s or f, depending on whether the jth virus is or is not located in our sample. The probability p that a virus in the vat will be in our sample may be taken as the ratio of the volume of the sample to the volume of the vat, $p = v/V$, if it is assumed that the viruses are dispersed uniformly in the vat. Assuming further that the viruses are independently dispersed in the vat, it follows by the binomial law that the probability $P[\{k\}]$ that the sample will contain exactly k viruses is given by

$$P[\{k\}] = \binom{mV}{k}\left(\frac{v}{V}\right)^k\left(1 - \frac{v}{V}\right)^{mV-k} \tag{3.8}$$

If it is assumed that the sample has a volume v less than 1% of the volume V of the vat, then by the Poisson approximation to the binomial law

$$P[\{k\}] = e^{-mv}\frac{(mv)^k}{k!}. \tag{3.9}$$

As an application of this result, let us consider a vat of vaccine that contains five viruses per 1000 cubic centimeters. Then $m = 0.005$. Let a sample of volume $v = 600$ cubic centimeters be taken. We are interested in determining the probability $P[\{0\}]$ that the sample will contain no viruses. This problem is of great importance in the design of a scheme to safety-test vaccine, for if the sample contains no viruses one might be led to pass as virus free the entire contents of the vat of vaccine from which the sample was drawn. By (3.9) we have

$$(3.10) \qquad P[\{0\}] = e^{-mv} = e^{-(0.005)(600)} = e^{-3} = 0.0498.$$

Let us attempt to interpret this result. If we desire to produce virus-free vaccine, we must design a production process so that the density m of viruses in the vaccine is 0. As a check that the production process is operating properly, we sample the vaccine produced. Now, (3.10) implies that when judging a given vat of vaccine it is not sufficient to rely merely on the sample from that vat, if we are taking samples of volume 600 cubic centimeters, since 5% of the samples drawn from vats with virus densities $m = 0.005$ viruses per cubic centimeter will yield the conclusion that no viruses are present in the vat. One way of decreasing this probability of a wrong decision might be to take into account the results of recent safety tests on similar vats of vaccine. ◀

Independent Trials with More Than 2 Possible Outcomes. In the foregoing we considered independent trials of a random experiment with just two possible outcomes. It is natural to consider next the independent trials of an experiment with several possible outcomes, say r possible outcomes, in which r is an integer greater than 2. For the sample description space of the outcomes of a particular trial we write $Z = \{s_1, s_2, \ldots, s_r\}$. We assume that we know positive numbers $p_1, p_2, \ldots, p_r$, whose sum is 1, such that at each trial p_k represents the probability that s_k will be the outcome of that trial. In symbols, there exist numbers $p_1, p_2, \ldots, p_r$ such that

$$(3.11) \qquad 0 < p_k < 1, \qquad \text{for } k = 1, 2, \ldots, r; \qquad p_1 + p_2 + \cdots + p_r = 1$$
$$P_Z[\{s_k\}] = p_k, \qquad \text{for } k = 1, 2, \ldots, r.$$

▶ **Example 3H.** Consider an experiment in which two fair dice are tossed. Consider three possible outcomes, s_1, s_2, and s_3, defined as follows: if the sum of the two dice is five or less, we say that s_1 is the outcome; if the sum of the two dice is six, seven, or eight, we say s_2 is the outcome; if the sum of the two dice is nine or more, we say s_3 is the outcome. Then $p_1 = \frac{5}{18}$, $p_2 = \frac{8}{18}$, $p_3 = \frac{5}{18}$. ◀

Let S be the sample description space of n independent repeated trials of the experiment described. There are r^n descriptions in S. The probability $P[\{(z_1, z_2, \ldots, z_n)\}]$ of any single-member event is equal to $p_1^{k_1} p_2^{k_2} \ldots p_r^{k_r}$, in which $k_1, k_2, \ldots, k_r$ denote, respectively, the number of occurrences of $s_1, s_2, \ldots, s_r$ among the components of the description $(z_1, z_2, \ldots, z_n)$.

Corresponding to the binomial law, we have the *multinomial law: the probability that in n trials the outcome s_1 will occur k_1 times, the outcome s_2 will occur k_2 times, . . . , the outcome s_r will occur k_r times, for any non-negative integers k_j satisfying the condition $k_1 + k_2 + \ldots + k_r = n$, is given by*

$$\frac{n!}{k_1!k_2!\cdots k_r!}p_1^{k_1}p_2^{k_2}\cdots p_r^{k_r}. \tag{3.12}$$

To prove (3.12), one must note only that the number of descriptions in S, which contain $k_1 s_1$'s, $k_2 s_2$'s, . . . , $k_r s_r$'s, is equal to the number of ways a set of size n can be partitioned into r ordered subsets of sizes $k_1, k_2, \ldots, k_r$, respectively, which is equal to $\binom{n}{k_1 k_2 \ldots k_r}$. Each of these descriptions has probability $p_1^{k_1}p_2^{k_2}\ldots p_r^{k_r}$. Consequently, (3.12) is proved. The name, "multinomial law" derives from the role played by the expressions given in (3.12) in the multinomial theorem [see (1.18) of Chapter 2]. The reader should note the similarity between (3.12) and (3.14) of Chapter 2; these two equations are in the same relationship to each other as (3.2) and (3.4) of Chapter 2.

THEORETICAL EXERCISES

3.1. Suppose one makes n independent trials of an experiment whose probability of success at each trial is p. Show that the conditional probability that any given trial will result in a success, given that there are k successes in the n trials, is equal to k/n.

3.2. Suppose one makes $m + n$ independent trials of an experiment whose probability of success at each trial is p. Let $q = 1 - p$.
(i) Show that for any $k = 0, 1, \ldots, n$ the conditional probability that exactly $m + k$ trials will result in success, given that the first m trials result in success, is equal to $\binom{n}{k}p^k q^{n-k}$.
(ii) Show that the conditional probability that exactly $m + k$ trials will result in success, given that at least m trials result in success, is equal to

$$\frac{\binom{m+n}{m+k}\left(\frac{p}{q}\right)^k}{\sum_{r=0}^{n}\binom{m+n}{m+r}\left(\frac{p}{q}\right)^r}. \tag{3.13}$$

3.3. Suppose one performed a sequence of independent Bernoulli trials (in which the probability of success at each trial is p) until the first success occurs. Show for any integer $n = 1, 2, \ldots$ that the probability that n will be the number of trials required to achieve the first success is pq^{n-1}. *Note:* Strictly speaking, this problem should be rephrased as follows.

Consider n independent Bernoulli trials, with probability p for success on any trial. What is the probability that the nth trial will be the first trial on which a success occurs? To show that the problem originally stated is equivalent to the reformulated problem requires the consideration of the theory of a countably infinite number of independent repeated Bernoulli trials; this is beyond the scope of this book.

3.4. **The behavior of the binomial probabilities.** Show that, as k goes from 0 to n, the terms $b(k; n, p)$ increase monotonically, then decrease monotonically, reaching their largest value (i) in the case that $(n + 1)p$ is not an integer, when k is equal to the integer m satisfying the inequalities

$$(n + 1)p - 1 < m < (n + 1)p \tag{3.14}$$

and (ii) in the case $(n + 1)p$ is an integer, when k is equal to either $(n + 1)p - 1$ or $(n + 1)p$. *Hint:* Use the fact that

$$\frac{b(k; n, p)}{b(k - 1; n, p)} = \frac{(n - k + 1)p}{kq} = 1 + \frac{(n + 1)p - k}{kq}. \tag{3.15}$$

3.5. Consider a series of n independent repeated Bernoulli trials at which the probability of success at each trial is p. Show that in order to have two successive integers, k_1 and k_2, between 0 and n, such that the probability of k_1 successes in the n trials will be equal to the probability of k_2 successes in the n trials, it is necessary and sufficient that $(n + 1)p$ be an integer.

3.6. Show that the probability [denoted by $P(r + 1)$, say] of at least $(r + 1)$ successes in $(n + 1)$ independent repeated Bernoulli trials, with probability p of success at each trial, is equal to

$$(r + 1)\binom{n + 1}{r + 1}\int_0^p x^r(1 - x)^{n-r}\,dx. \tag{3.16}$$

Hint: $P(r + 1)$ may be regarded as a function of p for r and n fixed. By differentiation, verify that

$$\frac{d}{dp}P(r + 1) = \frac{(n + 1)!}{r!(n - r)!}p^r q^{n-r}.$$

3.7. **The behavior of the Poisson probabilities.** Show that the probabilities of the Poisson probability law, given by (3.5), increase monotonically, then decrease monotonically as k increases, and reach their maximum when k is the largest integer not exceeding λ.

3.8. **The behavior of the multinomial probabilities.** Show that the probabilities of the multinomial probability law, given by (3.12), reach their maximum at $k_1, k_2, \ldots, k_r$, satisfying the inequalities, for $i = 1, 2, \ldots, r$,

$$np_i - 1 < k_i \le (n + r - 1)p_i. \tag{3.17}$$

Hint: Prove first that the maximum is attained at and only at values $k_1, \ldots, k_r$ satisfying $p_i k_j \le p_j(k_i + 1)$ for each pair of indices i and j. Add these inequalities for all j and also for all $i \ne j$. (This result is taken from W. Feller, *An Introduction to Probability Theory and its Applications*, second edition, New York, Wiley, 1957, p. 161, where it is ascribed to P. A. P. Moran.)

EXERCISES

3.1. Assuming that each child has probability 0.51 of being a boy, find the probability that a family of 4 children will have (i) exactly 1 boy, (ii) exactly 1 girl, (iii) at least one boy, (iv) at least 1 girl.

3.2. Find the number of children a couple should have in order that the probability of their having at least 2 boys will be greater than 0.75.

3.3. Assuming that each dart has probability 0.20 of hitting its target, find the probability that if one throws 5 darts at a target one will score (i) no hits, (ii) exactly 1 hit, (iii) at least 2 hits.

3.4. Assuming that each dart has probability 0.20 of hitting its target, find the number of darts one should throw at a target in order that the probability of at least 2 hits will be greater than 0.60.

3.5. Consider a family with 4 children, and assume that each child has probability 0.51 of being a boy. Find the conditional probability that all the children will be boys, given that (i) the eldest child is a boy, (ii) at least 1 of the children is a boy.

3.6. Assuming that each dart has probability 0.20 of hitting its target, find the conditional probability of obtaining 2 hits in 5 throws, given that one has scored an even number of hits in the 5 throws.

3.7. A certain manufacturing process yields electrical fuses, of which, in the long run, 15% are defective. Find the probability that in a sample of 10 fuses selected at random there will be (i) no defectives, (ii) at least 1 defective, (iii) no more than 1 defective.

3.8. A machine normally makes items of which 5% are defective. The practice of the producer is to check the machine every hour by drawing a sample of size 10, which he inspects. If the sample contains no defectives, he allows the machine to run for another hour. What is the probability that this practice will lead him to leave the machine alone when in fact it has shifted to producing items of which 10% are defective?

3.9. (Continuation of 3.8). How large a sample should be inspected to insure that if $p = 0.10$ the probability that the machine will not be stopped is less than or equal to 0.01?

3.10. Consider 3 friends who contract a disease; medical experience has shown that 10% of people contracting this disease do not recover. What is the probability that (i) none of the 3 friends will recover, (ii) all of them will recover?

3.11. Let the probability that a person aged x years will survive 1 year be denoted by p_x, whereas $q_x = 1 - p_x$ is the probability that he will die within a year. Consider a board of directors, consisting of a chairman and 5 members; all of the members are 60, the chairman is 65. Find the probability, in terms of q_{60} and q_{65}, that within a year (i) no members will die, (ii) not more than 1 member will die, (iii) neither a member nor the chairman will die, (iv) only the chairman will die. Evaluate these probabilities under the assumption that $q_{60} = 0.025$ and $q_{65} = 0.040$.

3.12. Consider a young man who is waiting for a young lady, who is late. To amuse himself while waiting, he decides to take a walk under the following set of rules. He tosses a coin (which we may assume is fair). If the coin falls heads, he walks 10 yards north; if the coin falls tails, he walks 10 yards south. He repeats this process every 10 yards and thus executes what is called a "random walk." What is the probability that after walking 100 yards he will be (i) back at his starting point, (ii) within 10 yards of his starting point, (iii) exactly 20 yards away from his starting point.

3.13. Do the preceding exercise under the assumption that the coin tossed by the young man is unfair and has probability 0.51 of falling heads (probability 0.49 of falling heads).

3.14. Let 4 persons play the game of "odd man out" with fair coins. What is the probability, for $n = 1, 2, \ldots$, that n plays will be required to conclude the game (that is, the nth play is the first play on which 1 of the players will have an outcome on his coin toss that is different from those of all the other players)?

3.15. Consider an experiment that consists of tossing 2 fair dice independently. Consider a sequence of n repeated independent trials of the experiment. What is the probability that the nth throw will be the first time that the sum of the 2 dice is a 7?

3.16. A man wants to open his door; he has 5 keys, only 1 of which fits the door. He tries the keys successively, choosing them (i) without replacement, (ii) with replacement, until he opens the door. For each integer $k = 1, 2, \ldots$, find the probability that the kth key tried will be the first to fit the door.

3.17. A man makes 5 independent throws of a dart at a target. Let p denote his probability of hitting the target at each throw. Given that he has made exactly 3 hits in the 5 throws, what is the probability that the first throw hit the target? Express your answer in terms as simple as you can.

3.18. Consider a loaded die; in 10 independent throws the probability that an even number will appear 5 times is twice the probability that an even number will appear 4 times. What is the probability that an even number will not appear at all in 10 independent throws of the die?

3.19. An accident insurance company finds that 0.001 of the population incurs a certain kind of accident each year. Assuming that the company has insured 10,000 persons selected randomly from the population, what is the probability that not more than 3 of the company's policyholders will incur this accident in a given year?

3.20. A certain airline finds that 4 per cent of the persons making reservations on a certain flight will not show up for the flight. Consequently, their policy is to sell to 75 persons reserved seats on a plane that has exactly 73 seats. What is the probability that for every person who shows up for the flight there will be a seat available?

3.21. Consider a flask containing 1000 cubic centimeters of vaccine drawn from a vat that contains on the average 5 live viruses in every 1000 cubic centimeters of vaccine. What is the probability that the flask contains (i) exactly 5 live viruses, (ii) 5 or more live viruses?

3.22. The items produced by a certain machine may be classified in 4 grades, A, B, C, and D. It is known that these items are produced in the following proportions:

Grade A	Grade B	Grade C	Grade D
0.3	0.4	0.2	0.1

What is the probability that there will be exactly 1 item of each grade in a sample of 4 items, selected at random from the output of the machine?

3.23. A certain door-to-door salesman sells 3 sizes of brushes, which he calls large, extra large, and giant. He estimates that among the persons he calls upon the probabilities are 0.4 that he will make no sale, 0.3 that he will sell a large brush, 0.1 that he will sell an extra large brush, and 0.2 that he will sell a giant brush. Find the probability that in 4 calls he will sell (i) no brushes, (ii) 4 large brushes, (iii) at least 1 brush of each kind.

3.24. Consider a man who claims to be able to locate hidden sources of water by use of a divining rod. To test his claim, he is presented with 10 covered cans, 1 at a time; he must decide, by means of his divining rod, whether each can contains water. What is the probability that the diviner will make at least 7 correct decisions just by chance? Do you think that the test described in this exercise is fairer than the test described in exercise 2.14 of Chapter 2? Will it make a difference if the diviner knows how many of the cans actually contain water?

3.25. In their paper "Testing the claims of a graphologist," *Journal of Personality*, Vol. 16 (1947), pp. 192–197, G. R. Pascal and B. Suttell describe an experiment designed to evaluate the ability of a professional graphologist. The graphologist claimed that she could distinguish the handwriting of abnormal from that of normal persons. The experimenters selected 10 persons who had been diagnosed as psychotics by at least 2 psychiatrists. For each of these persons a normal-control person was matched for age, sex, and education. Handwriting samples from each pair of persons were placed in a separate folder and presented to the graphologist, who was able to identify correctly the sample of the psychotic in 6 of the 10 pairs.

(i) What is the probability that she would have been correct on at least 6 pairs just by chance?

(ii) How many correct judgements would the graphologist need to make so that the probability of her getting at least that many correct by chance is 5% or less?

3.26. Two athletic teams play a series of games; the first team winning 4 games is the winner. The World Series is an example. Suppose that 1 of the teams is stronger than the other and has probability p of winning each game, independent of the outcomes of any other games. Assume that a game cannot end in a tie. Show that the probabilities that the series will end in 4, 5, 6, or 7 games are (i) if $p = \frac{2}{3}$, 0.21, 0.296, 0.274, and 0.22, respectively, and (ii) if $p = \frac{1}{2}$, 0.125, 0.25, 0.3125, and 0.3125, respectively.

3.27. Suppose that 9 people, chosen at random, are asked if they favor a certain proposal. Find the probability that a majority of the persons polled will favor the proposal, given that 45% of the population favor the proposal.

3.28. Suppose that (i) 2, (ii) 3 restaurants compete for the same 10 patrons. Find the number of seats each restaurant should have in order to have a probability greater than 95% that it can serve all patrons who come to it (assuming that all patrons arrive at the same time and choose, independently of one another, each restaurant with equal probability).

3.29. A fair die is to be thrown 9 times. What is the most probable number of throws on which the outcome is (i) a 6, (ii) an even number?

4. DEPENDENT TRIALS

In section 4 of Chapter 2 the notion of conditional probability was discussed for events defined on a sample description space on which a probability function was defined. However, an important use of the notion of conditional probability is *to set up a probability function on the subsets of a sample description space S, which consists of n trials that are dependent* (or, more correctly, nonindependent). In many applications of probability theory involving dependent trials one will state one's assumptions about the random phenomenon under consideration in terms of certain conditional probabilities that suffice to specify the probability model of the random phenomenon.

As in section 2, for $k = 1, 2, \ldots, n$, let $\mathscr{A}_k$ be the family of events on S which depend on the kth trial. Consider an event A that may be written as the intersection, $A = A_1A_2 \ldots A_n$, of events $A_1, A_2, \ldots, A_n$, which belong to $\mathscr{A}_1, \mathscr{A}_2, \ldots, \mathscr{A}_n$, respectively. Now suppose that a probability function $P[\cdot]$ has been defined on the subsets of S and suppose that $P[A] > 0$. Then, by the multiplicative rule given in theoretical exercise 1.4,

$$(4.1) \quad P[A] = P[A_1]P[A_2 \mid A_1]P[A_3 \mid A_1, A_2] \cdots P[A_n \mid A_1, A_2, \cdots, A_{n-1}].$$

Now, as shown in section 2, any event A that is a combinatorial product event may be written as the intersection of n events, each depending on only one trial. Further, as we pointed out there, a probability function defined on the subsets of a space S, consisting of n trials, is completely determined by its values on combinatorial product events.

Consequently, to know the value of $P[A]$ for any event A it suffices to know, for $k = 2, 3, \ldots, n$, the conditional probability $P[A_k \mid A_1, \ldots, A_{k-1}]$ of any event A_k depending on the kth trial, given any events $A_1, A_2, \ldots, A_{k-1}$ depending on the 1st, 2nd, ..., $(k-1)$st trials, respectively; one also must know $P[A_1]$ for any event A_1 depending on the first trial. In other words, if one assumes a knowledge of

$$
\begin{aligned}
&P[A_1] \\
&P[A_2 \mid A_1] \\
&P[A_3 \mid A_1, A_2] \\
&\quad\cdot \\
&\quad\cdot \\
&\quad\cdot \\
&P[A_n \mid A_1, A_2, \cdots, A_{n-1}]
\end{aligned}
\tag{4.2}
$$

for any events A_1 in $\mathscr{A}_1$, A_2 in $\mathscr{A}_2, \ldots, A_n$ in $\mathscr{A}_n$, one has thereby specified the value of $P[A]$ for any event A on S.

▶ **Example 4A.** Consider an urn containing M balls of which M_W are white. Let a sample of size $n \le M_W$ be drawn without replacement. Let us find the probability of the event that all the balls drawn will be white. The problem was solved in section 3 of Chapter 2; here, let us see how (4.2) may be used to provide insight into that solution. For $i = 1, \ldots, n$ let A_i be the event that the ball drawn on the ith draw is white. We are then seeking $P[A_1A_2 \ldots A_n]$. It is intuitively appealing that the conditional probability of drawing a white ball on the ith draw, given that white balls were drawn on the preceding $(i-1)$ draws, is described for $i = 2, \ldots, n$ by

$$P[A_i \mid A_1, A_2, \cdots, A_{i-1}] = \frac{M_W - (i-1)}{M - (i-1)}, \tag{4.3}$$

since just before the ith draw there are $M - (i-1)$ balls in the urn, of which $M_W - (i-1)$ are white. Let us assume that (4.3) is valid; more generally, *we assume a knowledge of all the probabilities in* (4.2) *by means of the assumption that, whatever the first* $(i-1)$ *choices, at the* i*th draw each of the remaining* $M - i + 1$ *elements will have probability* $1/(M - i + 1)$ *of being chosen.* Then, from (4.1) it follows that

$$P[A_1A_2 \cdots A_n] = \frac{M_W(M_W - 1) \cdots (M_W - n + 1)}{M(M-1) \cdots (M - n + 1)}, \tag{4.4}$$

which agrees with (3.1) of Chapter 2 for the case of $k = n$. ◀

Further illustrations of the specification of a probability function on the subsets of a space of n dependent trials by means of conditional probability functions of the form given in (4.2) are supplied in examples 4B and 4C.

▶ **Example 4B.** Consider two urns; urn I contains five white and three black balls, urn II, three white and seven black balls. One of the urns is selected at random, and a ball is drawn from it. Find the probability that the ball drawn will be white.

Solution: The sample description space of the experiment described consists of 2-tuples (z_1, z_2), in which z_1 is the number of the urn chosen and z_2 is the "name" of the ball chosen. The probability function $P[\cdot]$ on the subsets of S is specified by means of the functions listed in (4.2), with $n = 2$, which the assumptions stated in the problem enable us to compute. In particular, let C_1 be the event that urn I is chosen, and let C_2 be the event that urn II is chosen. Then $P[C_1] = P[C_2] = \frac{1}{2}$. Next, let B be the event that a white ball is chosen. Then $P[B \mid C_1] = \frac{5}{8}$, and $P[B \mid C_2] = \frac{3}{10}$. The events C_1 and C_2 are the complements of each other. Consequently, by (4.5) of Chapter 2,

$$(4.5) \qquad P[B] = P[B \mid C_1]P[C_1] + P[B \mid C_2]P[C_2] = \tfrac{37}{80}.$$ ◀

▶ **Example 4C. A case of hemophilia.*** The first child born to a certain woman was a boy who had hemophilia. The woman, who had a long family history devoid of hemophilia, was perturbed about having a second child. She reassured herself by reasoning as follows. "My son obviously did not inherit his hemophilia from me. Consequently, he is a mutant. The probability that my second child will have hemophilia, if he is a boy, is consequently the probability that he will be a mutant, which is a very small number m (equal to, say, 1/100,000)." Actually, what is the conditional probability that a second son will have hemophilia, given that the first son had hemophilia?

Solution: Let us write a 3-tuple (z_1, z_2, z_3) to describe the history of the mother and her two sons with regard to hemophilia. Let z_1 equal s or f, depending on whether the mother is or is not a hemophilia carrier. Let z_2 equal s or f, depending on whether the first son is or is not hemophilic. Let z_3 equal s or f, depending on whether the second son will or will not have hemophilia. On this sample description space, we define the events A_1, A_2, and A_3: A_1 is the event that the mother is a hemophilia carrier, A_2 is the event that the first son has hemophilia, and A_3 is the event that the second son will have hemophilia. To specify a probability function

* I am indebted to my esteemed colleague Lincoln E. Moses for the idea of this example.

on the subsets of S, we specify all conditional probabilities of the form given in (4.2):

$$
\begin{aligned}
&P[A_1] = 2m, \qquad P[A_1^c] = 1 - 2m,\\
&P[A_2 \mid A_1] = \tfrac{1}{2}, \qquad P[A_2^c \mid A_1] = \tfrac{1}{2},\\
&P[A_2 \mid A_1^c] = m, \qquad P[A_2^c \mid A_1^c] = 1 - m,\\
&P[A_3 \mid A_1, A_2] = P[A_3 \mid A_1, A_2^c] = \tfrac{1}{2},\\
&\quad P[A_3^c \mid A_1, A_2] = P[A_3^c \mid A_1, A_2^c] = \tfrac{1}{2},\\
&P[A_3 \mid A_1^c, A_2] = P[A_3 \mid A_1^c, A_2^c] = m,\\
&\quad P[A_3^c \mid A_1^c, A_2] = P[A_3^c \mid A_1^c, A_2^c] = 1 - m.
\end{aligned} \tag{4.6}
$$

In making these assumptions (4.6) we have used the fact that the woman has no family history of hemophilia. A boy usually carries an X chromosome and a Y chromosome; he has hemophilia if and only if, instead of an X chromsome, he has an X' chromosome which bears a gene causing hemophilia. Let m be the probability of mutation of an X chromosome into an X' chromosome. Now the mother carries two X chromosomes. Event A_1 can occur only if at least one of these X chromosomes is a mutant; this will happen with probability $1 - (1 - m)^2 \doteq 2m$, since m^2 is much smaller than $2m$. Assuming that the woman is a hemophilia carrier and exactly one of her chromosomes is X', it follows that her son will have probability $\frac{1}{2}$ of inheriting the X' chromosome.

We are seeking $P[A_3 \mid A_2]$. Now

$$
P[A_3 \mid A_2] = \frac{P[A_2A_3]}{P[A_2]}. \tag{4.7}
$$

To compute $P[A_2A_3]$, we use the formula

$$
\begin{aligned}
P[A_2A_3] &= P[A_1A_2A_3] + P[A_1^cA_2A_3]\\
&= P[A_1]P[A_2 \mid A_1]P[A_3 \mid A_2, A_1]\\
&\quad + P[A_1^c]P[A_2 \mid A_1^c]P[A_3 \mid A_2, A_1^c]\\
&= 2m(\tfrac{1}{2})\tfrac{1}{2} + (1 - 2m)mm\\
&\doteq \tfrac{1}{2}m,
\end{aligned} \tag{4.8}
$$

since we may consider $1 - 2m$ as approximately equal to 1 and m^2 as approximately equal to 0. To compute $P[A_2]$, we use the formula

$$
\begin{aligned}
P[A_2] &= P[A_2 \mid A_1]P[A_1] + P[A_2 \mid A_1^c]P[A_1^c]\\
&= \tfrac{1}{2}2m + m(1 - 2m)\\
&\doteq 2m.
\end{aligned} \tag{4.9}
$$

Consequently,

$$P[A_3 \mid A_2] = \frac{\frac{1}{2}m}{2m} = \frac{1}{4}. \tag{4.10}$$

Thus the conditional probability that the second son of a woman with no family history of hemophilia will have hemophilia, given that her first son has hemophilia, is approximately $\frac{1}{4}$! ◀

A very important use of the notion of conditional probability derives from the following extension of (4.5). Let $C_1, C_2, \ldots, C_n$ be n events, each of positive probability, which are mutually exclusive and are also *exhaustive* (that is, the union of all the events $C_1, C_2, \ldots, C_n$ is equal to the certain event). Then, for any event B one may express the unconditional probability $P[B]$ of B in terms of the conditional probabilities $P[B \mid C_1], \ldots, P[B \mid C_n]$ and the unconditional probabilities $P[C_1], \ldots, P[C_n]$:

$$P[B] = P[B \mid C_1]P[C_1] + \cdots + P[B \mid C_n]P[C_n] \tag{4.11}$$

if

$$C_1 \cup C_2 \cup \cdots \cup C_n = S, \qquad C_iC_j = \emptyset \qquad \text{for } i \neq j,$$
$$P[C_i] > 0.$$

Equation (4.11) follows immediately from the relation

$$P[B] = P[BC_1] + P[BC_2] + \cdots + P[BC_n] \tag{4.12}$$

and the fact that $P[BC_i] = P[B \mid C_i]P[C_i]$ for any event C_i.

▶ **Example 4D. On drawing a sample from a sample.** Consider a box containing five radio tubes selected at random from the output of a machine, which is known to be 20% defective on the average (that is, the probability that an item produced by the machine will be defective is 0.2). (i) Find the probability that a tube selected from the box will be defective. (ii) Suppose that a tube selected at random from the box is defective; what is the probability that a second tube selected at random from the box will be defective?

Solution: To describe the results of the experiment that consists in selecting five tubes from the output of the machine and then selecting one tube from among the five previously selected, we write a 6-tuple $(z_1, z_2, z_3, z_4, z_5, z_6)$; for $k = 1, 2, \ldots, 5$, z_k is equal to s or f, depending on whether the kth tube selected is defective or nondefective, whereas z_6 is equal to s or f, depending on whether the tube selected from those previously selected is defective or nondefective. For $j = 0, \ldots, 5$ let C_j denote the event that j defective tubes were selected from the output of the machine.

Assuming that the selections were independent, $P[C_j] = \binom{5}{j}(0.2)^j(0.8)^{5-j}$. Let B denote the event that the sixth tube selected from the box, is defective. We assume that $P[B \mid C_j] = j/5$; in words, each of the tubes in the box is equally likely to be chosen. By (4.11), it follows that

$$P[B] = \sum_{j=0}^{5} \frac{j}{5}\binom{5}{j}(0.2)^j(0.8)^{5-j}. \tag{4.13}$$

To evaluate the sum in (4.13), we write it as

$$\sum_{j=1}^{5} \frac{j}{5}\binom{5}{j}(0.2)^j(0.8)^{5-j} = (0.2)\sum_{j=1}^{5}\binom{4}{j-1}(0.2)^{j-1}(0.8)^{4-(j-1)} = 0.2, \tag{4.14}$$

in which we have used the easily verifiable fact that

$$\frac{j}{n}\binom{n}{j} = \binom{n-1}{j-1} \tag{4.15}$$

and the fact that the last sum in (4.14) is equal to 1 by the binomial theorem. Combining (4.13) and (4.14), we have $P[B] = 0.2$. In words, we have proved that *selecting an item randomly from a sample which has been selected randomly from a larger population is statistically equivalent to selecting the item from the larger population.* Note the fact that $P[B] = 0.2$ does not imply that the box containing five tubes will always contain one defective tube.

Let us next consider part (ii) of example 4D. To describe the results of the experiment that consists in selecting five tubes from the output of the machine and then selecting two tubes from among the five previously selected, we write a 7-tuple $(z_1, z_2, \ldots, z_7)$, in which z_6 and z_7 denote the tubes drawn from the box containing the first five tubes selected. Let $C_0, \ldots, C_5$ and B be defined as before. Let A be the event that the seventh tube is defective. We seek $P[A \mid B]$. Now, if two tubes, each of which has probability 0.2 of being defective, are drawn independently, the conditional probability that the second tube will be defective, given that the first tube is defective, is equal to the unconditional probability that the second tube will be defective, which is equal to 0.2. We now proceed to prove that $P[A \mid B] = 0.2$. In so doing, we are proving a special case of the principle that a sample of size 2, drawn without replacement from a sample of any size whose members are selected independently from a given population, has statistically the same properties as a sample of size 2 whose members are selected independently from the population! More general statements of this principle are given in the theoretical exercises of section

4, Chapter 4. We prove that $P[A \mid B] = 0.2$ under the assumption that $P[AB \mid C_j] = (j)_2/(5)_2$ for $j = 0, \ldots, 5$. Then, by (4.11),

$$P[AB] = \sum_{j=0}^{5} \frac{(j)_2}{(5)_2} \binom{5}{j} (0.2)^j (0.8)^{5-j}$$

$$= (0.2)^2 \sum_{j=2}^{5} \binom{3}{j-2} (0.2)^{j-2} (0.8)^{3-(j-2)} = (0.2)^2.$$

Consequently, $P[A \mid B] = P[AB]/P[B] = (0.2)^2/(0.2) = 0.2$. ◀

Bayes's Theorem. There is an interesting consequence to (4.11), which has led to much philosophical speculation and has been the source of much controversy. Let $C_1, C_2, \ldots, C_n$ be n mutually exclusive and exhaustive events, and let B be an event for which one knows the conditional probabilities $P[B \mid C_i]$ of B, given C_i, and also the absolute probabilities $P[C_i]$. One may then compute the conditional probability $P[C_i \mid B]$ of any one of the events C_i, given B, by the following formula:

$$P[C_i \mid B] = \frac{P[BC_i]}{P[B]} = \frac{P[B \mid C_i]P[C_i]}{\sum_{j=1}^{n} P[B \mid C_j]P[C_j]}. \tag{4.16}$$

The relation expressed by (4.16) is called "Bayes's theorem" or "Bayes's formula," after the English philosopher Thomas Bayes.* If the events C_i are called "causes," then Bayes's formula can be regarded as a formula for the probability that the event B, which has occurred, is the result of the "cause" C_i. In this way (4.16) has been interpreted as a formula for the probabilities of "causes" or "hypotheses." The difficulty with this interpretation, however, is that in many contexts one will rarely know the probabilities, especially the unconditional probabilities $P[C_i]$ of the "causes," which enter into the right-hand side of (4.16). However, Bayes's theorem has its uses, as the following examples indicate.†

▶ **Example 4E. Cancer diagnosis.** Suppose, contrary to fact, there were a diagnostic test for cancer with the properties that $P[A \mid C] = 0.95$, $P[A^c \mid C^c] = 0.95$, in which C denotes the event that a person tested has cancer and A denotes the event that the test states that the person tested

* A reprint of Bayes's original essay may be found in *Biometrika*, Vol. 46 (1958), pp. 293–315.

† The use of Bayes's formula to evaluate probabilities during the course of play of a bridge game is illustrated in Dan F. Waugh and Frederick V. Waugh, "On Probabilities in Bridge," *Journal of the American Statistical Association*, Vol. 48 (1953), pp. 79–87.

has cancer. Let us compute $P[C \mid A]$, the probability that a person who according to the test has cancer actually has it. We have

$$(4.17) \qquad P[C \mid A] = \frac{P[AC]}{P[A]} = \frac{P[A \mid C]P[C]}{P[A \mid C]P[C] + P[A \mid C^c]P[C^c]}.$$

Let us assume that the probability that a person taking the test actually has cancer is given by $P[C] = 0.005$. Then

$$(4.18) \qquad P[C \mid A] = \frac{(0.95)(0.005)}{(0.95)(0.005) + (0.05)(0.995)}$$

$$= \frac{0.00475}{0.00475 + 0.04975} = 0.087.$$

One should carefully consider the meaning of this result. On the one hand, the cancer diagnostic test is highly reliable, since it will detect cancer in 95% of the cases in which cancer is present. On the other hand, in only 8.7% of the cases in which the test gives a positive result and asserts cancer to be present is it actually true that cancer is present! (This example is continued in exercise 4.8.) ◀

▶ **Example 4F. Prior and posterior probability.** Consider an urn that contains a large number of coins. Not all of the coins are necessarily fair. Let a coin be chosen randomly from the urn and tossed independently 100 times. Suppose that in the 100 tosses heads appear 55 times. What is the probability that the coin selected is a fair coin (that is, the probability that the coin will fall heads at each toss is equal to $\frac{1}{2}$)?

Solution: To describe the results of the experiment we write a 101-tuple $(z_1, z_2, \ldots, z_{101})$. The components $z_2, \ldots, z_{101}$ are H or T, depending on whether the outcome of the respective toss is heads or tails. What are the possible values that may be assumed by the first component z_1? We assume that there is a set of N numbers, $p_1, p_2, \ldots, p_N$, each between 0 and 1, such that any coin in the urn has as its probability of falling heads some one of the numbers $p_1, p_2, \ldots, p_N$. Having selected a coin from the urn, we let z_1 denote the probability that the coin will fall heads; consequently, z_1 is one of the numbers $p_1, \ldots, p_N$. Now, for $j = 1, 2, \ldots, N$ let C_j be the event that the coin selected has probability p_j of falling heads, and let B be the event that the coin selected yielded 55 heads in 100 tosses. Let j_0 be the number, 1 to N, such that $p_{j_0} = \frac{1}{2}$. We are now seeking $P[C_{j_0} \mid B]$, the conditional probability that the coin selected is a fair coin, given that it yielded 55 heads in 100 tosses. In order to use (4.16) to evaluate $P[C_{j_0} \mid B]$, we require a knowledge of $P[C_j]$ and $P[B \mid C_j]$ for $j = 1, \ldots, N$. By the binomial law,

$$P[B \mid C_j] = \binom{100}{55}(p_j)^{55}(1 - p_j)^{45}. \tag{4.19}$$

The probabilities $P[C_j]$ cannot be computed but must be assumed. The probability $P[C_j]$ represents the proportion of coins in the urn which has probability p_j of falling heads. It is clear that the value we obtain for $P[C_{j_0} \mid B]$ depends directly on the values we assume for $P[C_1], \ldots, P[C_N]$. If the latter probabilities are unknown to us, then we must resign ourselves to not being able to compute $P[C_{j_0} \mid B]$. However, let us obtain a numerical answer for $P[C_{j_0} \mid B]$ under the assumption that $P[C_1] = \ldots = P[C_N] = 1/N$, so that a coin selected from the urn is equally likely to have any one of the probabilities $p_1, \ldots, p_N$. We then obtain that

$$P[C_{j_0} \mid B] = \frac{(1/N)\binom{100}{55}(p_{j_0})^{55}(1 - p_{j_0})^{45}}{(1/N)\sum_{j=1}^{N}\binom{100}{55}(p_j)^{55}(1 - p_j)^{45}}. \tag{4.20}$$

Let us next assume that $N = 9$, and $p_j = j/10$ for $j = 1, 2, \ldots, 9$. Then $j_0 = 5$, and

$$P[C_5 \mid B] = \frac{\binom{100}{55}(1/2)^{100}}{\sum_{j=1}^{9}\binom{100}{55}(j/10)^{55}[(10 - j)/10]^{45}} \tag{4.21}$$

$$= \frac{0.048475}{0.097664} = 0.496.$$

The probability $P[C_5] = \frac{1}{9}$ is called the prior (or a priori) probability of the event C_5; the conditional probability $P[C_5 \mid B] = 0.496$ is called the posterior (or a posteriori) probability of the event C_5. The prior probability is an unconditional probability that is known to us before any observations are taken. The posterior probability is a conditional probability that is of interest to us only if it is known that the conditioning event has occurred. ◀

Our next example illustrates a controversial use of Bayes's theorem.

▶ **Example 4G. Laplace's rule of succession.** Consider a coin that in n independent tosses yields k heads. What is the probability that n' subsequent independent tosses will yield k' heads? The problem may also be phrased in terms of drawing balls from an urn. Consider an urn that contains white and red balls in unknown proportions. In a sample of size n, drawn with replacement from the urn, k white balls appear. What is the probability that a sample of size n' drawn with replacement will contain k' white balls? A particular case of this problem, in which $k = n$ and $k' = n'$, can be interpreted as a simple form of the fundamental problem of inductive inference if one formulates the problem as follows: if n independent trials of an experiment have resulted in success, what is the probability that n' additional independent trials will result in success? Another reformulation is this: if the results of n independent experiments, performed to test a theory, agree with the theory, what is the probability that n' additional independent experiments will agree with the theory.

Solution: To describe the results of our observations, we write an $(n + n' + 1)$-tuple $(z_1, z_2, \ldots, z_{n+n'+1})$ in which the components $z_2, \ldots, z_{n+1}$ describe the outcomes of the coin tosses which have been made and the components $z_{n+2}, \ldots, z_{n+n'+1}$ describe the outcomes of the subsequent coin tosses. The first component z_1 describes the probability that the coin tossed has of falling heads; *we assume that there are N known numbers, $p_1, p_2, \ldots, p_N$, which z_1 can take as its value.* We have italicized this assumption to indicate that it is considered controversial. For $j = 1, 2, \ldots, N$ let C_j be the event that the coin tossed has probability p_j of falling heads. Let B be the event that the coin yields n heads in its first n tosses, and let A be the event that it yields n' heads in its subsequent n' tosses. We are seeking $P[A \mid B]$. Now

$$P[AB] = \sum_{j=1}^{N} P[AB \mid C_j]P[C_j] = \sum_{j=1}^{N} (p_j)^{n+n'}P[C_j], \tag{4.22}$$

whereas

$$P[B] = \sum_{j=1}^{N} (p_j)^n P[C_j]. \tag{4.23}$$

Let us now assume that p_j is equal to j/N and that $P[C_j] = 1/N$. Then

$$P[A \mid B] = \frac{(1/N)\sum_{j=1}^{N} (j/N)^{n+n'}}{(1/N)\sum_{j=1}^{N} (j/N)^n}. \tag{4.24}$$

The sums in (4.24) may be approximately evaluated in the case that N is large by means of the integral calculus. The sums can be regarded as approximating sums of Riemann integrals, and we have

$$\begin{aligned} \frac{1}{N}\sum_{j=1}^{N}\left(\frac{j}{N}\right)^{n+n'} &\doteq \int_0^1 x^{n+n'}\,dx = \frac{1}{n+n'+1}, \\ \frac{1}{N}\sum_{j=1}^{N}\left(\frac{j}{N}\right)^{n} &\doteq \int_0^1 x^{n}\,dx = \frac{1}{n+1}. \end{aligned} \tag{4.25}$$

Consequently, given that the first n tosses yielded a head, the conditional probability that n' subsequent tosses of the coin will yield a head, under *the assumption that the probability of the coin falling heads is equally likely to be any one of the numbers* $1/N, 2/N, \ldots, N/N$, and N *is large, is given by*

$$P[A \mid B] = \frac{n+1}{n+n'+1}. \tag{4.26}$$

Equation (4.26) is known as Laplace's general rule of succession. If we take $n' = 1$, then

$$P[A \mid B] = \frac{n+1}{n+2}. \tag{4.27}$$

Equation (4.27) is known as Laplace's special rule of succession.

Equation (4.27) has been interpreted by some writers on probability theory to imply that if a theory has been verified in n consecutive trials then the probability of its being verified on the $(n + 1)$st trial is $(n + 1)/(n + 2)$. That the rule has a certain appeal at first acquaintance may be seen from the following example:

Consider a tourist in a foreign city who scarcely understands the language. With trepidation, he selects a restaurant in which to eat. After ten meals taken there he has felt no ill effects. Consequently, he goes quite confidently to the restaurant the eleventh time in the knowledge that, according to the rule of succession, the probability is $\frac{11}{12}$ that he will not be poisoned by his next meal.

However, it is easy to exhibit applications of the rule that lead to absurd answers. A boy is 10 years old today. The rule says that, having lived ten years, he has probability $\frac{11}{12}$ of living one more year. On the other hand, his 80-year-old grandfather has probability 81/82 of living one more year! Yet, in fact, the boy has a greater probability of living one more year.

Laplace gave the following often-quoted application of the special rule of succession. "Assume," he says, "that history goes back 5000 years, that is, 1,826,213 days. The sun rose each day and so you can bet 1,826,214 against 1 that the sun will rise again tomorrow." However, before believing this assertion, ask yourself if you would believe the following consequence of the general rule of succession; the sun having risen on each of the last 1,826,213 days, the probability that it will rise on each of the next 1,826,214 days is $\frac{1}{2}$, which means that the probability is $\frac{1}{2}$ that on at least one of the next 1,826,214 days the sun will not rise. ◀

It is to be emphasized that Baye's formula and Laplace's rule of succession are true theorems, of mathematical probability theory. The foregoing examples do not in any way cast doubt on the validity of these theorems. Rather they serve to illustrate what may be called the *fundamental principle of applied probability theory*: before applying a theorem, one must carefully ponder whether the hypotheses of the theorem may be assumed to be satisfied.

THEORETICAL EXERCISES

4.1. An urn contains M balls, of which M_W are white (where $M_W \leq M$). Let a sample of size m (where $m \leq M_W$) be drawn from the urn with replacement [without replacement] and deposited in an empty urn. Let a sample of size n (where $n \leq m$) be drawn from the second urn without replacement. Show that for $k = 0, 1, \ldots, n$ the probability that the second sample will contain exactly k white balls continues to be given by (3.2) [(3.1)] of Chapter 2. The result shows that, as one might expect, drawing a sample of size n from a sample of larger size is statistically equivalent to drawing a sample of size n from the urn. An alternate statement of this theorem, and an outline of the proof, is given in theoretical exercise 4.1 of Chapter 4.

4.2. Consider a box containing N radio tubes selected at random from the output of a machine; the probability p that an item produced by the machine is defective is known.

(i) Let $k \leq n \leq N$ be integers. Show that the probability that n tubes selected at random from the box will have k defectives is given by $\binom{n}{k} p^k q^{n-k}$.

(ii) Suppose that m tubes are selected at random from the box and found to be defective. Show that the probability that n tubes selected at random from the remaining $N - m$ tubes in the box will contain k defectives is equal to $\binom{n}{k} p^k q^{n-k}$.

(iii) Suppose that $m + n$ tubes are selected at random from the box and tested. You are informed that at least m of the tubes are defective; show that the probability that exactly $m + k$ tubes are defective, where k is an integer from 0 to n, is given by (3.13). Express in words the conclusions implied by this exercise.

4.3. Consider an urn containing M balls, of which M_W are white. Let N be an integer such that $N \geq M_W$. Choose an integer n at random from the set $\{1, 2, \ldots, N\}$, and then choose a sample of size n without replacement from the urn. Show that the probability that all the balls in the sample will be white (letting $M_R = M - M_W$) is equal to

$$\frac{1}{N}\sum_{k=1}^{N} \frac{(M_W)_k}{(M)_k} = \frac{1}{N}\frac{M_W}{M_R + 1}.$$

4.4. **An application of Bayes's theorem.** Suppose that in answering a question on a multiple choice test an examinee either knows the answer or he guesses. Let p be the probability that he will know the answer, and let $1 - p$ be the probability that he will guess. Assume that the probability of answering a question correctly is unity for an examinee who knows the answer and $1/m$ for an examinee who guesses; m is the number of multiple choice alternatives. Show that the conditional probability that an examinee knew the answer to a question, given that he has correctly answered it, is equal to

$$\frac{mp}{1 + (m - 1)p}.$$

4.5. **Solution of a difference equation.** The difference equation

$$p_n = ap_{n-1} + b, \qquad n = 2, 3, \cdots,$$

in which a and b are given constants, arises in the theory of Markov dependent trials (see section 5). By mathematical induction, show that if a sequence of numbers $p_1, p_2, \ldots, p_n$ satisfies this difference equation, and if $a \neq 1$, then

$$p_n = \left(p_1 - \frac{b}{1 - a}\right)a^{n-1} + \frac{b}{1 - a}.$$

EXERCISES

4.1. Urn I contains 5 white and 7 black balls. Urn II contains 4 white and 2 black balls. Find the probability of drawing a white ball if (i) 1 urn is selected at random, and a ball is drawn from it, (ii) the 2 urns are emptied into a third urn from which 1 ball is drawn.

4.2. Urn I contains 5 white and 7 black balls. Urn II contains 4 white and 2 black balls. An urn is selected at random, and a ball is drawn from it. Given that the ball drawn is white, what is the probability that urn I was chosen?

4.3. A man draws a ball from an urn containing 4 white and 2 red balls. If the ball is white, he does not return it to the urn; if the ball is red, he does return it. He draws another ball. Let A be the event that the first ball drawn is white, and let B be the event that the second ball drawn is white. Answer each of the following statements, true or false. (i) $P[A] = \frac{2}{3}$, (ii) $P[B] = \frac{3}{5}$, (iii) $P[B \mid A] = \frac{3}{5}$, (iv) $P[A \mid B] = \frac{9}{14}$, (v) The events A and B are mutually exclusive. (vi) The events A and B are independent.

4.4. From an urn containing 6 white and 4 black balls, 5 balls are transferred into an empty second urn. From it 3 balls are transferred into an empty box. One ball is drawn from the box; it turns out to be white. What is the probability that exactly 4 of the balls transferred from the first to the second urn will be white?

4.5. Consider an urn containing 12 balls, of which 8 are white. Let a sample of size 4 be drawn with replacement (without replacement). Next, let a ball be selected randomly from the sample of size 4. Find the probability that it will be white.

4.6. Urn I contains 6 white and 4 black balls. Urn II contains 2 white and 2 black balls. From urn I 2 balls are transferred to urn II. A sample of size 2 is then drawn without replacement from urn II. What is the probability that the sample will contain exactly 1 white ball?

4.7. Consider a box containing 5 radio tubes selected at random from the output of a machine, which is known to be 20% defective on the average (that is, the probability that an item produced by the machine will be defective is 0.2). Suppose that 2 tubes are selected at random from the box and tested. You are informed that at least 1 of the tubes selected is defective; what is the probability that both tubes will be defective?

4.8. Let the events A and C be defined as in example 4E. Let $P[A \mid C] = P[A^c \mid C^c] = R$ and $P[C] = 0.005$. What value must R have in order that $P[C \mid A] = 0.95$? Interpret your answer.

4.9. In a certain college the geographical distribution of men students is as follows: 50% come from the East, 30% come from the Midwest, and 20% come from the Far West. The following proportions of the men students wear ties: 80% of the Easterners, 60% of the Midwesterners, and 40% of the Far Westerners. What is the probability that a student who wears a tie comes from the East? From the Midwest? From the Far West?

4.10. Consider an urn containing 10 balls, of which 4 are white. Choose an integer n at random from the set $\{1, 2, 3, 4, 5, 6\}$ and then choose a sample of size n without replacement from the urn. Find the probability that all the balls in the sample will be white.

4.11. Each of 3 boxes, identical in appearance, has 2 drawers. Box A contains a gold coin in each drawer; box B contains a silver coin in each drawer; box C contains a gold coin in 1 drawer and a silver coin in the other. A box is chosen, one of its drawers is opened, and a gold coin is found.

(i) What is the probability that the other drawer contains a silver coin? Write out the probability space of the experiment. Why is it fallacious to reason that the probability is $\frac{1}{2}$ that there will be a silver coin in the second drawer, since there are 2 possible types of coins, gold or silver, that may be found there?

(ii) What is the probability that the box chosen was box A? Box B? Box C?

4.12. Three prisoners, whom we may call A, B, and C, are informed by their jailer that one of them has been chosen at random to be executed, and the other 2 are to be freed. Prisoner A, who has studied probability theory, then reasons to himself that he has probability $\frac{1}{3}$ of being executed. He then asks the jailer to tell him privately which of his fellow prisoners will be set free, claiming that there would be no harm in divulging this information, since he already knows that at least 1 will go. The jailer (being an ethical fellow) refuses to reply to this question, pointing out that if A knew which of his fellows were to be set free then his probability of being executed would increase to $\frac{1}{2}$, since he would then be 1 of 2 prisoners, 1 of whom is to be executed. Show that the probability that A will be executed is still $\frac{1}{3}$, even if the jailer were to answer his question, assuming that, in the event that A is to be executed, the jailer is as likely to say that B is to be set free as he is to say that C is to be set free.

4.13. A male rat is either doubly dominant (AA) or heterozygous (Aa), owing to Mendelian properties, the probabilities of either being true is $\frac{1}{2}$. The male rat is bred to a doubly recessive (aa) female. If the male rat is doubly dominant, the offspring will exhibit the dominant characteristic; if heterozygous, the offspring will exhibit the dominant characteristic $\frac{1}{2}$ of the time and the recessive characteristic $\frac{1}{2}$ of the time. Suppose all of 3 offspring exhibit the dominant characteristic. What is the probability that the male is doubly dominant?

4.14. Consider an urn that contains 5 white and 7 black balls. A ball is drawn and its color is noted. It is then replaced; in addition, 3 balls of the color drawn are added to the urn. A ball is then drawn from the urn. Find the probability that (i) the second ball drawn will be black, (ii) both balls drawn will be black.

4.15. Consider a sample of size 3 drawn in the following manner. One starts with an urn containing 5 white and 7 red balls. At each trial a ball is drawn and its color is noted. The ball drawn is then returned to the urn, together with an additional ball of the same color. Find the probability that the sample will contain exactly (i) 0 white balls, (ii) 1 white ball, (iii) 3 white balls.

4.16. A certain kind of nuclear particle splits into 0, 1, or 2 new particles (which we call offsprings) with probabilities $\frac{1}{4}$, $\frac{1}{2}$, and $\frac{1}{4}$, respectively, and then dies. The individual particles act independently of each other. Given a particle, let X_1 denote the number of its offsprings, let X_2 denote the number of offsprings of its offsprings, and let X_3 denote the number of offsprings of the offsprings of its offsprings.

(i) Find the probability that $X_2 > 0$.

(ii) Find the conditional probability that $X_1 = 1$, given that $X_2 = 1$,

(iii) Find the probability that $X_3 = 0$.

4.17. A number, denoted by X_1, is chosen at random from the set of integers $\{1, 2, 3, 4\}$. A second number, denoted by X_2, is chosen at random from the set $\{1, 2, \ldots, X_1\}$.

(i) For each integer k, 1 to 4, find the conditional probability that $X_2 = 1$, given that $X_1 = k$.

(ii) Find the probability that $X_2 = 1$.

(iii) Find the conditional probability that $X_1 = 2$, given that $X_2 = 1$.

5. MARKOV DEPENDENT BERNOULLI TRIALS

Of interest in many problems of applied probability theory is the evolution in time of the state of a random phenomenon. For example, suppose one has two urns (I and II), each of which contains one white and one black ball. One ball is drawn simultaneously from each urn and placed in the other urn. One is often concerned with questions such as, what is the probability that after 100 repetitions of this procedure urn I will contain two white balls? The theory of Markov* chains is applicable to questions of this type.

The theory of Markov chains relates to every field of physical and social science (see the forthcoming book by A. T. Bharucha-Reid, *Introduction to the Theory of Markov Processes and their Applications*; for applications of the theory of Markov chains to the description of social or psychological phenomena, see the book by Kemeny and Snell cited in the next paragraph). There is an immense literature concerning the theory of Markov chains. In this section and the next we can provide only a brief introduction.

Excellent elementary accounts of this theory are to be found in the works of W. Feller, *An Introduction to Probability Theory and Its Applications*, second edition, Wiley, New York, 1957, and J. G. Kemeny and J. L. Snell, *Finite Markov Chains*, Van Nostrand, Princeton, New Jersey, 1959. The reader is referred to these books for proof of the assertions made in section 6.

The natural generalization of the notion of independent Bernoulli trials is the notion of *Markov dependent* Bernoulli trials. Given n trials of an experiment, which has only two possible outcomes (denoted by s or f, for "success" or "failure"), we recall that they are said to be independent Bernoulli trials if for any integer k (1 to $n-1$) and $k+1$ events A_1, $A_2, \ldots, A_{k+1}$, depending, respectively, on the first, second, ..., $(k+1)$st trials,

$$P[A_{k+1} \mid A_k, A_{k-1}, \cdots, A_1] = P[A_{k+1}]. \tag{5.1}$$

We define the trials as *Markov dependent Bernoulli trials* if, instead of (5.1), it holds that

$$P[A_{k+1} \mid A_k, A_{k-1}, \cdots, A_1] = P[A_{k+1} \mid A_k]. \tag{5.2}$$

In words, (5.2) says that at the kth trial the conditional probability of any event A_{k+1}, depending on the next trial, will not depend on what has happened in past trials but only on what is happening at the present time. One sometimes says that the trials have no memory.

* The theory of Markov chains derives its name from the celebrated Russian probabilist, A. A. Markov (1856–1922).

Suppose that the quantities

$$
\begin{aligned}
P(s, s) &= \text{probability of success on the } (k+1)\text{st trial,} \\
&\quad\ \text{given that there was success on the } k\text{th trial,} \\
P(f, s) &= \text{probability of success at the } (k+1)\text{st trial,} \\
&\quad\ \text{given that there was failure at the } k\text{th trial,} \\
P(f, f) &= \text{probability of failure at the } (k+1)\text{st trial,} \\
&\quad\ \text{given that there was failure at the } k\text{th trial,} \\
P(s, f) &= \text{probability of failure at the } (k+1)\text{st trial,} \\
&\quad\ \text{given that there was success at the } k\text{th trial,}
\end{aligned}
\tag{5.3}
$$

are independent of k. We then say that the trials are *Markov dependent repeated Bernoulli trials.*

▶ **Example 5A.** Let the weather be observed on n consecutive days. Let s describe a day on which rain falls, and let f describe a day on which no rain falls. Suppose one assumes that weather observations constitute a series of Markov dependent Bernoulli trials (or, in the terminology of section 6, a Markov chain with two states). Then $P(s, s)$ is the probability of rain tomorrow, given rain today; $P(s, f)$ is the probability of no rain tomorrow, given rain today; $P(f, f)$ is the probability of no rain tomorrow, given no rain today; and $P(f, s)$ is the probability of rain tomorrow, given no rain today. It is now natural to ask for such probabilities as that of rain the day after tomorrow, given no rain today; we denote this probability by $P_2(f, s)$ and obtain a formula for it. ◀

In the case of independent repeated Bernoulli trials the probability function $P[\cdot]$ on the sample description space of the n trials is completely specified once we have the probability p of success at each trial. In the case of Markov dependent repeated Bernoulli trials it suffices to specify the quantities

$$
\begin{aligned}
p_1(s) &= \text{probability of success at the first trial,} \\
p_1(f) &= \text{probability of failure at the first trial,}
\end{aligned}
\tag{5.4}
$$

as well as the conditional probabilities in (5.3). The probability of any event can be computed in terms of these quantities. For example, for $k = 1, 2, \ldots, n$ let

$$
\begin{aligned}
p_k(s) &= \text{probability of success at the } k\text{th trial,} \\
p_k(f) &= \text{probability of failure at the } k\text{th trial.}
\end{aligned}
\tag{5.5}
$$

The quantities $p_k(s)$ satisfy the following equations for $k = 2, 3, \ldots, n$:

$$(5.6) \qquad \begin{aligned} p_k(s) &= p_{k-1}(s)P(s, s) + p_{k-1}(f)P(f, s) \\ &= p_{k-1}(s)P(s, s) + [1 - p_{k-1}(s)][1 - P(f, f)] \\ &= p_{k-1}(s)[P(s, s) + P(f, f) - 1] + [1 - P(f, f)]. \end{aligned}$$

To justify (5.6), we reason as follows: if A_k is the event that there is success on the kth trial, then

$$(5.7) \qquad P[A_k] = P[A_{k-1}]P[A_k \mid A_{k-1}] + P[A^c_{k-1}]P[A_k \mid A^c_{k-1}].$$

From (5.7) and the fact that

$$(5.8) \quad P(s, f) = 1 - P(s, s) \qquad P(f, s) = 1 - P(f, f), \qquad p_k(f) = 1 - p_k(s)$$

one obtains (5.6).

Equation (5.6) constitutes a recursive relationship for $p_k(s)$, known as a difference equation. *Throughout this section we make the assumption that*

$$(5.9) \qquad |P(s, s) + P(f, f) - 1| < 1.$$

By using theoretical exercise 4.5, it follows from (5.6) and (5.9) that for $k = 1, 2, \ldots, n$

$$(5.10) \quad p_k(s) = \left[p_1(s) - \frac{1 - P(f, f)}{2 - P(s, s) - P(f, f)}\right][P(s, s) + P(f, f) - 1]^{k-1} + \left[\frac{1 - P(f, f)}{2 - P(s, s) - P(f, f)}\right].$$

By interchanging the role of s and f in (5.10), we obtain, similarly, for $k = 1, 2, \ldots, n$

$$(5.11) \quad p_k(f) = \left[p_1(f) - \frac{1 - P(s, s)}{2 - P(s, s) - P(f, f)}\right][P(s, s) + P(f, f) - 1]^{k-1} + \left[\frac{1 - P(s, s)}{2 - P(s, s) - P(f, f)}\right].$$

It is readily verifiable that the expressions in (5.10) and (5.11) sum to one, as they ought.

In many problems involving Markov dependent repeated Bernoulli trials we do not know the probability $p_1(s)$ of success at the first trial. We can only compute the quantities

$$
(5.12) \quad
\begin{aligned}
P_k(s, s) &= \text{conditional probability of success at the } (k+1)\text{st trial, given success at the first trial,} \\
P_k(s, f) &= \text{conditional probability of failure at the } (k+1)\text{st trial, given success at the first trial,} \\
P_k(f, f) &= \text{conditional probability of failure at the } (k+1)\text{st trial, given failure at the first trial,} \\
P_k(f, s) &= \text{conditional probability of success at the } (k+1)\text{st trial, given failure at the first trial.}
\end{aligned}
$$

Since

$$(5.13) \qquad P_k(s, f) = 1 - P_k(s, s), \qquad P_k(f, s) = 1 - P_k(f, f),$$

it suffices to obtain formulas for $P_k(s, s)$ and $P_k(f, f)$.

In the same way that we obtained (5.6) we obtain

$$
(5.14) \quad
\begin{aligned}
P_k(s, s) &= P_{k-1}(s, s)P(s, s) + P_{k-1}(s, f)P(f, s) \\
&= P_{k-1}(s, s)P(s, s) + [1 - P_{k-1}(s, s)][1 - P(f, f)] \\
&= P_{k-1}(s, s)[P(s, s) + P(f, f) - 1] + [1 - P(f, f)].
\end{aligned}
$$

By using theoretical exericse 4.5, it follows from (5.14) and (5.9) that for $k = 1, 2, \ldots, n$

$$
(5.15) \quad P_k(s, s) = \left[P_1(s, s) - \frac{1 - P(f, f)}{2 - P(s, s) - P(f, f)}\right] \times [P(s, s) + P(f, f) - 1]^{k-1} + \left[\frac{1 - P(f, f)}{2 - P(s, s) - P(f, f)}\right]
$$

which can be simplified to

$$
(5.16) \quad P_k(s, s) = \frac{1 - P(s, s)}{2 - P(s, s) - P(f, f)}[P(s, s) + P(f, f) - 1]^k + \left[\frac{1 - P(f, f)}{2 - P(s, s) - P(f, f)}\right].
$$

By interchanging the role of s and f, we obtain, similarly,

$$
(5.17) \quad P_k(f, f) = \frac{1 - P(f, f)}{2 - P(s, s) - P(f, f)}[P(s, s) + P(f, f) - 1]^k + \frac{1 - P(s, s)}{2 - P(s, s) - P(f, f)}
$$

By using (5.13), we obtain

$$(5.18)\quad P_k(s,f) = -\frac{1 - P(s, s)}{2 - P(s, s) - P(f,f)}[P(s, s) + P(f,f) - 1]^k + \frac{1 - P(s, s)}{2 - P(s, s) - P(f,f)},$$

$$(5.19)\quad P_k(f, s) = -\frac{1 - P(f,f)}{2 - P(s, s) - P(f,f)}[P(s, s) + P(f,f) - 1]^k + \frac{1 - P(f,f)}{2 - P(s, s) - P(f,f)}.$$

Equations (5.16) *to* (5.19) *represent the basic conclusions in the theory of Markov dependent Bernoulli trials* (in the case that (5.9) holds).

▶ **Example 5B.** Consider a communications system which transmits the digits 0 and 1. Each digit transmitted must pass through several stages, at each of which there is a probability p that the digit that enters will be unchanged when it leaves. Suppose that the system consists of three stages. What is the probability that a digit entering the system as 0 will be (i) transmitted by the third stage as 0, (ii) transmitted by each stage as 0 (that is, never changed from 0)? Evaluate these probabilities for $p = \frac{1}{3}$.

Solution: In observing the passage of the digit through the communications system, we are observing a 4-tuple (z_1, z_2, z_3, z_4), whose first component z_1 is 1 or 0, depending on whether the digit entering the system is 1 or 0. For $i = 2, 3, 4$ the component z_i is equal to 1 or 0, depending on whether the digit leaving the ith stage is 1 or 0. We now use the foregoing formulas, identifying s with 1, say, and 0 with f. Our basic assumption is that

$$(5.20)\qquad P(0, 0) = P(1, 1) = p.$$

The probability that a digit entering the system as 0 will be transmitted by the third stage as 0 is given by

$$(5.21)\quad \begin{aligned} P_3(0, 0) &= \frac{1 - P(0, 0)}{2 - P(0, 0) - P(1, 1)}[P(0, 0) + P(1, 1) - 1]^3 + \frac{1 - P(1, 1)}{2 - P(0, 0) - P(1, 1)} \\ &= \frac{1 - p}{2 - 2p}(2p - 1)^3 + \frac{1 - p}{2 - 2p} \\ &= \tfrac{1}{2}[1 + (2p - 1)^3]. \end{aligned}$$

If $p = \frac{1}{3}$, then $P_3(0, 0) = \frac{1}{2}[1 - (\frac{1}{3})^3] = \frac{13}{27}$. The probability that a digit entering the system as 0 will be transmitted by each stage as 0 is given by the product

$$P(0, 0)P(0, 0)P(0, 0) = p^3 = (\tfrac{1}{3})^3 = \tfrac{1}{27}. \quad ◀$$

▶ **Example 5C.** Suppose that the digit transmitted through the communications system described in example 5B (with $p = \frac{1}{3}$) is chosen by a chance mechanism; digit 0 is chosen with probability $\frac{1}{3}$ and digit 1, with probability $\frac{2}{3}$. What is the conditional probability that a digit transmitted by the third stage as 0 in fact entered the system as 0?

Solution: The conditional probability that a digit transmitted by the third stage as 0 entered the system as 0 is given by

$$\frac{p_1(0)P_3(0, 0)}{p_4(0)} \tag{5.22}$$

To justify (5.22), note that $p_1(0)P_3(0, 0)$ is the probability that the first digit is 0 and the fourth digit is 0, whereas $p_4(0)$ is the probability that the fourth digit is 0. Now, under the assumption that

$$P(1, 1) = P(0, 0) = \tfrac{1}{3}, \qquad p_1(0) = \tfrac{1}{3}, \qquad p_1(1) = \tfrac{2}{3},$$

it follows from (5.10) that

$$\begin{aligned} p_4(0) &= \left[p_1(0) - \frac{1 - P(1, 1)}{2 - P(0, 0) - P(1, 1)}\right][P(0, 0) + P(1, 1) - 1]^3 \\ &\quad + \frac{1 - P(1, 1)}{2 - P(0,0) - P(1,1)} \\ &= (\tfrac{1}{3} - \tfrac{1}{2})(-\tfrac{1}{3})^3 + \tfrac{1}{2} = \tfrac{41}{81} \end{aligned} \tag{5.23}$$

From (5.21) and (5.23) it follows that the conditional probability that a digit leaving the system as 0 entered the system as 0 is given by

$$\tfrac{1}{3} \cdot \tfrac{13}{27}/\tfrac{41}{81} = \tfrac{13}{41}. \quad ◀$$

▶ **Example 5D.** Let us use the considerations of examples 5B and 5C to solve the following problem, first proposed by the celebrated cosmologist A. S. Eddington (see W. Feller, "The Problem of n liars and Markov chains," *American Mathematical Monthly*, Vol. 58 (1951), pp. 606–608). "If A, B, C, D each speak the truth once in 3 times (independently), and A affirms that B denies that C declares that D is a liar, what is the probability that D was telling the truth?"

Solution: We consider a sample description space of 4-tuples (z_1, z_2, z_3, z_4) in which z_1 equals 0 or 1, depending on whether D is truthful or a liar, z_2 equals 0 or 1, depending on whether the statement made by C implies that

D is truthful or a liar, z_3 equals 0 or 1, depending on whether the statement made by B implies that D is truthful or a liar, and z_4 equals 0 or 1, depending on whether the statement made by A implies that D is truthful or a liar. The sample description space thus defined constitutes a series of Markov dependent repeated Bernoulli trials with

$$P(0, 0) = P(1, 1) = \tfrac{1}{3}, \qquad p_1(0) = \tfrac{1}{3}, \qquad p_1(1) = \tfrac{2}{3}.$$

The persons A, B, C, and D can be regarded as forming a communications system. We are seeking the conditional probability that the digit entering the system was 0 (which is equivalent to D being truthful), given that the digit transmitted by the third stage was 0 (if A affirms that B denies that C declares that D is a liar, then A is asserting that D is truthful). In view of example 5C, the required probability is $\frac{13}{41}$. ◀

Statistical Equilibrium. For large values of k the values of $p_k(s)$ and $p_k(f)$ are approximately given by

$$(5.24) \qquad \begin{aligned} p_k(s) &= \frac{1 - P(f,f)}{2 - P(s, s) - P(f,f)}, \\ p_k(f) &= \frac{1 - P(s, s)}{2 - P(s, s) - P(f,f)}. \end{aligned}$$

To justify (5.24), use (5.10) and (5.11) and the fact that (5.9) implies that

$$\lim_{k \to \infty} [P(s, s) + P(f, f) - 1]^{k-1} = 0.$$

Equation (5.24) has the following significant interpretation: *After a large number of Markov dependent repeated Bernoulli trials has been performed, one is in a state of statistical equilibrium, in the sense that the probability $p_k(s)$ of success on the kth trial is the same for all large values of k and indeed is functionally independent of the initial conditions, represented by $p_1(s)$.*

From (5.24) one sees that the trials are asymptotically fair in the sense that approximately

$$(5.25) \qquad p_k(s) = p_k(f) = \tfrac{1}{2}$$

for large values of k if and only if

$$(5.26) \qquad P(s, s) = P(f, f).$$

▶ **Example 5E.** If the communications system described in example 5B consists of a very large number of stages, then half the digits transmitted by the system will be 0's, irrespective of the proportion of 0's among the digits entering the system. ◀

EXERCISES

5.1. Consider a series of Markov dependent Bernoulli trials such that $P(s, s) = \frac{1}{3}$, $P(f, f) = \frac{1}{2}$. Find $P_3(s, f)$, $P_3(f, s)$.

5.2. Consider a series of Markov dependent Bernoulli trials such that $P(s, s) = \frac{1}{4}$, $P(f, f) = \frac{1}{2}$, $p_1(s) = \frac{1}{3}$. Find $p_3(s)$, $p_3(f)$.

5.3. Consider a series of Markov dependent Bernoulli trials such that $P(s, s) = \frac{1}{2}$, $P(f, f) = \frac{3}{4}$, $p_1(s) = \frac{1}{2}$. Find the conditional probability of a success at the first trial, given that there was a success at the fourth trial.

5.4. Consider a series of Markov dependent Bernoulli trials such that $P(s, s) = \frac{1}{2}$, $P(f, f) = \frac{3}{4}$. Find $\lim_{k \to \infty} p_k(s)$.

5.5. If A, B, C, and D each speak the truth once in 3 times (independently), and A affirms that B denies that C denies that D is a liar, what is the probability that D was telling the truth?

5.6. Suppose the probability is equal to p that the weather (rain or no rain) on any arbitrary day is the same as on the preceding day. Let p_1 be the probability of rain on the first day of the year. Find the probability p_n of rain on the nth day. Evaluate the limit of p_n as n tends to infinity.

5.7. Consider a game played as follows: a group of n persons is arranged in a line. The first person starts a rumor by telling his neighbor that the last person in line is a nonconformist. Each person in line then repeats this rumor to his neighbor; however, with probability $p > 0$, he reverses the sense of the rumor as it is told to him. What is the probability that the last person in line will be told he is a nonconformist if (i) $n = 5$, (ii) $n = 6$, (iii) n is very large?

5.8. Suppose you are confronted with 2 coins, A and B. You are to make n tosses, using the coin you prefer at each toss. You will be paid 1 dollar for each time the coin falls heads. Coin A has probability $\frac{1}{2}$ of falling heads, and coin B has probability $\frac{1}{4}$ of falling heads. Unfortunately, you are not told which of the coins is coin A. Consequently, you decide to toss the coins according to the following system. For the first toss you choose a coin at random. For all succeeding tosses you select the coin used on the preceding toss if it fell heads, and otherwise switch coins. What is the probability that coin A will be the coin tossed on the nth toss if (i) $n = 2$, (ii) $n = 4$, (iii) $n = 6$, (iv) n is very large? What is the probability that the coin tossed on the nth toss will fall heads if (i) $n = 2$, (ii) $n = 4$, (iii) $n = 6$, (iv) n is very large? *Hint:* On each trial let s denote the use of coin A and f denote the use of coin B.

5.9. A certain young lady is being wooed by a certain young man. The young lady tries not to be late for their dates too often. If she is late on 1 date, she is 90% sure to be on time on the next date. If she is on time, then there is a 60% chance of her being late on the next date. In the long run, how often is she late?

5.10. Suppose that people in a certain group may be classified into 2 categories (say, city dwellers and country dwellers; Republicans and Democrats; Easterners and Westerners; skilled and unskilled workers, and so on). Let us consider a group of engineers, some of whom are Easterners and some of whom are Westerners. Suppose that each person has a certain probability of changing his status: The probability that an Easterner will become a Westerner is 0.04, whereas the probability that a Westerner will become an Easterner is 0.01. In the long run, what proportion of the group will be (i) Easterners, (ii) Westerners, (iii) will move from East to West in a given year, (iv) will move from West to East in a given year? Comment on your answers.

6. MARKOV CHAINS

The notion of Markov dependence, defined in section 5 for Bernoulli trials, may be extended to trials with several possible outcomes. Consider n trials of an experiment with r possible outcomes $s_1, s_2, \ldots, s_r$, in which $r > 2$. For $k = 1, 2, \ldots, n$ and $j = 1, 2, \ldots, r$ let $A_k^{(j)}$ be the event that on the kth trial outcome s_j occurred. The trials are defined as *Markov dependent* if for any integer k from 1 to n and integers $j_1, j_2, \ldots, j_k$, from 1 to r, the events $A_1^{(j_1)}, A_2^{(j_2)}, \ldots, A_k^{(j_k)}$ satisfy the condition

$$P[A_k^{(j_k)} \mid A_{k-1}^{(j_{k-1})}, \cdots, A_1^{(j_1)}] = P[A_k^{(j_k)} \mid A_{k-1}^{(j_{k-1})}]. \tag{6.1}$$

In discussing Markov dependent trials with r possible outcomes, it is usual to employ an intuitively meaningful language. Instead of speaking of n trials of an experiment with r possible outcomes, we speak of observing at n times the *state of a system* which has r possible states. We number the states $1, 2, \ldots, r$ (or sometimes $0, 1, \ldots, r - 1$) and let $A_k^{(j)}$ be the event that the system is in state j at time k. If (6.1) holds, we say that the system is a *Markov chain* with r possible states. In words, (6.1) states that at any time the conditional probability of transition from one's present state to any other state does not depend on how one arrived in one's present state. One sometimes says that a Markov chain is a system without memory of the past.

Now suppose that for any states i and j the conditional probability

(6.2) $P(i, j) =$ conditional probability that the Markov chain is at time t in state j, given that at time $(t - 1)$ it was in state i,

is independent of t. The Markov chain is then said to be *homogeneous* (or time homogeneous). A homogeneous Markov chain with r states corresponds to the notion of Markov dependent repeated trials with r possible outcomes. In this section, by a Markov chain we mean a homogeneous Markov chain.

The m-step transition probabilities defined by

(6.3) $P_m(i,j) =$ the conditional probability that the Markov chain is at time $t + m$ in state j, given that at time t it was in state i,

are given recursively in terms of $P(i,j)$ by the system of equations, for $m = 2, 3, \ldots,$

$$P_m(i,j) = P_{m-1}(i, 1)P(1,j) + P_{m-1}(i, 2)P(2,j) + \cdots + P_{m-1}(i, r)P(r,j) = \sum_{k=1}^{r} P_{m-1}(i, k)P(k,j). \tag{6.4}$$

To justify (6.4), write it in the form

$$P[A_{m+1}^{(j)} \mid A_1^{(i)}] = \sum_{k=1}^{r} P[A_m^{(k)} \mid A_1^{(i)}]P[A_{m+1}^{(j)} \mid A_m^{(k)}];$$

recall that $A_m^{(j)}$ is the event that at time m the system is in state j.

One may similarly prove

$$P_m(i,j) = \sum_{k=1}^{r} P(i, k)P_{m-1}(k,j). \tag{6.5}$$

The unconditional probabilities,

(6.6) $p_n(j) =$ the probability that at time n the Markov chain is in state j,

are given for $n \geq 2$ in terms of the initial unconditional probabilities $p_1(j)$ by

$$p_n(j) = \sum_{k=1}^{r} p_1(k)P_{n-1}(k,j). \tag{6.7}$$

One proves (6.7) in exactly the same way that one proves (6.4) and (6.5). Similarly, one may show that

$$p_n(j) = \sum_{k=1}^{r} p_{n-1}(k)P(k,j). \tag{6.8}$$

The transition probabilities $P(i,j)$ of a Markov chain with r states are best exhibited in the form of a *matrix*.

$$P = \begin{bmatrix} P(1,1) & P(1,2) & P(1,3) & \ldots P(1,r-1) & P(1,r) \\ P(2,1) & P(2,2) & P(2,3) & \ldots P(2,r-1) & P(2,r) \\ \cdots & \cdots & \cdots & \cdots & \cdots \\ P(r-1,1) & P(r-1,2) & P(r-1,3) & \ldots P(r-1,r-1) & P(r-1,r) \\ P(r,1) & P(r,2) & P(r,3) & \ldots P(r,r-1) & P(r,r) \end{bmatrix} \tag{6.9}$$

The matrix P is said to be an $r \times r$ matrix, since it has r rows and r columns.

Given an $m \times r$ matrix A and an $r \times n$ matrix B,

$$A = \begin{bmatrix} a_{11} & a_{12} \dots a_{1r} \\ a_{21} & a_{22} \dots a_{2r} \\ \dots & \dots \\ a_{m1} & a_{m2} \dots a_{mr} \end{bmatrix} \qquad B = \begin{bmatrix} b_{11} & b_{12} \dots b_{1n} \\ b_{21} & b_{22} \dots b_{2n} \\ \dots & \dots \\ b_{r1} & b_{r2} \dots b_{rn} \end{bmatrix}.$$

we define the product $C = AB$ of the two matrices as the $m \times n$ matrix whose element c_{ij}, lying at the intersection of the ith row and the jth column, is given by

$$c_{ij} = a_{i1}b_{1j} + a_{i2}b_{2j} + \cdots + a_{ir}b_{rj} = \sum_{k=1}^{r} a_{ik}b_{kj}. \tag{6.10}$$

It should be noted that matrix multiplication is associative; $A(BC) = (AB)C$ for any matrices A, B, and C.

If we define the m-step transition probability matrix P_m of a Markov chain by

$$P_m = \begin{bmatrix} P_m(1, 1) & P_m(1, 2) \dots P_m(1, r) \\ P_m(2, 1) & P_m(2, 2) \dots P_m(2, r) \\ \dots & \dots \\ P_m(r, 1) & P_m(r, 2) \dots P_m(r, r) \end{bmatrix} \tag{6.11}$$

we see that (6.4) and (6.5) may be concisely expressed.

$$P_m = PP_{m-1} = P_{m-1}P, \qquad m = 2, 3, \cdots. \tag{6.12}$$

▶ **Example 6A.** If the transition probability matrix P of a Markov chain is given by

$$P = \begin{bmatrix} 0 & \frac{1}{3} & \frac{2}{3} \\ \frac{2}{3} & 0 & \frac{1}{3} \\ \frac{1}{3} & \frac{2}{3} & 0 \end{bmatrix},$$

then the chain consists of three states, since P is a 3×3 matrix, and the 2-step and 3-step transition probability matrices are given by

$$P_2 = PP = \begin{bmatrix} \frac{4}{9} & \frac{4}{9} & \frac{1}{9} \\ \frac{1}{9} & \frac{4}{9} & \frac{4}{9} \\ \frac{4}{9} & \frac{1}{9} & \frac{4}{9} \end{bmatrix}, \qquad P_3 = P_2P = \begin{bmatrix} \frac{1}{3} & \frac{2}{9} & \frac{4}{9} \\ \frac{4}{9} & \frac{1}{3} & \frac{2}{9} \\ \frac{2}{9} & \frac{4}{9} & \frac{1}{3} \end{bmatrix}.$$

If the initial unconditional probabilities are assumed to be given by

$$p_1 = (p_1(1), p_1(2), p_1(3)) = (\tfrac{1}{2}, \tfrac{1}{4}, \tfrac{1}{4}),$$

then

$$p_2 = (p_2(1), p_2(2), p_2(3)) = p_1P = (\tfrac{1}{4}, \tfrac{1}{3}, \tfrac{5}{12})$$
$$p_3 = (p_3(1), p_3(2), p_3(3)) = p_1P_2 = p_2P = (\tfrac{13}{36}, \tfrac{13}{36}, \tfrac{10}{36})$$
$$p_4 = (p_4(1), p_4(2), p_4(3)) = p_1P_3 = p_3P = (\tfrac{12}{36}, \tfrac{11}{36}, \tfrac{13}{36}). \quad ◀$$

We define a Markov chain with r states as *ergodic* if numbers $\pi_1, \pi_2, \ldots, \pi_r$ exist such that for any states i and j

$$\lim_{m\to\infty} P_m(i,j) = \pi_j. \tag{6.13}$$

In words, a Markov chain is ergodic if, as m tends to ∞, the m-step transition probabilities $P_m(i,j)$ tend to a limit that depends only on the final state j and not on the initial state i. If a Markov chain is ergodic, then after a large number of trials it achieves statistical equilibrium in the sense that the unconditional probabilities $p_n(j)$ tend to limits

$$\lim_{n\to\infty} p_n(j) = \pi_j, \tag{6.14}$$

which are the same, no matter what the values of the initial unconditional probabilities $p_1(j)$. To see that (6.13) implies (6.14), take the limit of both sides of (6.7) and use the fact that $\sum_{k=1}^{r} p_1(k) = 1$.

In view of (6.14), we call $\pi_1, \pi_2, \ldots, \pi_r$ the *stationary probabilities* of the Markov chain, since these represent the probabilities of being in the various states after one has achieved statistical equilibrium.

One of the important problems of the theory of Markov chains is to determine conditions under which a Markov chain is ergodic. A discussion of this problem is beyond the scope of this book. We state without proof the following theorem.

If there exists an integer m such that

$$P_m(i,j) > 0 \qquad \text{for all states } i \text{ and } j, \tag{6.15}$$

then the Markov chain with transition probability matrix P is ergodic.

It is sometimes possible to establish that a Markov chain is ergodic without having to exhibit an m-step transition probability matrix P_m, all the entries of which are positive. Given two states i and j in a Markov chain, we say that *one can reach i from j* if states $i_1, i_2, \ldots, i_N$ exist such that

$$0 < P(i, i_1)P(i_1, i_2) \cdots P(i_{N-1}, i_N)P(i_N, j). \tag{6.16}$$

Two states i and j are said to *communicate* if one can reach i from j and also j from i. The following theorem can be proved.

If all states in a Markov chain communicate and if a state i exists such that $P(i, i) > 0$, then the Markov chain is ergodic.

Having established that a Markov chain is ergodic, the next problem is to obtain the stationary probabilities π_j. It is clear from (6.4) that the stationary probabilities satisfy the system of linear equations,

$$\pi_j = \sum_{k=1}^{r} \pi_k P(k, j), \qquad j = 1, 2, \cdots, r. \tag{6.17}$$

Consequently, if a Markov chain is ergodic, then a solution of (6.17) that satisfies the conditions

$$\pi_j \geq 0 \qquad \text{for } j = 1, 2, \cdots, r; \qquad \sum_{j=1}^{r} \pi_j = 1 \tag{6.18}$$

exists. It may be shown that if a Markov chain with transition probability matrix P is ergodic then the solution of (6.17) satisfying (6.18) is unique and necessarily satisfies (6.13) and (6.14). Consequently, to find the stationary probabilities, we need solve only (6.17).

▶ **Example 6B.** The Markov chain considered in example 6A is ergodic, since $P_2(i, j) > 0$ for all states i and j. To compute the stationary probabilities π_1, π_2, π_3, we need only to solve the equations

$$\begin{aligned} \pi_1 &= \qquad\quad \tfrac{2}{3}\pi_2 + \tfrac{1}{3}\pi_3 \\ \pi_2 &= \tfrac{1}{3}\pi_1 \qquad\quad + \tfrac{2}{3}\pi_3 \\ \pi_3 &= \tfrac{2}{3}\pi_1 + \tfrac{1}{3}\pi_2 \end{aligned} \tag{6.19}$$

subject to (6.18). It is clear that

$$\pi_1 = \pi_2 = \pi_3 = \tfrac{1}{3}$$

is a solution of (6.19) satisfying (6.18). In the long run, the states 1, 2, and 3 are equally likely to be the state of the Markov chain. ◀

A matrix

$$A = \begin{bmatrix} a_{11} & a_{12} \cdots a_{1r} \\ a_{21} & a_{22} \cdots a_{2r} \\ \cdots & \cdots \cdots \\ a_{r1} & a_{r2} \cdots a_{rr} \end{bmatrix}$$

is defined as *stochastic* if the sum of the entries in any row is equal to 1; in symbols, A is stochastic if

$$\sum_{j=1}^{r} a_{ij} = 1 \qquad \text{for } i = 1, 2, \ldots, r. \tag{6.20}$$

The matrix A is defined as *doubly stochastic* if in addition the sum of the entries in any column is equal to 1; in symbols, A is doubly stochastic if (6.20) holds and also

$$\sum_{i=1}^{r} a_{ij} = 1 \qquad \text{for } j = 1, 2, \ldots, r. \tag{6.21}$$

It is clear that the transition probability matrix P of a Markov chain is stochastic. If P is doubly stochastic (as is the matrix in example 6A), then the stationary probabilities are given by

$$\pi_1 = \pi_2 = \cdots = \pi_r = \frac{1}{r}, \tag{6.22}$$

in which r is the number of states in the Markov chain. To prove (6.22) one need only verify that if P is doubly stochastic then (6.22) satisfies (6.17) and (6.18).

▶ **Example 6C. Random walk with retaining barriers.** Consider a straight line on which are marked off positions 0, 1, 2, 3, 4, and 5, arranged from left to right. A man (or an atomic particle, if one prefers physically significant examples) performs a random walk among the six positions by tossing a coin that has probability p (where $0 < p < 1$) of falling heads and acting in accordance with the following set of rules: if the coin falls heads, move one position to the right, if at 0, 1, 2, 3, or 4, and remain at 5, if at 5; if the coin falls tails, move one position to the left, if at 1, 2, 3, 4, or 5, and remain at 0, if at 0. The positions 0 and 5 are retaining barriers; one cannot move past them. In example 6D we consider the case in which positions 0 and 5 are absorbing barriers; if one reaches these positions, the walk stops. The transition probability matrix of the random walk with retaining barriers is given by

$$P = \begin{bmatrix} q & p & 0 & 0 & 0 & 0 \\ q & 0 & p & 0 & 0 & 0 \\ 0 & q & 0 & p & 0 & 0 \\ 0 & 0 & q & 0 & p & 0 \\ 0 & 0 & 0 & q & 0 & p \\ 0 & 0 & 0 & 0 & q & p \end{bmatrix}. \tag{6.23}$$

All states in this Markov chain communicate, since

$$0 < P(0, 1)P(1, 2)P(2, 3)P(3, 4)P(4, 5)P(5, 4)P(4, 3)P(3, 2)P(2, 1)P(1, 0).$$

The chain is ergodic, since $P(0, 0) > 0$. To find the stationary probabilities $\pi_0, \pi_1, \ldots, \pi_5$, we solve the system of equations:

$$\begin{aligned} \pi_0 &= q\pi_0 + q\pi_1 \\ \pi_1 &= p\pi_0 + q\pi_2 \\ \pi_2 &= p\pi_1 + q\pi_3 \\ \pi_3 &= p\pi_2 + q\pi_4 \\ \pi_4 &= p\pi_3 + q\pi_5 \\ \pi_5 &= p\pi_4 + p\pi_5. \end{aligned} \tag{6.24}$$

We solve these equations by successive substitution.

From the first equation we obtain

$$q\pi_1 = p\pi_0 \qquad \text{or} \qquad \pi_1 = \frac{p}{q}\,\pi_0.$$

By subtracting this result from the second equation in (6.24), we obtain

$$q\pi_2 = p\pi_1 \qquad \text{or} \qquad \pi_2 = \frac{p}{q}\,\pi_1 = \left(\frac{p}{q}\right)^2 \pi_0.$$

Similarly, we obtain

$$q\pi_3 = p\pi_2 \qquad \text{or} \qquad \pi_3 = \frac{p}{q}\,\pi_2 = \left(\frac{p}{q}\right)^3 \pi_0$$

$$q\pi_4 = p\pi_3 \qquad \text{or} \qquad \pi_4 = \frac{p}{q}\,\pi_3 = \left(\frac{p}{q}\right)^4 \pi_0$$

$$q\pi_5 = p\pi_4 \qquad \text{or} \qquad \pi_5 = \frac{p}{q}\,\pi_4 = \left(\frac{p}{q}\right)^5 \pi_0.$$

To determine π_0, we use the fact that

$$\begin{aligned} 1 = \pi_0 + \pi_1 + \cdots + \pi_5 &= \pi_0\left[1 + \frac{p}{q} + \left(\frac{p}{q}\right)^2 + \ldots + \left(\frac{p}{q}\right)^5\right] \\ &= \begin{cases} \pi_0 \dfrac{1 - (p/q)^6}{1 - (p/q)} & \text{if } p \neq q \\ 6\pi_0 & \text{if } p = q = \frac{1}{2}. \end{cases} \end{aligned}$$

We finally conclude that the stationary probabilities for the random walk with retaining barriers for $j = 0, 1, \ldots, 5$ are given by

$$\pi_j = \begin{cases} \left(\dfrac{p}{q}\right)^j \dfrac{1 - (p/q)}{1 - (p/q)^6} & \text{if } p \neq q \\ \frac{1}{6} & \text{if } p = q = \frac{1}{2}. \end{cases} \tag{6.25}$$

◀

If a Markov chain is ergodic, then the physical process represented by the Markov chain can continue indefinitely. Indeed, after a long time it achieves statistical equilibrium and probabilities $\pi_1, \ldots, \pi_r$ exist of being in the various states that depend only on the transition probability matrix P.

We next desire to study an important class of nonergodic Markov chains, namely those that possess absorbing states. A state j in a Markov chain is said to be *absorbing* if $P(j, i) = 0$ for all states $i \neq j$, so that it is impossible to leave an absorbing state. Equivalently, a state j is absorbing if $P(j,j) = 1$.

▶ **Example 6D. Random walk with absorbing barriers.** Consider a straight line on which positions 0, 1, 2, 3, 4, and 5, arranged from left to right, are marked off. Consider a man who performs a random walk among the six positions according to the following transition probability matrix:

$$P = \begin{bmatrix} 1 & 0 & 0 & 0 & 0 & 0 \\ q & 0 & p & 0 & 0 & 0 \\ 0 & q & 0 & p & 0 & 0 \\ 0 & 0 & q & 0 & p & 0 \\ 0 & 0 & 0 & q & 0 & p \\ 0 & 0 & 0 & 0 & 0 & 1 \end{bmatrix}. \tag{6.26}$$

In the Markov chain with transition probability matrix P, given by (6.26), the states 0 and 5 are absorbing states; consequently, this Markov chain is called a random walk with absorbing barriers. The model of a random walk with absorbing barriers describes the fortunes of gamblers with finite capital. Let two opponents, A and B, have 5 cents between them. Let A toss a coin, which has probability p of falling heads. On each toss he wins a penny if the coin falls heads and loses a penny if the coin falls tails. For $j = 0, \ldots, 5$ we define the chain to be in state j if A has j cents. ◀

Given a Markov chain with an absorbing state j, it is of interest to compute for each state i

(6.27) $u_j(i)$ = conditional probability of ever arriving at the absorbing state j, given that one started from state i.

We call $u_j(i)$ the probability of absorption in state j, given the initial state i, since one remains in j if one ever arrives there.

The probability $u_j(i)$ is defined on a sample description space consisting of a countably infinite number of trials. We do not in this book discuss the definition of probabilities on such sample spaces. Consequently, we cannot give a proof of the following basic theorem, which facilitates the computation of the absorption probabilities $u_j(i)$.

If j is an absorbing state in a Markov chain with states $\{1, 2, \ldots, r\}$, then the absorption probabilities $u_j(1), \ldots, u_j(r)$ are the unique solution to the system of equations:

$$
\begin{aligned}
&u_j(j) = 1 \\
(6.28) \qquad &u_j(i) = 0, \qquad \text{if } j \text{ cannot be reached from } i \\
&u_j(i) = \sum_{k=1}^{r} P(i, k)u_j(k), \qquad \text{if } j \text{ can be reached from } i.
\end{aligned}
$$

Equation (6.28) is proved as follows. The probability of going from state i to state j is the sum, over all states k, of the probability of going from i to j via k; this probability is the product of the probability $P(i, k)$ of going from i to k in one step and the probability $u_j(k)$ of then ever passing from k to j.

▶ **Example 6E. Probability of a gambler's ruin.** Let A and B play the coin-tossing game described in example 6D. If A's initial fortune is 3 cents and B's initial fortune is 2 cents, what is the probability that A's fortune will be 0 cents before it is 5 cents, and A will be ruined?

Solution: For $i = 0, 1, \ldots, 5$ let $u_0(i)$ be the probability that A's fortune will ever be 0, given that his initial fortune was i cents. In view of (6.28), the absorption probabilities $u_0(i)$ are the unique solution of the equations

$$
\begin{aligned}
&u_0(0) = 1 \\
&u_0(1) = qu_0(0) + pu_0(2) \quad \text{or} \quad q[u_0(1) - u_0(0)] = p[u_0(2) - u_0(1)] \\
&u_0(2) = qu_0(1) + pu_0(3) \quad \text{or} \quad q[u_0(2) - u_0(1)] = p[u_0(3) - u_0(2)] \\
(6.29) \quad &u_0(3) = qu_0(2) + pu_0(4) \quad \text{or} \quad q[u_0(3) - u_0(2)] = p[u_0(4) - u_0(3)] \\
&u_0(4) = qu_0(3) + pu_0(5) \quad \text{or} \quad q[u_0(4) - u_0(3)] = p[u_0(5) - u_0(4)] \\
&u_0(5) = 0.
\end{aligned}
$$

To solve these equations we note that, defining $c = u_0(1) - u_0(0)$,

$$
\begin{aligned}
u_0(2) - u_0(1) &= \frac{q}{p}[u_0(1) - u_0(0)] = \frac{q}{p}c \\
u_0(3) - u_0(2) &= \frac{q}{p}[u_0(2) - u_0(1)] = \left(\frac{q}{p}\right)^2 c \\
u_0(4) - u_0(3) &= \frac{q}{p}[u_0(3) - u_0(2)] = \left(\frac{q}{p}\right)^3 c \\
u_0(5) - u_0(4) &= \left(\frac{q}{p}\right)[u_0(4) - u_0(3)] = \left(\frac{q}{p}\right)^4 c.
\end{aligned}
$$

Therefore, there is a constant c such that (since $u_0(0) = 1$),

$$u_0(1) = 1 + c$$

$$u_0(2) = \frac{q}{p}c + 1 + c$$

$$u_0(3) = \left(\frac{q}{p}\right)^2 c + \left(\frac{q}{p}\right)c + 1 + c$$

$$u_0(4) = \left(\frac{q}{p}\right)^3 c + \left(\frac{q}{p}\right)^2 c + \left(\frac{q}{p}\right)c + 1 + c$$

$$u_0(5) = \left(\frac{q}{p}\right)^4 c + \left(\frac{q}{p}\right)^3 c + \left(\frac{q}{p}\right)^2 c + \left(\frac{q}{p}\right)c + 1 + c$$

To determine the constant c, we use the fact that $u_0(5) = 0$. We see that

$$1 + 5c = 0 \qquad \text{if } p = q = \tfrac{1}{2}$$

$$1 + c\left(\frac{1 - (q/p)^5}{1 - (q/p)}\right) = 0 \qquad \text{if } p \neq q$$

so that

$$\begin{aligned} c &= -\tfrac{1}{5} \qquad \text{if } p = q = \tfrac{1}{2} \\ &= -\frac{1 - (q/p)}{1 - (q/p)^5} \qquad \text{if } p \neq q. \end{aligned}$$

Consequently, for $i = 0, 1, \ldots, 5$

$$\text{(6.30)} \qquad \begin{aligned} u_0(i) &= 1 - \frac{i}{5} \qquad \text{if } p = q = \tfrac{1}{2} \\ &= 1 - \frac{1 - (q/p)^i}{1 - (q/p)^5} \qquad \text{if } p \neq q. \end{aligned}$$

In particular, the probability that A will be ruined, given his initial fortune is 3 cents, is given by

$$\begin{aligned} u_0(3) &= \tfrac{2}{5} \qquad \text{if } p = q = \tfrac{1}{2} \\ &= \frac{(q/p)^3 - (q/p)^5}{1 - (q/p)^5} = \frac{q^5 - p^2q^3}{q^5 - p^5} \qquad \text{if } p \neq q. \end{aligned}$$

◀

EXERCISES

6.1. Compute the 2-step and 3-step transition probability matrices for the Markov chains whose transition probability matrices are given by

(i) $P = \begin{bmatrix} \frac{1}{2} & \frac{1}{2} \\ \frac{1}{2} & \frac{1}{2} \end{bmatrix}$, (ii) $P = \begin{bmatrix} \frac{1}{2} & \frac{1}{2} & 0 \\ \frac{1}{2} & \frac{1}{2} & 0 \\ 0 & \frac{1}{2} & \frac{1}{2} \end{bmatrix}$

(iii) $P = \begin{bmatrix} \frac{1}{2} & \frac{1}{2} & 0 \\ \frac{1}{2} & \frac{1}{2} & 0 \\ 0 & 0 & 1 \end{bmatrix}$, (iv) $P = \begin{bmatrix} \frac{1}{2} & \frac{1}{2} & 0 \\ 0 & \frac{1}{2} & \frac{1}{2} \\ \frac{1}{2} & \frac{1}{2} & 0 \end{bmatrix}$

6.2. For each Markov chain in exercise 6.1, determine whether or not (i) it is ergodic, (ii) it has absorbing states.

6.3. Find the stationary probabilities for each of the following ergodic Markov chains:

(i) $\begin{bmatrix} \frac{2}{3} & \frac{1}{3} \\ \frac{1}{3} & \frac{2}{3} \end{bmatrix}$, (ii) $\begin{bmatrix} \frac{2}{3} & \frac{1}{3} \\ \frac{2}{3} & \frac{1}{3} \end{bmatrix}$, (iii) $\begin{bmatrix} 0.99 & 0.01 \\ 0.01 & 0.99 \end{bmatrix}$

6.4. Find the stationary probabilities for each of the following ergodic Markov chains:

(i) $\begin{bmatrix} \frac{1}{2} & \frac{1}{4} & \frac{1}{4} \\ \frac{1}{4} & \frac{1}{2} & \frac{1}{4} \\ \frac{1}{4} & \frac{1}{4} & \frac{1}{2} \end{bmatrix}$, (ii) $\begin{bmatrix} \frac{1}{4} & \frac{1}{4} & \frac{1}{2} \\ 0 & \frac{2}{3} & \frac{1}{3} \\ \frac{3}{4} & \frac{1}{4} & 0 \end{bmatrix}$, (iii) $\begin{bmatrix} \frac{1}{3} & \frac{1}{4} & \frac{5}{12} \\ \frac{1}{3} & \frac{1}{4} & \frac{5}{12} \\ \frac{1}{3} & \frac{1}{4} & \frac{5}{12} \end{bmatrix}$

6.5. Consider a series of independent repeated tosses of a coin that has probability $p > 0$ of falling heads. Let us say that at time n we are in state s_1, s_2, s_3, or s_4 depending on whether outcomes of tosses $n - 1$ and n were (H, H), (H, T), (T, H), or (T, T). Find the transition probability matrix P of this Markov chain. Also find P^2, P^3, P^4.

6.6. Random walk with retaining barriers. Consider a straight line on which positions 0, 1, 2, . . . , 7 are marked off. Consider a man who performs a random walk among the positions according to the following transition probability matrix:

$$P = \begin{bmatrix} q & p & 0 & 0 & 0 & 0 & 0 & 0 \\ q & 0 & p & 0 & 0 & 0 & 0 & 0 \\ 0 & q & 0 & p & 0 & 0 & 0 & 0 \\ 0 & 0 & q & 0 & p & 0 & 0 & 0 \\ 0 & 0 & 0 & q & 0 & p & 0 & 0 \\ 0 & 0 & 0 & 0 & q & 0 & p & 0 \\ 0 & 0 & 0 & 0 & 0 & q & 0 & p \\ 0 & 0 & 0 & 0 & 0 & 0 & q & p \end{bmatrix}$$

Prove that the Markov chain is ergodic. Find the stationary probabilities.

6.7. **Gambler's ruin.** Let two players A and B have 7 cents between them. Let A toss a coin, which has probability p of falling heads. On each toss he wins a penny if the coin falls heads and he loses a penny if the coin falls tails. If A's initial fortune is 3 cents, what is the probability that A's fortune will be 0 cents before it is 7 cents, and that A will be ruined.

6.8. Consider 2 urns, I and II, each of which contains 1 white and 1 red ball. One ball is drawn simultaneously from each urn and placed in the other urn. Let the probabilities that after n repetitions of this procedure urn I will contain 2 white balls, 1 white and 1 red, or 2 red balls be denoted by p_n, q_n, and r_n, respectively. Deduce formulas expressing p_{n+1}, q_{n+1}, and r_{n+1} in terms of p_n, q_n, and r_n. Show that p_n, q_n, and r_n tend to limiting values as n tends to infinity. Interpret these values.

6.9. In exercise 6.8 find the most probable number of red balls in urn I after (i) 2, (ii) 6 exchanges.

CHAPTER 4

Numerical-Valued Random Phenomena

In the foregoing we have considered mainly random phenomena whose sample description spaces were finite. We next consider random phenomena for which this is not necessarily the case. The simplest example of a random phenomenon whose sample description space is not necessarily finite is one which is numerical valued. The height of waves on a wind-swept sea, the number of alpha particles emitted from a radioactive source, the number of telephone calls arriving at a switchboard, the velocity of a particle in Brownian motion, the scores of students on an examination, the collar sizes of men, the dress sizes of women, and so on, constitute examples of numerical-valued random phenomena. In this chapter we discuss the notions and techniques used to treat numerical-valued random phenomena.

1. THE NOTION OF A NUMERICAL-VALUED RANDOM PHENOMENON

To introduce the notion of a numerical-valued random phenomenon, let us first consider a random phenomenon whose sample description space S is a set of real numbers; for example, the number of white balls in a sample of size n drawn from an urn or the number of hits in n independent throws of a dart. For the sample description space of each of these random phenomena one may take the set $\{0, 1, 2, \ldots, n\}$. However, it has already been indicated (in section 3 of Chapter 1) that one may make the sample description space S as large as one pleases, at the price of having a large number of sample descriptions in S to which zero probability is assigned. Consequently, we may take for the sample description space of these phenomena the set of all real numbers from $-\infty$ to ∞. The advantage of this procedure might be that it would render possible a unified theory of random phenomena whose sample description spaces are sets of real numbers.

There is still another advantage. Suppose one is measuring the weight of persons belonging to a certain group. One may measure the weight to the nearest pound, the nearest tenth of a pound, or the nearest hundredth of a pound. In the first case the space $S = \{$real numbers $x: x = k$ for some integer $k = 0, 1, 2, \ldots, 10^4\}$ would suffice as the sample description space; in the second case $S = \{$real numbers $x: x = k/10$ for some integer $k = 0, 1, 2, \ldots, 10^5\}$ would suffice; in the third case $S = \{$real numbers $x: x = k/100$ for some integer $k = 0, 1, 2, \ldots, 10^6\}$ would suffice. Nevertheless, it might be preferable in all three cases to take as one's sample description space the set of all numbers from $-\infty$ to ∞ and to develop the difference between the three cases in terms of the different probability functions adopted to describe the three random phenomena.

We are thus led to define the notion of a *numerical-valued random phenomenon* as a random phenomenon whose sample description space is the set R, consisting of all real numbers from $-\infty$ to ∞. The set R may be represented geometrically by a *real line*, which is an infinitely long line on which an origin and a unit distance have been marked off; then to every point on the line there corresponds a real number and to every real number there corresponds a point on the line.

We have previously defined an *event* as a set of sample descriptions; consequently, *events defined on numerical-valued random phenomena are sets of real numbers*. However, not every set of real numbers can be regarded as an event. There are certain sets of real numbers, defined by exceedingly involved limiting operations, that are nonprobabilizable, in the sense that for these sets it is not in general possible to answer, in a manner consistent with the axioms below, the question, "what is the probability that a given numerical-valued random phenomenon will have an observed value in the set?" Consequently, by the word "event" we mean not any set of real numbers but only a probabilizable set of real numbers. We do not possess at this stage in our discussion the notions with which to characterize the sets of real numbers that are probabilizable. We can point out only that it may be shown that the family (call it $\mathscr{F}$) of probabilizable sets always has the following properties:

(i) To $\mathscr{F}$ belongs any interval (an interval is a set of real numbers of the form $\{x: a < x < b\}$, $\{x: a < x \le b\}$, $\{x: a \le x < b\}$, or $\{x: a \le x \le b\}$, in which a and b may be finite or infinite numbers).

(ii) To $\mathscr{F}$ belongs the complement A^c of any set A belonging to $\mathscr{F}$.

(iii) To $\mathscr{F}$ belongs the union $\bigcup_{n=1}^{\infty} A_n$ of any sequence of sets $A_1, A_2, \ldots, A_n, \ldots$ belonging to $\mathscr{F}$.

If we desire to give a precise definition of the notion of an event at this stage in our discussion, we may do so as follows. There exists a smallest family of sets on the real line with the properties (i), (ii), and (iii). This family is denoted by $\mathscr{B}$, and any member of $\mathscr{B}$ is called a Borel set, after the great French mathematician and probabilist Émile Borel. Since $\mathscr{B}$ is the smallest family to possess properties (i), (ii), and (iii), it follows that $\mathscr{B}$ is contained in $\mathscr{F}$, the family of probabilizable sets. Thus every Borel set is probabilizable. Since the needs of mathematical rigor are fully met by restricting our discussion to Borel sets, in this book, *by an "event" concerning a numerical-valued random phenomena, we mean a Borel set of real numbers.*

We sum up the discussion of this section in a formal definition.

A *numerical-valued random phenomenon* is a random phenomenon whose sample description space is the set R (of all real numbers from $-\infty$ to ∞) on whose subsets is defined a function $P[\cdot]$, which to every Borel set of real numbers (also called an event) E assigns a nonnegative real number, denoted by $P[E]$, according to the following axioms:

AXIOM 1. $P[E] \geq 0$ for every event E.

AXIOM 2. $P[R] = 1$.

AXIOM 3. For any sequence of events $E_1, E_2, \ldots, E_n, \ldots$ which is mutually exclusive,

$$P\left[\bigcup_{n=1}^{\infty} E_n\right] = \sum_{n=1}^{\infty} P[E_n].$$

▶ **Example 1A.** Consider the random phenomenon that consists in observing the time one has to wait for a bus at a certain downtown bus stop. Let A be the event that one has to wait between 0 and 2 minutes, inclusive, and let B be the event that one has to wait between 1 and 3 minutes, inclusive. Assume that $P[A] = \frac{1}{2}$, $P[B] = \frac{1}{2}$, $P[AB] = \frac{1}{3}$. We can now answer all the usual questions about the events A and B. The conditional probability $P[B \mid A]$ that B has occurred given that A has occurred is $\frac{2}{3}$. The probability that neither the event A nor the event B has occurred is given by $P[A^cB^c] = 1 - P[A \cup B] = 1 - P[A] - P[B] + P[AB] = \frac{1}{3}$. ◀

EXERCISE

1.1. Consider the events A and B defined in example 1A. Assuming that $P[A] = P[B] = \frac{1}{2}$, $P[AB] = \frac{1}{3}$, find the probability for $k = 0, 1, 2$, that (i) exactly k, (ii) at least k, (iii) no more than k of the events A and B will occur.

2. SPECIFYING THE PROBABILITY FUNCTION OF A NUMERICAL-VALUED RANDOM PHENOMENON

Consider the probability function $P[\cdot]$ of a numerical-valued random phenomenon. The question arises concerning the convenient ways of stating the function without having actually to state the value of $P[E]$ for every set of real numbers E. In general, to state the function $P[\cdot]$, as with any function, one has to enumerate all the members of the domain of the function $P[\cdot]$, and for each of these members of the domain one states the value of the function. In special circumstances (which fortunately cover most of the cases encountered in practice) more convenient methods are available.

For many probability functions there exists a function $f(\cdot)$, defined for all real numbers x, from which $P[E]$ can, for any event E, be obtained by integration:

$$P[E] = \int_E f(x)\, dx. \tag{2.1}$$

Given a probability function $P[\cdot]$, which may be represented in the form of (2.1) in terms of some function $f(\cdot)$, we call the function $f(\cdot)$ the *probability density function* of the probability function $P[\cdot]$, and we say that the probability function $P[\cdot]$ is specified by the probability density function $f(\cdot)$.

A function $f(\cdot)$ must have certain properties in order to be a probability density function. To begin with, it must be sufficiently well behaved as a function so that the integral* in (2.1) is well defined. Next, letting $E = R$ in (2.1),

$$1 = P[R] = \int_R f(x)\, dx = \int_{-\infty}^{\infty} f(x)\, dx. \tag{2.2}$$

* We usually assume that the integral in (2.1) is defined in the sense of Riemann; to ensure that this is the case, we require that the function $f(\cdot)$ be defined and continuous at all but a finite number of points. The integral in (2.1) is then defined only for events E, which are either intervals or unions of a finite number of nonoverlapping intervals. In advanced probability theory the integral in (2.1) is defined by means of a theory of integration developed in the early 1900's by Henri Lebesgue. The function $f(\cdot)$ must then be a *Borel* function, by which is meant that for any real number c the set $\{x: f(x) < c\}$ is a Borel set. A function that is continuous at all but a finite number of points may be shown to be a Borel function. It may be shown that if a Borel function $f(\cdot)$ satisfies (2.2) and (2.3) then, for any Borel set B, the integral of $f(\cdot)$ over B exists as an integral defined in the sense of Lebesgue. If B is an interval, or a union of a finite number of nonoverlapping intervals, and if $f(\cdot)$ is continuous on B, then the integral of $f(\cdot)$ over B, defined in the sense of Lèbesgue, has the same value as the integral of $f(\cdot)$ over B, defined in the sense of Riemann. *Henceforth, in this book the word function (unless otherwise qualified) will mean a Borel function and the word set (of real numbers) will mean a Borel set.*

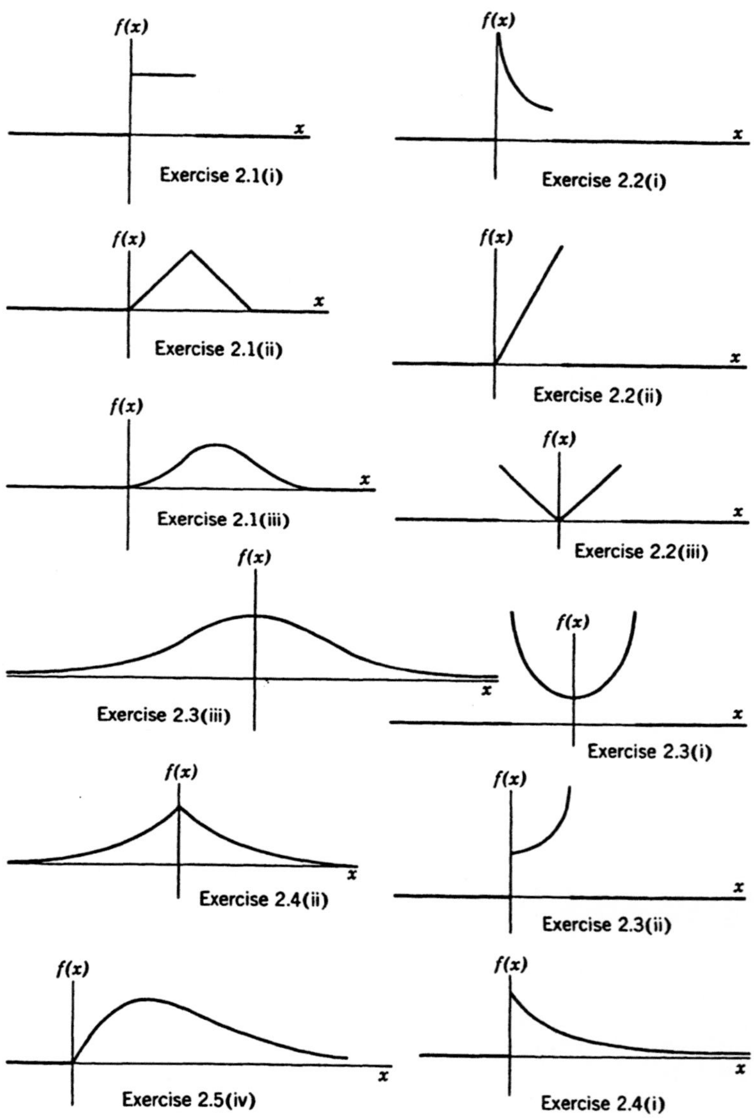

Fig. 2A. Graphs of the probability density functions given in the exercises indicated.

It is necessary that $f(\cdot)$ satisfy (2.2); in words, the integral of $f(\cdot)$ from $-\infty$ to ∞ must be equal to 1.

A function $f(\cdot)$ is said to be a probability density function if it satisfies (2.2) and, in addition, satisfies the condition*

(2.3) $$f(x) \geq 0 \qquad \text{for all } x \text{ in } R,$$

since a function $f(\cdot)$ satisfying (2.2) and (2.3) is the probability density function of a unique probability function $P[\cdot]$, namely the probability function with value $P[E]$ at any event E given by (2.1). Some typical probability density functions are illustrated in Fig. 2A.

▶ **Example 2A. Verifying that a function is a probability density function.** Suppose one is told that the time one has to wait for a bus on a certain street corner is a numerical-valued random phenomenon, with a probability function, specified by the probability density function $f(\cdot)$, given by

(2.4) $$\begin{aligned} f(x) &= 4x - 2x^2 - 1 \qquad 0 \leq x \leq 2 \\ &= 0 \qquad \text{otherwise.} \end{aligned}$$

The function $f(\cdot)$ is negative for various values of x; in particular, it is negative for $0 \leq x \leq \frac{1}{4}$ (prove this statement). Consequently, it is not possible for $f(\cdot)$ to be a probability density function. Next, suppose that the probability density function $f(\cdot)$ is given by

(2.5) $$\begin{aligned} f(x) &= 4x - 2x^2 \qquad 0 \leq x \leq 2 \\ &= 0 \qquad \text{otherwise.} \end{aligned}$$

The function $f(\cdot)$, given by (2.5), is nonnegative (prove this statement). However, its integral from $-\infty$ to ∞,

$$\int_{-\infty}^{\infty} f(x)\, dx = \frac{8}{3},$$

is not equal to 1. Consequently the function $f(\cdot)$, given by (2.5) is not a probability density function. However, the function $f(\cdot)$, given by

$$\begin{aligned} f(x) &= \tfrac{3}{8}(4x - 2x^2) \qquad 0 \leq x \leq 2 \\ &= 0 \qquad \text{otherwise,} \end{aligned}$$

is a probability density function. ◀

▶ **Example 2B. Computing probabilities from a probability density function.** Let us consider again the numerical-valued random phenomenon, discussed in example 1A, that consists in observing the time one has to

* For the purposes of this book we also require that a probability density function $f(\cdot)$ be defined and continuous at all but a finite number of points.

wait for a bus at a certain bus stop. Let us *assume* that the probability function $P[\cdot]$ of this phenomenon may be expressed by (2.1) in terms of the function $f(\cdot)$, whose graph is sketched in Fig. 2B. An algebraic formula for $f(\cdot)$ can be written as follows:

$$
\begin{aligned}
f(x) &= 0 && \text{for } x < 0 \\
&= (\tfrac{1}{9})(x + 1) && \text{for } 0 \leq x < 1 \\
&= (\tfrac{4}{9})(x - (\tfrac{1}{2})) && \text{for } 1 \leq x < (\tfrac{3}{2}) \\
&= (\tfrac{4}{9})((\tfrac{5}{2}) - x) && \text{for } (\tfrac{3}{2}) \leq x < 2 \\
&= (\tfrac{1}{9})(4 - x) && \text{for } 2 \leq x < 3 \\
&= (\tfrac{1}{9}) && \text{for } 3 \leq x < 6 \\
&= 0 && \text{for } 6 \leq x
\end{aligned}
\tag{2.6}
$$

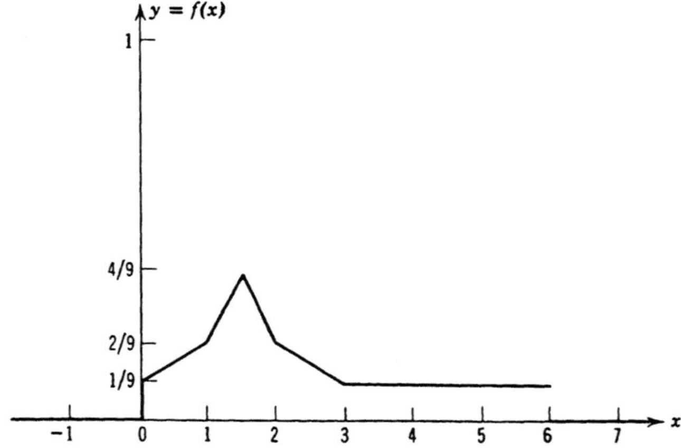

Fig. 2B. Graph of the probability density function $f(\cdot)$ defined by (2.6).

From (2.1) it follows that if $A = \{x: 0 \leq x \leq 2\}$ and $B = \{x: 1 \leq x \leq 3\}$ then

$$P[A] = \int_0^2 f(x)\,dx = \tfrac{1}{2}, \qquad P[B] = \int_1^3 f(x)\,dx = \tfrac{1}{2},$$

$$P[AB] = \int_1^2 f(x)\,dx = \tfrac{1}{3},$$

which agree with the values assumed in example 1A. ◀

▶ **Example 2C. The lifetime of a vacuum tube.** Consider the numerical-valued random phenomenon that consists in observing the total time a vacuum tube will burn from the moment it is first put into service. Suppose that the probability function $P[\cdot]$ of this phenomenon is expressed by (2.1) in terms of the function $f(\cdot)$ given by

$$\begin{aligned} f(x) &= 0 && \text{for } x < 0. \\ &= \frac{1}{1000} e^{(-x/1000)} && \text{for } x \geq 0. \end{aligned}$$

Let E be the event that the tube burns between 100 and 1000 hours, inclusive, and let F be the event that the tube burns more than 1000 hours. The events E and F may be represented as subsets of the real line: $E = \{x\colon 100 \leq x \leq 1000\}$ and $F = \{x\colon 1000 < x\}$. The probabilities of E and F are given by

$$\begin{aligned} P[E] &= \int_{100}^{1000} f(x)\,dx = \frac{1}{1000}\int_{100}^{1000} e^{-(x/1000)}\,dx = -e^{-(x/1000)}\Big|_{100}^{1000} \\ &= e^{-0.1} - e^{-1} = 0.537. \end{aligned}$$

$$\begin{aligned} P[F] &= \int_{1000}^{\infty} f(x)\,dx = \frac{1}{1000}\int_{1000}^{\infty} e^{-(x/1000)}\,dx = -e^{-(x/1000)}\Big|_{1000}^{\infty} \\ &= e^{-1} = 0.368. \end{aligned}$$ ◀

For many probability functions there exists a function $p(\cdot)$, defined for all real numbers x, but with value $p(x)$ equal to 0 for all x except for a finite or countably infinite set of values of x at which $p(x)$ is positive, such that from $p(\cdot)$ the value of $P[E]$ can be obtained for any event E by summation:

$$P[E] = \sum_{\substack{\text{over all} \\ \text{points } x \text{ in } E \\ \text{such that } p(x) > 0}} p(x) \tag{2.7}$$

In order that the sum in (2.7) may be meaningful, it suffices to impose the condition [letting $E = R$ in (2.7)] that

$$1 = \sum_{\substack{\text{over all} \\ \text{points } x \text{ in } R \\ \text{such that } p(x) > 0}} p(x) \tag{2.8}$$

Given a probability function $P[\cdot]$, which may be represented in the form (2.7), we call the function $p(\cdot)$ the *probability mass function* of the probability function $P[\cdot]$, and we say that the probability function $P[\cdot]$ is specified by the probability mass function $p(\cdot)$.

A function $p(\cdot)$, defined for all real numbers, is said to be a probability mass function if (i) *$p(x)$ equals zero for all x, except for a finite or countably infinite set of values of x for which $p(x) > 0$, and* (ii) *the infinite series in (2.8) converges and sums to 1.* Such a function is the probability mass function of a unique probability function $P[\cdot]$ defined on the subsets of the real line, namely the probability function with value $P[E]$ at any set E given by (2.7).

▶ **Example 2D. Computing probabilities from a probability mass function.** Let us consider again the numerical-valued random phenomenon considered in examples 1A and 2B. Let us *assume* that the probability function $P[\cdot]$

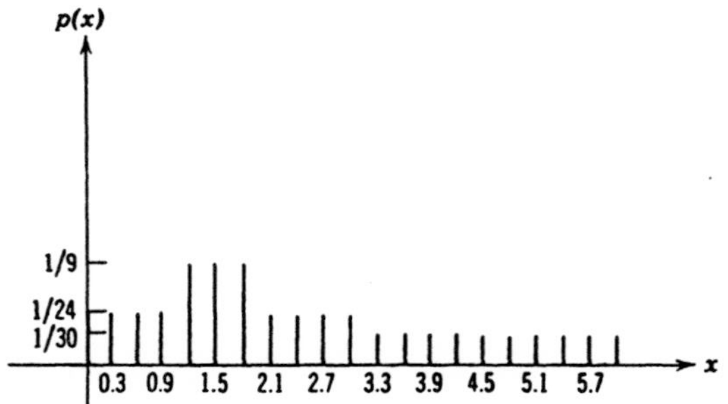

Fig. 2C. Graph of the probability mass function defined by (2.9).

of this phenomenon may be expressed by (2.7) in terms of the function $p(\cdot)$, whose graph is sketched in Fig. 2C. An algebraic formula for $p(\cdot)$ can be written as follows:

$$\begin{aligned}(2.9)\quad p(x) &= 0, &&\text{unless } x = (0.3)k \text{ for some } k = 0, 1, \cdots, 20\\ &= \tfrac{1}{24}, &&\text{for } x = 0, 0.3, 0.6, 0.9, 2.1, 2.4, 2.7, 3.0\\ &= \tfrac{1}{9}, &&\text{for } x = 1.2, 1.5, 1.8\\ &= \tfrac{1}{30}, &&\text{for } x = 3.3, 3.6, 3.9, 4.2, 4.5, 4.8, 5.1, 5.4, 5.7, 6.0.\end{aligned}$$

It then follows that

$$\begin{aligned}P[A] &= p(0) + p(0.3) + p(0.6) + p(0.9) + p(1.2) + p(1.5) + p(1.8)\\ &= \tfrac{1}{2},\\ P[B] &= p(1.2) + p(1.5) + p(1.8) + p(2.1) + p(2.4) + p(2.7) + p(3.0)\\ &= \tfrac{1}{2},\\ P[AB] &= p(1.2) + p(1.5) + p(1.8) = \tfrac{1}{3},\end{aligned}$$

which agree with the values assumed in example 1A. ◀

The terminology of "density function" and "mass function" comes from the following *physical representation of the probability function* $P[\cdot]$ of a numerical-valued random phenomenon. We imagine that a unit mass of some substance is distributed over the real line in such a way that the amount of mass over any set B of real numbers is equal to $P[B]$. The distribution of substance possesses a density, to be denoted by $f(x)$, at the point x, if for any interval containing the point x of length h (where h is a sufficiently small number) the mass of substance attached to the interval is equal to $hf(x)$. The distribution of substance possesses a mass, to be denoted by $p(x)$, at the point x, if there is a positive amount $p(x)$ of substance concentrated at the point.

We shall see in section 3 that a probability function $P[\cdot]$ always possesses a probability density function and a probability mass function. Consequently, in order for a probability function to be specified by either its probability density function or its probability mass function, it is necessary (and, from a practical point of view, sufficient) that one of these functions vanish identically.

EXERCISES

Verify that each of the functions $f(\cdot)$, given in exercises 2.1–2.5, is a probability density function (by showing that it satisfies (2.2) and (2.3)) and sketch its graph.* *Hint:* use freely the facts developed in the appendix to this section.

2.1. (i)
$$\begin{aligned} f(x) &= 1 && \text{for } 0 < x < 1 \\ &= 0 && \text{elsewhere.} \end{aligned}$$

(ii)
$$\begin{aligned} f(x) &= 1 - |1 - x| && \text{for } 0 < x < 2 \\ &= 0 && \text{elsewhere.} \end{aligned}$$

(iii)
$$\begin{aligned} f(x) &= (\tfrac{1}{2})x^2 && \text{for } 0 < x \le 1 \\ &= (\tfrac{1}{2})(x^2 - 3(x-1)^2) && \text{for } 1 \le x < 2 \\ &= (\tfrac{1}{2})(x^2 - 3(x-1)^2 + 3(x-2)^2) && \text{for } 2 \le x \le 3 \\ &= 0 && \text{elsewhere.} \end{aligned}$$

2.2. (i)
$$\begin{aligned} f(x) &= \frac{1}{2\sqrt{x}} && \text{for } 0 < x < 1 \\ &= 0 && \text{elsewhere.} \end{aligned}$$

(ii)
$$\begin{aligned} f(x) &= 2x && \text{for } 0 < x < 1 \\ &= 0 && \text{elsewhere.} \end{aligned}$$

(iii)
$$\begin{aligned} f(x) &= |x| && \text{for } |x| \le 1 \\ &= 0 && \text{elsewhere.} \end{aligned}$$

* The reader should note the *convention* used in the exercises of this book. When a function $f(\cdot)$ is defined by a single analytic expression for all x in $-\infty < x < \infty$, the fact that x varies between $-\infty$ and ∞ is not explicitly indicated.

2.3. (i) $f(x) = \dfrac{1}{\pi\sqrt{1 - x^2}}$ for $|x| < 1$

$= 0$ elsewhere.

(ii) $f(x) = \dfrac{2}{\pi} \dfrac{1}{\sqrt{1 - x^2}}$ for $0 < x < 1$

$= 0$ elsewhere.

(iii) $f(x) = \dfrac{1}{\pi} \dfrac{1}{1 + x^2}$

(iv) $f(x) = \dfrac{1}{\pi\sqrt{3}}\left(1 + \dfrac{x^2}{3}\right)^{-1}$

2.4. (i) $f(x) = e^{-x},$ $x \geq 0$

$= 0,$ $x < 0$

(ii) $f(x) = (\frac{1}{2})e^{-|x|}$

(iii) $f(x) = \dfrac{e^x}{(1 + e^x)^2}$

(iv) $f(x) = \dfrac{2}{\pi} \dfrac{e^x}{1 + e^{2x}}$

2.5. (i) $f(x) = \dfrac{1}{\sqrt{2\pi}} e^{-\frac{1}{2}x^2}$

(ii) $f(x) = \dfrac{1}{2\sqrt{2\pi}} e^{-\frac{1}{2}\left(\frac{x-2}{2}\right)^2}$

(iii) $f(x) = \dfrac{1}{\sqrt{2\pi x}} e^{-x/2}$ for $x > 0$

$= 0$ elsewhere.

(iv) $f(x) = \frac{1}{4}xe^{-x/2}$ for $x > 0$

$= 0$ elsewhere.

Show that each of the functions $p(\cdot)$ given in exercises 2.6 and 2.7 is a probability mass function [by showing that it satisfies (2.8)], and sketch its graph. *Hint:* use freely the facts developed in the appendix to this section.

2.6. (i) $p(x) = \frac{1}{3}$ for $x = 0$

$= \frac{2}{3}$ for $x = 1$

$= 0$ otherwise.

(ii) $p(x) = \dbinom{6}{x}\left(\dfrac{2}{3}\right)^x\left(\dfrac{1}{3}\right)^{6-x}$ for $x = 0, 1, \cdots, 6$

$= 0$ otherwise.

(iii) $p(x) = \dfrac{2}{3}\left(\dfrac{1}{3}\right)^{x-1}$ for $x = 1, 2, \cdots,$

$= 0$ otherwise.

(iv) $p(x) = e^{-2}\dfrac{2^x}{x!}$ for $x = 0, 1, 2, \cdots$

$= 0$ otherwise.

2.7. (i) $$p(x) = \frac{\binom{8}{x}\binom{4}{6-x}}{\binom{12}{6}} \quad \text{for } x = 0, 1, 2, 3, 4, 5, 6$$
$$= 0 \quad \text{otherwise.}$$

(ii) $$p(x) = \binom{1+x}{x}\left(\frac{2}{3}\right)^2\left(\frac{1}{3}\right)^x \quad \text{for } x = 0, 1, 2, \cdots$$
$$= 0 \quad \text{otherwise.}$$

(iii) $$p(x) = \frac{\binom{-8}{x}\binom{-4}{6-x}}{\binom{-12}{6}} \quad \text{for } x = 0, 1, 2, 3, 4, 5, 6$$
$$= 0 \quad \text{otherwise.}$$

2.8. The amount of bread (in hundreds of pounds) that a certain bakery is able to sell in a day is found to be a numerical-valued random phenomenon, with a probability function specified by the probability density function $f(\cdot)$, given by

$$\begin{aligned} f(x) &= Ax && \text{for } 0 \le x < 5 \\ &= A(10 - x) && \text{for } 5 \le x < 10 \\ &= 0 && \text{otherwise.} \end{aligned}$$

(i) Find the value of A which makes $f(\cdot)$ a probability density function.

(ii) Graph the probability density function.

(iii) What is the probability that the number of pounds of bread that will be sold tomorrow is (*a*) more than 500 pounds, (*b*) less than 500 pounds, (*c*) between 250 and 750 pounds?

(iv) Denote, respectively, by A, B, and C, the events that the number of pounds of bread sold in a day is (*a*) greater than 500 pounds, (*b*) less than 500 pounds, (*c*) between 250 and 750 pounds. Find $P[A \mid B]$, $P[A \mid C]$. Are A and B independent events? Are A and C independent events?

2.9. The length of time (in minutes) that a certain young lady speaks on the telephone is found to be a random phenomenon, with a probability function specified by the probability density function $f(\cdot)$, given by

$$\begin{aligned} f(x) &= Ae^{-x/5} && \text{for } x > 0 \\ &= 0 && \text{otherwise.} \end{aligned}$$

(i) Find the value of A that makes $f(\cdot)$ a probability density function.

(ii) Graph the probability density function.

(iii) What is the probability that the number of minutes that the young lady will talk on the telephone is (*a*) more than 10 minutes, (*b*) less than 5 minutes, (*c*) between 5 and 10 minutes?

(iv) For any real number b, let $A(b)$ denote the event that the young lady talks longer than b minutes. Find $P[A(b)]$. Show that, for $a > 0$ and $b > 0$, $P[A(a + b) \mid A(a)] = P[A(b)]$. In words, the conditional probability that a telephone conversation will last more than $a + b$ minutes, given that it has lasted at least a minutes, is equal to the unconditional probability that it will last more than b minutes.

2.10. The number of newspapers that a certain newsboy is able to sell in a day is found to be a numerical-valued random phenomenon, with a probability function specified by the probability mass function $p(\cdot)$, given by

$$\begin{aligned} p(x) &= Ax && \text{for } x = 1, 2, \cdots, 50 \\ &= A(100 - x) && \text{for } x = 51, 52, \cdots, 100 \\ &= 0 && \text{otherwise.} \end{aligned}$$

(i) Find the value of A that makes $p(\cdot)$ a probability mass function.

(ii) Sketch the probability mass function.

(iii) What is the probability that the number of newspapers that will be sold tomorrow is (*a*) more than 50, (*b*) less than 50, (*c*) equal to 50, (*d*) between 25 and 75, inclusive, (*e*) an odd number?

(iv) Denote, respectively, by A, B, C, and D, the events that the number of newspapers sold in a day is (*a*) greater than 50, (*b*) less than 50, (*c*) equal to 50, (*d*) between 25 and 75, inclusive. Find $P[A \mid B]$, $P[A \mid C]$, $P[A \mid D]$, $P[C \mid D]$. Are A and B independent events? Are A and D independent events? Are C and D independent events?

2.11. The number of times that a certain piece of equipment (say, a light switch) operates before having to be discarded is found to be a random phenomenon, with a probability function specified by the probability mass function $p(\cdot)$, given by

$$\begin{aligned} p(x) &= A(\tfrac{1}{3})^x && \text{for } x = 0, 1, 2, \cdots \\ &= 0 && \text{otherwise.} \end{aligned}$$

(i) Find the value of A which makes $p(\cdot)$ a probability mass function.

(ii) Sketch the probability mass function.

(iii) What is the probability that the number of times the equipment will operate before having to be discarded is (*a*) greater than 5, (*b*) an even number (regard 0 as even), (*c*) an odd number?

(iv) For any real number b, let $A(b)$ denote the event that the number of times the equipment operates is strictly greater than or equal to b. Find $P[A(b)]$. Show that, for any integers $a > 0$ and $b > 0$, $P[A(a + b) \mid A(a)] = P[A(b)]$. Express in words the meaning of this formula.

APPENDIX: THE EVALUATION OF INTEGRALS AND SUMS

If (2.1) and (2.7) are to be useful expressions for evaluating the probability of an event, then techniques must be available for evaluating sums and integrals. The purpose of this appendix is to state some of the notions and formulas with which the student should become familiar and to collect some important formulas that the reader should learn to use, even if he lacks the mathematical background to justify them.

To begin with, let us note the following principle. *If a function is defined by different analytic expressions over various regions, then to evaluate an integral whose integrand is this function one must express the integral as a sum of integrals corresponding to the different regions of definition of the function.* For example, consider the probability density function $f(\cdot)$ defined by

$$
\begin{aligned}
f(x) &= x && \text{for } 0 < x < 1 \\
&= 2 - x && \text{for } 1 < x < 2 \\
&= 0 && \text{elsewhere.}
\end{aligned} \tag{2.10}
$$

To prove that $f(\cdot)$ is a probability density function, we need to verify that (2.2) and (2.3) are satisfied. Clearly, (2.3) holds. Next,

$$
\begin{aligned}
\int_{-\infty}^{\infty} f(x)\, dx &= \int_{0}^{2} f(x)\, dx + \int_{-\infty}^{0} f(x)\, dx + \int_{2}^{\infty} f(x)\, dx \\
&= \int_{0}^{1} f(x)\, dx + \int_{1}^{2} f(x)\, dx + 0 \\
&= \frac{x^2}{2}\Big|_0^1 + \left(2x - \frac{x^2}{2}\right)\Big|_1^2 = \frac{1}{2} + \left(2 - \frac{3}{2}\right) = 1,
\end{aligned}
$$

and (2.2) has been shown to hold. It might be noted that the function $f(\cdot)$ in (2.10) can be written somewhat more concisely in terms of the absolute value notation:

$$
\begin{aligned}
f(x) &= 1 - |1 - x| && \text{for } 0 \le x \le 2 \\
&= 0 && \text{otherwise.}
\end{aligned} \tag{2.11}
$$

Next, in order to check his command of the basic techniques of integration, the reader should verify that the following formulas hold:

$$
\begin{gathered}
\int \frac{e^x}{(1 + e^x)^2}\, dx = \frac{-1}{1 + e^x}, \qquad \int \frac{e^x}{1 + e^{2x}}\, dx = \tan^{-1} e^x = \text{arc tan } e^x, \\
\int e^{-x-e^{-x}}\, dx = \int e^{-e^{-x}} e^{-x}\, dx = e^{-e^{-x}}.
\end{gathered} \tag{2.12}
$$

An important integration formula, obtained by integration by parts, is the following, for any real number t for which the integrals make sense:

$$
\int x^{t-1} e^{-x}\, dx = -x^{t-1} e^{-x} + (t - 1) \int x^{t-2} e^{-x}\, dx. \tag{2.13}
$$

Thus, for $t = 2$ we obtain

$$
\int x e^{-x}\, dx = -x e^{-x} + \int e^{-x}\, dx = -e^{-x}(x + 1). \tag{2.14}
$$

We next consider the Gamma function $\Gamma(\cdot)$, which plays an important role in probability theory. It is defined for every $t > 0$ by

$$\Gamma(t) = \int_0^\infty x^{t-1}e^{-x}\,dx. \tag{2.15}$$

The Gamma function is a generalization of the factorial function in the following sense. From (2.13) it follows that

$$\Gamma(t) = (t-1)\Gamma(t-1). \tag{2.16}$$

Therefore, for any integer r, $0 \leq r < t$,

$$\Gamma(t+1) = t\Gamma(t) = t(t-1)\cdots(t-r)\Gamma(t-r). \tag{2.17}$$

Since, clearly, $\Gamma(1) = 1$, it follows that for any integer $n \geq 0$

$$\Gamma(n+1) = n! \tag{2.18}$$

Next, it may be shown that for any integer $n > 0$

$$\Gamma\left(n + \frac{1}{2}\right) = \frac{1 \cdot 3 \cdot 5 \cdots (2n-1)}{2^n}\sqrt{\pi}, \tag{2.19}$$

which may be written for any even integer n

$$\Gamma\left(\frac{n+1}{2}\right) = \frac{1 \cdot 3 \cdot 5 \cdots (n-1)}{2^{n/2}}\sqrt{\pi}, \tag{2.20}$$

since

$$\Gamma(\tfrac{1}{2}) = \sqrt{\pi}. \tag{2.21}$$

We prove (2.21) by showing that $\Gamma(\frac{1}{2})$ is equal to another integral of whose value we have need. In (2.15), make the change of variable $x = \frac{1}{2}y^2$, and let $t = (n+1)/2$. Then, for any integer, $n = 0, 1, \ldots$, we have the formula

$$\Gamma\left(\frac{n+1}{2}\right) = \frac{1}{2^{(n-1)/2}}\int_0^\infty y^n e^{-\frac{1}{2}y^2}\,dy. \tag{2.22}$$

In view of (2.22), to establish (2.21) we need only show that

$$\Gamma\left(\frac{1}{2}\right) = \sqrt{2}\int_0^\infty e^{-\frac{1}{2}y^2}\,dy = \frac{1}{\sqrt{2}}\int_{-\infty}^\infty e^{-\frac{1}{2}y^2}\,dy = \sqrt{\pi}. \tag{2.23}$$

We prove (2.23) by proving the following *basic formula; for any* $u > 0$

$$\frac{1}{\sqrt{2\pi}}\int_{-\infty}^\infty e^{-\frac{1}{2}uy^2}\,dy = \frac{1}{\sqrt{u}}. \tag{2.24}$$

Equation (2.24) may be derived as follows. Let I be the value of the integral in (2.24). Then I^2 is a product of two single integrals. By the theorem for the evaluation of double integrals, it then follows that

$$I^2 = \frac{1}{2\pi}\int_{-\infty}^{\infty}\int_{-\infty}^{\infty} \exp\left[-\tfrac{1}{2}u(x^2 + y^2)\right] dx\, dy. \tag{2.25}$$

We now evaluate the double integral in (2.25) by means of a change of variables to polar coordinates. Then

$$I^2 = \frac{1}{2\pi}\int_0^{2\pi}\int_0^{\infty} e^{-\frac{1}{2}ur^2} r\, dr\, d\theta = \int_0^{\infty} e^{-\frac{1}{2}ur^2} r\, dr = \frac{1}{u},$$

so that $I = 1/\sqrt{u}$, which proves (2.24).

For large values of t there is an important asymptotic formula for the Gamma function, which is known as *Stirling's formula*. Taking $t = n + 1$, in which n is a positive integer, this formula can be written

$$\begin{aligned} \log n! &= \left(n + \frac{1}{2}\right)\log n - n + \frac{1}{2}\log 2\pi + \frac{r(n)}{12n}, \\ n! &= \left(\frac{n}{e}\right)^n \sqrt{2\pi n}\, e^{r(n)/12n}, \end{aligned} \tag{2.26}$$

in which $r(n)$ satisfies $1 - 1/(12n + 1) < r(n) < 1$. The proof of Stirling's formula may be found in many books. A particularly clear derivation is given by H. Robbins, "A Remark on Stirling's Formula," *American Mathematical Monthly*, Vol. 62 (1955), pp. 26–29.

We next turn to *the evaluation of sums and infinite sums*. The major tool in the evaluation of infinite sums is Taylor's theorem, which states that under certain conditions a function $g(x)$ may be expanded in a power series:

$$g(x) = \sum_{k=0}^{\infty} \frac{x^k}{k!} g^{(k)}(0), \tag{2.27}$$

in which $g^{(k)}(0)$ denotes the value at $x = 0$ of the kth derivative $g^{(k)}(x)$ of $g(x)$. Letting $g(x) = e^x$, we obtain

$$e^x = \sum_{k=0}^{\infty} \frac{x^k}{k!} = 1 + x + \frac{x^2}{2!} + \cdots + \frac{x^n}{n!} + \cdots, \quad -\infty < x < \infty. \tag{2.28}$$

Take next $g(x) = (1 - x)^n$, in which $n = 1, 2, \ldots$. Clearly

$$\begin{aligned} g^{(k)}(x) &= (-1)^k (n)_k (1 - x)^{n-k} \qquad \text{for } k = 0, 1, \cdots, n \\ &= 0 \qquad \text{for } k > n. \end{aligned} \tag{2.29}$$

Consequently, for $n = 1, 2, \ldots$

$$(1 - x)^n = \sum_{k=0}^{n} (-1)^k \binom{n}{k} x^k, \qquad -\infty < x < \infty, \tag{2.30}$$

which is a special case of the binomial theorem. One may deduce the binomial theorem from (2.30) by setting $x = (-b)/a$.

We obtain an important generalization of the binomial theorem by taking $g(x) = (1 - x)^t$, in which t is any real number. For any real number t and any integer $k = 1, 2, \ldots$ define the binomial coefficient

$$\binom{t}{k} = \frac{t(t-1)\cdots(t-k+1)}{k!} \qquad \text{for } k = 1, 2, \cdots \tag{2.31}$$
$$= 1 \qquad \text{for } k = 0.$$

Note that for any positive number n

$$\binom{-n}{k} = (-1)^k \frac{n(n+1)\cdots(n+k-1)}{k!} = (-1)^k \binom{n+k-1}{k}. \tag{2.32}$$

By Taylor's theorem, we obtain the important formula for all real numbers t and $-1 < x < 1$,

$$(1 - x)^t = \sum_{k=0}^{\infty} \binom{t}{k} (-x)^k. \tag{2.33}$$

For the case of n positive we may write, in view of (2.32),

$$(1 - x)^{-n} = \sum_{k=0}^{\infty} \binom{n+k-1}{k} x^k, \qquad |x| < 1. \tag{2.34}$$

Equation (2.34), with $n = 1$, is the familiar formula for the sum of a geometric series:

$$\sum_{k=0}^{\infty} x^k = 1 + x + x^2 + \cdots + x^n + \cdots = \frac{1}{1-x}, \qquad |x| < 1. \tag{2.35}$$

Equation (2.34) with $n = 2$ and 3 yields the formulas

$$\begin{aligned} &\sum_{k=0}^{\infty} (k+1)x^k = 1 + 2x + 3x^2 + \cdots = \frac{1}{(1-x)^2}, \qquad |x| < 1, \\ &\sum_{k=0}^{\infty} (k+2)(k+1)x^k = \frac{2}{(1-x)^3}, \qquad |x| < 1. \end{aligned} \tag{2.36}$$

From (2.33) we may obtain another important formula. By a comparison of the coefficients of x^n on both sides of the equation

$$(1 + x)^s (1 + x)^t = (1 + x)^{s+t},$$

we obtain *for any real numbers s and t and any positive integer n*

$$\binom{s}{0}\binom{t}{n} + \binom{s}{1}\binom{t}{n-1} + \cdots + \binom{s}{n}\binom{t}{0} = \binom{s+t}{n}. \tag{2.37}$$

If s and t are positive integers (2.37) could be verified by mathematical induction. A useful special case of (2.37) is when $s = t = n$; we then obtain (5.13) of Chapter 2.

THEORETICAL EXERCISES

2.1. Show that for any positive real numbers α, β, and t

$$\int_0^\infty x^{\beta-1}e^{-\alpha x}\,dx = \frac{\Gamma(\beta)}{\alpha^\beta}$$

(2.38)

$$\frac{\alpha^\beta}{\Gamma(\beta)}\int_0^\infty e^{-tx}x^{\beta-1}e^{-\alpha x}\,dx = \left(1 + \frac{t}{\alpha}\right)^{-\beta}.$$

2.2. Show for any $\sigma > 0$ and $n = 1, 2, \ldots$

$$2\int_0^\infty y^n e^{-\frac{1}{2}(y/\sigma)^2}\,dy = (2\sigma^2)^{(n+1)/2}\Gamma\left(\frac{n+1}{2}\right). \tag{2.39}$$

2.3. The integral

$$B(m, n) = \int_0^1 x^{m-1}(1 - x)^{n-1}\,dx, \tag{2.40}$$

which converges if m and n are positive, defines a function of m and n, called the *beta function*. Show that the beta function is symmetrical in its arguments, $B(m, n) = B(n, m)$, and may be expressed [letting $x = \sin^2\theta$ and $x = 1/(1 + y)$, respectively] by

$$\begin{aligned} B(m, n) &= 2\int_0^{\pi/2} \sin^{2m-1}\theta\cos^{2n-1}\theta\,d\theta \\ &= \int_0^\infty \frac{y^{n-1}}{(1 + y)^{m+n}}\,dy. \end{aligned} \tag{2.41}$$

Show finally that the beta and gamma functions are connected by the relation

$$B(m, n) = \frac{\Gamma(m)\Gamma(n)}{\Gamma(m + n)}. \tag{2.42}$$

Hint: By changing to polar coordinates, we have

$$\begin{aligned} \Gamma(m)\Gamma(n) &= 4\int_0^\infty\int_0^\infty x^{2m-1}e^{-x^2}y^{2n-1}e^{-y^2}\,dx\,dy \\ &= 4\int_0^{\pi/2} d\theta\cos^{2m-1}\theta\sin^{2n-1}\theta\int_0^\infty dr e^{-r^2}r^{2m+2n-1}. \end{aligned}$$

2.4. Use (2.41) and (2.42), with $m = n = \frac{1}{2}$, to prove (2.23).

2.5. Prove that the integral defining the gamma function converges for any real number $t > 0$.

2.6. Prove that the integral defining the beta function converges for any real numbers m and n, such that $m > 0$ and $n > 0$.

2.7 Taylor's theorem with remainder. Show that if the function $g(\cdot)$ has a continuous nth derivative in some interval containing the origin then for x in this interval

$$g(x) = g(0) + xg'(0) + \frac{x^2}{2!}g''(0) + \cdots + \frac{x^{n-1}}{(n-1)!}g^{(n-1)}(0) + \frac{x^n}{(n-1)!}\int_0^1 dt(1-t)^{n-1}g^{(n)}(xt). \tag{2.43}$$

Hint: Show, for $k = 2, 3, \ldots, n$, that

$$-\frac{x^k}{(k-1)!}\int_0^1 g^{(k)}(xt)(1-t)^{k-1}\,dt + \frac{x^{k-1}}{(k-2)!}\int_0^1 g^{(k-1)}(xt)(1-t)^{k-2}\,dt = \frac{x^{k-1}}{(k-1)!}g^{(k-1)}(0).$$

2.8 Lagrange's form of the remainder in Taylor's theorem. Show that if $g(\cdot)$ has a continuous nth derivative in the closed interval from 0 to x, where x may be positive or negative, then

$$\int_0^1 g^{(n)}(xt)(1-t)^{n-1}\,dt = \frac{1}{n}g^{(n)}(\theta x) \tag{2.44}$$

for some number θ in the interval $0 < \theta < 1$.

3. DISTRIBUTION FUNCTIONS

To describe completely a numerical-valued random phenomenon, one needs only to state its probability function. The probability function $P[\cdot]$ is a function of sets and for this reason is somewhat unwieldy to treat analytically. It would be preferable if there were a function of points (that is, a function of real numbers x), which would suffice to determine completely the probability function. In the case of a probability function, specified by a probability density function or by a probability mass function, the density and mass functions provide a point function that determines the probability function. Now it may be shown that for any numerical-valued random phenomenon whatsoever there exists a point function, called the *distribution function*, which suffices to determine the probability function in the sense that the probability function may be reconstructed from the distribution function. The distribution function thus provides a point function that contains all the information necessary to describe the probability properties of the random phenomenon. Consequently, to study the general properties of numerical valued random phenomena without restricting ourselves to those whose probability functions are specified by either a probability density function or by a probability mass function, it suffices to study the general properties of distribution functions.

The distribution function $F(\cdot)$ of a numerical valued random phenomenon is defined as having as its value, at any real number x, the probability that an observed value of the random phenomenon will be less than or equal to the number x. In symbols, for any real number x,

$$F(x) = P[\{\text{real numbers } x'\colon \quad x' \leq x\}]. \tag{3.1}$$

Before discussing the general properties of distribution functions, let us consider the distribution functions of numerical valued random phenomena, whose probability functions are specified by either a probability mass function or a probability density function. If the probability function is specified by a probability mass function $p(\cdot)$, then the corresponding distribution function $F(\cdot)$ for any real number x is given by

$$F(x) = \sum_{\substack{\text{points } x' \leq x \\ \text{such that } p(x') > 0}} p(x'). \tag{3.2}$$

Equation (3.2) follows immediately from (3.1) and (2.7). If the probability function is specified by a probability density function $f(\cdot)$, then the corresponding distribution function $F(\cdot)$ for any real number x is given by

$$F(x) = \int_{-\infty}^{x} f(x')\,dx'. \tag{3.3}$$

Equation (3.3) follows immediately from (3.1) and (2.1).

We may classify numerical valued random phenomena by classifying their distribution functions. To begin with, consider a random phenomenon whose probability function is specified by its probability mass function, so that its distribution function $F(\cdot)$ is given by (3.2). The graph $y = F(x)$ then appears as it is shown in Fig. 3A; it consists of a sequence of horizontal line segments, each one higher than its predecessor. The points at which one moves from one line to the next are called the *jump points* of the distribution function $F(\cdot)$; they occur at all points x at which the probability mass function $p(x)$ is positive. We define a *discrete distribution function* as one that is given by a formula of the form of (3.2), in terms of a probability mass function $p(\cdot)$, or equivalently as one whose graph (Fig. 3A) consists only of jumps and level stretches. The term "discrete" connotes the fact that the numerical valued random phenomenon corresponding to a discrete distribution function could be assigned, as its sample description space, the set consisting of the (at most countably infinite number of) points at which the graph of the distribution function jumps.

Let us next consider a numerical valued random phenomenon whose probability function is specified by a probability density function, so that

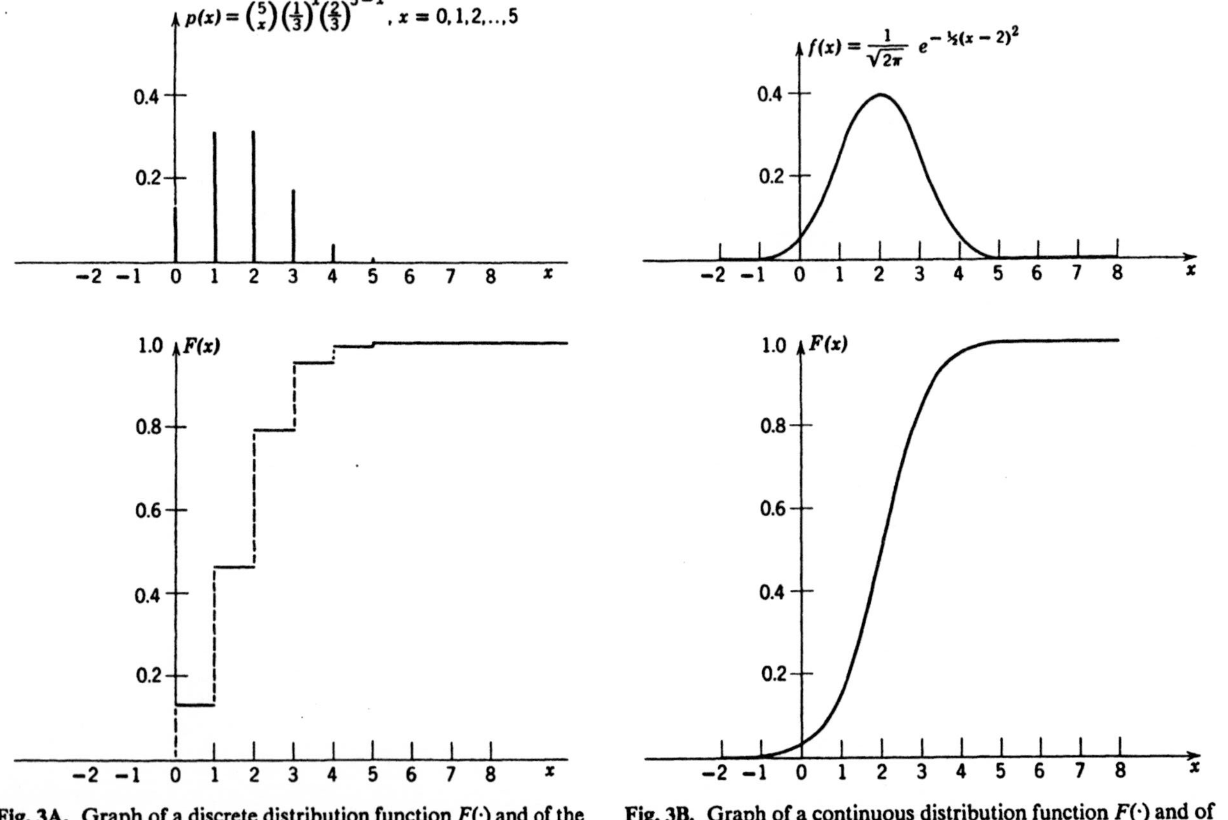

Fig. 3A. Graph of a discrete distribution function $F(\cdot)$ and of the probability mass function $p(\cdot)$ in terms of which $F(\cdot)$ is given by (3.2).

Fig. 3B. Graph of a continuous distribution function $F(\cdot)$ and of the probability density function $f(\cdot)$ in terms of which $F(\cdot)$ is given by (3.3).

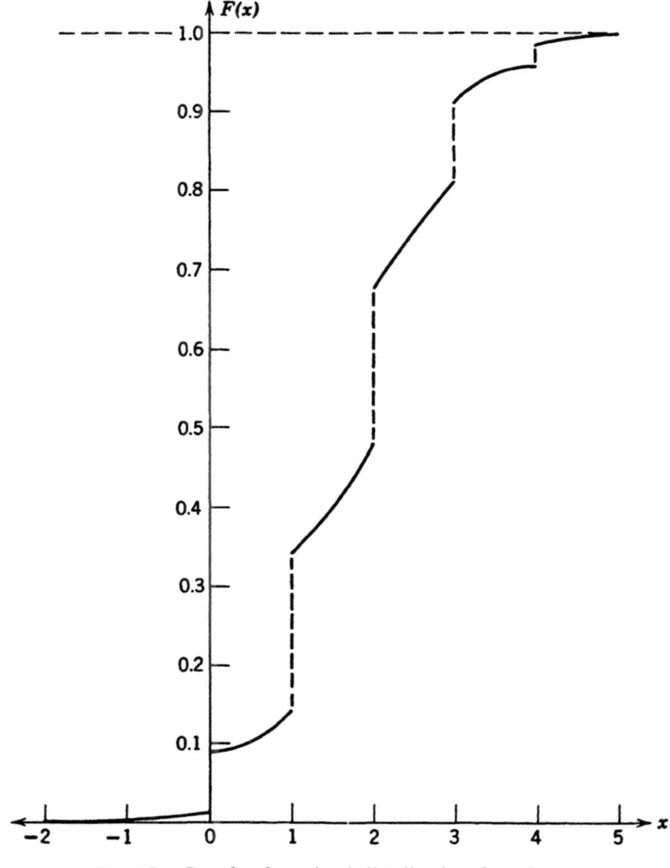

Fig. 3C. Graph of a mixed distribution function.

its distribution function $F(\cdot)$ is given by (3.3). The graph $y = F(x)$ then appears (Fig. 3B) as an unbroken curve. The function $F(\cdot)$ is continuous. However, even more is true; the derivative $F'(x)$ exists at all points (except perhaps for a finite number of points) and is given by

$$F'(x) = \frac{d}{dx} F(x) = f(x). \tag{3.4}$$

We define a *continuous distribution function* as one that is given by a formula of the form of (3.3) in terms of a probability density function.

Most of the distribution functions arising in practice are either discrete or continuous. Nevertheless, it is important to realize that there are distribution functions, such as the one whose graph is shown in Fig. 3C, that are neither discrete nor continuous. Such distribution functions are called *mixed.* A distribution function $F(\cdot)$ is called mixed if it can be written as a linear combination of two distribution functions, denoted by $F^d(\cdot)$ and $F^c(\cdot)$, which are discrete and continuous, respectively, in the following way: for any real number x

$$F(x) = c_1F^d(x) + c_2F^c(x), \tag{3.5}$$

in which c_1 and c_2 are constants between 0 and 1, whose sum is one. The distribution function $F(\cdot)$, graphed in Fig. 3C, is mixed, since $F(x) = \frac{3}{5}F^d(x) + \frac{2}{5}F^c(x)$, in which $F^d(\cdot)$ and $F^c(\cdot)$ are the distribution functions graphed in Fig. 3A and 3B, respectively.

Any numerical valued random phenomenon possesses a probability mass function $p(\cdot)$ defined as follows: for any real number x

$$p(x) = P[\{\text{real numbers } x' : x' = x\}] = P[\{x\}]. \tag{3.6}$$

Thus $p(x)$ represents the probability that the random phenomenon will have an observed value equal to x. In terms of the representation of the probability function as a distribution of a unit mass over the real line, $p(x)$ represents the mass (if any) concentrated at the point x. It may be shown that $p(x)$ represents the size of the jump at x in the graph of the distribution function $F(\cdot)$ of the numerical valued random phenomenon. Consequently, $p(x) = 0$ for all x if and only if $F(\cdot)$ is continuous.

We now introduce the following notation. Given a numerical valued random phenomenon, we write X to denote the observed value of the random phenomenon. For any real numbers a and b we write $P[a \le X \le b]$ to mean the probability that an observed value X of the numerical valued random phenomenon lies in the interval a to b. It is important to keep in mind that $P[a \le X \le b]$ represents an informal notation for $P[\{x: a \le x \le b\}]$.

Some writers on probability theory call a number X determined by the outcome of a random experiment (as is the observed value X of a numerical valued random phenomenon) a *random variable.* In Chapter 7 we give a rigorous definition of the notion of random variable in terms of the notion of function, and show that the observed value X of a numerical valued random phenomenon can be regarded as a random variable. For the present we have the following definition:

A quantity X is said to be a random variable (*or, equivalently, X is said to be an observed value of a numerical valued random phenomenon*) *if for every real number x there exists a probability* (*which we denote by* $P[X \le x]$) *that X is less than or equal to x.*

Given an observed value X of a numerical valued random phenomenon with distribution function $F(\cdot)$ and probability mass function $p(\cdot)$, we have the following formulas for any real numbers a and b (in which $a < b$):

$$
\begin{aligned}
&P[a < X \le b] = P[\{x: a < x \le b\}] = F(b) - F(a) \\
(3.7)\quad &P[a \le X \le b] = P[\{x: a \le x \le b\}] = F(b) - F(a) + p(a) \\
&P[a \le X < b] = P[\{x: a \le x < b\}] = F(b) - F(a) + p(a) - p(b) \\
&P[a < X < b] = P[\{x: a < x < b\}] = F(b) - F(a) - p(b).
\end{aligned}
$$

To prove (3.7), define the events A, B, C, and D:

$$A = \{X \le a\}, \qquad B = \{X \le b\}, \qquad C = \{X = a\}, \qquad D = \{X = b\}.$$

Then (3.7) merely expresses the facts that (since $A \subset B$, $C \subset A$, $D \subset B$)

$$
\begin{aligned}
P[BA^c] &= P[B] - P[A] \\
(3.8)\quad P[BA^c \cup C] &= P[B] - P[A] + P[C] \\
P[BA^cD^c \cup C] &= P[B] - P[A] + P[C] - P[D] \\
P[BA^cD^c] &= P[B] - P[A] - P[D].
\end{aligned}
$$

The use of (3.7) in solving probability problems posed in terms of distribution functions is illustrated in example 3A.

▶ **Example 3A.** Suppose that the duration in minutes of long distance telephone calls made from a certain city is found to be a random phenomenon, with a probability function specified by the distribution function $F(\cdot)$, given by

$$
\begin{aligned}
(3.9)\qquad F(x) &= 0 \qquad \text{for } x < 0 \\
&= 1 - \tfrac{1}{2}e^{-(x/3)} - \tfrac{1}{2}e^{-[x/3]} \qquad \text{for } x \ge 0,
\end{aligned}
$$

in which the expression $[y]$ is defined for any real number $y \ge 0$ as the largest integer less than or equal to y. What is the probability that the duration in minutes of a long distance telephone call is (i) more than six minutes, (ii) less than four minutes, (iii) equal to three minutes? What is the conditional probability that the duration in minutes of a long distance telephone call is (iv) less than nine minutes, given that it is more than five minutes, (v) more than five minutes, given that it is less than nine minutes?

Solution: The distribution function given by (3.9) is neither continuous nor discrete but mixed. Its graph is given in Fig. 3D. For the sake of brevity, we write X for the duration in minutes of a telephone call and $P[X > 6]$ as an abbreviation in mathematical symbols of the verbal statement "the probability that a telephone call has a duration strictly greater than six minutes." The intuitive statement $P[X > 6]$ is identified in our model with $P[\{x': x' > 6\}]$, the value at the set $\{x': x' > 6\}$ of the probability function $P[\cdot]$ corresponding to the distribution function $F(\cdot)$ given by (3.9). Consequently,

$$P[X > 6] = 1 - F(6) = \tfrac{1}{2}e^{-2} + \tfrac{1}{2}e^{-[2]} = e^{-2} = 0.135.$$

Next, the probability that the duration of a call will be less than four minutes (or, more concisely written, $P[X < 4]$) is equal to $F(4) - p(4)$, in which $p(4)$ is the jump in the distribution function $F(\cdot)$ at $x = 4$. A glance at the graph of $F(\cdot)$, drawn in Fig. 3D, reveals that the graph is unbroken at $x = 4$. Consequently, $p(4) = 0$, and

$$P[X < 4] = 1 - \tfrac{1}{2}e^{-(4/3)} - \tfrac{1}{2}e^{-[4/3]} = 1 - \tfrac{1}{2}e^{-(4/3)} - \tfrac{1}{2}e^{-1} = 0.684.$$

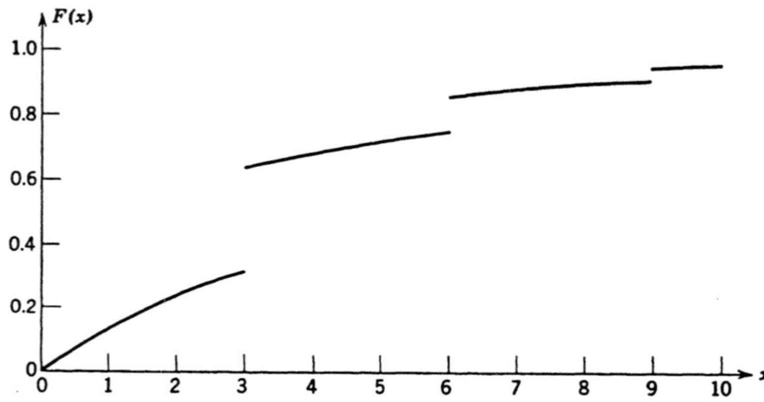

Fig. 3D. Graph of the distribution function given by (3.9).

The probability $P[X = 3]$ that an observed value X of the duration of a call is equal to 3 is given by

$$\begin{aligned} P[X = 3] = p(3) &= (1 - \tfrac{1}{2}e^{-(3/3)} - \tfrac{1}{2}e^{-[3/3]}) - (1 - \tfrac{1}{2}e^{-(3/3)} - \tfrac{1}{2}e^{-[2/3]}) \\ &= \tfrac{1}{2}(1 - e^{-1}) = 0.316, \end{aligned}$$

in which $p(3)$ is the jump in the graph of $F(\cdot)$ at $x = 3$. Solutions to parts (iv) and (v) of the example may be obtained similarly:

$$\begin{aligned} P[X < 9 \mid X > 5] &= \frac{P[5 < X < 9]}{P[X > 5]} = \frac{F(9) - p(9) - F(5)}{1 - F(5)} \\ &= \frac{\tfrac{1}{2}(e^{-(5/3)} + e^{-1} - e^{-3} - e^{-2})}{\tfrac{1}{2}(e^{-(5/3)} + e^{-1})} = \frac{0.187}{0.279} = 0.670, \\ P[X > 5 \mid X < 9] &= \frac{P[5 < X < 9]}{P[X < 9]} = \frac{F(9) - p(9) - F(5)}{F(9) - p(9)} \\ &= \frac{\tfrac{1}{2}(e^{-(5/3)} + e^{-1} - e^{-3} - e^{-2})}{\tfrac{1}{2}(2 - e^{-3} - e^{-2})} = \frac{0.187}{0.908} = 0.206. \end{aligned}$$

◀

In section 2 we gave the conditions a function must satisfy in order to be a probability density function or a probability mass function. The question naturally arises as to the conditions a function must satisfy in order to be a distribution function. In advanced studies of probability theory it is shown that the properties a function $F(\cdot)$ must have in order to be a distribution function are the following: (i) $F(\cdot)$ must be nondecreasing in the sense that for any real numbers a and b

$$F(a) \leq F(b) \qquad \text{if } a < b; \tag{3.10}$$

(ii) the limits of $F(x)$, as x tends to either plus or minus infinity, must exist and be given by

$$\lim_{x \to -\infty} F(x) = 0, \qquad \lim_{x \to \infty} F(x) = 1; \tag{3.11}$$

(iii) at any point x the limit from the right $\lim_{b \to x+} F(b)$, which is defined as the limit of $F(b)$, as b tends to x through values greater than x, must be equal to $F(x)$,

$$\lim_{b \to x+} F(b) = F(x) \tag{3.12}$$

so that at any point x the graph of $F(x)$ is unbroken as one approaches x from the right; (iv) at any point x, the limit from the left, written $F(x-)$ or $\lim_{a \to x-} F(a)$, which is defined as the limit of $F(a)$ as a tends to x through values less than x, must be equal to $F(x) - p(x)$; in symbols,

$$F(x-) = \lim_{a \to x-} F(a) = F(x) - p(x), \tag{3.13}$$

where we define $p(x)$ as the probability that the observed value of the random phenomenon is equal to x. Note that $p(x)$ represents the size of the jump in the graph of $F(x)$ at x.

From these facts it follows that the graph $y = F(x)$ of a typical distribution function $F(\cdot)$ has as its asymptotes the lines $y = 0$ and $y = 1$. The graph is nondécreasing. However, it need not increase at every point but rather may be level (horizontal) over certain intervals. The graph need not be unbroken [that is, $F(\cdot)$ need not be continuous] at all points, but there is at most a countable infinity of points at which the graph has a break; at these points it jumps upward and possesses limits from the right and the left, satisfying (3.12) and (3.13).

The foregoing mathematical properties of the distribution function of a numerical valued random phenomenon serve to characterize completely such functions. It may be shown that for any function possessing the first three properties listed there is a unique set function $P[\cdot]$, defined on the Borel sets of the real line, satisfying axioms 1–3 of section 1 and the condition that for any finite real numbers a and b, at which $a \leq b$,

$$P[\{\text{real numbers } x\colon \; a < x \leq b\}] = F(b) - F(a). \tag{3.14}$$

From this fact it follows that to specify the probability function it suffices to specify the distribution function.

The fact that a distribution function is continuous does not imply that it may be represented in terms of a probability density function by a formula such as (3.3). If this is the case, it is said to be *absolutely continuous.* There also exists another kind of continuous distribution function, called *singular continuous*, whose derivative vanishes at almost all points. This is a somewhat difficult notion to picture, and examples have been constructed only by means of fairly involved analytic operations. From a practical point of view, one may act as if singular distribution functions do not exist, since examples of these functions are rarely, if ever, encountered in practice. It may be shown that any distribution function may be represented in the form

$$F(x) = c_1 F^d(x) + c_2 F^{ac}(x) + c_3 F^{sc}(x), \tag{3.15}$$

in which $F^d(\cdot)$, $F^{ac}(\cdot)$, and $F^{sc}(\cdot)$, respectively, are discrete, absolutely continuous, and singular continuous, and c_1, c_2, and c_3 are constants between 0 and 1, inclusive, the sum of which is 1. If it is assumed that the coefficient c_3 vanishes for any distribution function encountered in practice, it follows that in order to study the properties of a distribution function it suffices to study those that are discrete or continuous.

THEORETICAL EXERCISES

3.1. Show that the probability mass function $p(\cdot)$ of a numerical valued random phenomenon can be positive at no more than a countable infinity of points. *Hint:* For $n = 2, 3, \ldots$, define E_n as the set of points x at which $p(x) > (1/n)$. The size of E_n is less than n, for if it were greater than n it would follow that $P[E_n] > 1$. Thus each of the sets E_n is of finite size. Now the set E of points x at which $p(x) > 0$ is equal to the union $E_2 \cup E_3 \cup \ldots \cup E_n \cup \ldots$, since $p(x) > 0$ if and only if, for some integer n, $p(x) > (1/n)$. The set E, being a union of a countable number of sets of finite size, is therefore proved to have at most a countable infinity of members.

EXERCISES

3.1–3.7. For $k = 1, 2, \ldots, 7$, exercise 3.k is to sketch the distribution function corresponding to each probability density function or probability mass function given in exercise 2.k.

3.8. In the game of "odd man out" (described in section 3 of Chapter 3) the number of trials required to conclude the game, if there are 5 players, is a numerical valued random phenomenon, with a probability function specified by the distribution function $F(\cdot)$, given by

$$\begin{aligned} F(x) &= 0 && \text{for } x < 1 \\ &= 1 - (\tfrac{11}{16})^{[x]} && \text{for } x \geq 1, \end{aligned}$$

in which $[x]$ denotes the largest integer less than or equal to x.

(i) Sketch the distribution function.

(ii) Is the distribution function discrete? If so, give a formula for its probability mass function.

(iii) What is the probability that the number of trials required to conclude the game will be (*a*) more than 3, (*b*) less than 3, (*c*) equal to 3, (*d*) between 2 and 5, inclusive.

(iv) What is the conditional probability that the number of trials required to conclude the game will be (*a*) more than 5, given that it is more than 3 trials, (*b*) more than 3, given that it is more than 5 trials?

3.9. Suppose that the amount of money (in dollars) that a person in a certain social group has saved is found to be a random phenomenon, with a probability function specified by the distribution function $F(\cdot)$, given by

$$\begin{aligned} F(x) &= \tfrac{1}{2}e^{-(x/50)^2} && \text{for } x \leq 0 \\ &= 1 - \tfrac{1}{2}e^{-(x/50)^2} && \text{for } x \geq 0. \end{aligned}$$

Note that a negative amount of savings represents a debt.

(i) Sketch the distribution function.

(ii) Is the distribution function continuous? If so, give a formula for its probability density function.

(iii) What is the probability that the amount of savings possessed by a person in the group will be (*a*) more than 50 dollars, (*b*) less than −50 dollars, (*c*) between −50 dollars and 50 dollars, (*d*) equal to 50 dollars?

(iv) What is the conditional probability that the amount of savings possessed by a person in the group will be (*a*) less than 100 dollars, given that it is more than 50 dollars, (*b*) more than 50 dollars, given that it is less than 100 dollars?

3.10. Suppose that the duration in minutes of long-distance telephone calls made from a certain city is found to be a random phenomenon, with a probability function specified by the distribution function $F(\cdot)$, given by

$$\begin{aligned} F(x) &= 0 && \text{for } x \leq 0 \\ &= 1 - \tfrac{2}{3}e^{-(x/3)} - \tfrac{1}{3}e^{-[x/3]} && \text{for } x > 0. \end{aligned}$$

(i) Sketch the distribution function.

(ii) Is the distribution function continuous? Discrete? Neither?

(iii) What is the probability that the duration in minutes of a long-distance telephone call will be (*a*) more than 6 minutes, (*b*) less than 4 minutes, (*c*) equal to 3 minutes, (*d*) between 4 and 7 minutes?

(iv) What is the conditional probability that the duration of a long-distance telephone call will be (*a*) less than 9 minutes, given that it has lasted more than 5 minutes, (*b*) less than 9 minutes, given that it has lasted more than 15 minutes?

3.11. Suppose that the time in minutes that a man has to wait at a certain subway station for a train is found to be a random phenomenon, with a probability function specified by the distribution function $F(\cdot)$, given by

$$\begin{aligned} F(x) &= 0 && \text{for } x \leq 0 \\ &= \tfrac{1}{2}x && \text{for } 0 \leq x \leq 1 \\ &= \tfrac{1}{2} && \text{for } 1 \leq x \leq 2 \\ &= \tfrac{1}{4}x && \text{for } 2 \leq x \leq 4 \\ &= 1 && \text{for } x \geq 4. \end{aligned}$$

(i) Sketch the distribution function.

(ii) Is the distribution function continuous? If so, give a formula for its probability density function.

(iii) What is the probability that the time the man will have to wait for a train will be (*a*) more than 3 minutes, (*b*) less than 3 minutes, (*c*) between 1 and 3 minutes?

(iv) What is the conditional probability that the time the man will have to wait for a train will be (*a*) more than 3 minutes, given that it is more than 1 minute, (*b*) less than 3 minutes, given that it is more than 1 minute?

3.12. Consider a numerical valued random phenomenon with distribution function

$$\begin{aligned} F(x) &= 0 && \text{for } x \leq 0 \\ &= (\tfrac{1}{4})x && \text{for } 0 < x < 1 \\ &= \tfrac{1}{3} && \text{for } 1 \leq x \leq 2 \\ &= (\tfrac{1}{6})x && \text{for } 2 < x \leq 3 \\ &= \tfrac{1}{2} && \text{for } 3 < x \leq 4 \\ &= (\tfrac{1}{8})x && \text{for } 4 < x \leq 8 \\ &= 1 && \text{for } 8 < x. \end{aligned}$$

What is the conditional probability that the observed value of the random phenomenon will be between 2 and 5, given that it is between 1 and 6, inclusive.

4. PROBABILITY LAWS

The notion of the probability law of a random phenomenon is introduced in this section in order to provide a concise and intuitively meaningful language for describing the probability properties of a random phenomenon.

In order to describe a numerical valued random phenomenon, it is necessary and sufficient to state its probability function $P[\cdot]$; this is equivalent to stating for any Borel set B of real numbers the probability that an observed value of the random phenomenon will be in the Borel set B. However, other functions exist, a knowledge of which is equivalent to a knowledge of the probability function. The distribution function is one such function, for between probability functions and distribution functions there is a one-to-one correspondence. Similarly, between discrete distribution functions and probability mass functions and between continuous distribution functions and probability density functions one-to-one correspondences exist. Thus we have available different, but equivalent, representations of the same mathematical concept, which we may call *the probability law (or sometimes the probability distribution) of the numerical valued random phenomenon.*

A probability law is called *discrete* if it corresponds to a discrete distribution function and *continuous* if it corresponds to a continuous distribution function.

For example, suppose one is considering the numerical valued random phenomenon that consists in observing the number of hits in five independent tosses of a dart at a target, where the probability at each toss of hitting the target is some constant p. To describe the phenomenon, one needs to know, by definition, the probability function $P[\cdot]$, which for any set E of real numbers is given by

$$P[E] = \sum_{k \text{ in } E\{0, 1, \ldots, 5\}} \binom{5}{k} p^k q^{5-k}. \tag{4.1}$$

It should be recalled that $E\{0, 1, \ldots, 5\}$ represents the intersection of the sets E and $\{0, 1, \ldots, 5\}$.

Equivalently, one may describe the phenomenon by stating its distribution function $F(\cdot)$; this is done by giving the value of $F(x)$ at any real number x,

$$F(x) = \sum_{k=0}^{[x]} \binom{5}{k} p^k q^{5-k}. \tag{4.2}$$

It should be recalled that $[x]$ denotes the largest integer less than or equal to x.

Equivalently, since the distribution function is discrete, one may describe the phenomenon by stating its probability mass function $p(\cdot)$, given by

$$\begin{aligned} p(x) &= \binom{5}{x} p^x q^{5-x} \qquad \text{for } x = 0, 1, \ldots, 5 \\ &= 0 \qquad \text{otherwise.} \end{aligned} \tag{4.3}$$

Equations (4.1), (4.2), and (4.3) constitute equivalent representations, or statements, of the same concept, which we call the probability law of the random phenomenon. This particular probability law is discrete.

We next note that *probability laws may be classified into families on the basis of similar functional form.* For example, consider the function $b(\cdot\,; n, p)$ defined for any $n = 1, 2, \ldots$ and $0 \le p \le 1$ by

$$b(x; n, p) = \binom{n}{x} p^x q^{n-x} \qquad \text{for } x = 0, 1, \cdots, n$$
$$= 0 \qquad \text{otherwise.}$$

For fixed values of n and p the function $b(\cdot\,; n, p)$ is a probability mass function and thus defines a probability law. The probability laws determined by $b(\cdot\,; n_1, p_1)$ and $b(\cdot\,; n_2, p_2)$ for two different sets of values n_1, p_1 and n_2, p_2 are different. Nevertheless, the common functional form of the two functions $b(\cdot\,; n_1, p_1)$ and $b(\cdot\,; n_2, p_2)$ enables us to treat simultaneously the two probability laws that they determine. We call n and p *parameters*, and $b(\cdot\,; n, p)$ the probability mass function of the binomial probability law with parameters n and p.

We next list some frequently occurring discrete probability laws, to be followed by a list of some frequently occurring continuous probability laws.

The *Bernoulli* probability law with parameter p, where $0 \le p \le 1$, is specified by the probability mass function

$$p(x) = p \qquad \text{if } x = 1$$
$$= 1 - p = q \qquad \text{if } x = 0$$
$$= 0 \qquad \text{otherwise.} \tag{4.4}$$

An example of a numerical valued random phenomena obeying the Bernoulli probability law with parameter p is the outcome of a Bernoulli trial in which the probability of success is p, if instead of denoting success and failure by s and f, we denote them by 1 and 0, respectively.

The *binomial* probability law with parameters n and p, where $n = 1, 2, \ldots$, and $0 \le p \le 1$, is specified by the probability mass function

$$p(x) = \binom{n}{x} p^x (1 - p)^{n-x} \qquad \text{for } x = 0, 1, \ldots, n$$
$$= 0 \qquad \text{otherwise.} \tag{4.5}$$

An important example of a numerical valued random phenomenon obeying the binomial probability law with parameters n and p is the number of successes in n independent repeated Bernoulli trials in which the probability of success at each trial is p.

The *Poisson* probability law with parameter λ, where $\lambda > 0$, is specified by the probability mass function

$$p(x) = e^{-\lambda} \frac{\lambda^x}{x!} \qquad \text{for } x = 0, 1, 2, \ldots$$
$$= 0 \qquad \text{otherwise.} \tag{4.6}$$

In section 3 of Chapter 3 it was seen that the Poisson probability law provides under certain conditions an approximation to the binomial probability law. In section 3 of Chapter 6 we discuss random phenomena that obey the Poisson probability law.

The *geometric* probability law with parameter p, where $0 \leq p \leq 1$, is specified by the probability mass function

$$(4.7) \qquad \begin{aligned} p(x) &= p(1-p)^{x-1} \qquad \text{for } x = 1, 2, \ldots \\ &= 0 \qquad \text{otherwise.} \end{aligned}$$

An important example of a numerical valued random phenomenon obeying the geometric probability law with parameter p is the number of trials required to obtain the first success in a sequence of independent repeated Bernoulli trials in which the probability of success at each trial is p.

The *hypergeometric* probability law with parameters N, n, and p (where N may be any integer $1, 2, \ldots, n$ is an integer in the set $1, 2, \ldots, N$ and $p = 0, 1/N, 2/N, \ldots, 1$) is specified by the probability mass function, letting $q = 1 - p$,

$$(4.8) \qquad \begin{aligned} p(x) &= \frac{\binom{Np}{x}\binom{Nq}{n-x}}{\binom{N}{n}} \qquad \text{for } x = 0, 1, \ldots, n \\ &= 0 \qquad \text{otherwise.} \end{aligned}$$

The hypergeometric probability law may also be defined by using (2.31), for any value of p in the interval $0 \leq p \leq 1$. An example of a random phenomenon obeying the hypergeometric probability law is given by the number of white balls contained in a sample of size n drawn without replacement from an urn containing N balls, of which Np are white.

The *negative binomial* probability law with parameters r and p, where $r = 1, 2, \ldots$ and $0 \leq p \leq 1$, is specified by the probability mass function, letting $q = 1 - p$,

$$(4.9) \qquad \begin{aligned} p(x) &= \binom{r+x-1}{x} p^r q^x = \binom{-r}{x} p^r (-q)^x \qquad \text{for } x = 0, 1, \ldots \\ &= 0 \qquad \text{otherwise.} \end{aligned}$$

An example of a random phenomenon obeying the negative binomial probability law with parameters r and p is the number of failures encountered in a sequence of independent repeated Bernoulli trials (with probability p of success at each trial) before the rth success. Note that the number of trials required to achieve the rth success is equal to r plus the number of failures encountered before the rth success is met.

Some important continuous probability laws are the following.

The *uniform* probability law over the interval a to b, where a and b are any finite real numbers such that $a < b$, is specified by the probability density function

$$(4.10) \qquad f(x) = \frac{1}{b-a} \qquad \text{for } a < x < b$$
$$= 0 \qquad \text{otherwise.}$$

Examples of random phenomena obeying a uniform probability law are discussed in section 5.

The *normal* probability law with parameters m and σ, where $-\infty < m < \infty$ and $\sigma > 0$, is specified by the probability density function

$$(4.11) \qquad f(x) = \frac{1}{\sigma\sqrt{2\pi}} e^{-\frac{1}{2}\left(\frac{x-m}{\sigma}\right)^2}, \qquad -\infty < x < \infty.$$

The role played by the normal probability law in probability theory is discussed in Chapter 6. In section 6 we introduce certain functions that are helpful in the study of the normal probability law.

The *exponential* probability law with parameter λ, in which $\lambda > 0$, is specified by the probability density function

$$(4.12) \qquad f(x) = \lambda e^{-\lambda x} \qquad \text{for } x > 0$$
$$= 0 \qquad \text{otherwise.}$$

The *gamma* probability law with parameters r and λ, in which $r = 1, 2, \ldots$ and $\lambda > 0$, is specified by the probability density function

$$(4.13) \qquad f(x) = \frac{\lambda}{(r-1)!} (\lambda x)^{r-1} e^{-\lambda x} \qquad \text{for } x \geq 0$$
$$= 0 \qquad \text{otherwise.}$$

The exponential and gamma probability laws are discussed in Chapter 6.

The *Cauchy* probability law with parameters α and β; in which $-\infty < \alpha < \infty$ and $\beta > 0$, is specified by the probability density function

$$(4.14) \qquad f(x) = \frac{1}{\pi\beta\left\{1 + \left(\frac{x-\alpha}{\beta}\right)^2\right\}}, \qquad -\infty < x < \infty.$$

Student's distribution with parameter $n = 1, 2, \ldots$ (also called Student's t-distribution with n degrees of freedom) is specified by the probability density function

$$(4.15) \qquad f(x) = \frac{1}{\sqrt{n\pi}} \frac{\Gamma[(n+1)/2]}{\Gamma(n/2)} \left(1 + \frac{x^2}{n}\right)^{-(n+1)/2}$$

It should be noted that Student's distribution with parameter $n = 1$ coincides with the Cauchy probability law with parameters $\alpha = 0$ and $\beta = 1$.

The χ^2 distribution with parameters $n = 1, 2, \ldots$ and $\sigma > 0$ is specified by the probability density function

$$(4.16) \qquad f(x) = \frac{1}{2^{n/2}\sigma^n\Gamma(n/2)} x^{(n/2)-1}e^{-(x/2\sigma^2)} \qquad \text{for } x > 0$$
$$= 0 \qquad \text{for } x < 0$$

The symbol χ is the Greek letter chi, and one sometimes writes chi-square for χ^2. The χ^2 distribution with parameters n and $\sigma = 1$ is called in statistics the χ^2 distribution with n degrees of freedom. The χ^2 distribution with parameters n and σ coincides with the gamma distribution with parameters $r = n/2$ and $\lambda = 1/(2\sigma^2)$ [to define the gamma probability law for non-integer r, replace $(r - 1)!$ in (4.13) by $\Gamma(r)$].

The χ distribution with parameters $n = 1, 2, \ldots$ and $\sigma > 0$ is specified by the probability density function

$$(4.17) \qquad f(x) = \frac{2(n/2)^{n/2}}{\sigma^n\Gamma(n/2)} x^{n-1}e^{-(n/2\sigma^2)x^2} \qquad \text{for } x > 0$$
$$= 0 \qquad \text{for } x < 0.$$

The χ distribution with parameters n and $\sigma = 1$ is often called the chi distribution with n degrees of freedom. (The relation between the χ^2 and χ distributions is given in exercise 8.1 of Chapter 7).

The *Rayleigh* distribution with parameter $\alpha > 0$ is specified by the probability density function

$$(4.18) \qquad f(x) = \frac{1}{\alpha^2} xe^{-\frac{1}{2}(x/\alpha)^2} \qquad \text{for } x > 0$$
$$= 0 \qquad \text{for } x < 0.$$

The Rayleigh distribution coincides with the χ distribution with parameters $n = 2$ and $\sigma = \alpha\sqrt{2}$.

The *Maxwell* distribution with parameter $\alpha > 0$ is specified by the probability density function

$$(4.19) \qquad f(x) = \frac{4}{\sqrt{\pi}} \frac{1}{\alpha^3} x^2e^{-x^2/\alpha^2} \qquad \text{for } x > 0$$
$$= 0 \qquad \text{for } x < 0$$

The Maxwell distribution with parameter α coincides with the χ distribution with parameter $n = 3$ and $\sigma = \alpha\sqrt{3/2}$.

The F distribution with parameters $m = 1, 2, \ldots$ and $n = 1, 2, \ldots$ is specified by the probability density function

$$(4.20) \qquad f(x) = \frac{\Gamma[(m+n)/2]}{\Gamma(m/2)\Gamma(n/2)} (m/n)^{m/2} \frac{x^{(m/2)-1}}{[1 + (m/n)x]^{(m+n)/2}} \qquad \text{for } x > 0$$
$$= 0 \qquad \text{for } x < 0.$$

The *beta* probability law with parameters a and b, in which a and b are positive real numbers, is specified by the probability density function

$$(4.21) \qquad f(x) = \frac{1}{B(a, b)} x^{a-1}(1 - x)^{b-1} \qquad 0 < x < 1$$
$$= 0 \qquad \text{elsewhere.}$$

THEORETICAL EXERCISES

4.1. The probability law of the number of white balls in a sample drawn without replacement from an urn of random composition. Consider an urn containing N balls. Suppose that the number of white balls in the urn is a numerical valued random phenomenon obeying (i) a binomial probability law with parameters N and p, (ii) a hypergeometric probability law with parameters M, N, and p. [For example, suppose that the balls in the urn constitute a sample of size N drawn with replacement (without replacement) from a box containing M balls, of which a proportion p is white.] Let a sample of size n be drawn without replacement from the urn. Show that the number of white balls in the sample obeys either a binomial probability law with parameters n and p, or a hypergeometric probability law with parameters M, n, and p, depending on whether the number of white balls in the urn obeys a binomial or a hypergeometric probability law.

Hint: Establish the conditions under which the following statements are valid:

$$\binom{N}{m} = \binom{N-k}{m-k} \frac{(N)_k}{(m)_k};$$

$$\frac{\binom{m}{k}\binom{N-m}{n-k}}{\binom{N}{n}} = \frac{\binom{n}{k}\binom{N-n}{m-k}}{\binom{N}{m}};$$

$$\sum_{m=0}^{N} \frac{\binom{m}{k}\binom{N-m}{n-k}}{\binom{N}{n}} p(m) = \sum_{m=k}^{N-n+k} \frac{\binom{n}{k}\binom{N-n}{m-k}}{\binom{N}{m}} p(m)$$

where

$$p(m) = \binom{N}{m} p^m q^{N-m},$$

$$p(m) = \frac{\binom{Mp}{m}\binom{Mq}{N-m}}{\binom{M}{N}} = \frac{\binom{N}{m}\binom{M-N}{Mp-m}}{\binom{M}{Mp}}.$$

Finally, use the fact that

$$\frac{\binom{Mp}{k}\binom{Mq}{n-k}}{\binom{M}{n}} = \frac{\binom{n}{k}\binom{M-n}{Mp-k}}{\binom{M}{Mp}}.$$

EXERCISES

4.1. Give formulas for, and identify, the probability law of each of the following numerical valued random phenomena:

(i) The number of defectives in a sample of size 20, chosen without replacement from a batch of 200 articles, of which 5% are defective.

(ii) The number of baby boys in a series of 30 independent births, assuming the probability at each birth that a boy will be born is 0.51.

(iii) The minimum number of babies a woman must have in order to give birth to a boy (ignore multiple births, assume independence, and assume the probability at each birth that a boy will be born is 0.51).

(iv) The number of patients in a group of 35 having a certain disease who will recover if the long-run frequency of recovery from this disease is 75% (assume that each patient has an independent chance to recover).

In exercises 4.2–4.9 consider an urn containing 12 balls, numbered 1 to 12. Further, the balls numbered 1 to 8 are white, and the remaining balls are red. Give a formula for the probability law of the numerical valued random phenomenon described.

4.2. The number of white balls in a sample of size 6 drawn from the urn without replacement.

4.3. The number of white balls in a sample of size 6 drawn from the urn with replacement.

4.4. The smallest number occurring on the balls in a sample of size 6, drawn from the urn without replacement (see theoretical exercise 5.1 of Chapter 2.)

4.5. The second smallest number occurring in a sample of size 6, drawn from the urn without replacement.

4.6. The minimum number of balls that must be drawn, when sampling without replacement, to obtain a white ball.

4.7. The minimum number of balls that must be drawn, when sampling with replacement, to obtain a white ball.

4.8. The minimum number of balls that must be drawn, when sampling without replacement, to obtain 2 white balls.

4.9. The minimum number of balls that must be drawn, when sampling with replacement, to obtain 2 white balls.

5. THE UNIFORM PROBABILITY LAW

The notion of the uniform probability law (or uniform distribution) over the interval a to b, in which a and b are finite real numbers, is best defined in the following manner. Consider a numerical valued random phenomenon whose values can lie only in a certain finite interval S; that is $S = \{\text{real numbers } x: a \leq x \leq b\}$ for some finite numbers a and b. The random phenomenon is said to obey a uniform probability law over the finite interval S if the value $P[B]$ of its probability function, at any interval B, satisfies the relation

$$(5.1)\qquad \begin{aligned} P[B] &= \frac{\text{length of } B}{\text{length of } S} \qquad \text{if } B \text{ is a subset of } S \\ &= 0 \qquad \text{if } B \text{ and } S \text{ have no points in common.} \end{aligned}$$

It should be noted that knowing $P[B]$ at intervals suffices to determine it on any Borel set B of real numbers.

From (5.1) one sees that *the notion of a uniform distribution represents an extension of the notion of a finite sample description space S with equally likely descriptions*, since in this case the probability $P[A]$ of any event A on S is given by the formula

$$(5.2)\qquad P[A] = \frac{\text{size of } A}{\text{size of } S}.$$

There are many random phenomena for which it appears plausible to assume a uniform probability law. For example, suppose one is tossing a dart at a line marked 0 to 1. If one is always sure to land on the line and if one feels that *any two intervals on the line of equal length have an equal chance of being hit*, then one is led to conclude that the place at which the dart hits the line has a probability function satisfying (5.1), with S denoting the interval 0 to 1.

The distribution function $F(\cdot)$ of a random phenomenon, which obeys a uniform probability law over the interval a to b is obtained from (5.1):

$$(5.3)\qquad \begin{aligned} F(x) &= 0 \qquad && \text{if } x \leq a \\ &= \frac{x-a}{b-a} && \text{if } a \leq x \leq b \\ &= 1 && \text{if } x \geq b. \end{aligned}$$

By differentiation, the probability density function may be obtained:

$$(5.4) \qquad f(x) = \frac{1}{b-a} \quad \text{if } a < x < b$$
$$= 0 \quad \text{otherwise.}$$

From (5.4) it follows that the definition of a uniform probability law given by (5.1) coincides with the definition given by (4.10). (See Fig. 5A.)

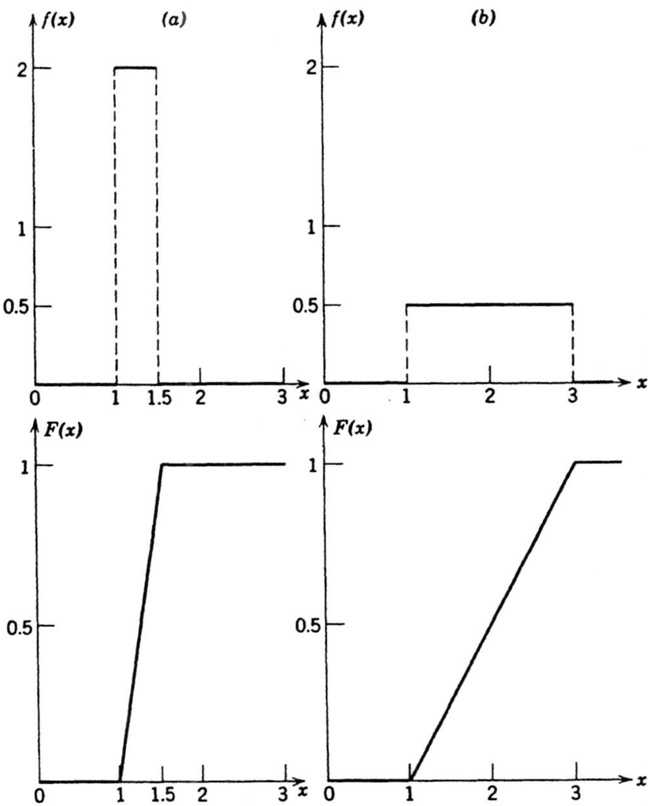

Fig. 5A. Probability density function and distribution function of the uniform probability law over (*a*) the interval 1 to 1.5, (*b*) the interval 1 to 3.

◀ **Example 5A. Waiting time for a train.** Between 7 A.M. and 8 A.M. trains leave a certain station at 3, 5, 8, 10, 13, 15, 18, 20, . . . minutes past the hour. What is the probability that a person arriving at the station will have to wait less than a minute for a train, assuming that the person's time of arrival at the station obeys a uniform probability law over the interval of time (i) 7 A.M. to 8 A.M., (ii) 7:15 A.M. to 7:30 A.M., (iii) 7:02 A.M. to 7:15 A.M., (iv) 7:03 A.M. to 7:15 A.M., (v) 7:04 A.M. to 7:15 A.M.?

Solution: We must first find the set B of real numbers in which the person's arrival must lie in order for his waiting time to be less than 1 minute. One sees that B is the set of real numbers consisting of the intervals 2 to 3, 4 to 5, 7 to 8, 9 to 10, and so on. (See Fig. 5B.) The probability that the person will wait less than a minute for a train is given by $P[B]$, which is equal to, in the various cases, (i) $\frac{24}{60} = \frac{2}{5}$, (ii) $\frac{6}{15} = \frac{2}{5}$, (iii) $\frac{6}{13}$, (iv) $\frac{5}{12}$, (v) $\frac{5}{11}$. ◀

Fig. 5B. In order that the person discussed in example 5A will have to wait less than one minute for a train, his time of arrival X at the station (measured in minutes after 7 A.M.) must lie in the set B, consisting of the shaded intervals.

▶ **Example 5B. The probability law of the second digit in the decimal expansion of the square root of a randomly chosen number.** A number is chosen from the interval 0 to 1 by a random mechanism that obeys a uniform probability law over the interval. What is the probability that the second decimal place of the square root of the number will be the digit 3? Is the digit k for $k = 0, 1, \ldots, 9$?

Solution: For $k = 0, 1, \ldots, 9$ let B_k be the set of numbers on the unit interval whose square roots have a second decimal equal to the digit k. A number x belongs to B_k if and only if $\sqrt{x}$ satisfies for some $m = 0, 1, \ldots, 9$

$$m + \frac{k}{10} \leq 10\sqrt{x} < m + \frac{k+1}{10}$$

or

$$\frac{1}{100}\left(m + \frac{k}{10}\right)^2 \leq x < \frac{1}{100}\left(m + \frac{k+1}{10}\right)^2. \tag{5.5}$$

The length of the interval described by (5.5) is

$$\frac{1}{100}\left(m + \frac{k+1}{10}\right)^2 - \frac{1}{100}\left(m + \frac{k}{10}\right)^2 = \frac{1}{10{,}000}(20m + 2k + 1).$$

Hence the probability of the set B_k is given by

$$P[B_k] = \frac{1}{10,000} \sum_{m=0}^{9} (20m + 2k + 1) = 0.091 + 0.002k.$$

In particular, $P[B_3] = 0.097$. ◀

EXERCISES

5.1. The time, measured in minutes, required by a certain man to travel from his home to a train station is a random phenomenon obeying a uniform probability law over the interval 20 to 25. If he leaves his home promptly at 7:05 A.M., what is the probability that he will catch a train that leaves the station promptly at 7:28 A.M.?

5.2. A radio station broadcasts the correct time every hour on the hour between the hours of 6 A.M. and 12 midnight. What is the probability that a listener will have to wait less than 10 minutes to hear the correct time if the time at which he tunes in is distributed uniformly over (chosen randomly from) the interval (i) 6 A.M. to 12 midnight, (ii) 8 A.M. to 6 P.M., (iii) 7:30 A.M. to 5:30 P.M., (iv) 7:30 A.M. to 5 P.M?

5.3. The circumference of a wheel is divided into 37 arcs of equal length, which are numbered 0 to 36 (this is the principle of construction of a roulette wheel). The wheel is twirled. After the wheel comes to rest, the point on the wheel located opposite a certain fixed marker is noted. Assume that the point thus chosen obeys a uniform probability law over the circumference of the wheel. What is the probability that the point thus chosen will lie in an arc (i) with a number 1 to 10, inclusive, (ii) with an odd number, (iii) numbered 0?

5.4. A parachutist lands on the line connecting 2 towns, A and B. Suppose that the point at which he lands obeys a uniform probability law over the line. What is the probability that the ratio of his distance from A to his distance from B will be (i) greater than 3, (ii) equal to 3, (iii) greater than R, where R is a given real number?

5.5. An angle θ is chosen from the interval $-\pi/2$ to $\pi/2$ by a random mechanism that obeys a uniform probability law over the interval. A line is then drawn on an (x, y)-plane through the point $(0, 1)$ at the angle θ with the y-axis. What is the probability, for any positive number z, that the x-coordinate of the point at which the line intersects the x-axis will be less than z?

5.6. A number is chosen from the interval 0 to 1 by a random mechanism that obeys a uniform probability law over the interval. What is the probability that (i) its first decimal will be a 3, (ii) its second decimal will be a 3, (iii) its first 2 decimals will be 3's, (iv) any specified decimal will be a 3, (v) any 2 specified decimals will be 3's?

5.7. A number is chosen from the interval 0 to 1 by a random mechanism that obeys a uniform probability law over the interval. What is the probability that (i) the first decimal of its square root will be a 3, (ii) the negative of its logarithm (to the base e) will be less than 3?

6. THE NORMAL DISTRIBUTION AND DENSITY FUNCTIONS

A fundamental role in probability theory is played by the functions $\phi(\cdot)$ and $\Phi(\cdot)$, defined as follows: for any real number x

$$\phi(x) = \frac{1}{\sqrt{2\pi}} e^{-\frac{1}{2}x^2} \tag{6.1}$$

$$\Phi(x) = \int_{-\infty}^{x} \phi(y)\, dy = \frac{1}{\sqrt{2\pi}} \int_{-\infty}^{x} e^{-\frac{1}{2}y^2}\, dy. \tag{6.2}$$

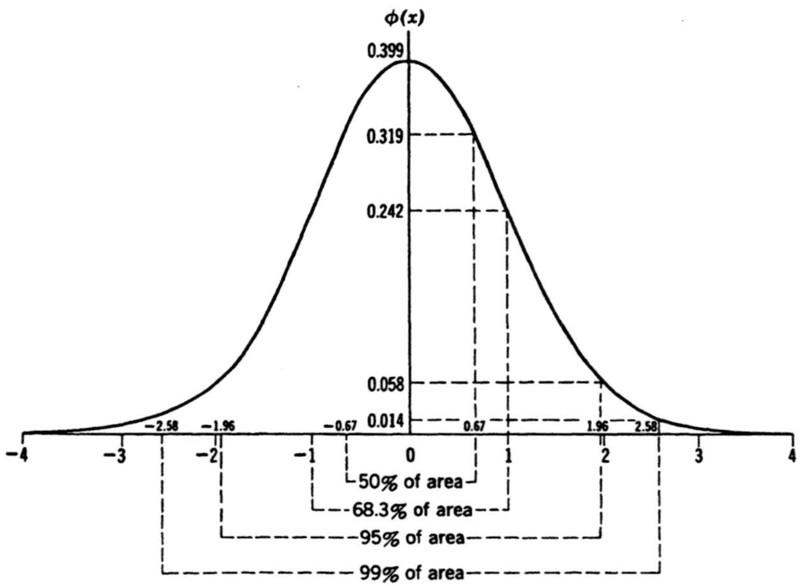

Fig. 6A. Graph of the normal density function $\phi(x)$.

Because of their close relation to normal probability laws $\phi(\cdot)$ is called the *normal density function* and $\Phi(\cdot)$, the *normal distribution function*. These functions are graphed in Figs. 6A and 6B, respectively. The graph of $\phi(\cdot)$ is a symmetric bell-shaped curve. The graph of $\Phi(\cdot)$ is an *S*-shaped curve. It suffices to know these functions for positive x, in order to know them for all x, in view of the relations (see theoretical exercise 6.3)

$$\phi(-x) = \phi(x) \tag{6.3}$$

$$\Phi(-x) = 1 - \Phi(x). \tag{6.4}$$

A table of $\Phi(x)$ for positive values of x is given in Table I (see p. 441).

The function $\phi(x)$ is positive for all x. Further, from (2.24)

$$\int_{-\infty}^{\infty} \phi(x)\, dx = 1, \tag{6.5}$$

so that $\phi(\cdot)$ is a probability density function.

The importance of the function $\Phi(\cdot)$ arises from the fact that probabilities concerning random phenomena obeying a normal probability law with parameters m and σ are easily computed, since they may be expressed in terms of the tabulated function $\Phi(\cdot)$. More precisely, consider a random

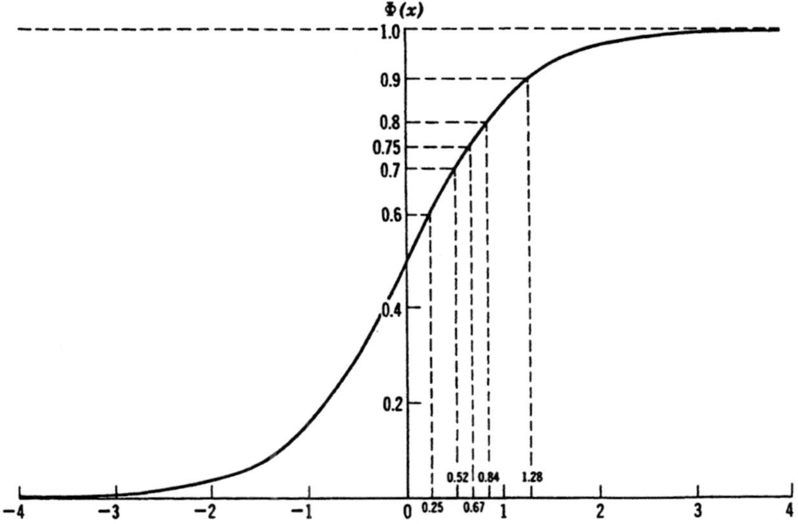

Fig. 6B. Graph of the normal distribution function $\Phi(x)$.

phenomenon whose probability law is specified by the probability density function $f(\cdot)$, given by (4.11). The corresponding distribution function is given by

$$\begin{aligned} F(x) &= \frac{1}{\sqrt{2\pi}\sigma} \int_{-\infty}^{x} e^{-\frac{1}{2}\left(\frac{y-m}{\sigma}\right)^2} dy \\ &= \frac{1}{\sqrt{2\pi}} \int_{-\infty}^{(x-m)/\sigma} e^{-\frac{1}{2}y^2}\, dy = \Phi\left(\frac{x-m}{\sigma}\right) \end{aligned} \tag{6.6}$$

Consequently, if X is an observed value of a numerical valued random phenomenon obeying a normal probability law with parameters m and σ, *then for any real numbers a and b (finite or infinite, in which $a < b$),*

$$(6.7)\quad P[a \le X \le b] = F(b) - F(a) = \Phi\left(\frac{b-m}{\sigma}\right) - \Phi\left(\frac{a-m}{\sigma}\right)$$

▶ **Example 6A. "Grading on the curve."** The properties of the normal distribution function provide the basis for the system of "grading on the curve" used in assigning final grades in large courses in American universities. Under this system, the letters A, B, C, D are used as passing grades. Of the students with passing grades, 15% receive A, 35% receive B, 35% receive C, and 15% receive D. The system is based on the assumption that the score X, which each student obtains on the examinations in the course, is an observed value of a numerical valued random phenomenon obeying a normal probability law with parameters m and σ (which the instructor can estimate, given the scores of many students). From (6.7) it follows that

$$(6.8)\quad \begin{aligned} P[0 \le X - m \le \sigma] &= P[-\sigma \le X - m \le 0] \\ &= \Phi(1) - \Phi(0) = 0.3413 \\ P[\sigma \le X - m] &= P[X - m \le -\sigma] \\ &= 1 - \Phi(1) = 0.1587 \end{aligned}$$

Therefore, if one assigns the letter A to a student whose score X is greater than $m + \sigma$, one would expect 0.1587 (approximately 15%) of the students to receive a grade A. Similarly, 0.3413 (approximately 35%) of the students receive a grade of B if B is assigned to a student with a score X between m and $m + \sigma$; approximately 35% receive C if C is assigned to a student with a score between $m - \sigma$ and m; and approximately 15% receive D if D is assigned to a student with a score less than $m - \sigma$. ◀

The following example illustrates the use of (6.7) in solving problems involving random phenomena obeying normal probability laws.

▶ **Example 6B.** Consider a random phenomenon obeying the normal probability law with parameters $m = 2$ and $\sigma = 2$. The probability that an observed value X of the random phenomenon will have a value between 0 and 3 is given by

$$\begin{aligned} P[0 \le X \le 3] &= \frac{1}{2\sqrt{2\pi}} \int_0^3 e^{-\frac{1}{2}\left(\frac{x-2}{2}\right)^2} dx = \Phi\left(\frac{3-2}{2}\right) - \Phi\left(\frac{0-2}{2}\right) \\ &= \Phi\left(\frac{1}{2}\right) - \Phi(-1) = \Phi\left(\frac{1}{2}\right) + \Phi(1) - 1 = 0.533; \end{aligned}$$

the probability that an observed value X of the random phenomenon will have a value between -1 and 1 is given by

$$P[|X| \leq 1] = \Phi\left(\frac{1-2}{2}\right) - \Phi\left(\frac{-1-2}{2}\right) = \Phi\left(-\frac{1}{2}\right) - \Phi\left(-\frac{3}{2}\right)$$

$$= \Phi\left(\frac{3}{2}\right) - \Phi\left(\frac{1}{2}\right) = 0.242.$$

The conditional probability that an observed value X of the random phenomenon will have a value between -1 and 1, given that it has a value between 0 and 3, is given by

$$P[-1 \leq X \leq 1 \mid 0 \leq X \leq 3] = \frac{P[0 \leq X \leq 1]}{P[0 \leq X \leq 3]}$$

$$= \frac{\Phi[(1-2)/2] - \Phi[(0-2)/2]}{0.533} = \frac{0.150}{0.533} = 0.281. \blacktriangleleft$$

The most widely available tables of the normal distribution are the *Tables of the Normal Probability Functions*, National Bureau of Standards, Applied Mathematics Series 23, Washington, 1953, which tabulate

$$Q(x) = \frac{1}{\sqrt{2\pi}} e^{-\frac{1}{2}x^2}, \qquad P(x) = \frac{1}{\sqrt{2\pi}} \int_{-x}^{x} e^{-\frac{1}{2}y^2}\, dy$$

to 15 decimals for $x = 0.0000\ (0.0001)\ 1.0000\ (0.001)\ 7.800$.

THEORETICAL EXERCISES

6.1. One of the properties of the normal density functions which make them convenient to work with mathematically is the following identity. Verify algebraically that for any real numbers x, m_1, m_2, σ_1, and σ_2 (among which σ_1 and σ_2 are positive)

$$\text{(6.9)} \qquad \exp\left[-\tfrac{1}{2}\left(\frac{x-m_1}{\sigma_1}\right)^2\right] \exp\left[-\tfrac{1}{2}\left(\frac{x-m_2}{\sigma_2}\right)^2\right]$$
$$= \exp\left[-\tfrac{1}{2}\left(\frac{x-m}{\sigma}\right)^2\right] \exp\left[-\tfrac{1}{2}\,\frac{(m_1-m_2)^2}{\sigma_1^2+\sigma_2^2}\right],$$

where

$$\text{(6.10)} \qquad m = \frac{m_1\sigma_2^2 + m_2\sigma_1^2}{\sigma_1^2+\sigma_2^2}, \qquad \sigma^2 = \frac{\sigma_1^2\sigma_2^2}{\sigma_1^2+\sigma_2^2}.$$

6.2. Although it is not possible to obtain an explicit formula for the normal distribution function $\Phi(\cdot)$, in terms of more familiar functions, it is possible to obtain various inequalities on $\Phi(x)$. Show the following inequality, which is particularly useful for large values of x: for any $x > 0$

$$1 - \Phi(x) = \int_x^\infty \phi(y)\,dy \leq \frac{1}{x\sqrt{2\pi}} e^{-\frac{1}{2}x^2} \tag{6.11}$$

Hint: Use the fact that $\int_x^\infty ye^{-\frac{1}{2}y^2}\,dy = e^{-\frac{1}{2}x^2}$.

6.3. Prove (6.3) and (6.4). *Hint*: Verify that

$$\int_{-\infty}^{-x} \phi(y)\,dy = \int_x^\infty \phi(y)\,dy.$$

EXERCISES

6.1. Let X be the observed value of a numerical valued random phenomenon obeying a normal probability law with parameters (i) $m = 0$ and $\sigma = 1$, (ii) $m = 0$ and $\sigma = 2$. For α in $0 < \alpha < 1$ define $J(\alpha)$ and $K(\alpha)$ so that

$$P[X > J(\alpha)] = \alpha, \qquad P[|X| < K(\alpha)] = \alpha.$$

Find $J(\alpha)$ and $K(\alpha)$ for $\alpha = 0.05, 0.10, 0.50, 0.90, 0.95, 0.99$.

6.2. Suppose that the life in hours of a electronic tube manufactured by a certain process is normally distributed with parameters $m = 160$ hours and σ. What is the maximum allowable value for σ, if the life X of a tube is to have probability 0.80 of being between 120 and 200 hours?

6.3. Assume that the height in centimeters of a man aged 21 is a random phenomenon obeying a normal probability law with parameters $m = 170$ and $\sigma = 5$. What is the conditional probability that the height of a man aged 21 will be greater than 170 centimeters, given that it is greater than 160 centimeters?

6.4. A shirt manufacturer determines by observation that the circumference of the neck of a college man is a random phenomenon approximately obeying a normal probability law with parameters $m = 14.25$ inches and $\sigma = 0.50$ inches. For the purpose of determining how many shirts of a manufacturer's total production should have various collar sizes, compute for each of the sizes (measured in inches), 14, 14.25, 14.50, 14.75, 15.00, 15.25, 15.50, 15.75, and 16.00, the probability that a college man will wear a shirt collar of the given size, assuming that his collar size is the smallest size more than $\frac{3}{4}$ of an inch larger than the circumference of his neck.

6.5. A machine produces bolts in a length (in inches) found to obey a normal probability law with parameters $m = 10$ and $\sigma = 0.10$. The specifications for the bolt call for items with a length (in inches) equal to 10.05 ± 0.12. A bolt not meeting these specifications is called defective.

(i) What is the probability that a bolt produced by this machine will be defective?

(ii) If the machine were adjusted so that the length of bolts produced by it is normally distributed with parameters $m = 10.10$ and $\sigma = 0.10$, what is the probability that a bolt produced by the machine will be defective? (iii) If the machine is adjusted so that the lengths of bolts produced by it are normally distributed with parameters $m = 10.05$ and $\sigma = 0.06$, what is the probability a bolt produced by the machine will be defective?

6.6. Let

$$g(x) = \frac{1}{2\sqrt{2\pi}} \exp\left[-\tfrac{1}{2}\left(\frac{x-1}{2}\right)^2\right], \qquad G(x) = \int_{-\infty}^{x} g(x')\,dx'.$$

Tabulate $g(x)$ and $G(x)$ for $x = 0, \pm 1, \pm 2, \pm 3$. Compare these functions with $\phi(x)$ and $\Phi(x)$, by plotting $\phi(x)$ and $g(x)$ on one graph and $\Phi(x)$ and $G(x)$ on a second graph.

6.7. Tabulate

$$H(x) = \frac{1}{2\sqrt{2\pi}} \int_{-x}^{x} \exp\left[-\tfrac{1}{2}\left(\frac{y-1}{2}\right)^2\right] dy \qquad \text{for } x = 0, 1, 2, 3.$$

Give a probabilistic meaning to $H(x)$.

7. NUMERICAL n-TUPLE VALUED RANDOM PHENOMENA

In many cases the result of a random experiment is not expressed by a single quantity but by a family of simultaneously observed quantities. Thus, to describe the outcome of the tossing of a pair of distinguishable dice, one requires a 2-tuple (x_1, x_2), in which x_1 denotes the number obtained on the first die and x_2 denotes the number obtained on the second die. Similarly, to describe the geographical location of an object (such as a ship), one requires a 2-tuple (x_1, x_2), whose components represent the latitude and longitude of the ship, respectively. One may want to describe the prices of some commodity (such as wheat or International Business Machines' common stock) on the first day of each month of a given year; to do this, one requires a 12-tuple $(x_1, x_2, \ldots, x_{12})$ whose components $x_1, x_2, \ldots, x_{12}$ represent the price on the first day of January, February, March, . . . , November, and December, respectively. On the other hand, for some integer n one may want to describe the price of each of n commodities on a list (bread, milk, meats, shoes, electricity, etc.) on July 1 of a given year; to do this, one requires an n-tuple $(x_1, x_2, \ldots, x_n)$, whose components $x_1, x_2, \ldots, x_n$ represent the price on July 1 of the first commodity on the list, the second commodity on the list, and so on, up to the nth commodity on the list.

We are thus led to the notion of a *numerical n-tuple valued random phenomenon*, which we define as a random phenomenon whose sample description space is the set R_n consisting of all n-tuples $(x_1, x_2, \ldots, x_n)$ in which the components $x_1, x_2, \ldots, x_n$ are real numbers from $-\infty$ to ∞. In this section we indicate the notation that is used to discuss numerical n-tuple valued random phenomena. We begin by considering the case of $n = 2$.

A numerical 2-tuple valued random phenomenon is described by stating its probability function $P[\cdot]$, whose value $P[B]$ at any set B of 2-tuples of real numbers represents the probability that an observed occurrence of the random phenomenon will have a description lying in the set B. It is useful to think of the probability function $P[\cdot]$ as representing a distribution of a unit mass of some substance, which we call *probability*, over a 2-dimensional plane on which rectangular coordinates have been marked off, as in Fig. 7A. For any (probabilizable) set B of 2-tuples $P[B]$ states the weight of the probability substance distributed over the set B.

In order to know for all (probabilizable) sets B of 2-tuples the value $P[B]$ of the probability function $P[\cdot]$ of a numerical 2-tuple valued random phenomenon, it suffices to know it for all real numbers x_1 and x_2 for the sets

$$B_{x_1,x_2} = \{\text{2-tuples } (x_1', x_2'): x_1' \leq x_1, x_2' \leq x_2\}. \tag{7.1}$$

In words, B_{x_1,x_2} is the set consisting of all 2-tuples (x_1', x_2') whose first component x_1' is less than the specified real number x_1 and whose second component x_2' is less than the specified real number x_2. We are thus led to introduce the distribution function $F(.\,,.)$ of the numerical 2-tuple valued random phenomenon, which is a function of two variables, defined for all real numbers x_1 and x_2 by the equation

$$F(x_1, x_2) = P[B_{x_1,x_2}]. \tag{7.2}$$

The quantity $F(x_1, x_2)$ represents the probability that an observed occurrence of the random phenomenon under consideration will have as its description a 2-tuple whose first component is less than or equal to x_1 and whose second component is less than or equal to x_2. In terms of the unit mass of probability distributed over the plane of Fig. 7A, $F(x_1, x_2)$ is equal to the weight of the probability substance lying over the "infinitely extended rectangle," which consists of all 2-tuples (x_1', x_2'), such that $x_1' \leq x_1$ and $x_2' \leq x_2$, which corresponds to the shaded area in Fig. 7A.

The probability assigned to any rectangle in the plane may also be expressed in terms of the distribution function $F(.\,,.)$: for any real numbers a_1 and a_2 and any positive numbers h_1 and h_2

$$\begin{aligned} (7.3)\quad P[\{(x_1', x_2'):\ & a_1 < x_1' \leq a_1 + h_1, a_2 < x_2' \leq a_2 + h_2\}] \\ &= F(a_1 + h_1, a_2 + h_2) + F(a_1, a_2) - F(a_1 + h_1, a_2) - F(a_1, a_2 + h_2). \end{aligned}$$

As in the case of numerical valued random phenomena, the most important cases of numerical 2-tuple valued random phenomena are those in which the probability function is specified either by a probability mass function or a probability density function.

Given a numerical 2-tuple valued random phenomenon, we define its *probability mass function*, denoted by $p(.\,,.)$, as a function of two variables, defined for all real numbers x_1 and x_2 by the equation

$$p(x_1, x_2) = P[\{(x_1', x_2'):\ x_1' = x_1, x_2' = x_2\}]. \tag{7.4}$$

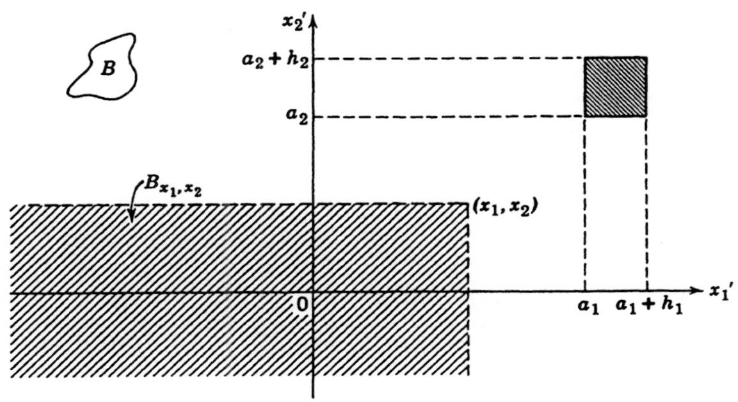

Fig. 7A. The set R_2 of all 2-tuples (x_1', x_2') of real numbers, represented as a 2-dimensional plane on which a rectangular coordinate system has been imposed.

The quantity $p(x_1, x_2)$ represents the probability that an observed occurrence of the random phenomenon under consideration will have as its description a 2-tuple whose first component is equal to x_1 and whose second component is equal to x_2. It may be shown that there is only a finite or countably infinite number of points at which $p(x_1, x_2) > 0$.

We define a numerical 2-tuple valued random phenomenon as obeying a *discrete probability law* if the sum of its probability mass function, over the points (x_1, x_2) where $p(x_1, x_2) > 0$, is equal to 1. Equivalently, the random phenomenon obeys a discrete probability law if its probability function $P[\cdot]$ is specified by its probability mass function $p(.\,,.)$ by the formula for any set B of 2-tuples:

$$P[B] = \sum_{\substack{\text{over } (x_1,x_2) \text{ lying in } B \\ \text{such that } p(x_1,x_2)>0}} p(x_1, x_2) \tag{7.5}$$

In terms of the unit of probability mass distributed over the plane of Fig. 7A by the probability function $P[\cdot]$, a numerical 2-tuple valued random phenomenon obeys a *discrete probability law* if in order to distribute the corresponding unit probability mass one needs only attach a positive probability mass at each of a finite or countably infinite number of points.

We next consider numerical 2-tuple valued random phenomena whose probability functions $P[\cdot]$ may be specified in terms of a function $f(.\,,.)$ of two variables, which we call its *probability density function.* For every (probabilizable) set B of 2-tuples

$$P[B] = \iint_B f(x_1, x_2)\, dx_1\, dx_2. \tag{7.6}$$

Equivalently, its distribution function $F(.\,,.)$ satisfies, for every pair of real numbers x_1 and x_2

$$F(x_1, x_2) = \int_{-\infty}^{x_1} \int_{-\infty}^{x_2} f(x_1', x_2')\, dx_1'\, dx_2'. \tag{7.7}$$

Consequently, the probability density function may be obtained from the distribution function $F(.\,,.)$ by differentiation:

$$f(x_1, x_2) = \frac{\partial^2}{\partial x_1\, \partial x_2} F(x_1, x_2) \tag{7.8}$$

at all 2-tuples (x_1, x_2), where the second-order mixed partial derivative $\partial^2/(\partial x_1\, \partial x_2)F(x_1, x_2)$ exists.

In the case of numerical 2-tuple valued random phenomena it remains true, *from a practical point of view*, that the only random phenomena whose distribution functions $F(.\,,.)$ are continuous, regarded as a function of two variables, are those whose distribution functions are specified by a probability density function. Consequently, we shall say that a numerical 2-tuple valued random phenomenon obeys a *continuous probability law* if its probability function and distribution function are specified by a probability density function.

All the notions of this section extend immediately to numerical n-tuple valued random phenomena by reading n-tuple for 2-tuple and $(x_1, x_2, \ldots, x_n)$ for (x_1, x_2) in the foregoing discussion. In place of (7.1) to (7.8) read the following equations:

$$\begin{aligned} &B_{x_1, x_2, \cdots, x_n} \\ &\quad = \{n\text{-tuples } (x_1', x_2', \cdots, x_n'):\ \ x_1' \le x_1, x_2' \le x_2, \cdots, x_n' \le x_n\}. \end{aligned} \tag{7.1'}$$

$$F(x_1, x_2, \cdots, x_n) = P[B_{x_1, x_2, \cdots, x_n}]. \tag{7.2'}$$

$$(7.3') \quad P[\{(x_1', x_2', \cdots, x_n'): \ a_1 < x_1' \le a_1 + h_1, \ a_2 < x_2' \le a_2 + h_2, \cdots, a_n < x_n' \le a_n + h_n\}]$$
$$= F(a_1 + h_1, a_2 + h_2, \cdots, a_n + h_n)$$
$$- F(a_1, a_2 + h_2, \cdots, a_n + h_n) - \cdots - F(a_1 + h_1, \cdots, a_{n-1} + h_{n-1}, a_n)$$
$$+ \cdots\cdots\cdots\cdots\cdots\cdots$$
$$+ (-1)^n F(a_1, a_2, \cdots, a_n).$$

$$(7.4') \quad p(x_1, x_2, \cdots, x_n) = P[\{(x_1', x_2', \cdots, x_n'): \ x_1' = x_1, x_2' = x_2, \cdots, x_n' = x_n\}].$$

$$(7.5') \quad P[B] = \sum_{\substack{\text{over } (x_1, x_2, \ldots, x_n) \text{ lying in } B \\ \text{such that } p(x_1, x_2, \ldots, x_n) > 0}} p(x_1, x_2, \cdots, x_n).$$

$$(7.6') \quad P[B] = \int\int \cdots \int_B f(x_1, x_2, \cdots, x_n)\, dx_1\, dx_2 \cdots dx_n.$$

$$(7.7') \quad F(x_1, x_2, \cdots, x_n) = \int_{-\infty}^{x_1} \int_{-\infty}^{x_2} \cdots \int_{-\infty}^{x_n} f(x_1', x_2', \cdots, x_n')\, dx_1'\, dx_2' \cdots dx_n'.$$

$$(7.8') \quad f(x_1, x_2, \cdots, x_n) = \frac{\partial^n}{\partial x_1\, \partial x_2 \cdots \partial x_n} F(x_1, x_2, \cdots, x_n).$$

There are many other notions that arise in connection with numerical n-tuple valued random phenomena, but they are best formulated in terms of random variables and consequently are discussed in Chapter 7.

EXERCISES

7.1. Let, for some finite constants a, b, c, and K,

$$f(x_1, x_2) = Ke^{-(ax_1^2 + 2bx_1x_2 + cx_2^2)}$$

Show that in order for $f(x_1, x_2)$ to be the probability density function of a 2-tuple valued random phenomenon it is necessary and sufficient that the constants a, b, c, and K satisfy the conditions $a > 0$, $c > 0$, $b^2 - ac < 0$, $K = (1/\pi)\sqrt{ac - b^2}$.

7.2. An urn contains M balls, numbered 1 to M. Two balls are drawn, 1 after the other, with replacement (without replacement). Consider the 2-tuple valued random phenomenon (x_1, x_2), in which x_1 is the number on the first ball drawn, and x_2 is the number of the second ball drawn. Find the probability mass function of this 2-tuple valued random phenomenon and show that its probability law is discrete.

7.3. Consider a square sheet of tin, 20 inches wide, that contains 10 rows and 10 columns of circular holes, each 1 inch in diameter, with centers evenly spaced at a distance 2 inches apart.

(i) What is the probability that a particle of sand (considered as a point) blown against the tin sheet will fall upon 1 of the holes and thus pass through?

(ii) What is the probability that a ball of diameter $\frac{1}{2}$ inch thrown upon the sheet will pass through 1 of the holes without touching the tin sheet? Assume an appropriate uniform probability law.

CHAPTER 5

Mean and Variance of a Probability Law

It has been emphasized that in order to describe a numerical valued random phenomenon one must specify its probability function $P[\cdot]$ or, equivalently, its distribution function $F(\cdot)$. In the special case in which the random phenomenon obeys a discrete or a continuous probability law its probability function is determined by a knowledge of the probability mass function $p(\cdot)$ or of the probability density function $f(\cdot)$. Thus, to describe a numerical valued random phenomenon, certain functions must be specified. It is desirable to be able to summarize some of the outstanding features of the probability law of a numerical valued random phenomenon by specifying only a few numbers rather than an entire function. Such numbers are provided by the expectation of various functions $g(\cdot)$ with respect to the probability law of the random phenomenon.

1. THE NOTION OF AN AVERAGE

In order to motivate our definition of the notion of expectation, let us first discuss the meaning of the word "average." Given a set of n quantities, which we denote by $x_1, x_2, \ldots, x_n$, we define their average, often denoted by $\bar{x}$, as the sum of the quantities divided by n; in symbols

$$\bar{x} = \frac{x_1 + x_2 + \cdots + x_n}{n} = \frac{1}{n} \sum_{i=1}^{n} x_i. \tag{1.1}$$

The quantity $\bar{x}$ is also called the *arithmetic mean* of the numbers $x_1, x_2, \ldots, x_n$.

For example, consider the scores on an examination of a class of 20 students:

$$\{10, 10, 10, 10, 9, 9, 9, 9, 9, 8, 8, 8, 8, 8, 7, 7, 6, 5, 5, 5\} \tag{1.2}$$

The average of these scores is $160/20 = 8$.

Very often, a set of n numbers, $x_1, x_2, \ldots, x_n$, which is to be averaged, may be described in the following way. There are k real numbers, which we may denote by $x_1', x_2', \ldots, x_k'$, and k integers, $n_1, n_2, \ldots, n_k$ (whose sum is n), such that the set of numbers $\{x_1, x_2, \ldots, x_n\}$ consists of n_1 repetitions of the number x_1', n_2 repetitions of the number x_2', and so on, up to n_k repetitions of the number x_k'. Thus the set of scores in (1.2) may be described by the following table:

(1.3)

Possible values x_i' in the set	10	9	8	7	6	5
Number n_i of occurrences of x_i' in the set	4	5	5	2	1	3

In terms of this notation, the average $\bar{x}$ defined by (1.1) may be written

$$\bar{x} = \frac{1}{n} \sum_{i=1}^{k} x_i' n_i. \tag{1.4}$$

We may go one step further. Let us define the quantity

$$f(x_i') = \frac{n_i}{n} \tag{1.5}$$

that represents the fraction of the set of numbers $\{x_1, x_2, \ldots, x_n\}$, which is equal to the number x_i'. Then (1.4) becomes

$$\bar{x} = \sum_{i=1}^{k} x_i' f(x_i'). \tag{1.6}$$

In words, we may read (1.6) as follows: *the average $\bar{x}$ of a set of numbers, $x_1, x_2, \ldots, x_n$, is equal to the sum, over the set of numbers, $x_1', x_2', \ldots, x_k'$, which occur in the set $\{x_1, x_2, \ldots, x_n\}$, of the product of the value of x_i' and the fraction $f(x_i')$; $f(x_i')$ is the fraction of numbers in the set $\{x_1, x_2, \ldots, x_n\}$ which are equal to x_i'.*

The question naturally arises as to the meaning to be assigned to the average of a set of numbers. It seems clear that the average of a set of numbers is computed for the purpose of *summarizing the data* represented by the set of numbers, so as to better comprehend it. Given the examination scores of a large number of students, it is difficult to form an opinion as to how well the students performed, except perhaps by forming averages.

However, it is also clear that the average of a set of numbers, as defined by (1.1) or (1.6), does not serve to summarize the data completely. Consider a second group of twenty students who, in the same examination on which the scores in (1.2) were obtained, gave the following performance:

(1.7)

Scores x_i'	10	9	8	7	6	5
Number n_i of students scoring the score x_i'	3	5	6	2	3	1

The average of this set of scores is 8, as it would have been if the scores had been

(1.8)

Scores x_i'	10	9	8	7	6	5
Number n_i of students scoring the score x_i'	3	3	8	3	3	0

Consequently, if we are to summarize these collections of data, we shall require more than the average, in the sense of (1.6), to do it.

The average, in the sense of (1.6), is a measure of what might be called the mid-point, or *mean*, of the data, about which the numbers in the data are, loosely speaking, "centered." More precisely, the mean $\bar{x}$ represents the center of gravity of a long rod on which masses $f(x_1'), \ldots, f(x_k')$ have been placed at the points $x_1', \ldots, x_k'$, respectively.

Perhaps another characteristic of the data for which one should have a measure is its *spread* or *dispersion* about the mean. Of course, it is not clear how this measure should be defined.

The dispersion might be defined as the average of the absolute value of the deviation of each number in the set from the mean $\bar{x}$; in symbols,

$$\text{absolute dispersion} = \sum_{i=1}^{k} |x_i' - \bar{x}| f(x_i'). \tag{1.9}$$

The value of the expression (1.9) for the data in (1.3), (1.7), and (1.8) is equal to 1.3, 1.1, and 0.9, respectively, where in each case the mean $\bar{x} = 8$.

Another possible measure of the spread of the data is the average of the squares of the deviation from the mean $\bar{x}$ of each number x_i' in the set; in symbols,

$$\text{square dispersion} = \sum_{i=1}^{k} (x_i' - \bar{x})^2 f(x_i'), \tag{1.10}$$

which has the values 2.7, 2.0, and 1.5 for the data in (1.3), (1.7), and (1.8), respectively.

Next, one may desire a measure for the symmetry of the distribution of the scores about their mean, for which purpose one might take the average of the cubes of the deviation of each number in the set from the mid-point $\bar{x}$ (= 8); in symbols,

$$\sum_{i=1}^{k} (x_i' - \bar{x})^3 f(x_i'), \tag{1.11}$$

which has the values −2.7, −1.2, and 0 for the data in (1.3), (1.7), and (1.8), respectively.

From the foregoing discussion *one conclusion emerges clearly.* Given data $\{x_1, x_2, \ldots, x_n\}$, there are many kinds of averages one can define, depending on the particular aspect of the data in which one is interested. Consequently, we cannot speak of the average of a set of numbers. Rather, we must consider some function $g(x)$ of a real variable x; for example, $g(x) = x$, $g(x) = (x - 8)^2$, or $g(x) = (x - 8)^3$. We then *define the average of the function $g(x)$ with respect to a set of numbers $\{x_1, x_2, \ldots, x_n\}$ as*

$$\frac{1}{n} \sum_{j=1}^{n} g(x_j) = \sum_{i=1}^{k} g(x_i') f(x_i'), \tag{1.12}$$

in which the numbers $x_1', \ldots, x_k'$ occur in the proportions $f(x_1'), \ldots, f(x_k')$ in the set $\{x_1, x_2, \ldots, x_n\}$.

EXERCISES

In each of the following exercises find the average with respect to the data given for these functions: (i) $g(x) = x$; (ii) $g(x) = (x - \bar{x})^2$, in which $\bar{x}$ is the answer obtained to question (i); (iii) $g(x) = (x - \bar{x})^3$; (iv) $g(x) = (x - \bar{x})$; (v) $g(x) = |x - \bar{x}|$. *Hint:* First compute the number of times each number appears in the data.

1.1. The number of rainy days in a certain town during the month of January for the years 1950–1959 was as follows:

Year	1950	1951	1952	1953	1954	1955	1956	1957	1958	1959
Number of rainy days in January	8	9	21	16	16	9	13	9	8	21

1.2. Record the last digits of the last 20 telephone numbers appearing on the first page of your local telephone directory.

1.3. Ten light bulbs were subjected to a forced life test. Their lifetimes were found to be (to the nearest 10 hours)

850, 1090, 1150, 940, 1150, 960, 1040, 920, 1040, 960.

1.4. An experiment consists of drawing 2 balls without replacement from an urn containing 6 balls, numbered 1 to 6, and recording the sum of the 2 numbers drawn. In 30 repetitions of the experiment the sums recorded were (compare example 4A of Chapter 2)

7 9 5 8 5 7 4 6 3 5 9 11 9 4 9
11 7 10 4 8 5 6 10 9 5 7 9 10 10 3.

2. EXPECTATION OF A FUNCTION WITH RESPECT TO A PROBABILITY LAW

Consider a numerical valued random phenomenon, with probability function $P[\cdot]$. The probability function $P[\cdot]$ determines a distribution of a unit mass on the real line, the amount of which lying on any (Borel) set B of real numbers is equal to $P[B]$. In order to summarize the characteristics of $P[\cdot]$ by a few numbers, we define in this section the notion of the *expectation of a continuous function* $g(x)$ *of a real variable* x, *with respect to the probability function* $P[\cdot]$, *to be denoted by* $E[g(x)]$. It will be seen that the expectation $E[g(x)]$ has much the same properties as the average of $g(x)$, with respect to a set of numbers.

For the case in which the probability function $P[\cdot]$ is specified by a probability mass function $p(\cdot)$, we define, in analogy with (1.12),

$$E[g(x)] = \sum_{\substack{\text{over all } x \text{ such} \\ \text{that } p(x) > 0}} g(x)p(x). \tag{2.1}$$

The sum written in (2.1) may involve the summation of a countably infinite number of terms and therefore is not always meaningful. For reasons made clear in section 1 of Chapter 8 the expectation $E[g(x)]$ is said to exist if

$$E[|g(x)|] = \sum_{\substack{\text{over all } x \text{ such} \\ \text{that } p(x) > 0}} |g(x)|p(x) < \infty. \tag{2.2}$$

In words, *the expectation* $E[g(x)]$, *defined in* (2.1), *exists if and only if the infinite series defining* $E[g(x)]$ *is absolutely convergent.* A test for convergence of an infinite series is given in theoretical exercise 2.1.

For the case in which the probability function $P[\cdot]$ is specified by a probability density function $f(\cdot)$, we define

$$E[g(x)] = \int_{-\infty}^{\infty} g(x)f(x)\,dx. \tag{2.3}$$

The integral written in (2.3) is an improper integral and therefore is not always meaningful. Before one can speak of the expectation $E[g(x)]$, one must verify its existence. The expectation $E[g(x)]$ is said to exist if

$$E[|g(x)|] = \int_{-\infty}^{\infty} |g(x)| f(x)\, dx < \infty. \tag{2.4}$$

In words, *the expectation $E[g(x)]$ defined in* (2.3) *exists if and only if the improper integral defining $E[g(x)]$ is absolutely convergent.* In the case in which the functions $g(\cdot)$ and $f(\cdot)$ are continuous for all (but a finite number of values of) x, the integral in (2.3) may be defined as an improper Riemann* integral by the limit

$$\int_{-\infty}^{\infty} g(x)f(x)\, dx = \lim_{\substack{a \to -\infty \\ b \to \infty}} \int_a^b g(x)f(x)\, dx. \tag{2.5}$$

A useful tool for determining whether or not the expectation $E[g(x)]$, given by (2.3), exists is the test for convergence of an improper integral given in theoretical exercise 2.1.

A discussion of the definition of the expectation in the case in which the probability function must be specified by the distribution function is given in section 6.

The expectation $E[g(x)]$ is sometimes called the *ensemble average* of the function $g(x)$ in order to emphasize that the expectation (or ensemble average) is a theoretically computed quantity. It is not an average of an observed set of numbers, as was the case in section 1. We shall later consider averages with respect to observed values of random phenomena, and these will be called sample averages.

A special terminology is introduced to describe the expectation $E[g(x)]$ of various functions $g(x)$.

We call $E[x]$, the expectation of the function $g(x) = x$ with respect to a probability law, the *mean of the probability law.* For a discrete probability law, with probability mass function $p(\cdot)$,

$$E[x] = \sum_{\substack{\text{over all } x \text{ such} \\ \text{that } p(x) > 0}} x\, p(x). \tag{2.6}$$

For a continuous probability law, with probability density function $f(\cdot)$,

$$E[x] = \int_{-\infty}^{\infty} x\, f(x)\, dx. \tag{2.7}$$

* For the benefit of the reader acquainted with the theory of Lebesgue integration, let it be remarked that if the integral in (2.3) is defined as an integral in the sense of Lebesgue then the notion of expectation $E[g(x)]$ may be defined for a Borel function $g(x)$.

It may be shown that the mean of a probability law has the following meaning. Suppose one makes a sequence $X_1, X_2, \ldots, X_n, \ldots$ of independent observations of a random phenomenon obeying the probability law and forms the successive arithmetic means

$$A_1 = X_1, \qquad A_2 = \frac{1}{2}(X_1 + X_2),$$

$$A_3 = \frac{1}{3}(X_1 + X_2 + X_3), \cdots, \qquad A_n = \frac{1}{n}(X_1 + X_2 + \cdots + X_n), \cdots.$$

These successive arithmetic means, $A_1, A_2, \ldots, A_n$, will (with probability one) tend to a limiting value if and only if the mean of the probability law is finite. Further, this limiting value will be precisely the mean of the probability law.

We call $E[x^2]$, the expectation of the function $g(x) = x^2$ with respect to a probability law, the *mean square of the probability law*. This notion is not to be confused with the *square mean of the probability law*, which is the square $(E[x])^2$ of the mean and which we denote by $E^2[x]$. For a discrete probability law, with probability mass function $p(\cdot)$,

$$E[x^2] = \sum_{\substack{\text{over all } x \text{ such} \\ \text{that } p(x) > 0}} x^2 p(x). \tag{2.8}$$

For a continuous probability law, with probability density function $f(\cdot)$,

$$E[x^2] = \int_{-\infty}^{\infty} x^2 f(x)\, dx. \tag{2.9}$$

More generally, for any integer $n = 1, 2, 3, \ldots$, we call $E[x^n]$, the expectation of $g(x) = x^n$ with respect to a probability law, the *nth moment of the probability law*. Note that the first moment and the mean of a probability law are the same; also, the second moment and the mean square of a probability law are the same.

Next, for any real number c, and integer $n = 1, 2, 3, \ldots$, we call $E[(x - c)^n]$ the *nth moment of the probability law about the point c*. Of especial interest is the case in which c is equal to the mean $E[x]$. We call $E[(x - E[x])^n]$ the *nth moment of the probability law about its mean* or, more briefly, the *nth central moment of the probability law*.

The second central moment $E[(x - E[x])^2]$ is especially important and is called the *variance of the probability law*. Given a probability law, we shall use the symbols m and σ^2 to denote, respectively, its mean and variance; consequently,

$$m = E[x], \qquad \sigma^2 = E[(x - m)^2]. \tag{2.10}$$

The square root σ of the variance is called the *standard deviation* of the probability law. The intuitive meaning of the variance is discussed in section 4.

▶ **Example 2A.** The *normal* probability law with parameters m and σ is specified by the probability density function $f(\cdot)$, given by (4.11) of Chapter 4. Its mean is equal to

$$(2.11)\quad E[x] = \frac{1}{\sigma\sqrt{2\pi}} \int_{-\infty}^{\infty} x e^{-\frac{1}{2}\left(\frac{x-m}{\sigma}\right)^2} dx = \frac{1}{\sqrt{2\pi}} \int_{-\infty}^{\infty} (m + \sigma y) e^{-\frac{1}{2}y^2}\, dy,$$

where we have made the change of variable $y = (x - m)/\sigma$. Now

$$(2.12)\quad \int_{-\infty}^{\infty} e^{-\frac{1}{2}y^2}\, dy = \int_{-\infty}^{\infty} y^2 e^{-\frac{1}{2}y^2}\, dy = \sqrt{2\pi}, \qquad \int_{-\infty}^{\infty} y e^{-\frac{1}{2}y^2}\, dy = 0.$$

Equation (2.12) follows from (2.20) and (2.22) of Chapter 4 and the fact that for any integrable function $h(y)$

$$(2.13)\quad \begin{aligned} &\int_{-\infty}^{\infty} h(y)\, dy = 0 \qquad \text{if } h(-y) = -h(y) \\ &\int_{-\infty}^{\infty} h(y)\, dy = 2\int_{0}^{\infty} h(y)\, dy \qquad \text{if } h(-y) = h(y). \end{aligned}$$

From (2.12) and (2.13) it follows that the mean $E[x]$ is equal to m. Next, the variance is equal to

$$(2.14)\quad \begin{aligned} E[(x - m)^2] &= \frac{1}{\sigma\sqrt{2\pi}} \int_{-\infty}^{\infty} (x - m)^2 e^{-\frac{1}{2}\left(\frac{x-m}{\sigma}\right)^2} dx \\ &= \sigma^2 \frac{1}{\sqrt{2\pi}} \int_{-\infty}^{\infty} y^2 e^{-\frac{1}{2}y^2}\, dy = \sigma^2. \end{aligned}$$

Notice that *the parameters m and σ in the normal probability law were chosen equal to the mean and standard deviation of the probability law.* ◀

The operation of taking expectations has certain basic properties with which one may perform various formal manipulations. To begin with, we have the following properties for any constant c and any functions $g(x)$, $g_1(x)$, and $g_2(x)$ whose expectations exist:

$$(2.15)\quad E[c] = c.$$

$$(2.16)\quad E[cg(x)] = cE[g(x)].$$

$$(2.17)\quad E[g_1(x) + g_2(x)] = E[g_1(x)] + E[g_2(x)].$$

$$(2.18)\quad E[g_1(x)] \le E[g_2(x)] \qquad \text{if } g_1(x) \le g_2(x) \qquad \text{for all } x.$$

$$(2.19)\quad |E[g(x)]| \le E[|g(x)|].$$

In words, the first three of these properties may be stated as follows: the expectation of a constant c [that is, of the function $g(x)$, which is equal to c for every value of x] is equal to c; the expectation of the product of a constant and a function is equal to the constant multiplied by the expectation of the function; the expectation of a function which is the sum of two functions is equal to the sum of the expectations of the two functions.

Equations (2.15) to (2.19) are immediate consequences of the definition of expectation. We write out the details only for the case in which the expectations are taken with respect to a continuous probability law with probability density function $f(\cdot)$. Then, by the properties of integrals,

$$E[c] = \int_{-\infty}^{\infty} cf(x)\,dx = c\int_{-\infty}^{\infty} f(x)\,dx = c,$$

$$E[cg(x)] = \int_{-\infty}^{\infty} cg(x)f(x)\,dx = c\int_{-\infty}^{\infty} g(x)f(x)\,dx = cE[g(x)],$$

$$\begin{aligned} E[g_1(x) + g_2(x)] &= \int_{-\infty}^{\infty} (g_1(x) + g_2(x))f(x)\,dx \\ &= \int_{-\infty}^{\infty} g_1(x)f(x)\,dx + \int_{-\infty}^{\infty} g_2(x)f(x)\,dx = E[g_1(x)] + E[g_2(x)], \end{aligned}$$

$$E[g_2(x)] - E[g_1(x)] = \int_{-\infty}^{\infty} [g_2(x) - g_1(x)]f(x)\,dx \geq 0.$$

Equation (2.19) follows from (2.18), applied first with $g_1(x) = g(x)$ and $g_2(x) = |g(x)|$ and then with $g_1(x) = -|g(x)|$ and $g_2(x) = g(x)$.

▶ **Example 2B.** To illustrate the use of (2.15) to (2.19), we note that $E[4] = 4$, $E[x^2 - 4x] = E[x^2] - 4E[x]$, and $E[(x - 2)^2] = E[x^2 - 4x + 4] = E[x^2] - 4E[x] + 4$. ◀

We next derive an extremely important expression for the *variance* of a probability law:

$$\sigma^2 = E[(x - E[x])^2] = E[x^2] - E^2[x]. \tag{2.20}$$

In words, *the variance of a probability law is equal to its mean square, minus its square mean.* To prove (2.20), we write, letting $m = E[x]$,

$$\begin{aligned} \sigma^2 &= E[x^2 - 2mx + m^2] = E[x^2] - 2mE[x] + m^2 \\ &= E[x^2] - 2m^2 + m^2 = E[x^2] - m^2. \end{aligned}$$

In the remainder of this section we compute the mean and variance of various probability laws. A tabulation of the results obtained is given in Tables 3A and 3B at the end of section 3.

▶ **Example 2C.** The Bernoulli probability law with parameter p, in which $0 \le p \le 1$, is specified by the probability mass function $p(\cdot)$, given by $p(0) = 1 - p$, $p(1) = p$, $p(x) = 0$ for $x \neq 0$ or 1. Its mean, mean square, and variance, letting $q = 1 - p$, are given by

$$E[x] = 0 \cdot q + 1 \cdot p = p$$

$$E[x^2] = 0^2 \cdot q + 1^2 \cdot p = p \tag{2.21}$$

$$\sigma^2 = E[x^2] - m^2 = p - p^2 = pq.$$ ◀

▶ **Example 2D.** The binomial probability law with parameters n and p is specified by the probability mass function given by (4.5) of Chapter 4. Its mean is given by

$$\begin{aligned} E[x] &= \sum_{k=0}^{n} kp(k) = \sum_{k=0}^{n} k\binom{n}{k} p^k q^{n-k} \\ &= np\sum_{k=1}^{n} \binom{n-1}{k-1} p^{k-1} q^{(n-1)-(k-1)} = np(p+q)^{n-1} = np. \end{aligned} \tag{2.22}$$

Its mean square is given by

$$E[x^2] = \sum_{k=0}^{n} k^2 \binom{n}{k} p^k q^{n-k}. \tag{2.23}$$

To evaluate $E[x^2]$, we write $k^2 = k(k-1) + k$. Then

$$E[x^2] = \sum_{k=2}^{n} k(k-1)\binom{n}{k} p^k q^{n-k} + E[x]. \tag{2.24}$$

Since $k(k-1)\binom{n}{k} = n(n-1)\binom{n-2}{k-2}$, the sum in (2.24) is equal to

$$n(n-1)p^2 \sum_{k=2}^{n} \binom{n-2}{k-2} p^{k-2} q^{(n-2)-(k-2)} = n(n-1)p^2(p+q)^{n-2}.$$

Consequently, $E[x^2] = n(n-1)p^2 + np$, so that

$$E[x^2] = npq + n^2p^2, \qquad \sigma^2 = E[x^2] - E^2[x] = npq. \tag{2.25}$$ ◀

▶ **Example 2E.** The hypergeometric probability law with parameters N, n, and p is specified by the probability mass function $p(\cdot)$ given by (4.8) of Chapter 4. Its mean is given by

$$E[x] = \sum_{k=0}^{n} kp(k) = \frac{1}{\binom{N}{n}} \sum_{k=1}^{n} k \binom{Np}{k} \binom{Nq}{n-k} \tag{2.26}$$

$$= \frac{Np}{\binom{N}{n}} \sum_{k=1}^{n} \binom{a-1}{k-1} \binom{b}{n-k},$$

in which we have let $a = Np$, $b = Nq$. Now, letting $j = k - 1$ and using (2.37) of Chapter 4, the last sum written is equal to

$$\sum_{j=0}^{n-1} \binom{a-1}{j} \binom{b}{n-1-j} = \binom{a+b-1}{n-1} = \binom{N-1}{n-1}.$$

Consequently,

$$E[x] = Np \frac{\binom{N-1}{n-1}}{\binom{N}{n}} = np. \tag{2.27}$$

Next, we evaluate $E[x^2]$ by first evaluating $E[x(x-1)]$ and then using the fact that $E[x^2] = E[x(x-1)] + E[x]$. Now

$$\binom{N}{n} E[x(x-1)] = \sum_{k=2}^{n} k(k-1) \binom{a}{k} \binom{b}{n-k}$$

$$= a(a-1) \sum_{k=2}^{n} \binom{a-2}{k-2} \binom{b}{n-k}$$

$$= a(a-1) \binom{a+b-2}{n-2} = Np(Np-1) \binom{N-2}{n-2} \tag{2.28}$$

$$E[x^2] = np(Np-1) \frac{n-1}{N-1} + np = \frac{np}{N-1} (Np(n-1) + N - n)$$

$$\sigma^2 = \frac{np}{N-1} (N - n + pN(n-1) - pn(N-1)) = npq \frac{N-n}{N-1}.$$

Notice that the mean of the hypergeometric probability law is the same as that of the corresponding binomial probability law, whereas the variances differ by a factor that is approximately equal to 1 if the ratio n/N is a small number. ◀

▶ **Example 2F.** The uniform probability law over the interval a to b has probability density function $f(\cdot)$ given by (4.10) of Chapter 4. Its mean, mean square, and variance are given by

$$E[x] = \int_{-\infty}^{\infty} xf(x)\,dx = \frac{1}{b-a}\int_a^b x\,dx = \frac{b^2-a^2}{2(b-a)} = \frac{b+a}{2}$$

$$(2.29)\quad E[x^2] = \int_{-\infty}^{\infty} x^2 f(x)\,dx = \frac{1}{b-a}\int_a^b x^2\,dx = \frac{1}{3}(b^2+ba+a^2)$$

$$\sigma^2 = E[x^2] - E^2[x] = \tfrac{1}{12}(b-a)^2.$$

Note that the variance of the uniform probability law depends only on the length of the interval, whereas the mean is equal to the mid-point of the interval. The higher moments of the uniform probability law are also easily obtained:

$$(2.30)\qquad E[x^n] = \frac{1}{b-a}\int_a^b x^n\,dx = \frac{b^{n+1}-a^{n+1}}{(n+1)(b-a)}. \qquad ◀$$

▶ **Example 2G.** The Cauchy probability law with parameters $\alpha = 0$ and $\beta = 1$ is specified by the probability density function

$$(2.31)\qquad f(x) = \frac{1}{\pi}\frac{1}{1+x^2}.$$

The mean $E[x]$ of the Cauchy probability law does not exist, since

$$(2.32)\qquad E[|x|] = \frac{1}{\pi}\int_{-\infty}^{\infty} |x|\frac{1}{1+x^2}\,dx = \infty.$$

However, for $r < 1$ the rth absolute moments

$$(2.33)\qquad E[|x|^r] = \frac{1}{\pi}\int_{-\infty}^{\infty} |x|^r\frac{1}{1+x^2}\,dx$$

do exist, as one may see by applying theoretical exercise 2.1. ◀

THEORETICAL EXERCISES

2.1. **Test for convergence or divergence of infinite series and improper integrals.** Prove the following statements. Let $h(x)$ be a continuous function. If, for some real number $r > 1$, the limits

$$(2.34)\qquad \lim_{x\to\infty} x^r|h(x)|, \qquad \lim_{x\to-\infty} |x|^r|h(x)|$$

both exist and are finite, then

$$(2.35) \qquad \int_{-\infty}^{\infty} h(x)\,dx, \qquad \sum_{k=-\infty}^{\infty} h(k)$$

converge absolutely; if, for some $r \le 1$, *either* of the limits in (2.34) exist and is not equal to 0, then the expressions in (2.35) fail to converge absolutely.

2.2. Pareto's distribution with parameters r and A, in which r and A are positive, is defined by the probability density function

$$(2.36) \qquad \begin{aligned} f(x) &= rA^r \frac{1}{x^{r+1}} && \text{for } x \ge A \\ &= 0 && \text{for } x < A. \end{aligned}$$

Show that Pareto's distribution possesses a finite nth moment if and only if $n < r$. Find the mean and variance of Pareto's distribution in the cases in which they exist.

2.3. "Student's" t-distribution with parameter $\nu > 0$ is defined as the continuous probability law specified by the probability density function

$$(2.37) \qquad f(x) = \frac{1}{\sqrt{\nu\pi}} \frac{\Gamma[(\nu+1)/2]}{\Gamma(\nu/2)} \left(1 + \frac{x^2}{\nu}\right)^{-(\nu+1)/2}$$

Note that "Student's" t-distribution with parameter $\nu = 1$ coincides with the Cauchy probability law given by (2.31). Show that for "Student's" t-distribution with parameter ν (i) the nth moment $E[x^n]$ exists only for $n < \nu$, (ii) if $n < \nu$ and n is odd, then $E[x^n] = 0$, (iii) if $n < \nu$ and n is even, then

$$(2.38) \qquad E[x^n] = \nu^{n/2} \frac{\Gamma[(n+1)/2]\Gamma[(\nu-n)/2]}{\Gamma(1/2)\Gamma(\nu/2)}$$

Hint: Use (2.41) and (2.42) in Chapter 4.

2.4. **A characterization of the mean.** Consider a probability law with finite mean m. Define, for every real number a, $h(a) = E[(x-a)^2]$. Show that $h(a) = E[(x-m)^2] + (m-a)^2$. Consequently $h(a)$ is minimized at $a = m$, and its minimum value is the variance of the probability law.

2.5. **A geometrical interpretation of the mean of a probability law.** Show that for a continuous probability law with probability density function $f(\cdot)$ and distribution function $F(\cdot)$

$$(2.39) \qquad \begin{aligned} \int_0^{\infty} [1 - F(x)]\,dx &= \int_0^{\infty} dx \int_x^{\infty} dy f(y) = \int_0^{\infty} y f(y)\,dy, \\ -\int_{-\infty}^{0} F(x)\,dx &= -\int_{-\infty}^{0} dx \int_{-\infty}^{x} dy f(y) = -\int_0^{-\infty} y f(y)\,dy. \end{aligned}$$

Consequently the mean m of the probability law may be written

$$(2.40) \qquad m = \int_{-\infty}^{\infty} y f(y)\,dy = \int_0^{\infty} [1 - F(x)]\,dx - \int_{-\infty}^{0} F(x)\,dx.$$

These equations may be interpreted geometrically. Plot the graph

$y = F(x)$ of the distribution function on an (x, y)-plane, as in Fig. 2A, and define the areas I and II as indicated: I is the area to the right of the y-axis bounded by $y = 1$ and $y = F(x)$; II is the area to the left of the y-axis bounded by $y = 0$ and $y = F(x)$. Then the mean m is equal to

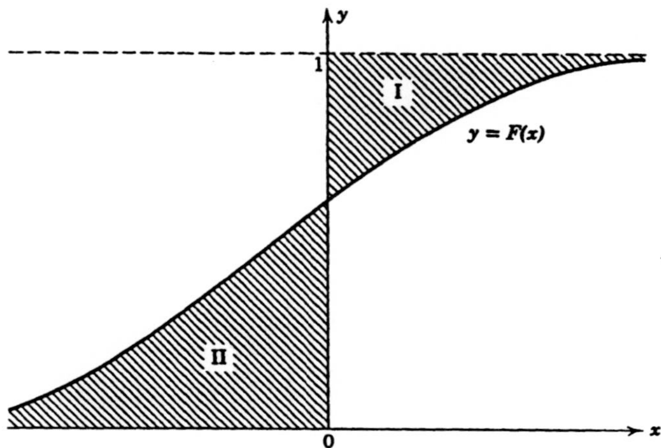

Fig. 2A. The mean of a probability law with distribution function $F(\cdot)$ is equal to the shaded area to the right of the y-axis, minus the shaded area to the left of the y-axis.

area I, minus area II. Although we have proved this assertion only for the case of a continuous probability law, it holds for any probability law.

2.6. **A geometrical interpretation of the higher moments.** Show that the nth moment $E[x^n]$ of a continuous probability law with distribution function $F(\cdot)$ can be expressed for $n = 1, 2, \ldots$

$$E[x^n] = \int_{-\infty}^{\infty} x^n f(x)\, dx = \int_0^{\infty} dy\, ny^{n-1}[1 - F(y) + (-1)^n F(-y)]. \tag{2.41}$$

Use (2.41) to interpret the nth moment in terms of area.

2.7. **The relation between the moments and central moments of a probability law.** Show that from a knowledge of the moments of a probability law one may obtain a knowledge of the central moments, and conversely. In particular, it is useful to have expressions for the first 4 central moments in terms of the moments. Show that

$$\begin{aligned} E[(x - E[x])^3] &= E[x^3] - 3E[x]E[x^2] + 2E^3[x] \\ E[(x - E[x])^4] &= E[x^4] - 4E[x]E[x^3] + 6E^2[x]E[x^2] - 3E^4[x]. \end{aligned} \tag{2.42}$$

2.8. **The square mean is less than or equal to the mean square.** Show that

$$|E[x]| \leq E[|x|] \leq E^{1/2}[x^2]. \tag{2.43}$$

Give an example of a probability law whose mean square $E[x^2]$ is equal to its square mean.

2.9. The mean is not necessarily greater than or equal to the variance. The binomial and the Poisson are probability laws having the property that their mean m is greater than or equal to their variance σ^2 (show this); this circumstance has sometimes led to the belief that for the probability law of a random variable assuming only nonnegative values it is always true that $m \geq \sigma^2$. Prove this is not the case by showing that $m < \sigma^2$ for the probability law of the number of failures up to the first success in a sequence of independent repeated Bernoulli trials.

2.10. The median of a probability law. The mean of a probability law provides a measure of the "mid-point" of a probability distribution. Another such measure is provided by the *median of a probability law*, denoted by m_e, which is defined as a number such that

$$\lim_{x \to m_e -} F(x) = F(m_e - 0) \leq \frac{1}{2} \leq F(m_e + 0) = \lim_{x \to m_e +} F(x). \tag{2.44}$$

If the probability law is continuous, the median m_e may be defined as a number satisfying $\int_{-\infty}^{m_e} f(x)\, dx = \frac{1}{2}$. Thus m_e is the projection on the x-axis of the point in the (x, y)-plane at which the line $y = \frac{1}{2}$ intersects the curve $y = F(x)$. A more probabilistic definition of the median m_e is as a number such that $P[X < m_e] \leq \frac{1}{2} \geq P[X > m_e]$, in which X is an observed value of a random phenomenon obeying the given probability law. There may be an interval of points that satisfies (2.44); if this is the case, we take the mid-point of the interval as the median. Show that one may *characterize the median* m_e as a number at which the function $h(a) = E[|x - a|]$ achieves its minimum value; this is therefore $E[|x - m_e|]$. *Hint:* Although the assertion is true in general, show it only for a continuous probability law. Show, and use the fact, that for any number a

$$E[|x - a|] = E[|x - m_e|] + 2\int_a^{m_e} (x - a) f(x)\, dx. \tag{2.45}$$

2.11. The mode of a continuous or discrete probability law. For a continuous probability law with probability density function $f(x)$ a mode of the probability law is defined as a number m_0 at which the probability density has a relative maximum; assuming that the probability density function is twice differentiable, a point m_0 is a mode if $f'(m_0) = 0$ and $f''(m_0) < 0$. Since the probability density function is the derivative of the distribution function $F(\cdot)$, these conditions may be stated in terms of the distribution function: a point m_0 is a mode if $F''(m_0) = 0$ and $F'''(m_0) < 0$. Similarly, for a discrete probability law with probability mass function $p(\cdot)$ a mode of the probability law is defined as a number m_0 at which the probability mass function has a relative maximum; more precisely, $p(m_0) \geq p(x)$ for x equal to the largest probability mass point less than m_0 and for x equal to the smallest probability mass point larger than m_0. A probability law is said to be (i) *unimodal* if it possesses just 1 mode, (ii) *bimodal* if it possesses exactly 2 modes, and so on. Give examples of continuous and discrete probability laws which are (*a*) unimodal, (*b*) bimodal. Give examples of continuous and discrete probability laws for which the mean, median, and mode (*c*) coincide, (*d*) are all different.

2.12. The interquartile range of a probability law. Possible measures exist of the dispersion of a probability distribution, in addition to the variance, which one may consider (especially if the variance is infinite). The most important of these is the *interquartile range* of the probability law, defined as follows: for any number p, between 0 and 1, define the p percentile $\mu(p)$ of the probability law as the number satisfying $F(\mu(p) - 0) \le p \le F(\mu(p) + 0)$. Thus $\mu(p)$ is the projection on the x-axis of the point in the (x, y)-plane at which the line $y = p$ intersects the curve $y = F(x)$. The 0.5 percentile is usually called the median. The *interquartile range*, defined as the difference $\mu(0.75) - \mu(0.25)$, may be taken as a measure of the dispersion of the probability law.

(i) Show that the ratio of the interquartile range to the standard deviation is (a), for the normal probability law with parameters m and σ, 1.3490, (b), for the exponential probability law with parameter λ, $\log_e 3 = 1.099$, (c), for the uniform probability law over the interval a to b, $\sqrt{3}$.

(ii) Show that the Cauchy probability law specified by the probability density function $f(x) = [\pi(1 + x^2)]^{-1}$ possesses neither a mean nor a variance. However, it possesses a median and an interquartile range given by $m_e = \mu(\frac{1}{2}) = 0$, $\mu(\frac{3}{4}) - \mu(\frac{1}{4}) = 2$.

EXERCISES

In exercises 2.1 to 2.7, compute the mean and variance of the probability law specified by the probability density function, probability mass function, or distribution function given.

2.1. (i)
$$\begin{aligned} f(x) &= 2x && \text{for } 0 < x < 1 \\ &= 0 && \text{elsewhere.} \end{aligned}$$

(ii)
$$\begin{aligned} f(x) &= |x| && \text{for } |x| \le 1 \\ &= 0 && \text{elsewhere.} \end{aligned}$$

(iii)
$$\begin{aligned} f(x) &= \frac{1 + .8x}{2} && |x| < 1, \\ &= 0 && \text{elsewhere.} \end{aligned}$$

2.2. (i)
$$\begin{aligned} f(x) &= 1 - |1 - x| && \text{for } 0 < x < 2 \\ &= 0 && \text{elsewhere.} \end{aligned}$$

(ii)
$$\begin{aligned} f(x) &= \frac{1}{2\sqrt{x}} && \text{for } 0 < x < 1 \\ &= 0 && \text{elsewhere.} \end{aligned}$$

2.3. (i)
$$f(x) = \frac{1}{\pi\sqrt{3}}\left(1 + \frac{x^2}{3}\right)^{-1}$$

(ii)
$$f(x) = \frac{2}{\pi\sqrt{3}}\left(1 + \frac{x^2}{3}\right)^{-2}$$

(iii)
$$f(x) = \frac{8/3}{\pi\sqrt{3}}\left(1 + \frac{x^2}{3}\right)^{-3}$$

2.4. (i) $f(x) = \frac{1}{2\sqrt{2\pi}} e^{-\frac{1}{2}\left(\frac{x-2}{2}\right)^2}$

(ii) $f(x) = \sqrt{(2/\pi)}e^{-\frac{1}{2}x^2}$ for $x > 0$
$= 0$ elsewhere.

2.5. (i) $p(x) = \frac{1}{3}$ for $x = 0$
$= \frac{2}{3}$ for $x = 1$,
$= 0$ elsewhere.

(ii) $p(x) = \binom{6}{x}\left(\frac{2}{3}\right)^x\left(\frac{1}{3}\right)^{6-x}$ for $x = 0, 1, \cdots, 6$
$= 0$ elsewhere.

(iii) $p(x) = \frac{\binom{8}{x}\binom{4}{6-x}}{\binom{12}{6}}$ for $x = 0, 1, \cdots, 6$
$= 0$ elsewhere.

2.6. (i) $p(x) = \frac{2}{3}\left(\frac{1}{3}\right)^{x-1}$ for $x = 1, 2, \cdots$
$= 0$ otherwise.

(ii) $p(x) = e^{-2}\frac{2^x}{x!}$ for $x = 0, 1, 2, \cdots$
$= 0$ otherwise.

2.7. (i) $F(x) = 0$ for $x < 0$
$= x^2$ for $0 \le x \le 1$
$= 1$ for $x > 1$.

(ii) $F(x) = 0$ for $x < 0$
$= x^{\frac{1}{2}}$ for $0 \le x \le 1$
$= 1$ for $x > 1$.

2.8. Compute the means and variances of the probability laws obeyed by the numerical valued random phenomena described in exercise 4.1 of Chapter 4.

2.9. For what values of r does the probability law, specified by the following probability density function, possess (i) a finite mean, (ii) a finite variance:

$$f(x) = \frac{r-1}{2|x|^r}, \qquad |x| > 1$$
$$= 0 \qquad \text{otherwise.}$$

3. MOMENT-GENERATING FUNCTIONS

The evaluation of expectations requires the use of operations of summation and integration for which completely routine methods are not available. We now discuss a method of evaluating the moments of a probability law, which, when available, requires the performance of only one summation or integration, after which all the moments of the probability law can be obtained by routine differentiation.

The *moment-generating function* of a probability law is a function $\psi(\cdot)$, defined for all real numbers t by

$$\psi(t) = E[e^{tx}]. \tag{3.1}$$

In words, $\psi(t)$ is the expectation of the exponential function e^{tx}.

In the case of a discrete probability law, specified by a probability mass function $p(\cdot)$, the moment-generating function is given by

$$\psi(t) = \sum_{\substack{\text{over all points } x \text{ such} \\ \text{that } p(x) > 0}} e^{tx} p(x). \tag{3.2}$$

In the case of a continuous probability law, specified by a probability density function $f(\cdot)$, the moment-generating function is given by

$$\psi(t) = \int_{-\infty}^{\infty} e^{tx} f(x)\, dx. \tag{3.3}$$

Since, for fixed t, the integrand e^{tx} is a positive function of x, it follows that $\psi(t)$ is either finite or infinite. We say that a probability law *possesses a moment-generating function* if there exists a positive number T such that $\psi(t)$ is finite for $|t| \le T$. It may then be shown that all moments of the probability law exist and may be expressed in terms of the successive derivatives at $t = 0$ of the moment-generating function [see (3.5)]. We have already shown that there are probability laws without finite means. Consequently, probability laws that do not possess moment-generating functions also exist. It may be seen in Chapter 9 that for every probability law one can define a function, called the characteristic function, that always exists and can be used as a moment-generating function to obtain those moments that do exist.

If a moment-generating function $\psi(t)$ exists for $|t| \le T$ (for some $T > 0$), then one may form its successive derivatives by successively differentiating under the integral or summation sign. Consequently, we obtain

$$\begin{aligned}
\psi'(t) &= \frac{d}{dt}\psi(t) = E\left[\frac{\partial}{\partial t} e^{tx}\right] = E[xe^{tx}] \\
\psi''(t) &= \frac{d^2}{dt^2}\psi(t) = E\left[\frac{\partial}{\partial t} xe^{tx}\right] = E[x^2e^{tx}]. \\
\psi^{(3)}(t) &= \frac{d^3}{dt^3}\psi(t) = E\left[\frac{\partial}{\partial t} x^2e^{tx}\right] = E[x^3e^{tx}] \\
&\cdots \\
\psi^{(n)}(t) &= \frac{d^n}{dt^n}\psi(t) = E[x^ne^{tx}].
\end{aligned} \tag{3.4}$$

Letting $t = 0$, we obtain

$$\psi'(0) = E[x]$$

$$\psi''(0) = E[x^2]$$

(3.5) $$\psi^{(3)}(0) = E[x^3]$$

$$\cdots$$

$$\psi^{(n)}(0) = E[x^n].$$

If the moment-generating function $\psi(t)$ is finite for $|t| \leq T$ (for some $T > 0$), it then possesses a power-series expansion (valid for $|t| < T$).

(3.6) $$\psi(t) = 1 + E[x]t + E[x^2]\frac{t^2}{2!} + \cdots + E[x^n]\frac{t^n}{n!} + \cdots.$$

To prove (3.6), use the definition of $\psi(t)$ and the fact that

(3.7) $$e^{tx} = 1 + xt + x^2\frac{t^2}{2!} + \cdots + x^n\frac{t^n}{n!} + \cdots.$$

In view of (3.6), if one can readily obtain the power-series expansion of $\psi(t)$, then one can readily obtain the nth moment $E[x^n]$ for any integer n, since $E[x^n]$ is the coefficient of $t^n/n!$ in the power-series expansion of $\psi(t)$.

▶ **Example 3A.** The Bernoulli probability law with parameter p has a moment-generating function for $-\infty < t < \infty$.

(3.8) $$\psi(t) = e^{t0} \cdot q + e^{t1} \cdot p = pe^t + q$$

with derivatives

(3.9) $$\psi'(t) = pe^t, \qquad E[x] = \psi'(0) = p,$$
$$\psi''(t) = pe^t, \qquad E[x^2] = \psi''(0) = p.$$ ◀

▶ **Example 3B.** The binomial probability law with parameters n and p has a moment-generating function for $-\infty < t < \infty$.

(3.10) $$\psi(t) = \sum_{k=0}^{n} e^{tk}p(k) = \sum_{k=0}^{n}\binom{n}{k}(pe^t)^k q^{n-k} = (pe^t + q)^n,$$

with derivatives

$$\psi'(t) = npe^t(pe^t + q)^{n-1}, \qquad E[x] = \psi'(0) = np,$$

(3.11) $$\psi''(t) = npe^t(pe^t + q)^{n-1} + n(n-1)p^2e^{2t}(pe^t + q)^{n-2},$$

$$E[x^2] = np + n(n-1)p^2 = npq + n^2p^2.$$ ◀

TABLE 3A

SOME FREQUENTLY ENCOUNTERED DISCRETE PROBABILITY LAWS WITH THEIR MOMENTS AND GENERATING FUNCTIONS

Probability Law	Parameters	Probability Mass Function $p(\cdot)$	Mean $m = E[x]$	Variance $\sigma^2 = E[x^2] - E^2[x]$
Bernoulli	$0 \le p \le 1$	$p(x) = p \quad x = 1$ $= q \quad x = 0$ $= 0 \quad$ otherwise	p	pq
Binomial	$n = 1, 2, \cdots$ $0 \le p \le 1$	$p(x) = \binom{n}{x} p^x q^{n-x} \quad x = 0, 1, 2, \cdots, n$ $= 0 \quad$ otherwise	np	npq
Poisson	$\lambda > 0$	$p(x) = e^{-\lambda} \frac{\lambda^x}{x!} \quad x = 0, 1, 2, \cdots$ $= 0 \quad$ otherwise	λ	λ
Geometric	$0 \le p \le 1$	$p(x) = pq^{x-1} \quad x = 1, 2, \cdots$ $= 0 \quad$ otherwise	$\frac{1}{p}$	$\frac{q}{p^2}$
Negative binomial	$r > 0$ $0 \le p \le 1$	$p(x) = \binom{r + x - 1}{x} p^r q^x$ $= \binom{-r}{x} p^r(-q)^x, \quad x = 0, 1, 2, \cdots$ $= 0 \quad$ otherwise	$\frac{rq}{p} = rP$ if $P = \frac{q}{p}$	$\frac{rq}{p^2} = rPQ$ if $Q = \frac{1}{p}$
Hypergeometric	$N = 1, 2, \cdots$ $n = 1, 2, \cdots, N$ $p = 0, \frac{1}{N}, \frac{2}{N}, \cdots, 1$	$p(x) = \frac{\binom{Np}{x}\binom{Nq}{n-x}}{\binom{N}{n}} \quad x = 0, 1, \cdots, n$ $= 0 \quad$ otherwise	np	$npq\left(\frac{N-n}{N-1}\right)$

TABLE 3A (*Continued*).

SOME FREQUENTLY ENCOUNTERED DISCRETE PROBABILITY LAWS WITH THEIR MOMENTS AND GENERATING FUNCTIONS

Probability Law	Moment-Generating Function $\psi(t) = E[e^{tx}]$	Characteristic Function $\phi(u) = E[e^{iux}]$	Third Central Moment $E[(x - E[x])^3]$	Fourth Central Moment $E[(x - E[x])^4]$
Bernoulli	$pe^t + q$	$pe^{iu} + q$	$pq(q - p)$	$3p^2q^2 + pq(1 - 6pq)$
Binomial	$(pe^t + q)^n$	$(pe^{iu} + q)^n$	$npq(q - p)$	$3n^2p^2q^2 + pqn(1 - 6pq)$
Poisson	$e^{\lambda(e^t - 1)}$	$e^{\lambda(e^{iu} - 1)}$	λ	$\lambda + 3\lambda^2$
Geometric	$\dfrac{pe^t}{1 - qe^t}$	$\dfrac{pe^{iu}}{1 - qe^{iu}}$	$\dfrac{q}{p^2}\left(1 + 2\dfrac{q}{p}\right)$	$\dfrac{q}{p^2}\left(1 + 9\dfrac{q}{p^2}\right)$
Negative binomial	$\left(\dfrac{P}{1 - qe^t}\right)^r = (Q - Pe^t)^{-r}$	$\left(\dfrac{P}{1 - qe^{iu}}\right)^r = (Q - Pe^{iu})^{-r}$	$\dfrac{rq}{p^2}\left(1 + 2\dfrac{q}{p}\right) = rPQ(Q + P)$	$\dfrac{rq}{p^2}\left(1 + (6 + 3r)\dfrac{q}{p^2}\right) = 3r^2P^2Q^2 + rPQ(1 + 6PQ)$
Hypergeometric	see M. G. Kendall, *Advanced Theory of Statistics*, Charles Griffin, London, 1948, p. 127.			

TABLE 3B

SOME FREQUENTLY ENCOUNTERED CONTINUOUS PROBABILITY LAWS WITH THEIR MOMENTS AND GENERATING FUNCTIONS

Probability Law	Parameters	Probability Density Function $f(\cdot)$	Mean $m = E[x]$	Variance $\sigma^2 = E[x^2] - E^2(x)$
Uniform over interval a to b	$-\infty < a < b < \infty$	$f(x) = \frac{1}{b-a} \quad a < x < b$ $= 0 \quad$ otherwise	$\frac{a+b}{2}$	$\frac{(b-a)^2}{12}$
Normal	$-\infty < m < \infty$ $\sigma > 0$	$f(x) = \frac{1}{\sigma\sqrt{2\pi}} e^{-\frac{1}{2}\left(\frac{x-m}{\sigma}\right)^2}$	m	σ^2
Exponential	$\lambda > 0$	$f(x) = \lambda e^{-\lambda x} \quad x > 0$ $= 0 \quad$ otherwise	$\frac{1}{\lambda}$	$\frac{1}{\lambda^2}$
Gamma	$r > 0$	$f(x) = \frac{\lambda}{\Gamma(r)} (\lambda x)^{r-1} e^{-\lambda x} \quad x > 0$ $= 0 \quad$ otherwise	$\frac{r}{\lambda}$	$\frac{r}{\lambda^2}$

TABLE 3B (*Continued*).

SOME FREQUENTLY ENCOUNTERED CONTINUOUS PROBABILITY LAWS WITH THEIR MOMENTS AND GENERATING FUNCTIONS

Probability Law	Moment-Generating Function $\psi(t) = E[e^{tx}]$	Characteristic Function $\phi(u) = E[e^{iux}]$	Third Central Moment $E[(x - E[x])^3]$	Fourth Central Moment $E[(x - E[x])^4]$
Uniform over interval a to b	$\dfrac{e^{tb} - e^{ta}}{t(b - a)}$	$\dfrac{e^{iub} - e^{iua}}{iu(b - a)}$	0	$\dfrac{(b - a)^4}{80}$
Normal	$e^{tm + \frac{1}{2}t^2\sigma^2}$	$e^{ium - \frac{1}{2}u^2\sigma^2}$	0	$3\sigma^4$
Exponential	$\dfrac{\lambda}{\lambda - t} = \left(1 - \dfrac{t}{\lambda}\right)^{-1}$	$\left(1 - \dfrac{iu}{\lambda}\right)^{-1}$	$\dfrac{2}{\lambda^3}$	$\dfrac{9}{\lambda^4}$
Gamma	$\left(1 - \dfrac{t}{\lambda}\right)^{-r}$	$\left(1 - \dfrac{iu}{\lambda}\right)^{-r}$	$\dfrac{2r}{\lambda^3}$	$\dfrac{6r + 3r^2}{\lambda^4}$

▶ **Example 3C.** The Poisson probability law with parameter λ has a moment-generating function for all t.

$$(3.12) \qquad \psi(t) = \sum_{k=0}^{\infty} e^{tk} p(k) = e^{-\lambda} \sum_{k=0}^{\infty} \frac{(\lambda e^t)^k}{k!} = e^{-\lambda} e^{\lambda e^t} = e^{\lambda(e^t - 1)},$$

with derivatives

$$(3.13) \qquad \begin{aligned} \psi'(t) &= e^{\lambda(e^t-1)} \lambda e^t, \qquad & E[x] &= \psi'(0) = \lambda, \\ \psi''(t) &= e^{\lambda(e^t-1)} \lambda^2 e^{2t} + \lambda e^t e^{\lambda(e^t-1)}, \qquad & E[x^2] &= \psi''(0) = \lambda^2 + \lambda. \end{aligned}$$

Consequently, the variance $\sigma^2 = E[x^2] - E^2[x] = \lambda$. Thus *for the Poisson probability law the mean and the variance are equal.* ◀

▶ **Example 3D.** The geometric probability law with parameter p has a moment-generating function for t such that $qe^t < 1$.

$$(3.14) \qquad \psi(t) = \sum_{k=1}^{\infty} e^{tk} p(k) = pe^t \sum_{k=1}^{\infty} (qe^t)^{k-1} = pe^t \frac{1}{1 - qe^t}.$$

From (3.14) one may show that the mean and variance of the geometric probability law are given by

$$(3.15) \qquad m = E[x] = \frac{1}{p}, \qquad \sigma^2 = E[x^2] - E^2[x] = \frac{q}{p^2}.$$ ◀

▶ **Example 3E.** The normal probability law with mean m and variance σ has a moment-generating function for $-\infty < t < \infty$.

$$(3.16) \qquad \begin{aligned} \psi(t) &= \frac{1}{\sqrt{2\pi}\sigma} \int_{-\infty}^{\infty} e^{tx} e^{-\frac{1}{2}\left(\frac{x-m}{\sigma}\right)^2} dx = e^{mt} \frac{1}{\sqrt{2\pi}} \int_{-\infty}^{\infty} e^{ty\sigma} e^{-\frac{1}{2}y^2} dy \\ &= e^{mt + (\frac{1}{2})\sigma^2 t^2} \frac{1}{\sqrt{2\pi}} \int_{-\infty}^{\infty} e^{-\frac{1}{2}(y - \sigma t)^2} dy = e^{mt + (\frac{1}{2})\sigma^2 t^2}. \end{aligned}$$

From (3.16) one may show that the central moments of the normal probability law are given by

$$(3.17) \qquad \begin{aligned} E[(x - m)^n] &= 0 \qquad \text{if } n = 3, 5, \cdots, \\ &= 1 \cdot 3 \cdot 5 \cdots (n-1)\sigma^n \qquad \text{if } n = 2, 4, \cdots. \end{aligned}$$

An alternate method of deriving (3.17) is by use of (2.22) in Chapter 4. ◀

▶ **Example 3F.** The exponential probability law with parameter λ has a moment-generating function for $t < \lambda$.

$$(3.18) \qquad \psi(t) = \lambda \int_0^{\infty} e^{tx} e^{-\lambda x} dx = \frac{\lambda}{\lambda - t} = \left(1 - \frac{t}{\lambda}\right)^{-1}.$$

One may show from (3.18) that *for the exponential probability law the mean m and the standard deviation σ are equal, and are equal to the reciprocal of the parameter λ.* ◀

▶ **Example 3G. The lifetime of a radioactive atom.** It is shown in section 4 of Chapter 6 that the time between emissions of particles by a radioactive atom obeys an exponential probability law with parameter λ. By example 3F, the mean time between emissions is $1/\lambda$. The time between emissions is called the lifetime of the atom. The half life of the atom is defined as the time T such that the probability is $\frac{1}{2}$ that the lifetime will be greater than T. Since the probability that the lifetime will be greater than a given number t is $e^{-\lambda t}$, it follows that T is the solution of $e^{-\lambda T} = \frac{1}{2}$, or $T = \log_e 2/\lambda$. In words, the half life T is equal to the mean $1/\lambda$ multiplied by $\log_e 2$. ◀

THEORETICAL EXERCISES

3.1. Generating function of moments about a point. Define the moment-generating function of a probability law about a point c as a function $\psi_c(\cdot)$ defined for all real numbers t by $\psi_c(t) = E[e^{t(x-c)}]$. Show that $\psi_c(t)$ may be obtained from $\psi(t)$ by $\psi_c(t) = e^{-ct}\psi(t)$. The nth moment $E[(x - c)^n]$ of the probability law about the point c is given by $E[(x - c)^n] = \psi_c^{(n)}(0)$ and may be read off as the coefficient of $t^n/n!$ in the power-series expansion of $\psi_c(t)$.

3.2. The factorial moment-generating function. $\Phi(u)$ of a probability law is defined for all u such that $|u| < 1$ by

$$\Phi(u) = E[(1 + u)^x] = E[e^{x \log (1+u)}] = \psi[\log (1 + u)].$$

Its nth derivative evaluated at $u = 0$

$$\Phi^{(n)}(0) = E[x(x - 1) \cdots (x - n + 1)]$$

is called the nth factorial moment of the probability law. From a knowledge of the first n factorial moments of a probability law one may obtain a knowledge of the first n moments of the probability law, and conversely. Thus, for example,

$$(3.19) \quad E[x(x - 1)] = E[x^2] - E[x], \qquad E[x^2] = E[x(x - 1)] + E[x].$$

Equation (3.19) was implicitly used in calculating certain second moments and variances in section 2. Show that the first n moments of two distinct probability laws coincide if and only if their first n factorial moments coincide. *Hint*: Consult M. Kendall, *The Advanced Theory of Statistics*, Vol. I, Griffin, London, 1948, p. 58.

3.3. The factorial moment-generating function of the probability law of the number of matches in the matching problem. The number of matches obtained by distributing, 1 to an urn, M balls, numbered 1 to M, among

M urns, numbered 1 to M, has a probability law specified by the probability mass function

$$p(m) = \frac{1}{m!}\sum_{k=0}^{M-m}(-1)^k\frac{1}{k!} \qquad m = 0, 1, 2, \ldots, M \tag{3.20}$$
$$= 0 \qquad \text{otherwise.}$$

Show that the corresponding moment-generating function may be written

$$\psi(t) = \sum_{m=0}^{M} e^{tm}p(m) = \sum_{r=0}^{M}\frac{1}{r!}(e^t - 1)^r. \tag{3.21}$$

Consequently the factorial moment-generating function of the number of matches may be written

$$\Phi(u) = \sum_{r=0}^{M}\frac{1}{r!}u^r. \tag{3.22}$$

3.4. *The first M moments of the number of matches in the problem of matching M balls in M urns coincide with the first M moments of the Poisson probability law with parameter* $\lambda = 1$. Show that the factorial moment-generating function of the Poisson law with parameter λ is given by

$$\Phi_{P}(u) = e^{\lambda u} = \sum_{r=0}^{\infty}\frac{\lambda^r}{r!}u^r. \tag{3.23}$$

By comparing (3.22) and (3.23), it follows that the first M factorial moments, and, consequently, the first M moments of the probability law of the number of matches and the Poisson probability law with parameter 1, coincide.

EXERCISES

Compute the moment generating function, mean, and variance of the probability law specified by the probability density function, probability mass function, or distribution function given.

3.1. (i)
$$f(x) = e^{-x} \quad \text{for } x \geq 0$$
$$= 0 \quad \text{elsewhere.}$$

(ii)
$$f(x) = e^{-(x-5)} \quad \text{for } x \geq 5$$
$$= 0 \quad \text{elsewhere.}$$

3.2. (i)
$$f(x) = \frac{1}{\sqrt{2\pi x}}e^{-x/2} \quad \text{for } x > 0$$
$$= 0 \quad \text{elsewhere.}$$

(ii)
$$f(x) = \tfrac{1}{4}xe^{-x/2} \quad \text{for } x > 0$$
$$= 0 \quad \text{elsewhere.}$$

3.3. (i)
$$p(x) = \tfrac{2}{3}(\tfrac{1}{3})^{x-1} \quad \text{for } x = 1, 2, \cdots$$
$$= 0 \quad \text{elsewhere.}$$

(ii)
$$p(x) = e^{-2}\frac{2^x}{x!} \quad \text{for } x = 0, 1, \cdots$$
$$= 0 \quad \text{elsewhere.}$$

3.4. (i) $$F(x) = \Phi\left(\frac{x-2}{2}\right)$$

(ii) $$\begin{aligned} F(x) &= 0 && \text{for } x < 0 \\ &= 1 - e^{-x/5} && \text{for } x \geq 0. \end{aligned}$$

3.5. Find the mean, variance, third central moment, and fourth central moment of the number of matches when (i) 4 balls are distributed in 4 urns, 1 to an urn, (ii) 3 balls are distributed in 3 urns, 1 to an urn.

3.6. Find the factorial moment-generating function of the (i) binomial, (ii) Poisson, (iii) geometric probability laws and use it to obtain their means, variances, and third and fourth central moments.

4. CHEBYSHEV'S INEQUALITY

From a knowledge of the mean and variance of a probability law one cannot in general determine the probability law. In the circumstance that the functional form of the probability law is known up to several unspecified parameters (for example, a probability law may be assumed to be a normal distribution with parameters m and σ), it is often possible to relate the parameters and the mean and variance. One may then use a knowledge of the mean and variance to determine the probability law. In the case in which the functional form of the probability law is unknown one can obtain crude estimates of the probability law, which suffice for many purposes, from a knowledge of the mean and variance.

For any probability law with finite mean m and finite variance σ^2, define the quantity $Q(h)$, for any $h > 0$, as the probability assigned to the interval $\{x:\ m - h\sigma < x \leq m + h\sigma\}$ by the probability law. In terms of a distribution function $F(\cdot)$ or a probability density function $f(\cdot)$,

$$Q(h) = F(m + h\sigma) - F(m - h\sigma) = \int_{m-h\sigma}^{m+h\sigma} f(x)\,dx. \tag{4.1}$$

Let us compute $Q(h)$ in certain cases. For the normal probability law with mean m and standard deviation σ

$$Q(h) = \frac{1}{\sqrt{2\pi}\sigma}\int_{m-h\sigma}^{m+h\sigma} e^{-\frac{1}{2}\left(\frac{y-m}{\sigma}\right)^2}\,dy = \Phi(h) - \Phi(-h). \tag{4.2}$$

For the exponential law with mean $1/\lambda$

$$\begin{aligned} Q(h) &= e^{-1}(e^{h} - e^{-h}) && \text{for } h \leq 1 \\ &= 1 - e^{-(1+h)} && \text{for } h \geq 1. \end{aligned} \tag{4.3}$$

For the uniform distribution over the interval a to b, for $h < \sqrt{3}$,

$$(4.4) \qquad Q(h) = \frac{1}{b-a} \int_{\frac{b+a}{2} - h\frac{b-a}{\sqrt{12}}}^{\frac{b+a}{2} + h\frac{b-a}{\sqrt{12}}} dx = \frac{h}{\sqrt{3}}.$$

For the other frequently encountered probability laws one cannot so readily evaluate $Q(h)$. Nevertheless, the function $Q(h)$ is still of interest, since it is possible to obtain a lower bound for it, which does not depend on the probability law under consideration. This lower bound, known as Chebyshev's inequality, was named after the great Russian probabilist P. L. Chebyshev (1821–1894).

Chebyshev's inequality. For any distribution function $F(\cdot)$ and any $h \geq 0$

$$(4.5) \qquad Q(h) = F(m + h\sigma) - F(m - h\sigma) \geq 1 - \frac{1}{h^2}.$$

Note that (4.5) is trivially true for $h < 1$, since the right-hand side is then negative.

We prove (4.5) for the case of a continuous probability law with probability density function $f(\cdot)$. It may be proved in a similar manner (using Stieltjes integrals, introduced in section 6) for a general distribution function. The inequality (4.5) may be written in the continuous case

$$(4.6) \qquad \int_{m-h\sigma}^{m+h\sigma} f(x)\, dx \geq 1 - \frac{1}{h^2}.$$

To prove (4.6), we first obtain the inequality

$$(4.7) \qquad \sigma^2 \geq \int_{-\infty}^{m-h\sigma} (x - m)^2 f(x)\, dx + \int_{m+h\sigma}^{\infty} (x - m)^2 f(x)\, dx$$

that follows, since the variance σ^2 is equal to the sum of the two integrals on the right-hand side of (4.7), plus the nonnegative quantity $\int_{m-h\sigma}^{m+h\sigma} (x - m)^2 f(x)\, dx$. Now for $x \leq m - h\sigma$, it holds that $(x - m)^2 \geq h^2\sigma^2$. Similarly, $x \geq m + h\sigma$ implies $(x - m)^2 \geq h^2\sigma^2$. By replacing $(x - m)^2$ by these lower bounds in (4.7), we obtain

$$(4.8) \qquad \sigma^2 \geq \sigma^2 h^2 \left[\int_{-\infty}^{m-h\sigma} f(x)\, dx + \int_{m+h\sigma}^{\infty} f(x)\, dx \right].$$

The sum of the two integrals in (4.8) is equal to $1 - Q(h)$. Therefore (4.8) implies that $1 - Q(h) \leq (1/h^2)$, and (4.5) is proved.

In Fig. 4A the function $Q(h)$, given by (4.2), (4.3), and (4.4), and the lower bound for $Q(h)$, given by Chebyshev's inequality, are plotted.

In terms of the observed value X of a numerical valued random phenomenon, Chebyshev's inequality may be reformulated as follows. The quantity $Q(h)$ is then essentially equal to $P[|X - m| \leq h\sigma]$; in words, $Q(h)$ is equal to the probability that an observed value of a numerical valued random phenomenon, with distribution function $F(\cdot)$, will lie in an

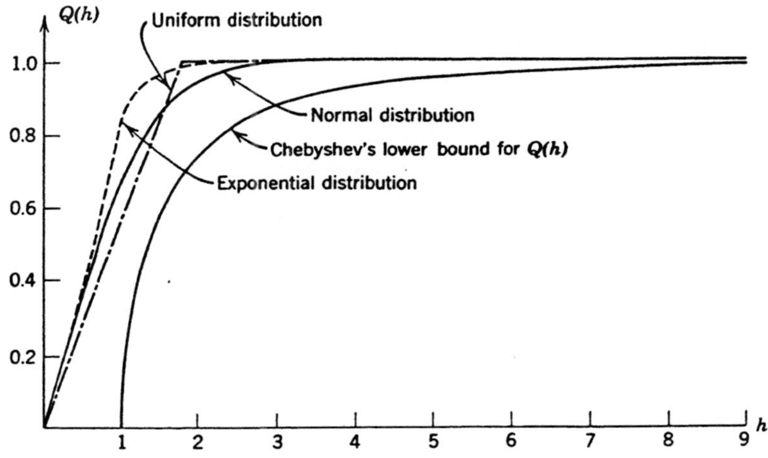

Fig. 4A. Graphs of the function $Q(h)$.

interval centered at the mean and of length $2h$ standard deviations. Chebyshev's inequality may be reformulated: for any $h > 0$

$$(4.9) \qquad P[|X - m| \leq h\sigma] \geq 1 - \frac{1}{h^2}, \qquad P[|X - m| > h\sigma] \leq \frac{1}{h^2}.$$

Chebyshev's inequality (with $h = 4$) states that the probability is at least 0.9375 that an observed value X will lie within four standard deviations of the mean, whereas the probability is at least 0.99 that an observed value X will lie within ten standard deviations of the mean. Thus, in terms of the standard deviation σ (and consequently in terms of the variance σ^2), we can state intervals in which, with very high probability, an observed value of a numerical valued random phenomenon may be expected to lie. It may be remarked that it is this fact that renders the variance a measure of the *spread*, or *dispersion*, of the probability mass that a probability law distributes over the real line.

Generalizations of Chebyshev's inequality. As a practical tool for using the lower-order moments of a probability law for obtaining inequalities on its distribution function, Chebyshev's inequality can be improved upon if various additional facts about the distribution function are known. Expository surveys of various generalizations of Chebyshev's inequality are given by H. J. Godwin, "On generalizations of Tchebychef's inequality," *Journal of the American Statistical Association*, Vol. 50 (1955), pp. 923–945, and by C. L. Mallows, "Generalizations of Tchebycheff's inequalities," *Journal of the Royal Statistical Society*, Series B, Vol. 18 (1956), pp. 139–176 (with discussion).

EXERCISES

4.1. Use Chebyshev's inequality to determine how many times a fair coin must be tossed in order that the probability will be at least 0.90 that the ratio of the observed number of heads to the number of tosses will lie between 0.4 and 0.6.

4.2. Suppose that the number of airplanes arriving at a certain airport in any 20-minute period obeys a Poisson probability law with mean 100. Use Chebyshev's inequality to determine a lower bound for the probability that the number of airplanes arriving in a given 20-minute period will be between 80 and 120.

4.3. Consider a group of N men playing the game of "odd man out" (that is, they repeatedly perform the experiment in which each man independently tosses a fair coin until there is an "odd" man, in the sense that either exactly 1 of the N coins falls heads or exactly 1 of the N coins falls tails). Find, for (i) $N = 4$, (ii) $N = 8$, the exact probability that the number of repetitions of the experiment required to conclude the game will be within 2 standard deviations of the mean number of repetitions required to conclude the game. Compare your answer with the lower bound given by Chebyshev's inequality.

4.4. For Pareto's distribution, defined in theoretical exercise 2.2, compute and graph the function $Q(h)$, for $A = 1$ and $r = 3$ and 4, and compare it with the lower bound given by Chebyshev's inequality.

5. THE LAW OF LARGE NUMBERS FOR INDEPENDENT REPEATED BERNOULLI TRIALS

Consider an experiment with two possible outcomes, denoted by success and failure. Suppose, however, that the probability p of success at each trial is unknown. According to the frequency interpretation of probability, p represents the relative frequency of successes in an indefinitely prolonged series of trials. Consequently, one might think that in order to determine p one must only perform a long series of trials and take as the value of p the observed relative frequency of success. The question arises: can one justify this procedure, not by appealing to the frequency interpretation of probability theory, but by appealing to the mathematical theory of probability?

The mathematical theory of probability is a logical construct, consisting of conclusions logically deduced from the axioms of probability theory. These conclusions are applicable to the world of real experience in the sense that they are conclusions about real phenomena, which are *assumed* to satisfy the axioms. We now show that one can reach a conclusion within the mathematical theory of probability that may be interpreted to justify the frequency interpretation of probability (and consequently may be used to justify the procedure described for estimating p). This result is known as the law of large numbers, since it applies to the outcome of a large number of trials. The law of large numbers we are about to investigate may be considerably generalized. Consequently, the version to be discussed is called the *Bernoulli law of large numbers*, as it was first discovered by Jacob Bernoulli and published in his posthumous book *Ars conjectandi* (1713).

The Bernoulli Law of Large Numbers. Let S_n be the observed number of successes in n independent repeated Bernoulli trials, with probability p of success at each trial. Let

$$f_n = \frac{S_n}{n} \tag{5.1}$$

denote the relative frequency of successes in the n trials. Then, for any positive number ϵ, no matter how small, it follows that

$$\lim_{n\to\infty} P[|f_n - p| \leq \epsilon] = 1, \tag{5.2}$$

$$\lim_{n\to\infty} P[|f_n - p| > \epsilon] = 0. \tag{5.3}$$

In words, (5.2) and (5.3) state that as the number n of trials tends to infinity the relative frequency of successes in n trials tends to the true probability p of success at each trial, in the probabilistic sense that any nonzero difference ϵ between f_n and p becomes less and less probable of observation as the number of trials is increased indefinitely.

Bernoulli proved (5.3) by a tedious evaluation of the probability in (5.3). Using Chebyshev's inequality, one can give a very simple proof of (5.3). By using the fact that the probability law of S_n has mean np and variance npq, one may prove that the probability law of f_n has mean p and variance $[p(1 - p)]/n$. Consequently, for any $\epsilon > 0$

$$P[|f_n - p| > \epsilon] \leq \frac{p(1 - p)}{n\epsilon^2}. \tag{5.4}$$

Now, for any value of p in the interval $0 \leq p \leq 1$

$$p(1-p) \leq \tfrac{1}{4}, \tag{5.5}$$

using the fact that $4p(1-p) - 1 = -(2p-1)^2 \leq 0$. Consequently, for any $\epsilon > 0$

$$P[|f_n - p| > \epsilon] \leq \frac{1}{4n\epsilon^2} \to 0 \qquad \text{as } n \to \infty, \tag{5.6}$$

no matter what the true value of p. To prove (5.2), one uses (5.3) and the fact that

$$P[|f_n - p| \leq \epsilon] = 1 - P[|f_n - p| > \epsilon]. \tag{5.7}$$

It is shown in section 5 of Chapter 8 that the foregoing method of proof, using Chebyshev's inequality, permits one to prove that if $X_1, X_2, \ldots, X_n, \ldots$ is a sequence of independent observations of a numerical valued random phenomenon whose probability law has mean m then for any $\epsilon > 0$

$$\lim_{n \to \infty} P\left[\left|\frac{X_1 + X_2 + \cdots + X_n}{n} - m\right| > \epsilon\right] = 0. \tag{5.8}$$

The result given by (5.8) is known as the law of large numbers.

The Bernoulli law of large numbers states that to estimate the unknown value of p, as an estimate of p, the observed relative frequency f_n of successes in n trials can be employed; this estimate becomes perfectly correct as the number of trials becomes infinitely large. In practice, a finite number of trials is performed. Consequently, the number of trials must be determined, in order that, with high probability, the observed relative frequency be within a preassigned distance ϵ from p. In symbols, to any number α one desires to find n so that

$$P[|f_n - p| \leq \epsilon \mid p] \geq \alpha \qquad \text{for all } p \text{ in } 0 \leq p \leq 1 \tag{5.9}$$

where we write $P[\cdot \mid p]$ to indicate that the probability is being calculated under the assumption that p is the true probability of success at each trial.

One may obtain an expression for the value of n that satisfies (5.9) by means of Chebyshev's inequality. Since

$$P[|f_n - p| \leq \epsilon] \geq 1 - \frac{1}{4n\epsilon^2} \qquad \text{for all } p \text{ in } 0 \leq p \leq 1, \tag{5.10}$$

it follows that (5.9) is satisfied if n is chosen so that

$$n \geq \frac{1}{4\epsilon^2(1-\alpha)}. \tag{5.11}$$

▶ **Example 5A.** How many trials of an experiment with two outcomes, called A and B, should be performed in order that the probability be 95% or better that the observed relative frequency of occurrences of A will differ from the probability p of occurrence of A by no more than 0.02? Here $\alpha = 0.95$, $\epsilon = 0.02$. Therefore, the number n of trials should be chosen so that $n \geq 12{,}500$. ◀

The estimate of n given by (5.11) can be improved upon. In section 2 of Chapter 6 we prove the normal approximation to the binomial law. In particular, it is shown that if p is the probability of success at each trial then the number S_n of successes in n independent repeated Bernoulli trials approximately satisfies, for any $h > 0$,

$$P\left[\frac{|S_n - np|}{\sqrt{npq}} \leq h\right] = 2\Phi(h) - 1. \tag{5.12}$$

Consequently, the relative frequency of successes satisfies, for any $\epsilon > 0$,

$$P[|f_n - p| \leq \epsilon] \doteq 2\Phi(\epsilon\sqrt{n/pq}) - 1. \tag{5.13}$$

To obtain (5.13) from (5.12), let $h = \epsilon\sqrt{n/pq}$.

Define $K(\alpha)$ as the solution of the equation

$$2\Phi(K(\alpha)) - 1 = \int_{-K(\alpha)}^{K(\alpha)} \phi(y)\, dy = \alpha. \tag{5.14}$$

A table of selected values of $K(\alpha)$ is given in Table 5A.

TABLE 5A

α	$K(\alpha)$
0.50	0.675
0.6827	1.000
0.90	1.645
0.95	1.960
0.9546	2.000
0.99	2.576
0.9973	3.000

From (5.13) we may obtain the conclusion that

$$P[|f_n - p| \leq \epsilon] \geq \alpha \qquad \text{if } \epsilon\sqrt{(n/pq)} \geq K(\alpha). \tag{5.15}$$

To justify (5.15), note that $\epsilon\sqrt{(n/pq)} > K(\alpha)$ implies that the right-hand side of (5.13) is greater than the left-hand side of (5.14).

Since $pq \leq (\frac{1}{4})$ for all p, we finally obtain from (5.15) that (5.9) *will hold if*

$$n \geq \frac{K^2(\alpha)}{4\epsilon^2}. \tag{5.16}$$

▶ **Example 5B.** If $\alpha = 0.95$ and $\epsilon = 0.02$, then according to (5.16) n should be chosen so that $n \geq 2500$. Thus the number of trials required for f_n to be within 0.02 of p with probability greater than 95% is approximately 2500, which is $\frac{1}{5}$ of the number of trials that Chebyshev's inequality states is required. ◀

EXERCISES

5.1. A sample is taken to find the proportion p of smokers in a certain population. Find a sample size so that the probability is (i) 0.95 or better, (ii) 0.99 or better that the observed proportion of smokers will differ from the true proportion of smokers by less than (*a*) 1%, (*b*) 10%.

5.2. Consider an urn that contains 10 balls numbered 0 to 9, each of which is equally likely to be drawn; thus choosing a ball from the urn is equivalent to choosing a number 0 to 9; this experiment is sometimes described by saying a *random digit* has been chosen. Let n balls be chosen with replacement.

(i) What does the law of large numbers tell you about occurrences of 9's in the n drawings.

(ii) How many drawings must be made in order that, with probability 0.95 or better, the relative frequency of occurrence of 9's will be between 0.09 and 0.11?

5.3. If you wish to estimate the proportion of engineers and scientists who have studied probability theory and you wish your estimate to be correct, within 2%, with probability 0.95 or better, how large a sample should you take (i) if you feel confident that the true proportion is less than 0.2, (ii) if you have no idea what the true proportion is.

5.4. The law of large numbers, in popular terminology, is called the law of averages. Comment on the following advice. When you toss a fair coin to decide a bet, let your companion do the calling. "Heads" is called 7 times out of 10. The simple law of averages gives the man who listens a tremendous advantage.

6. MORE ABOUT EXPECTATION

In this section we define the expectation of a function with respect to (i) a probability law specified by its distribution function, and (ii) a numerical n-tuple valued random phenomenon.

Stieltjes Integral. In section 2 we defined the expectation of a continuous function $g(x)$ with respect to a probability law, which is specified by a probability mass function or by a probability density function. We now consider the case of a general probability law, which is specified by its distribution function $F(\cdot)$.

In order to define the expectation with respect to a probability law specified by a distribution function $F(\cdot)$, we require a generalization of the notion of integral, which goes under the name of the *Stieltjes integral*. Given a continuous function $g(x)$, a distribution function $F(\cdot)$, and a half-open interval $(a, b]$ on the real line (that is, $(a, b]$ consists of all the points strictly greater than a and less than or equal to b), we define the Stieltjes integral of $g(\cdot)$, with respect to $F(\cdot)$ over $(a, b]$, written $\int_{a+}^{b} g(x)\, dF(x)$, as follows. We start with a partition of the interval $(a, b]$ into n subintervals $(x_{i-1}, x_i]$, in which $x_0, x_1, \ldots, x_n$ are $(n + 1)$ points chosen so that $a = x_0 < x_1 < \ldots < x_n = b$. We then choose a set of points $x_1', x_2', \ldots.$ x_n', one in each subinterval, so that $x_{i-1} < x_i' < x_i$ for $i = 1, 2, \ldots, n$, We define

$$\int_{a+}^{b} g(x)\, dF(x) = \operatorname*{limit}_{n\to\infty} \sum_{i=1}^{n} g(x_i')[F(x_i) - F(x_{i-1})] \tag{6.1}$$

in which the limit is taken over all partitions of the interval $(a, b]$, as the maximum length of subinterval in the partition tends to 0.

It may be shown that if $F(\cdot)$ is specified by a probability density function $f(\cdot)$, then

$$\int_{a+}^{b} g(x)\, dF(x) = \int_{a}^{b} g(x) f(x)\, dx, \tag{6.2}$$

whereas if $F(\cdot)$ is specified by a probability mass function $p(\cdot)$ then

$$\int_{a+}^{b} g(x)\, dF(x) = \sum_{\substack{\text{over all } x \text{ such that} \\ a<x\le b \text{ and } p(x)>0}} g(x)p(x). \tag{6.3}$$

The Stieltjes integral of the continuous function $g(\cdot)$, with respect to the distribution function $F(\cdot)$ over the whole real line, is defined by

$$\int_{-\infty}^{\infty} g(x)\, dF(x) = \lim_{\substack{a\to-\infty \\ b\to\infty}} \int_{a+}^{b} g(x)\, dF(x). \tag{6.4}$$

The discussion in section 2 in regard to the existence and finiteness of integrals over the real line applies also to Stieltjes integrals. We say that $\int_{-\infty}^{\infty} g(x)\, dF(x)$ exists if and only if $\int_{-\infty}^{\infty} |g(x)|\, dF(x)$ is finite. Thus only absolutely convergent Stieltjes integrals are to be invested with sense.

We now define the expectation of a continuous function $g(\cdot)$, with respect to a probability law specified by a distribution function $F(\cdot)$, as the Stieltjes integral of $g(\cdot)$, with respect to $F(\cdot)$ over the infinite real line; in symbols,

$$E[g(x)] = \int_{-\infty}^{\infty} g(x)\, dF(x). \tag{6.5}$$

Stieltjes integrals are only of theoretical interest. They provide a compact way of defining, and working with, the properties of expectation. In practice, one evaluates a Stieltjes integral by breaking it up into a sum of an ordinary integral and an ordinary summation by means of the following theorem: if there exists a probability density function $f(\cdot)$, a probability mass function $p(\cdot)$, and constants c_1 and c_2, whose sum is 1, such that for every x

$$F(x) = c_1 \int_{-\infty}^{x} f(x')\, dx' + c_2 \sum_{\substack{\text{over all } x' \leq x \text{ such} \\ \text{that } p(x') > 0}} p(x'), \tag{6.6}$$

then for any continuous function $g(\cdot)$

$$\int_{-\infty}^{\infty} g(x)\, dF(x) = c_1 \int_{-\infty}^{\infty} g(x) f(x)\, dx + c_2 \sum_{\substack{\text{over all } x \text{ such} \\ \text{that } p(x) > 0}} g(x) p(x). \tag{6.7}$$

In giving the proofs of various propositions about probability laws we most often confine ourselves to the case in which the probability law is specified by a probability density function, for here we may employ only ordinary integrals. However, the properties of Stieltjes integrals are very much the same as those of ordinary Riemann integrals; consequently, the proofs we give are immediately translatable into proofs of the general case that require the use of Stieltjes integrals.

Expectations with Respect to Numerical n-Tuple Valued Random Phenomena. The foregoing ideas extend immediately to a numerical n-tuple valued random phenomenon. Given the distribution function $F(x_1, x_2, \ldots, x_n)$ of such a random phenomenon and any continuous function $g(x_1, \ldots, x_n)$ of n real variables, we define the expectation of the function with respect to the random phenomenon by

$$E[g(x_1, x_2, \cdots, x_n)] = \iint_{R_n} \cdots \int g(x_1, x_2, \cdots, x_n)\, dF(x_1, x_2, \cdots, x_n) \tag{6.8}$$

in which the integral is a Stieltjes integral over the space R_n of all n-tuples $(x_1, x_2, \ldots, x_n)$ of real numbers. We shall not write out here the definition of this integral.

We note that (6.2) and (6.3) generalize. If the distribution function $F(x_1, x_2, \ldots, x_n)$ is specified by a probability density function $f(x_1, x_2, \ldots, x_n)$ so that (7.7′) of Chapter 4 holds, then

$$(6.9)\quad E[g(x_1, x_2, \cdots, x_n)] = \underbrace{\int_{-\infty}^{\infty}\int_{-\infty}^{\infty}\cdots\int_{-\infty}^{\infty}}_{n \text{ integrals}} g(x_1, x_2, \cdots, x_n) f(x_1, x_2, \cdots, x_n)\, dx_1\, dx_2 \cdots dx_n$$

If the distribution function $F(x_1, x_2, \ldots, x_n)$ is specified by a probability mass function $p(x_1, x_2, \ldots, x_n)$, so that (7.8′) of Chapter 4 holds, then

$$(6.10)\quad E[g(x_1, x_2, \cdots, x_n)] = \sum_{\substack{\text{over all } (x_1,x_2,\cdots,x_n) \\ \text{such that } p(x_1,x_2,\cdots,x_n) > 0}} g(x_1, x_2, \cdots, x_n) p(x_1, x_2, \cdots, x_n).$$

EXERCISES

6.1. Compute the mean, variance, and moment-generating function of each of the probability laws specified by the following distribution functions. (Recall that $[x]$ denotes the largest integer less than or equal to x.)

(i) $F(x) = 0$ for $x < 0$
$= 1 - \frac{1}{3}e^{-(x/3)} - \frac{2}{3}e^{-[x/3]}$ for $x \geq 0$.

(ii) $F(x) = 0$ for $x < 0$
$= 8\int_0^x ye^{-4y}\,dy + \frac{e^{-2}}{2}\sum_{k=0}^{[x]}\frac{2^k}{k!}$ for $x \geq 0$.

(iii) $F(x) = 0$ for $x < 1$
$= 1 - \frac{1}{2x^2} - \frac{1}{2^{[x]}}$ for $x \geq 1$.

(iv) $F(x) = 0$ for $x < 1$
$= 1 - \frac{2}{3x} - \frac{1}{3^{[x]}}$ for $x \geq 1$.

6.2. Compute the expectation of the function $g(x_1, x_2) = x_1x_2$ with respect to the probability laws of the numerical 2-tuple valued random phenomenon specified by the following probability density functions or probability mass functions:

(i) $f(x_1, x_2) = \exp(-2|x_1| - 2|x_2|)$

(ii) $f(x_1, x_2) = \frac{1}{(b_1 - a_1)(b_2 - a_2)}$ if $a_1 \leq x_1 \leq b_1$ and $a_2 \leq x_2 \leq b_2$
$= 0$ otherwise.

(iii) $f(x_1, x_2) = \frac{1}{2\pi\sqrt{1 - \rho^2}} \exp\left[-(x_1^2 + x_2^2 + 2\rho x_1 x_2)/2(1 - \rho^2)\right]$

in which $|\rho| < 1$.

(iv) $p(x_1, x_2) = \frac{1}{36}$ if $x_1 = 1, 2, 3, \cdots, 6$ and $x_2 = 1, 2, \cdots, 6$

$= 0$ otherwise.

(v) $p(x_1, x_2) = \binom{6}{x_1}\binom{6}{x_2}\left(\frac{2}{3}\right)^{x_1+x_2}\left(\frac{1}{3}\right)^{12-x_1-x_2}$ for x_1 and x_2 equal to nonnegative integers

$= 0$ otherwise.

(vi) $p(x_1, x_2) = \binom{12}{x_1\ x_2\ 12 - x_1 - x_2}(\frac{1}{4})^{x_1}(\frac{1}{3})^{x_2}(\frac{5}{12})^{12-x_1-x_2}$ for x_1 and x_2 equal to nonnegative integers

$= 0$ otherwise.

CHAPTER 6

Normal, Poisson, and Related Probability Laws

In applied probability theory the binomial, normal, and Poisson probability laws play a central role. In this chapter we discuss the reasons for the importance of the normal and Poisson probability laws.

1. THE IMPORTANCE OF THE NORMAL PROBABILITY LAW

The normal distribution function and the normal probability laws have played a significant role in probability theory since the early eighteenth century, and it is important to understand from what this signifiance derives.

To begin with, there are random phenomena that obey a normal probability law precisely. One example of such a phenomenon is the velocity in any given direction of a molecule (with mass M) in a gas at absolute temperature T (which, according to Maxwell's law of velocities, obeys a normal probability law with parameters $m = 0$ and $\sigma^2 = M/kT$, where k is the physical constant called Boltzmann's constant). However, with the exception of certain physical phenomena, there are not many random phenomena that obey a normal probability law precisely. Rather, the normal probability laws derive their importance from the fact that under various conditions they closely approximate many other probability laws.

The normal distribution function was first encountered (in the work of de Moivre, 1733) as a means of giving an approximate evaluation of the distribution function of the binomial probability law with parameters n and p for large values of n. This fact is a special case of the famed central limit theorem of probability theory (discussed in Chapters 8 and 10) which describes a very general class of random phenomena whose distribution functions may be approximated by the normal distribution function.

A normal probability law has many properties that make it easy to manipulate. Consequently, for mathematical convenience one may often, in practice, assume that a random phenomenon obeys a normal probability law if its true probability law is specified by a probability density function of a shape similar to that of the normal density function, in the sense that it possesses a single peak about which it is approximately symmetrical. For example, the height of a human being appears to obey a probability law possessing an approximately bell-shaped probability density function. Consequently, one might assume that this quantity obeys a normal probability law in certain respects. However, care must be taken in using this approximation; for example, it is conceivable for a normally distributed random quantity to take values between -10^6 and -10^{100}, although the probability of its doing so may be exceedingly small. On the other hand, no man's height can assume such a large negative value. In this sense, it is incorrect to state that a man's height is approximately distributed in accordance with a normal probability law. One may, nevertheless, insist on regarding a man's height as obeying approximately a normal probability law, in order to take advantage of the computational simplicity of the normal distribution. As long as the justification of this approximation is kept clearly in mind, there does not seem to be too much danger in employing it.

There is another sense in which a random phenomenon may approximately obey a normal probability law. It may happen that the random phenomenon, which as measured does not obey a normal probability law, can, by a numerical transformation of the measurement, be cast into a random phenomenon that does obey a normal probability law. For example, the cube root of the weight of an animal may obey a normal probability law (since the cube root of weight may be proportional to height) in a case in which the weight does not.

Finally, the study of the normal density function is important even for the study of a random phenomenon that does not obey a normal probability law, for under certain conditions its probability density function may be expanded in an infinite series of functions whose terms involve the successive derivatives of the normal density function.

2. THE APPROXIMATION OF THE BINOMIAL PROBABILITY LAW BY THE NORMAL AND POISSON PROBABILITY LAWS

Some understanding of the kinds of random phenomena that obey the normal probability law can be obtained by examining the manner in which the normal density function and the normal distribution function first arose in probability theory as means of approximately evaluating probabilities associated with the binomial probability law.

The following theorem was stated by de Moivre in 1733 for the case $p = \frac{1}{2}$ and proved for arbitrary values of p by Laplace in 1812.

The probability that a random phenomenon obeying the binomial probability law with parameters n and p will have an observed value lying between a and b, inclusive, for any two integers a and b, is given approximately by

$$\sum_{k=a}^{b} \binom{n}{k} p^k q^{n-k} \doteq \frac{1}{\sqrt{2\pi}} \int_{\frac{a-np-\frac{1}{2}}{\sqrt{npq}}}^{\frac{b-np+\frac{1}{2}}{\sqrt{npq}}} e^{-\frac{1}{2}y^2}\, dy = \Phi\left(\frac{b - np + \frac{1}{2}}{\sqrt{npq}}\right) - \Phi\left(\frac{a - np - \frac{1}{2}}{\sqrt{npq}}\right) \tag{2.1}$$

Before indicating the proof of this theorem, let us explain its meaning and usefulness by the following examples.

▶ **Example 2A.** Suppose that $n = 6000$ tosses of a fair die are made. The probability that exactly k of the tosses will result in a "three" is given by $\binom{6000}{k}\left(\frac{1}{6}\right)^k\left(\frac{5}{6}\right)^{6000-k}$. The probability that the number of tosses on which a "three" will occur is between 980 and 1030, inclusive, is given by the sum

$$\sum_{k=980}^{1030} \binom{6000}{k}\left(\frac{1}{6}\right)^k\left(\frac{5}{6}\right)^{6000-k} \tag{2.2}$$

It is clearly quite laborious to evaluate this sum directly. Fortunately, by (2.1), the sum in (2.2) is approximately equal to

$$\frac{1}{\sqrt{2\pi}} \int_{\frac{980-1000-\frac{1}{2}}{28.87}}^{\frac{1030-1000+\frac{1}{2}}{28.87}} e^{-\frac{1}{2}y^2}\, dy = \Phi(1.06) - \Phi(-0.71) = 0.617.$$

◀

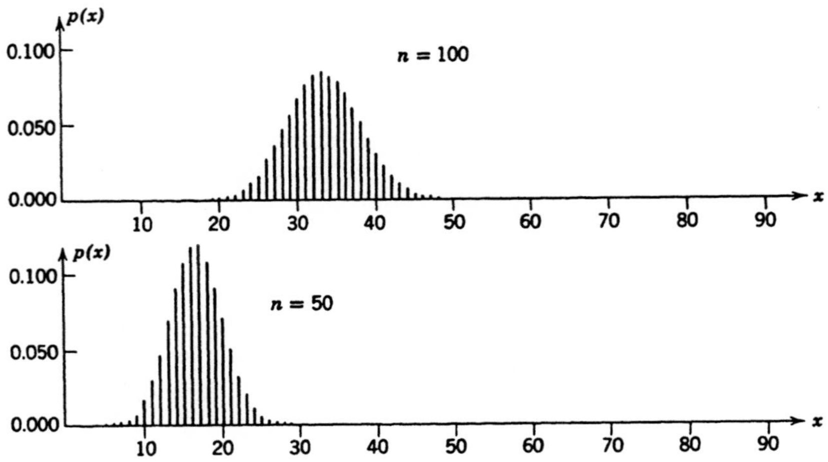

Fig. 2A. Graphs of the binomial probability mass function $p(x)$ for $p = \frac{1}{3}$ and $n = 50$ and 100.

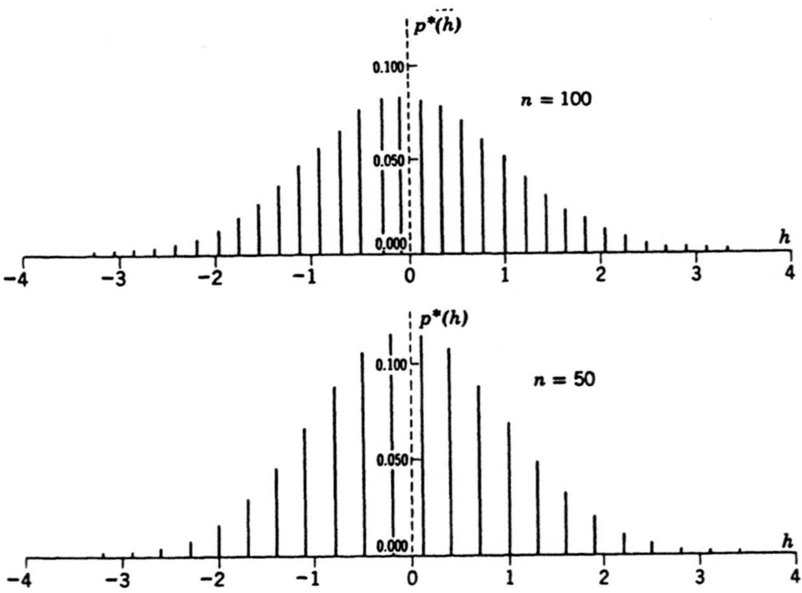

Fig. 2B. Graphs of the probability mass function $p^*(h)$ for $p = \frac{1}{3}$ and $n = 50$ and 100.

▶ **Example 2B.** In 40,000 independent tosses of a coin heads appeared 20,400 times. Find the probability that if the coin were fair one would observe in 40,000 independent tosses (i) 20,400 or more heads, (ii) between 19,600 and 20,400 heads.

Solution: Let X be the number of heads in 40,000 independent tosses of a fair coin. Then X obeys a binomial probability law with mean $m = np = 20{,}000$, variance $\sigma^2 = npq = 10{,}000$, and standard deviation $\sigma = 100$. According to the normal approximation to the binomial probability law, X approximately obeys a normal probability law with parameters $m = 20{,}000$ and $\sigma = 100$ [in making this statement we are ignoring the terms in (2.1) involving $\frac{1}{2}$, which are known as a "continuity" correction]. Since 20,400 is four standard deviations more then the mean of the probability law of X, the probability is approximately 0 that one would observe a value of X more than 20,400. Similarly, the probability is 1 that one would observe a value of X between 19,600 and 20,400. ◀

In order to have a convenient language in which to discuss the proof of (2.1), let us suppose that we are observing the number X of successes in n independent repeated Bernoulli trials with probability p of success at each trial. Next, to each outcome X let us compute the quantity

$$h = \frac{X - np}{\sqrt{npq}}, \tag{2.3}$$

which represents the deviation of X from np divided by $\sqrt{npq}$. Recall that the quantities np and $\sqrt{npq}$ are equal, respectively, to the *mean* and *standard deviation* of the binomial probability law. The deviation h, defined by (2.3), is a random quantity obeying a discrete probability law specified by a probability mass function $p^*(h)$, which may be given in terms of the probability mass function $p(x)$ by

$$p^*(h) = p(h\sqrt{npq} + np). \tag{2.4}$$

In words, (2.4) expresses the fact that for any given real number h the probability that the deviation (of the number of successes from np, divided by $\sqrt{npq}$) will be equal to h is the same as the probability that the number of successes will be equal to $h\sqrt{npq} + np$.

The advantage in considering the probability mass function $p^*(h)$ over the original probability mass function $p(x)$ can be seen by comparing their graphs, which are given in Figs. 2A and 2B for $n = 50$ and 100 and $p = \frac{1}{3}$. The graph of the function $p(x)$ becomes flatter and flatter and spreads out more and more widely along the x-axis as the number n of trials increases.

The graphs of the functions $p^*(h)$, on the other hand, are very similar to each other, for different values of n, and seem to approach a limit as n becomes infinite. It might be thought that it is possible to find a smooth curve that would be a good approximation to $p^*(h)$, at least when the number of trials n is large. We now show that this is indeed possible. More precisely, we show that if h is a real number for which $p^*(h) > 0$ then, approximately,

$$p^*(h) \doteq \frac{1}{\sqrt{npq}} \frac{1}{\sqrt{2\pi}} e^{-\frac{1}{2}h^2}, \tag{2.5}$$

in the sense that

$$\lim_{n \to \infty} \frac{\sqrt{2\pi npq}\, p(h\sqrt{npq} + np)}{e^{-\frac{1}{2}h^2}} = 1. \tag{2.6}$$

To prove (2.6), we first obtain the approximate expression for $p(x)$; for $k = 0, 1, \ldots, n$

$$p(k) = \frac{1}{\sqrt{2\pi}} \sqrt{\frac{n}{k(n-k)}} \left(\frac{np}{k}\right)^k \left(\frac{nq}{n-k}\right)^{n-k} e^R, \tag{2.7}$$

in which $|R| < \frac{1}{12}\left(\frac{1}{n} + \frac{1}{k} + \frac{1}{n-k}\right)$. Equation (2.7) is an immediate consequence of the approximate expression for the binomial coefficient $\binom{n}{k}$; for any integers n and $k = 0, 1, \ldots, n$

$$\binom{n}{k} = \frac{n!}{k!(n-k)!} = \frac{1}{\sqrt{2\pi}} \sqrt{\frac{n}{k(n-k)}} \left(\frac{n}{k}\right)^k \left(\frac{n}{n-k}\right)^{n-k} e^R. \tag{2.8}$$

Equation (2.8), in turn, is an immediate consequence of the approximate expression for $m!$ given by Stirling's formula; for any integer $m = 1, 2, \ldots$

$$m! = \sqrt{2\pi m}\, m^m e^{-m} e^{r(m)}, \qquad 0 < r(m) < \frac{1}{12m}. \tag{2.9}$$

In (2.7) let $k = np + h\sqrt{npq}$. Then $n - k = nq - h\sqrt{npq}$, and

$$\frac{k(n-k)}{n} = n(p + h\sqrt{pq/n})(q - h\sqrt{pq/n}) \doteq npq. \tag{2.10}$$

Then, using the expansion $\log_e(1 + x) = x - \frac{1}{2}x^2 + \theta x^3$ for some θ such that $|\theta| < 1$, valid for $|x| < \frac{1}{4}$, we obtain that

$$
\begin{aligned}
(2.11)\quad & -\log_e\{\sqrt{2\pi(npq)}p^*(h)\} \\
&= -\log_e\left\{\left(\frac{np}{k}\right)^k\left(\frac{nq}{n-k}\right)^{n-k}\right\} \\
&= (np + h\sqrt{npq})\log(1 + h\sqrt{q/np}) \\
&\quad + (nq - h\sqrt{npq})\log(1 - h\sqrt{p/nq}) \\
&= (np + h\sqrt{npq})\left[h\sqrt{(q/np)} - \frac{h^2}{2}\frac{q}{np} + \text{terms in } \frac{1}{n^{3/2}}\right] \\
&\quad + (nq - h\sqrt{npq})\left[-h\sqrt{(p/nq)} - \frac{h^2}{2}\frac{p}{nq} + \text{terms in } \frac{1}{n^{3/2}}\right] \\
&= \left(h\sqrt{npq} - q\frac{h^2}{2} + qh^2 + \text{terms in } \frac{1}{n^{1/2}}\right) \\
&\quad + \left(-h\sqrt{npq} - p\frac{h^2}{2} + ph^2 + \text{terms in } \frac{1}{n^{1/2}}\right) \\
&= \frac{1}{2}\cdot h^2 + \text{terms in } \frac{1}{n^{1/2}}.
\end{aligned}
$$

If we ignore all terms that tend to 0 as n tends to infinity in (2.11), we obtain the desired conclusion, namely (2.6).

From (2.6) one may obtain a proof of (2.1). However, in this book we give only a heuristic geometric proof that (2.6) implies (2.1). For an elementary rigorous proof of (2.1) the reader should consult J. Neyman, *First Course in Probability and Statistics*, New York, Henry Holt, 1950, pp. 234–242. In Chapter 10 we give a rigorous proof of (2.1) by using the method of characteristic functions.

A geometric derivation of (2.1) from (2.6) is as follows. First plot $p^*(h)$ in Fig. 2B as a function of h; note that $p^*(h) = 0$ for all points h, except those that may be represented in the form

$$(2.12)\qquad h = (k - np)/\sqrt{npq}$$

for some integer $k = 0, 1, \ldots, n$. Next, as in Fig. 2C, plot $p^*(h)$ by a series of rectangles of height $(1/\sqrt{2\pi})e^{-\frac{1}{2}h^2}$, centered at all points h of the form of (2.12).

From Fig. 2C we obtain (2.1). It is clear that

$$\sum_{k=a}^{b} p(k) = \sum_{\substack{\text{over } h \text{ of the form of} \\ (2.12) \text{ for } k = a, \cdots, b}} p^*(h),$$

which is equal to the sum of the areas of the rectangles in Fig. 2C centered at the points h of the form of (2.12), corresponding to the integers k from a to b, inclusive. Now, the sum of the area of these rectangles is an approximating sum to the integral of the function $(1/\sqrt{2\pi})e^{-\frac{1}{2}h^2}$ between the limits $(a - np - \frac{1}{2})/\sqrt{npq}$ and $(b - np + \frac{1}{2})/\sqrt{npq}$. We have thus obtained the approximate formula (2.1).

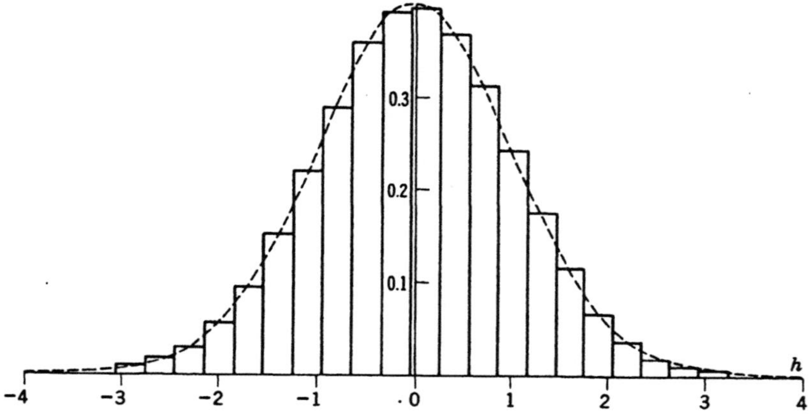

Fig. 2C. The normal approximation to the binomial probability law. The continuous curve represents the normal density function. The area of each rectangle represents the approximate value given by (2.5) for the value of the probability mass function $p^*(h)$ at the mid-point of the base of the rectangle.

It should be noted that we have available two approximations to the probability mass function $p(x)$ of the binomial probability law. From (2.5) and (2.6) it follows that

$$\binom{n}{x} p^x q^{n-x} \doteq \frac{1}{\sqrt{2\pi npq}} \exp\left(-\frac{1}{2}\frac{(x-np)^2}{npq}\right), \tag{2.13}$$

whereas from (2.1) one obtains, setting $a = b = x$,

$$\binom{n}{x} p^x q^{n-x} \doteq \frac{1}{\sqrt{2\pi}} \int_{\frac{x-np-\frac{1}{2}}{\sqrt{npq}}}^{\frac{x-np+\frac{1}{2}}{\sqrt{npq}}} e^{-\frac{1}{2}y^2}\, dy. \tag{2.14}$$

In using any approximation formula, such as that given by (2.1), it is important to have available "remainder terms" for the determination of the accuracy of the approximation formula. Analytic expressions for the remainder terms involved in the use of (2.1) are to be found in J. V. Uspensky, *Introduction to Mathematical Probability*, McGraw-Hill, New York, 1937, p. 129, and W. Feller, "On the normal approximation to the binomial distribution," *Annals of Mathematical Statistics*, Vol. 16, (1945), pp. 319–329. However, these expressions do not lead to conclusions that are easy to state. A booklet entitled *Binomial, Normal, and Poisson Probabilities*, by Ed. Sinclair Smith (published by the author in 1953 at Bel Air, Maryland), gives extensive advice on how to compute expeditiously binomial probabilities with 3-decimal accuracy. Smith (p. 38) states that (2.1) gives 2-decimal accuracy or better if $np > 37$. The accuracy of the approximation is much better for p close to 0.5, in which case 2-decimal accuracy is obtained with n as small as 3.

In treating problems in this book, the student will not be seriously wrong if he uses the normal approximation to the binomial probability law in cases in which $np(1 - p) \geq 10$.

Extensive tables of the binomial distribution function

$$F_B(x; n, p) = \sum_{k=0}^{x} \binom{n}{k} p^k q^{n-k}, \qquad x = 0, 1, \ldots, n \tag{2.15}$$

have become available in recent years. The *Tables of the Binomial Probability Distribution*, National Bureau of Standards, Applied Mathematics Series 6, Washington, 1950, give $1 - F_B(x; n, p)$ to seven decimal places for $p = 0.01\ (0.01)\ 0.50$ and $n = 2(1)\ 49$. These tables are extended in H. G. Romig, *50–100 Binomial Tables*, Wiley, New York, 1953, in which $F_B(x; n, p)$ is tabulated for $n = 50(5)\ 100$ and $p = 0.01\ (0.01)\ 0.50$. A more extensive tabulation of $F_B(x; n, p)$ for $n = 1(1)\ 50(2)\ 100(10)\ 200(20)\ 500(50)\ 1000$ and $p = 0.01\ (0.01)\ 50$ and also p equal to certain other fractional values is available in *Tables of the Cumulative Binomial Probability Distribution*, Harvard University Press, 1955.

The Poisson Approximation to the Binomial Probability Law. The Poisson approximation, whose proof and usefulness was indicated in section 3 of Chapter 3, states that

$$\binom{n}{k} p^k q^{n-k} \doteq e^{-np} \frac{(np)^k}{k!} \tag{2.16}$$

$$\sum_{k=0}^{m} \binom{n}{k} p^k q^{n-k} \doteq \sum_{k=0}^{m} e^{-np} \frac{(np)^k}{k!}. \tag{2.17}$$

The Poisson approximation applies when the binomial probability law is very far from being bell shaped; this is true, say, when $p \leq 0.1$.

It may happen that p is very small, so that the Poisson approximation may be used; but n is so large that (2.14) holds, and the normal approximation may be used. This implies that for large values of $\lambda = np$ the Poisson law and the normal law approximate each other. In theoretical exercise 2.1 it is shown directly that the *Poisson probability law with parameter λ may be approximated by the normal probability law for large values of λ.*

▶ **Example 2C. A telephone trunking problem.** Suppose you are designing the physical premises of a newly organized research laboratory. Since there will be a large number of private offices in the laboratory, there will also be a large number n of individual telephones, each connecting to a central laboratory telephone switchboard. The question arises: how many outside lines will the switchboard require to establish a fairly high probability, say 95%, that any person who desires the use of an outside telephone line (whether on the outside of the laboratory calling in or on the inside of the laboratory calling out) will find one immediately available?

Solution: We begin by regarding the problem as one involving independent Bernoulli trials. We suppose that for each telephone in the laboratory, say the jth telephone, there is a probability p_j that an outside line will be required (either as the result of an incoming call or an outgoing call). One could estimate p_j by observing in the course of an hour how many minutes h_j an outside line is engaged, and estimating p_j by the ratio $h_j/60$. In order to have *repeated* Bernoulli trials, we assume $p_1 = p_2 = \ldots = p_n = p$. We next assume independence of the n events $A_1, A_2, \ldots, A_n$, in which A_j is the event that the jth telephone demands an outside line at the moment of time at which we are regarding the laboratory. The probability that exactly k outside lines will be in demand at a given moment is, by the binomial law, given by $\binom{n}{k} p^k q^{n-k}$.

Consequently, if we let K denote the number of outside lines connected to the laboratory switchboard and make the assumption that they are all free at the moment at which we are regarding the laboratory, then the probability that a person desiring an outside line will find one available is the same as the probability that the number of outside lines demanded at that moment is less than or equal to K, which is equal to

$$(2.18) \quad \sum_{k=0}^{K} \binom{n}{k} p^k (1-p)^{n-k} \doteq e^{-np} \sum_{k=0}^{K} \frac{(np)^k}{k!}$$

$$\doteq \Phi\left(\frac{K - np + \frac{1}{2}}{\sqrt{np(1-p)}}\right) - \Phi\left(\frac{-np - \frac{1}{2}}{\sqrt{np(1-p)}}\right),$$

where the first equality sign in (2.18) holds if the Poisson approximation to the binomial applies and the second equality sign holds if the normal approximation to the binomial applies.

Define, for any $\lambda > 0$ and integer $n = 0, 1, \ldots,$

$$F_P(n; \lambda) = \sum_{k=0}^{n} e^{-\lambda} \frac{\lambda^k}{k!}, \tag{2.19}$$

which is essentially the distribution function of the Poisson probability law with parameter λ. Next, define for P, such that $0 \leq P \leq 1$, the symbol $\mu(P)$ to denote the P-percentile of the normal distribution function, defined by

$$\Phi(\mu(P)) = \int_{-\infty}^{\mu(P)} \phi(x)\, dx = P. \tag{2.20}$$

One may give the following *expressions for the minimum number K of outside lines that should be connected to the laboratory switchboard in order to have a probability greater than a preassigned level P_0 that all demands for outside lines can be handled.* Depending on whether the Poisson or the normal approximation applies, K is the smallest integer such that

$$F_P(K; np) \geq P_0 \tag{2.21}$$

$$K \geq \mu(P_0)\sqrt{np(1-p)} + np - \tfrac{1}{2}. \tag{2.22}$$

In writing (2.22), we are approximating $\Phi[(-np - \frac{1}{2})/\sqrt{npq}]$ by 0, since

$$\frac{np}{\sqrt{npq}} \geq \sqrt{npq} \geq 4 \qquad \text{if } npq \geq 16$$

The value of $\mu(P)$ can be determined from Table I (see p. 441). In particular,

$$\mu(0.95) = 1.645, \qquad \mu(0.99) = 2.326. \tag{2.23}$$

The solution K of the inequality (2.21) can be read from tables prepared by E. C. Molina (published in a book entitled *Poisson's Exponential Binomial Limit*, Van Nostrand, New York, 1942) which tabulate, to six decimal places, the function

$$1 - F_P(K; \lambda) = \sum_{k=K+1}^{\infty} e^{-\lambda} \frac{\lambda^k}{k!} \tag{2.24}$$

for about 300 values of λ in the interval $0.001 \leq \lambda \leq 100$.

The value of K, determined by (2.21) and (2.22), is given in Table 2A for $p = \frac{1}{30}, \frac{1}{10}, \frac{1}{3}$, $n = 90, 900$, and $P_0 = 0.95, 0.99$. ◀

TABLE 2A

THE VALUES OF K DETERMINED BY (2.21) AND (2.22)

	p	$\frac{1}{30}$		$\frac{1}{10}$		$\frac{1}{3}$	
Approximation		Poisson	Normal	Poisson	Normal	Poisson	Normal
$P_0 = 0.95$	$n = 90$	6	5.3	14	13.2	39	36.9
	$n = 900$	39	38.4	106	104.3		322.8
$P_0 = 0.99$	$n = 90$	8	6.5	17	15.1	43	39.9
	$n = 900$	43	42.0	113	110.4		332.4

THEORETICAL EXERCISES

2.1. Normal approximation to the Poisson probability law. Consider a random phenomenon obeying a Poisson probability law with parameter λ. To an observed outcome X of the random phenomenon, compute $h = (X - \lambda)/\sqrt{\lambda}$, which represents the deviation of X from λ, divided by $\sqrt{\lambda}$. The quantity h is a random quantity obeying a discrete probability law specified by a probability mass function $p^*(h)$, which may be given in terms of the probability function $p(x)$ by $p^*(h) = p(h\sqrt{\lambda} + \lambda)$. In the same way that (2.6), (2.1), and (2.13) are proved show that for fixed values of a, b, and k, the following differences tend to 0 as λ tends to infinity:

$$\sqrt{\lambda}p^*(h) - \frac{1}{\sqrt{2\pi}} e^{-\frac{1}{2}h^2} \to 0$$

$$\sum_{k=a}^{b} e^{-\lambda} \frac{\lambda^k}{k!} - \frac{1}{\sqrt{2\pi}} \int_{\frac{a-\lambda-\frac{1}{2}}{\sqrt{\lambda}}}^{\frac{b-\lambda+\frac{1}{2}}{\sqrt{\lambda}}} e^{-\frac{1}{2}y^2}\, dy \to 0 \tag{2.25}$$

$$e^{-\lambda} \frac{\lambda^k}{k!} - \frac{1}{\sqrt{2\pi}} \int_{\frac{k-\lambda-\frac{1}{2}}{\sqrt{\lambda}}}^{\frac{k-\lambda+\frac{1}{2}}{\sqrt{\lambda}}} e^{-\frac{1}{2}y^2}\, dy \to 0$$

2.2. A competition problem. Suppose that m restaurants compete for the same n patrons. Show that the number of seats that each restaurant should have to order to have a probability greater than P_0 that it can serve all patrons who come to it (assuming that all the patrons arrive at the same time and choose, independently of one another, each restaurant with probability $p = 1/m$) is given by (2.22), with $p = 1/m$. Compute K for $m = 2, 3, 4$ and $P_0 = 0.75$ and 0.95. Express in words how the size of a restaurant (represented by K) depends on the size of its market (represented by n), the number of its competitors (represented by m), and the share of the market it desires (represented by P_0).

EXERCISES

2.1. In 10,000 independent tosses of a coin 5075 heads were observed. Find approximately the probability of observing (i) exactly 5075 heads, (ii) 5075 or more heads if the coin (*a*) is fair, (*b*) has probability 0.51 of falling heads.

2.2. Consider a room in which 730 persons are assembled. For $i = 1, 2, \ldots, 730$, let A_i be the event that the ith person was born on January 1. Assume that the events $A_1, \ldots, A_{730}$ are independent and that each event has probability equal to 1/365. Find approximately the probability that (i) exactly 2, (ii) 2 or more persons were born on January 1. Compare the answers obtained by using the normal and Poisson approximations to the binomial law.

2.3. Plot the probability mass function of the binomial probability law with parameters $n = 10$ and $p = \frac{1}{2}$ against its normal approximation. In your opinion, is the approximation close enough for practical purposes?

2.4. Consider an urn that contains 10 balls, numbered 0 to 9, each of which is equally likely to be drawn; thus choosing a ball from the urn is equivalent to choosing a number 0 to 9, and one sometimes describes this experiment by saying that a *random digit* has been chosen. Now let n balls be chosen with replacement. Find the probability that among the n numbers thus chosen the number 7 will appear between $(n - 3\sqrt{n})/10$ times and $(n + 3\sqrt{n})/10$ times, inclusive, if (i) $n = 10$, (ii) $n = 100$, (iii) $n = 10{,}000$. Compute the answers exactly or by means of the normal and Poisson approximations to the binomial probability law.

2.5. Find the probability that in 3600 independent repeated trials of an experiment, in which the probability of success of each trial is p, the number of successes is between $3600p - 20$ and $3600p + 20$, inclusive, if (i) $p = \frac{1}{2}$, (ii) $p = \frac{1}{3}$.

2.6. A certain corporation has 90 junior executives. Assume that the probability is $\frac{1}{10}$ that an executive will require the services of a secretary at the beginning of the business day. If the probability is to be 0.95 or greater that a secretary will be available, how many secretaries should be hired to constitute a pool of secretaries for the group of 90 executives?

2.7. Suppose that (i) 2, (ii) 3 restaurants compete for the same 800 patrons. Find the number of seats that each restaurant should have in order to have a probability greater than 95% that it can serve all patrons who come to it (assuming that all patrons arrive at the same time and choose, independently of one another, each restaurant with equal probability).

2.8. At a certain men's college the probability that a student selected at random on a given day will require a hospital bed is 1/5000. If there are 8000 students, how many beds should the hospital have so that the probability that a student will be turned away for lack of a bed is less than 1% (in other words, find K so that $P[X > K] \leq 0.01$, where X is the number of students requiring beds).

2.9. Consider an experiment in which the probability of success at each trial is p. Let X denote the successes in n independent trials of the experiment. Show that

$$P[|X - np| \leq (1.96)\sqrt{npq}] \doteq 95\%.$$

Consequently, if $p = 0.5$, with probability approximately equal to 0.95, the observed number X of successes in n independent trials will satisfy the inequalities

$$(0.5)n - (0.98)\sqrt{n} \leq X \leq 0.5n + (0.98)\sqrt{n}. \tag{2.26}$$

Determine how large n should be, under the assumption that (i) $p = 0.4$, (ii) $p = 0.6$, (iii) $p = 0.7$, to have a probability of 5% that the observed number X of successes in the n trials will satisfy (2.26).

2.10. In his book *Natural Inheritance*, p. 63, F. Galton in 1889 described an apparatus known today as Galton's *quincunx*. The apparatus consists of a board in which nails are arranged in rows, the nails of a given row being placed below the mid-points of the intervals between the nails in the row above. Small steel balls of equal diameter are poured into the apparatus through a funnel located opposite the central pin of the first row. As they run down the board, the balls are "influenced" by the nails in such a manner that, after passing through the last row, they take up positions deviating from the point vertically below the central pin of the first row. Let us call this point $x = 0$. Assume that the distance between 2 neighboring pins is taken to be 1 and that the diameter of the balls is slightly smaller than 1. Assume that in passing from one row to the next the abscissa (x-coordinate) of a ball changes by either $\frac{1}{2}$ or $-\frac{1}{2}$, each possibility having equal probability. To each opening in a row of nails, assign as its abscissa the mid-point of the interval between the 2 nails. If there is an even number of rows of nails, then the openings in the last row will have abscissas $0, \pm 1, \pm 2, \ldots$. Assuming that there are 36 rows of nails, find for $k = 0, \pm 1, \pm 2, \ldots, \pm 10$ the probability that a ball inserted in the funnel will pass through the opening in the last row, which has abscissa k.

2.11. Consider a liquid of volume V, which contains N bacteria. Let the liquid be vigorously shaken and part of it transferred to a test tube of volume v. Suppose that (i) the probability p that any given bacterium will be transferred to the test tube is equal to the ratio of the volumes v/V and that (ii) the appearance of 1 particular bacterium in the test tube is independent of the appearance of the other $N - 1$ bacteria. Consequently, the number of bacteria in the test tube is a numerical valued random phenomenon obeying a binomial probability law with parameters N and $p = v/V$. Let $m = N/V$ denote the average number of bacteria per unit volume. Let the volume v of the test tube be equal to 3 cubic centimeters.

(i) Assume that the volume v of the test tube is very small compared to the volume V of liquid, so that $p = v/V$ is a small number. In particular, assume that $p = 0.001$ and that the bacterial density $m = 2$ bacteria per cubic centimeter. Find approximately the probability that the number of bacteria in the test tube will be greater than 1.

(ii) Assume that the volume v of the test tube is comparable to the volume V of the liquid. In particular, assume that $V = 12$ cubic centimeters and $N = 10{,}000$. What is the probability that the number of bacteria in the test tube will be between 2400 and 2600, inclusive?

2.12. Suppose that among 10,000 students at a certain college 100 are red-haired.

(i) What is the probability that a sample of 100 students, selected with replacement, will contain at least one red-haired student?

(ii) How large is a random sample, drawn with replacement, if the probability of its containing a red-haired student is 0.95?

It would be more realistic to assume that the sample is drawn without replacement. Would the answers to (i) and (ii) change if this assumption were made? *Hint:* State conditions under which the hypergeometric law is approximated by the Poisson law.

2.13. Let S be the observed number of successes in n independent repeated Bernoulli trials with probability p of success at each trial. For each of the following events, find (i) its exact probability calculated by use of the binomial probability law, (ii) its approximate probability calculated by use of the normal approximation, (iii) the percentage error involved in using (ii) rather than (i).

	n	p	the event that		n	p	the event that
(i)	4	0.3	$S \le 2$	(viii)	49	0.2	$S \le 4$
(ii)	9	0.7	$S \ge 6$	(ix)	49	0.2	$S \ge 8$
(iii)	9	0.7	$2 \le S \le 8$	(x)	49	0.2	$S \le 16$
(iv)	16	0.4	$2 \le S \le 10$	(xi)	100	0.5	$S \le 10$
(v)	16	0.2	$S \le 2$	(xii)	100	0.5	$S \ge 40$
(vi)	25	0.9	$S \le 20$	(xiii)	100	0.5	$S = 50$
(vii)	25	0.3	$5 \le S \le 10$	(xiv)	100	0.5	$S \le 60$

3. THE POISSON PROBABILITY LAW

The Poisson probability law has become increasingly important in recent years as more and more random phenomena to which the law applies have been studied. In physics the random emission of electrons from the filament of a vacuum tube, or from a photosensitive substance under the influence of light, and the spontaneous decomposition of radioactive atomic nuclei lead to phenomena obeying a Poisson probability law. This law arises frequently in the fields of operations research and management science, since *demands for service*, whether upon the cashiers or salesmen of a department store, the stock clerk of a factory, the runways of an airport, the cargo-handling facilities of a port, the maintenance man of a machine shop, and the trunk lines of a telephone exchange, and also *the rate at which service is rendered*, often lead to random phenomena either exactly or approximately obeying a Poisson probability law. Such random phenomena also arise in connection with the occurrence of accidents, errors, breakdowns, and other similar calamities.

The kinds of random phenomena that lead to a Poisson probability law can best be understood by considering the kinds of phenomena that lead to a binomial probability law. The usual situation to which the binomial probability law applies is one in which n independent occurrences of some experiment are observed. One may then determine (i) the number of trials on which a certain event occurred and (ii) the number of trials on which the event did not occur. There are random events, however, that do not occur as the outcomes of definite trials of an experiment but rather at random points in time or space. For such events one may count the number of occurrences of the event in a period of time (or space). However, it makes no sense to speak of the number of nonoccurrences of such an event in a period of time (or space). For example, suppose one observes the number of airplanes arriving at a certain airport in an hour. One may report how many airplanes arrived at the airport; however, it makes no sense to inquire how many airplanes did not arrive at the airport. Similarly, if one is observing the number of organisms in a unit volume of some fluid, one may count the number of organisms present, but it makes no sense to speak of counting the number of organisms not present.

We next indicate some conditions under which one may expect that the number of occurrences of a random event occurring in time or space (such as the presence of an organism at a certain point in 3-dimensional space, or the arrival of an airplane at a certain point in time) obeys a Poisson probability law. We make the basic assumption that there exists a positive quantity μ such that, for any small positive number h and any time interval of length h,

(i) the probability that exactly one event will occur in the interval is approximately equal to μh, in the sense that it is equal to $\mu h + r_1(h)$, and $r_1(h)/h$ tends to 0 as h tends to 0;

(ii) the probability that exactly zero events occur in the interval is approximately equal to $1 - \mu h$, in the sense that it is equal to $1 - \mu h + r_2(h)$, and $r_2(h)/h$ tends to 0 as h tends to 0; and,

(iii) the probability that two or more events occur in the interval is equal to a quantity $r_3(h)$ such that the quotient $r_3(h)/h$ tends to 0 as the length h of the interval tends to 0.

The parameter μ may be interpreted as the *mean rate at which events occur* per unit time (or space); consequently, we refer to μ as the *mean rate of occurrence* (of events).

▶ **Example 3A.** Suppose one is observing the times at which automobiles arrive at a toll collector's booth on a toll bridge. Let us suppose that we are informed that the mean rate μ of arrival of automobiles is given by $\mu = 1.5$ automobiles per minute. The foregoing assumption then states that in a time period of length $h = 1$ second $= (\frac{1}{60})$ minute, exactly one car will arrive with approximate probability $\mu h = (1.5)(\frac{1}{60}) = \frac{1}{40}$, whereas exactly zero cars will arrive with approximate probability $1 - \mu h = \frac{39}{40}$. ◀

In addition to the assumption concerning the existence of the parameter μ with the properties stated, we also make the assumption that if an interval of time is divided into n subintervals and, for $i = 1, \ldots, n$, A_i denotes the event that at least one event of the kind we are observing occurs in the ith subinterval then, for any integer n, $A_1, \ldots, A_n$ are independent events.

We now show, under these assumptions, that *the number of occurrences of the event in a period of time (or space) of length (or area or volume) t obeys a Poisson probability law with parameter μt*; more precisely, *the probability that exactly k events occur in a time period of length t is equal to*

$$e^{-\mu t}\frac{(\mu t)^k}{k!}. \tag{3.1}$$

Consequently, we may describe briefly a sequence of events occurring in time (or space), and which satisfy the foregoing assumptions, by saying that the events *obey a Poisson probability law at the rate of μ events per unit time (or unit space).*

Note that if X is the number of events occurring in a time interval of length t, then X obeys a Poisson probability law with mean μt. Consequently, μ is the mean rate of occurrence of events per unit time, in the sense that the number of events occurring in a time interval of length 1 obeys a Poisson probability law with mean μ.

To prove (3.1), we divide the time period of length t into n time periods of length $h = t/n$. Then the probability that k events will occur in the time t is approximately equal to the probability that exactly one event has occurred in exactly k of the n subintervals of time into which the original interval was divided. By the foregoing assumptions, this is equal to the probability of scoring exactly k successes in n independent repeated Bernoulli trials in which the probability of success at each trial is $p = h\mu = (\mu t)/n$; this is equal to

$$\binom{n}{k}\left(\frac{\mu t}{n}\right)^k\left(1 - \frac{\mu t}{n}\right)^{n-k}. \tag{3.2}$$

Now (3.2) is only an approximation to the probability that k events will occur in time t. To get an exact evaluation, we must let the number of subintervals increase to infinity. Then (3.2) tends to (3.1) since rewriting (3.2)

$$\frac{1}{k!}(\mu t)^k\left(1-\frac{\mu t}{n}\right)^{n-k}\frac{(n)_k}{n^k} \rightarrow \frac{1}{k!}(\mu t)^k e^{-\mu t}$$

as $n \rightarrow \infty$.

It should be noted that the foregoing derivation of (3.1) is not completely rigorous. To give a rigorous proof of (3.1), one must treat the random phenomenon under consideration as a stochastic process. A sketch of such proof, using differential equations, is given in section 5.

▶ **Example 3B.** It is known that bacteria of a certain kind occur in water at the rate of two bacteria per cubic centimeter of water. Assuming that this phenomenon obeys a Poisson probability law, what is the probability that a sample of two cubic centimeters of water will contain (i) no bacteria, (ii) at least two bacteria?

Solution: Under the assumptions made, it follows that the number of bacteria in a two-cubic-centimeter sample of water obeys a Poisson probability law with parameter $\mu t = (2)(2) = 4$, in which μ denotes the rate at which bacteria occur in a unit volume and t represents the volume of the sample of water under consideration. Consequently, the probability that there will be no bacteria in the sample is equal to e^{-4}, and the probability that there will be two or more bacteria in the sample is equal to $1 - 5e^{-4}$. ◀

▶ **Example 3C. Misprints.** In a certain published book of 520 pages 390 typographical errors occur. What is the probability that four pages, selected randomly by the printer as examples of his work, will be free from errors?

Solution: The problem as stated is incapable of mathematical solution. However, let us recast the problem as follows. Assume that typographical errors occur in the work of a certain printer in accordance with the Poisson probability law at the rate of $390/520 = \frac{3}{4}$ errors per page. The number of errors in four pages then obeys a Poisson probability law with parameter $(\frac{3}{4})\,4 = 3$; consequently, the probability is e^{-3} that there will be no errors in the four pages. ◀

▶ **Example 3D. Shot noise in electron tubes.** The sensitivity attainable with electronic amplifiers and apparatus is inherently limited by the spontaneous current fluctuations present in such devices, usually called *noise*. One source of noise in vacuum tubes is shot noise, which is due to the random emission of electrons from the heated cathode. Suppose that the potential difference between the cathode and the anode is so great that all electrons emitted by the cathode have such high velocities that there is no accumulation of electrons between the cathode and the anode (and thus no space charge). If we consider an emission of an electron from the cathode as an event, then the assumptions preceding (3.1) may be shown as satisfied (see W. B. Davenport, Jr. and W. L. Root, *An Introduction to the Theory of Random Signals and Noise*, McGraw-Hill, New York, 1958, pp. 112–119). Consequently, the number of electrons emitted from the cathode in a time interval of length t obeys a Poisson probability law with parameter λt, in which λ is the mean rate of emission of electrons from the cathode. ◀

The Poisson probability law was first published in 1837 by Poisson in his book *Recherches sur la probabilité des jugements en matière criminelle et en matière civile.* In 1898, in a work entitled *Das Gesetz der kleinen Zahlen*, Bortkewitz described various applications of the Poisson distribution. However until 1907 the Poisson distribution was regarded as more of a curiosity than a useful scientific tool, since the applications made of it were to such phenomena as the suicides of women and children and deaths from the kick of a horse in the Prussian army. Because of its derivation as a limit of the binomial law, the Poisson law was usually described as the probability law of the number of successes in a very large number of independent repeated trials, each with a very small probability of success.

In 1907 the celebrated statistician W. S. Gosset (writing, as was his wont, under the pseudonym "Student") deduced the Poisson law as the probability law of the number of minute corpuscles to be found in sample drops of a liquid, under the assumption that the corpuscles are distributed at random throughout the liquid; see "Student," "On the error of counting with a Haemocytometer," *Biometrika*, Vol. 5, p. 351. In 1910 the Poisson law was shown to fit the number of "α-particles discharged per $\frac{1}{8}$-minute or $\frac{1}{4}$-minute interval from a film of polonium"; see Rutherford and Geiger, "The probability variations in the distribution of α-particles," *Philosophical Magazine*, Vol. 20, p. 700.

Although one is able to state assumptions under which a random phenomenon will obey a Poisson probability law with some parameter λ, the value of the constant λ cannot be deduced theoretically but must be determined empirically. The determination of λ is a statistical problem. The following procedure for the determination of λ can be justified on various grounds. Given events occurring in time, choose an interval of length t. Observe a large number N of time intervals of length t. For each integer $k = 0, 1, 2, \ldots$ let N_k be the number of intervals in which exactly k events have occurred. Let

$$T = 0 \cdot N_0 + 1 \cdot N_1 + 2 \cdot N_2 + \cdots + k \cdot N_k + \cdots \tag{3.3}$$

be the total number of events observed in the N intervals of length t. Then the ratio T/N represents the observed average number of events happening per time interval of length t. As an estimate $\hat{\lambda}$ of the value of the parameter λ, we take

$$\hat{\lambda} = \frac{T}{N} = \frac{1}{N} \sum_{k=0}^{\infty} kN_k. \tag{3.4}$$

If we believe that the random phenomenon under observation obeys a Poisson probability law with parameter $\hat{\lambda}$, then we may compute the probability $p(k; \hat{\lambda})$ that in a time interval of length t exactly k successes will occur.

▶ **Example 3E. Vacancies in the United States Supreme Court.** W. A. Wallis, writing on "The Poisson Distribution and the Supreme Court," *Journal of the American Statistical Association*, Vol. 31 (1936), pp. 376–380, reports that vacancies in the United States Supreme Court, either by death or resignation of members, occurred as follows during the 96 years, 1837 to 1932:

k = number of vacancies during the year	N_k = number of years with k vacancies
0	59
1	27
2	9
3	1
over 3	0

Since $T = 27 + 2 \cdot 9 + 1 \cdot 3 = 48$ and $N = 96$, it follows from (3.4) that $\hat{\lambda} = 0.5$. If it is believed that vacancies in the Supreme Court occur in accord with a Poisson probability law at a mean rate of 0.5 a year, then it follows that the probability is equal to e^{-2} that during his four-year term of office the next president will make no appointments to the Supreme Court.

The foregoing data also provide a method of testing the hypothesis that vacancies in the Supreme Court obey a Poisson probability at the rate of 0.5 vacancies per year. If this is the case, then the probability that in a year there will be k vacancies is given by

$$p(k; 0.5) = e^{-0.5} \frac{(0.5)^k}{k!}, \qquad k = 0, 1, 2, \cdots.$$

The expected number of years in N years in which k vacancies occur, which is equal to $Np(k; 0.5)$, may be computed and compared with the observed number of years in which k vacancies have occurred; refer to Table 3A.

TABLE 3A

Number of Vacancies k	Probability $p(k; 0.5)$ of k Vacancies	Number of Years out of 96 in which k Vacancies Occur: Expected Number $(96)p(k; 0.5)$	Observed Number N_k
0	0.6065	58.224	59
1	0.3033	29.117	27
2	0.0758	7.277	9
3	0.0126	1.210	1
over 3	0.0018	0.173	0

The observed and expected numbers may then be compared by various statistical criteria (such as the χ^2-test for goodness of fit) to determine whether the observations are compatible with the hypothesis that the number of vacancies obeys a Poisson probability law at a mean rate of 0.5. ◀

The Poisson, and related, probability laws arise in a variety of ways in the mathematical theory of queues (waiting lines) and the mathematical theory of inventory and production control. We give a very simple example of an inventory problem. It should be noted that to make the following example more realistic one must take into account the costs of the various actions available.

▶ **Example 3F. An inventory problem.** Suppose a retailer discovers that the number of items of a certain kind demanded by customers in a given time period obeys a Poisson probability law with known parameter λ. What stock K of this item should the retailer have on hand at the beginning of the time period in order to have a probability 0.99 that he will be able to supply immediately all customers who demand the item during the time period under consideration?

Solution: The problem is to find the number K, such that the probability is 0.99 that there will be K or less occurrences during the time period of the event when the item is demanded. Since the number of occurrences of this event obeys a Poisson probability law with parameter λ, we seek the integer K such that

$$(3.5) \qquad \sum_{k=0}^{K} e^{-\lambda}\frac{\lambda^k}{k!} \geq 0.99, \qquad \sum_{k=K+1}^{\infty} e^{-\lambda}\frac{\lambda^k}{k!} \leq 0.01.$$

The solution K of the second inequality in (3.5) can be read from Molina's tables (E. C. Molina, *Poisson's Exponential Binomial Limit*, Van Nostrand,

New York, 1942). If λ is so large that the normal approximation to the Poisson law may be used, then (3.5) may be solved explicitly for K. Since the first sum in (3.5) is approximately equal to

$$\Phi\left(\frac{K - \lambda + \frac{1}{2}}{\sqrt{\lambda}}\right),$$

K should be chosen so that $(K - \lambda + \frac{1}{2})/\sqrt{\lambda} = 2.326$ or

$$K = 2.326\sqrt{\lambda} + \lambda - \tfrac{1}{2}. \tag{3.6}$$

◀

THEORETICAL EXERCISES

3.1. **A problem of aerial search.** State conditions for the validity of the following assertion: if N ships are distributed at random over a region of the ocean of area A, and if a plane can search over Q square miles of ocean per hour of flight, then the number of ships sighted by a plane in a flight of T hours obeys a Poisson probability law with parameter $\lambda = NQT/A$.

3.2. **The number of matches approximately obeys a Poisson probability law.** Consider the number of matches obtained by distributing M balls, numbered 1 to M, among M urns in such a way that each urn contains exactly 1 ball. Show that the probability of exactly m matches tends to $e^{-1}(1/m!)$, as M tends to infinity, so that for large M the number of matches approximately obeys a Poisson probability law with parameter 1.

EXERCISES

State carefully the probabilistic assumptions under which you solve the following problems. Keep in mind the empirically observed fact that the occurrence of accidents, errors, breakdowns, and so on, in many instances appear to obey Poisson probability laws.

3.1. The incidence of polio during the years 1949–1954 was approximately 25 per 100,000 population. In a city of 40,000 what is the probability of having 5 or fewer cases? In a city of 1,000,000 what is the probability of having 5 or fewer cases? State your assumptions.

3.2. A manufacturer of wool blankets inspects the blankets by counting the number of defects. (A defect may be a tear, an oil spot, etc.) From past records it is known that the mean number of defects per blanket is 5. Calculate the probability that a blanket will contain 2 or more defects.

3.3. Bank tellers in a certain bank make errors in entering figures in their ledgers at the rate of 0.75 error per page of entries. What is the probability that in 4 pages there will be 2 or more errors?

3.4. Workers in a certain factory incur accidents at the rate of 2 accidents per week. Calculate the probability that there will be 2 or fewer accidents during (i) 1 week, (ii) 2 weeks; (iii) calculate the probability that there will be 2 or fewer accidents in each of 2 weeks.

3.5. A radioactive source is observed during 4 time intervals of 6 seconds each. The number of particles emitted during each period are counted. If the particles emitted obey a Poisson probability law, at a rate of 0.5 particles emitted per second, find the probability that (i) in each of the 4 time intervals 3 or more particles will be emitted, (ii) in at least 1 of the 4 time intervals 3 or more particles will be emitted.

3.6. Suppose that the suicide rate in a certain state is 1 suicide per 250,000 inhabitants per week.

(i) Find the probability that in a certain town of population 500,000 there will be 6 or more suicides in a week.

(ii) What is the expected number of weeks in a year in which 6 or more suicides will be reported in this town.

(iii) Would you find it surprising that during 1 year there were at least 2 weeks in which 6 or more suicides were reported?

3.7. Suppose that customers enter a certain shop at the rate of 30 persons an hour.

(i) What is the probability that during a 2-minute interval either no one will enter the shop or at least 2 persons will enter the shop.

(ii) If you observed the number of persons entering the shop during each of 30 2-minute intervals, would you find it surprising that 20 or more of these intervals had the property that either no one or at least 2 persons entered the shop during that time?

3.8. Suppose that the telephone calls coming into a certain switchboard obey a Poisson probability law at a rate of 16 calls per minute. If the switchboard can handle at most 24 calls per minute, what is the probability, using a normal approximation, that in 1 minute the switchboard will receive more calls than it can handle (assume all lines are clear).

3.9. In a large fleet of delivery trucks the average number inoperative on any day because of repairs is 2. Two standby trucks are available. What is the probability that on any day (i) no standby trucks will be needed, (ii) the number of standby trucks is inadequate.

3.10. Major motor failures occur among the buses of a large bus company at the rate of 2 a day. Assuming that each motor failure requires the services of 1 mechanic for a whole day, how many mechanics should the bus company employ to insure that the probability is at least 0.95 that a mechanic will be available to repair each motor as it fails? (More precisely, find the smallest integer K such that the probability is greater than or equal to 0.95 that K or fewer motor failures will occur in a day.)

3.11. Consider a restaurant located in the business section of a city. How many seats should it have available if it wishes to serve at least 95% of all those who desire its services in a given hour, assuming that potential customers (each of whom takes at least an hour to eat) arrive in accord with the following schemes:

(i) 1000 persons pass by the restaurant in a given hour, each of whom has probability 1/100 of desiring to eat in the restaurant (that is, each person passing by the restaurant enters the restaurant once in every 100 times);

(ii) persons, each of whom has probability 1/100 of desiring to eat in the restaurant, pass by the restaurant at the rate of 1000 an hour;

(iii) persons, desiring to be patrons of the restaurant, arrive at the restaurant at the rate of 10 an hour.

3.12. Flying-bomb hits on London. The following data (R. D. Clarke, "An application of the Poisson distribution," *Journal of the Institute of Actuaries*, Vol. 72 (1946), p. 48) give the number of flying-bomb hits recorded in each of 576 small areas of $t = \frac{1}{4}$ square kilometers each in the south of London during World War II.

k = number of flying-bomb hits per area	N_k = number of areas with k hits
0	229
1	211
2	93
3	35
4	7
5 or over	1

Using the procedure in example 3E, show that these observations are well fitted by a Poisson probability law.

3.13. For each of the following numerical valued random phenomena state conditions under which it may be expected to obey, either exactly or approximately, a Poisson probability law: (i) the number of telephone calls received at a given switchboard per minute; (ii) the number of automobiles passing a given point on a highway per minute; (iii) the number of bacterial colonies in a given culture per 0.01 square millimeter on a microscope slide; (iv) the number of times one receives 4 aces per 75 hands of bridge; (v) the number of defective screws per box of 100.

4. THE EXPONENTIAL AND GAMMA PROBABILITY LAWS

It has already been seen that the geometric and negative binomial probability laws arise in response to the following question: through how many trials need one wait in order to achieve the rth success in a sequence of independent repeated Bernoulli trials in which the probability of success at each trial is p? In the same way, exponential and gamma probability laws arise in response to the question: how long a time need one wait if one is observing a sequence of events occurring in time in accordance with a Poisson probability law at the rate of μ events per unit time in order to observe the rth occurrence of the event?

▶ **Example 4A.** How long will a toll collector at a toll station at which automobiles arrive at the mean rate $\mu = 1.5$ automobiles per minute have to wait before he collects the rth toll for any integer $r = 1, 2, \ldots$? ◀

We now show that the waiting time to the rth event in a series of events happening in accordance with a Poisson probability law at the rate of μ events per unit of time (or space) obeys a *gamma probability law* with parameter r and μ; consequently, it has probability density function

$$f(t) = \frac{\mu}{(r-1)!}(\mu t)^{r-1}e^{-\mu t} \qquad t \geq 0 \tag{4.1}$$
$$= 0 \qquad t < 0.$$

In particular, the waiting time to the first event obeys the *exponential* probability law with parameter μ (or equivalently, the gamma probability law with parameters $r = 1$ and μ) with probability density function

$$f(t) = \mu e^{-\mu t} \qquad t \geq 0 \tag{4.2}$$
$$= 0 \qquad t < 0.$$

To prove (4.1), first find the distribution function of the time of occurrence of the rth event. For $t \geq 0$, let $F_r(t)$ denote the probability that the time of occurrence of the rth event will be less than or equal to t. Then $1 - F_r(t)$ represents the probability that the time of occurrence of the rth event will be greater than t. Equivalently, $1 - F_r(t)$ is the probability that the number of events occurring in the time from 0 to t is less than r; consequently,

$$1 - F_r(t) = \sum_{k=0}^{r-1} \frac{1}{k!}(\mu t)^k e^{-\mu t}. \tag{4.3}$$

By differentiating (4.3) with respect to t, one obtains (4.1).

▶ **Example 4B.** Consider a baby who cries at random times at a mean rate of six distinct times per hour. If his parents respond only to every second time, what is the probability that ten or more minutes will elapse between two responses of the parents to the baby?

Solution: From the assumptions given (which may not be entirely realistic) the length T in hours of the time interval between two responses obeys a gamma probability law with parameters $r = 2$ and $\mu = 6$, Consequently,

$$P\left[T \geq \frac{1}{6}\right] = \int_{1/6}^{\infty} 6(6t)e^{-6t}\,dt = 2e^{-1}, \tag{4.4}$$

in which the integral has been evaluated by using (4.3). If the parents responded only to every third cry of the baby, then

$$P\left[T \geq \frac{1}{6}\right] = \int_{1/6}^{\infty} \frac{6}{2!}(6t)^2 e^{-6t}\, dt = \frac{5}{2} e^{-1}.$$

More generally, if the parents responded only to every rth cry of the baby, then

$$P\left[T \geq \frac{1}{6}\right] = \int_{1/6}^{\infty} \frac{6}{(r-1)!}(6t)^{r-1} e^{-6t}\, dt \tag{4.5}$$

$$= e^{-1}\left\{1 + \frac{1}{1!} + \frac{1}{2!} + \cdots + \frac{1}{(r-1)!}\right\}. \quad ◀$$

The exponential and gamma probability laws are of great importance in applied probability theory, since recent studies have indicated that in addition to describing the lengths of waiting times they also describe such numerical valued random phenomena as the life of an electron tube, the time intervals between successive breakdowns of an electronic system, the time intervals between accidents, such as explosions in mines, and so on.

The exponential probability law may be characterized in a manner that illuminates its applicability as a law of waiting times or as a law of time to failure. Let T be the observed waiting time (or time to failure). By definition, T obeys an exponential probability law with parameter λ if and only if for every $a > 0$

$$P[T \geq a] = 1 - F(a) = \int_a^{\infty} \lambda e^{-\lambda t}\, dt = e^{-\lambda a}. \tag{4.6}$$

It then follows that for any positive numbers a and b

$$P[T > a + b \mid T > b] = e^{-\lambda a} = P[T > a]. \tag{4.7}$$

In words, (4.7) says that, given an item of equipment that has served b or more time units, its conditional probability of serving $a + b$ or more time units is the same as its original probability, when first put into service of serving a or more time units. Another way of expressing (4.7) is to say that if the time to failure of a piece of equipment obeys an exponential probability law then the equipment is not subject to wear or to fatigue.

The converse is also true, as we now show. If the time to failure of an item of equipment obeys (4.7), then it obeys an exponential probability law. More precisely, *let $F(x)$ be the distribution function of the time to failure and assume that $F(x) = 0$ for $x < 0$, $F(x) < 1$ for $x > 0$, and*

$$\frac{1 - F(x + y)}{1 - F(y)} = 1 - F(x) \qquad \text{for } x, y > 0. \tag{4.8}$$

Then necessarily, for some constant $\lambda > 0$,

$$1 - F(x) = e^{-\lambda x} \qquad \text{for } x > 0. \tag{4.9}$$

If we define $g(x) = \log_e [1 - F(x)]$, then the foregoing assertion follows from a more general theorem.

THEOREM. If a function $g(x)$ satisfies the functional equation

$$g(x + y) = g(x) + g(y), \qquad x, y > 0 \tag{4.10}$$

and is bounded in the interval 0 to 1,

$$|g(x)| \leq M, \qquad 0 < x \leq 1, \tag{4.11}$$

for some constant M, then the function $g(x)$ is given by

$$g(x) = g(1)x, \qquad x > 0. \tag{4.12}$$

Proof: Suppose that (4.12) were not true. Then the function $G(x) = g(x) - g(1)x$ would not vanish identically in x. Let $x_0 > 0$ be a point such that $G(x_0) \neq 0$. Now it is clear that $G(x)$ satisfies the functional equation in (4.10). Therefore, $G(2x_0) = G(x_0) + G(x_0)$, and, for any integer n, $G(nx_0) = nG(x_0)$. Consequently, $\lim_{n\to\infty} |G(nx_0)| = \infty$. We now show that this cannot be true, since the function $G(x)$ satisfies the inequality $|G(x)| \leq 2M$ for all x, in which M is the constant given in (4.11). To prove this, note that $G(1) = 0$. Since $G(x)$ satisfies the functional equation in (4.10) it follows that, for any integer n, $G(n) = 0$ and $G(n + x) = G(x)$ for $0 < x \leq 1$. Thus $G(x)$ is a function that is periodic, with period 1. By (4.11), $G(x)$ satisfies the inequality $|G(x)| \leq 2M$ for $0 < x \leq 1$. Being periodic with period 1, it therefore satisfies this inequality for all x. The proof of the theorem is now complete.

For references to the history of the foregoing theorem, and a generalization, the reader may consult G. S. Young, "The Linear Functional Equation," *American Mathematical Monthly*, Vol. 65 (1958), pp. 37–38.

EXERCISES

4.1. Consider a radar set of a type whose failure law is exponential. If radar sets of this type have a failure rate $\lambda = 1$ set/1000 hours, find a length T of time such that the probability is 0.99 that a set will operate satisfactorily for a time greater than T.

4.2. The lifetime in hours of a radio tube of a certain type obeys an exponential law with parameter (i) $\lambda = 1000$, (ii) $\lambda = 1/1000$. A company producing these tubes wishes to guarantee them a certain lifetime. For how many hours should the tube be guaranteed to function, to achieve a probability of 0.95 that it will function at least the number of hours guaranteed?

4.3. Describe the probability law of the following random phenomenon: the number N of times a fair die is tossed until an even number appears (i) for the first time, (ii) for the second time, (iii) for the third time.

4.4. A fair coin is tossed until heads appears for the first time. What is the probability that 3 tails will appear in the series of tosses?

4.5. The customers of a certain newsboy arrive in accordance with a Poisson probability law at a rate of 1 customer per minute. What is the probability that 5 or more minutes have elapsed since (i) his last customer arrived, (ii) his next to last customer arrived?

4.6. Suppose that a certain digital computer, which operates 24 hours a day, suffers breakdowns at the rate of 0.25 per hour. We observe that the computer has performed satisfactorily for 2 hours. What is the probability that the machine will not fail within the next 2 hours?

4.7. Assume that the probability of failure of a ball bearing at any revolution is constant and equal to p. What is the probability that the ball bearing will fail on or before the nth revolution? If $p = 10^{-4}$, how many revolutions will be reached before 10% of such ball bearings fail? More precisely, find K so that $P[X > K] \leq 0.1$, where X is the number of revolutions to failure.

4.8. A lepidopterist wishes to estimate the frequency with which an unusual form of a certain species of butterfly occurs in a particular district. He catches individual specimens of the species until he has obtained exactly 5 butterflies of the form desired. Suppose that the total number of butterflies caught is equal to 25. Find the probability that 25 butterflies would have to be caught in order to obtain 5 of a desired form, if the relative frequency p of occurrence of butterflies of the desired form is given by (i) $p = \frac{1}{5}$, (ii) $p = \frac{1}{6}$.

4.9. Consider a shop at which customers arrive at random at a rate of 30 per hour. What fraction of the time intervals between successive arrivals will be (i) longer than 2 minutes, (ii) shorter than 4 minutes, (iii) between 1 and 3 minutes.

5. BIRTH AND DEATH PROCESSES

In this section we indicate briefly how one may derive the Poisson probability law, and various related probability laws, by means of differential equations. The process to be examined is treated in the literature of stochastic processes under the name "birth and death."

Consider a population, such as the molecules present in a certain subvolume of gas, the particles emitted by a radioactive source, biological organisms of a certain kind present in a certain environment, persons waiting in a line (queue) for service, and so on. Let X_t be the size of the population at a given time t. The probability law of X_t is specified by its probability mass function,

$$p(n; t) = P[X_t = n] \qquad n = 0, 1, 2, \cdots. \tag{5.1}$$

A differential equation for the probability mass function of X_t may be found under assumptions similar in spirit to, but somewhat more general than, those made in deriving (3.1). In reading the following discussion the reader should attempt to formulate explicitly for himself the assumptions that are being made. A rigorous treatment of this discussion is given by W. Feller, *An Introduction to Probability Theory and its Applications*, Wiley, 1957, pp. 397–411.

Let $r_0(h)$, $r_1(h)$, and $r_2(h)$ be functions defined for $h > 0$ with the property that

$$\lim_{h \to 0} \frac{r_0(h)}{h} = \lim_{h \to 0} \frac{r_1(h)}{h} = \lim_{h \to 0} \frac{r_2(h)}{h} = 0.$$

Assume that the probability is $r_2(h)$ that in the time from t to $t + h$ the population size will change by two or more. For $n \geq 1$ the event that $X_{t+h} = n$ (n members in the population at time $t + h$) can then essentially happen in any one of three mutually exclusive ways: (i) the population size at time t is n and undergoes no change in the time from t to $t + h$; (ii) the population size at time t is $n - 1$ and increases by one in the time from t to $t + h$; (iii) the population size at time t is $n + 1$ and decreases by one in the time from t to $t + h$. For $n = 0$, the event that $X_{t+h} = 0$ can happen only in ways (i) and (iii). Now let us introduce quantities λ_n and μ_n, defined as follows; $\lambda_n h + r_1(h)$ for any time t and positive value of h is the conditional probability that the population size will increase by one in the time from t to $t + h$, given that the population had size n at time t, whereas $\mu_n h + r_0(h)$ is the conditional probability that the population size will decrease by one in the time from t to $t + h$, given that the population had size n at time t. In symbols, λ_n and μ_n are such that, for any time t and small $h > 0$,

$$(5.2) \qquad \begin{aligned} \lambda_n h &\doteq P[X_{t+h} - X_t = 1 \mid X_t = n], \qquad & n \geq 0 \\ \mu_n h &\doteq P[X_{t+h} - X_t = -1 \mid X_t = n], \qquad & n \geq 1; \end{aligned}$$

the approximation in (5.2) is such that the difference between the two sides of each equation tends to 0 faster than h, as h tends to 0. In writing the next equations we omit terms that tend to 0 faster than h, as h tends to 0, since these terms vanish in deriving the differential equations in (5.10) and (5.11). The reader may wish to verify this statement for himself.

The event (i) then has probability,

$$(5.3) \qquad p(n; t)(1 - \lambda_n h - \mu_n h);$$

the event (ii) has probability

$$(5.4) \qquad p(n - 1; t)\lambda_{n-1} h;$$

the event (iii) has probability

$$(5.5) \qquad p(n + 1; t)\mu_{n+1} h.$$

Consequently, one obtains for $n \geq 1$

$$(5.6)\quad p(n; t+h) = p(n;t)(1 - \lambda_n h - \mu_n h) + p(n-1;t)\lambda_{n-1}h + p(n+1;t)\mu_{n+1}h.$$

For $n = 0$ one obtains

$$(5.7)\quad p(0;t+h) = p(0;t)(1-\lambda_0 h) + p(1;t)\mu_1 h.$$

It may be noted that if there is a maximum possible value N for the population size then (5.6) holds only for $1 \leq n \leq N-1$, whereas for $n = N$ one obtains

$$(5.8)\quad p(N;t) = p(N;t)(1-\mu_N h) + p(N-1;t)\lambda_{N-1}h.$$

Rearranging (5.6), one obtains

$$(5.9)\quad \frac{p(n;t+h) - p(n;t)}{h} = -(\lambda_n + \mu_n)p(n;t) + \lambda_{n-1}p(n-1;t) + \mu_{n+1}p(n+1;t).$$

Letting h tend to 0, one finally obtains for $n \geq 1$

$$(5.10)\quad \frac{\partial}{\partial t}p(n;t) = -(\lambda_n + \mu_n)p(n;t) + \lambda_{n-1}p(n-1;t) + \mu_{n+1}p(n+1;t).$$

Similarly, for $n = 0$ one obtains

$$(5.11)\quad \frac{\partial}{\partial t}p(0;t) = -\lambda_0 p(0;t) + \mu_1 p(1;t).$$

The question of the existence and uniqueness of solutions of these equations is nontrivial and is not discussed here.

We solve these equations only in the case that

$$(5.12)\quad \begin{aligned} \lambda_0 = \lambda_1 = \lambda_2 = \cdots = \lambda_n = \cdots = \lambda \\ \mu_1 = \mu_2 = \mu_3 = \cdots = \mu_n = \cdots = 0, \end{aligned}$$

which corresponds to the assumptions made before (3.1). Then (5.11) becomes

$$(5.13)\quad \frac{\partial}{\partial t}p(0;t) = -\lambda p(0;t),$$

which has solution (under the assumption $p(0;0) = 0$)

$$(5.14)\quad p(0;t) = e^{-\lambda t}.$$

Next (5.10) for the case $n = 1$ becomes

$$\frac{\partial}{\partial t} p(1;t) = -\lambda p(1;t) + \lambda p(0;t), \tag{5.15}$$

which has solution (under the assumption $p(1;0) = 0$)

$$\begin{aligned} p(1;t) &= \lambda e^{-\lambda t} \int_0^t e^{\lambda t'} p(0;t')\,dt' \\ &= \lambda t e^{-\lambda t}. \end{aligned} \tag{5.16}$$

Proceeding inductively, one obtains (assuming $p(n;0) = 0$)

$$p(n;t) = \frac{(\lambda t)^n}{n!} e^{-\lambda t}, \tag{5.17}$$

so that the size X_t of the population at time t obeys a Poisson probability law with mean λt.

THEORETICAL EXERCISES

5.1. The Yule process. Consider a population whose numbers can (by splitting or otherwise) give birth to new members but cannot die. Assume that the probability is approximately equal to λh that in a short time interval of length h a member will create a new member. More precisely, in the model of section 5, assume that

$$\lambda_n = n\lambda, \qquad \mu_n = 0.$$

If at time 0 the population size is k, show that the probability that the population size at time t is equal to n is given by

$$p(n;t) = \binom{n-1}{n-k} e^{-k\lambda t}(1 - e^{-\lambda t})^{n-k}, \qquad n \geq k. \tag{5.18}$$

Show that the probability law defined by (5.18) has mean m and variance σ^2 given by

$$m = ke^{\lambda t}, \qquad \sigma^2 = ke^{\lambda t}(e^{\lambda t} - 1). \tag{5.19}$$

CHAPTER 7

Random Variables

It has been stressed in the foregoing chapters that the probability of a random event can be discussed only with reference to a sample description space on which a probability function has been defined. However, in many applications of probability theory the terminology of sample description spaces does not explicitly enter (although, as we shall see, the notion is always implicitly present). Rather, many applications of probability theory are based on *the notion of a random variable*. This chapter gives a rigorous definition of the notion of a random variable and presents the main concepts and techniques used to treat random variables.

1. THE NOTION OF A RANDOM VARIABLE

In applications of probability theory one usually has to deal simultaneously with several random phenomena. In section 7 of Chapter 4, we indicated one way of treating several random phenomena by means of the notion of a numerical n-tuple valued random phenomenon. However, this is not a very satisfactory method, for it requires one to fix in advance the number n of random phenomena to be considered in a given context. Further, it provides no convenient way of generating, by means of various algebraic and analytic operations, new random phenomena from known random phenomena. These difficulties are avoided by using random variables. Random variables are usually denoted by capital letters, especially the letters X, Y, Z, U, V, and W. To these letters numerical subscripts may be added, so that X_1, X_2, ... are random variables. For the purpose of defining the terminology we consider a random variable which we denote by X.

The notion of a random variable is intimately related to the notion of a function, as the following definitions indicate.

THE DEFINITION OF A FUNCTION. An object X, or $X(\cdot)$, is said to be a function defined on a space S if for every member s of S there is a real number, denoted by $X(s)$, which is called the value of the function X at s.

THE DEFINITION OF A RANDOM VARIABLE. An object X is said to be a random variable if (i) it is a real valued function defined on a sample description space on a family of whose subsets a probability function $P[\cdot]$ has been defined, and (ii) for every Borel set B of real numbers the set $\{s\colon\ X(s) \text{ is in } B\}$ belongs to the domain of $P[\cdot]$.

A random variable then is a function defined on the outcome of a random phenomenon; consequently, the value of a random variable is a random phenomenon and indeed is a numerical valued random phenomenon. Conversely, every numerical valued random phenomenon can be interpreted as the value of a random variable X; namely, the random variable X defined on the real line for every real number x by $X(x) = x$.

One of the major difficulties students have with the notion of a random variable is that objects that are random variables are not always defined in a manner to make this fact explicit. However, we have previously encountered a similar situation with regard to the notion of a random event. We have defined a random event as a set on a sample description space on which a probability function is defined. In every day discourse random events are defined verbally, so that in order to discuss a random event one must first formulate the event in a mathematical manner as a *set*. Similarly, with regard to random variables, one must learn how to recognize, and formulate mathematically as *functions*, verbally described objects that are random variables.

▶ **Example 1A. The number of white balls in a sample is a random variable.** Let us consider the object X defined as follows: X is the number of white balls in a sample of size 2 drawn without replacement from an urn containing 6 balls, of which 4 are white. The sample description space S of the experiment of drawing the sample may be taken as the set of 30 ordered 2-tuples given in (3.1) of Chapter 1, in which the white balls have been numbered 1 to 4 and the remaining 2 balls, 5 and 6. To render S a probability space, we need to define a probability function upon its subsets; let us do so by assuming all descriptions equally likely. The number X of white balls in the sample drawn can be regarded as a function on this probability space, for if the sample description s is known then the value of X is known.

$$
\begin{aligned}
(1.1)\quad X(s) &= 0 \quad \text{if } s = (5, 6), (6, 5)\\
&= 1 \quad \text{if } s = (1, 5), (1, 6), (2, 5), (2, 6), (3, 5), (3, 6), (4, 5), (4, 6)\\
&\qquad\qquad (5, 1), (6, 1), (5, 2), (6, 2), (5, 3), (6, 3), (5, 4), (6, 4)\\
&= 2 \quad \text{if } s = (1, 2), (1, 3), (1, 4), (2, 1), (2, 3), (2, 4), (3, 1), (3, 2)\\
&\qquad\qquad (3, 4), (4, 1), (4, 2), (4, 3).
\end{aligned}
$$

◀

EXERCISE

1.1. Show that the following quantities are random variables by explaining how they may be defined as functions on a probability space:

(i) The sum of 2 dice that are tossed independently.

(ii) The number of times a coin is tossed until a head appears for the first time.

(iii) The second digit in the decimal expansion of a number chosen on the unit interval in accordance with a uniform probability law.

(iv) The absolute value of a number chosen on the real line in accordance with a normal probability law.

(v) The number of urns that contain balls bearing the same number, when 52 balls, numbered 1 to 52, are distributed, one to an urn, among 52 urns, numbered 1 to 52.

(vi) The distance from the origin of a 2-tuple (x_1, x_2) in the plane chosen in accordance with a known probability law, specified by the probability density function $f(x_1, x_2)$.

2. DESCRIBING A RANDOM VARIABLE

Although, by definition, a random variable X is a function on a probability space, in probability theory we are rarely concerned with the functional form of X, for we are not interested in computing the value $X(s)$ that the function X assumes at any individual member s of the sample description space S on which X is defined. Indeed, we do not usually wish to know the space S on which X is defined. Rather, we are interested in the probability that an observed value of the random variable X will lie in a given set B. We are interested in a random variable as a mechanism that gives rise to a numerical valued random phenomenon, and the questions we shall ask about a random variable X are precisely the same as those asked about numerical valued random phenomena. Similarly, the techniques we use to describe random variables are precisely the same as those used to describe numerical valued random phenomena.

To begin with, we define the *probability function of a random variable* X, denoted by $P_X[\cdot]$, as a set function defined for every Borel set B of real numbers, whose value $P_X[B]$ is the probability that X is in B. We sometimes write the intuitively meaningful expression $P[X \text{ is in } B]$ for the mathematically correct expression $P_X[B]$. Similarly, we adopt the following expressions for any real numbers a, b, and x:

$$\begin{aligned} P[a < X \leq b] &= P_X[\{\text{real numbers } x\colon \ a < x \leq b\}] \\ P[X \leq x] &= P_X[\{\text{real numbers } x'\colon \ x' \leq x\}] \\ P[X = x] &= P_X[\{\text{real numbers } x'\colon \ x' = x\}] = P_X[\{x\}]. \end{aligned} \tag{2.1}$$

One obtains the probability function $P_X[\cdot]$ of the random variable X from the probability function $P[\cdot]$, which exists on the sample description space S on which X is defined as a function, by means of the following *basic formula: for any Borel set B of real numbers*

$$P_X[B] = P[\{s\colon \ X(s) \text{ is in } B\}]. \tag{2.2}$$

Equation (2.2) represents the definition of $P_X[B]$; it is clear that it embodies the intuitive meaning of $P_X[B]$ given above, since the function X will have an observed value lying in the set B if and only if the observed value s of the underlying random phenomenon is such that $X(s)$ is in B.

▶ **Example 2A. The probability function of the number of white balls in a sample.** To illustrate the use of (2.2), let us compute the probability function of the random variable X defined by (1.1). Assuming equally likely descriptions on S, one determines for any set B of real numbers that the value of $P_X[B]$ depends on the intersection of B with the set $\{0, 1, 2\}$:

$P_X[B] = 0$	$\frac{1}{15}$	$\frac{8}{15}$	$\frac{6}{15}$	$\frac{9}{15}$	$\frac{7}{15}$	$\frac{14}{15}$	1
if $B\{0, 1, 2\} = \emptyset$	$\{0\}$	$\{1\}$	$\{2\}$	$\{0, 1\}$	$\{0, 2\}$	$\{1, 2\}$	$\{0, 1, 2\}$ ◀

We may represent the probability function $P_X[\cdot]$ of a random variable as a distribution of a unit mass over the real line in such a way that the amount of mass over any set B of real numbers is equal to the value $P_X[B]$ of the probability function of X at B. We have seen in Chapter 4 that a distribution of probability mass may be specified in various ways by means of probability mass functions, probability density functions, and distribution functions. We now introduce these notions in connection with random variables. However, the reader should bear constantly in mind that, as mathematical functions defined on the real line, these notions have the same mathematical properties, whether they arise from random variables or from numerical valued random phenomena.

The *probability law of a random variable X is defined as a probability function $P[\cdot]$ over the real line that coincides with the probability function $P_X[\cdot]$ of the random variable X.* By definition, *probability theory is concerned with the statements that can be made about a random variable, knowing only its probability law.* Consequently, a proposition stated about a probability function $P[\cdot]$ is, from the point of view of probability theory, a proposition stated about all random variables $X, Y, \ldots,$ whose probability functions $P_X[\cdot], P_Y[\cdot], \ldots$ coincide with $P[\cdot]$.

Two random variables X and Y are said to be *identically distributed* if their probability functions are equal; that is, $P_X[B] = P_Y[B]$ for all Borel sets B.

The *distribution function* of a random variable X, denoted by $F_X(\cdot)$, is defined for any real number x by

$$F_X(x) = P[X \le x]. \tag{2.3}$$

The distribution function $F_X(\cdot)$ of a random variable possesses all the properties stated in section 3 of Chapter 4 for the distribution function of a numerical valued random phenomenon. The distribution function of X uniquely determines the probability function of X.

The distribution function may be used to classify random variables into types. *A random variable X is said to be discrete or continuous, depending on whether its distribution function $F_X(\cdot)$ is discrete or continuous.*

The *probability mass function* of a random variable X, denoted by $p_X(\cdot)$, is a function whose value $p_X(x)$ at any real number x represents the probability that the observed value of the random variable X will be equal to x; in symbols,

$$p_X(x) = P[X = x] = P_X[\{x': \ x' = x\}]. \tag{2.4}$$

A real number x for which $p_X(x)$ is positive is called a *probability mass point* of the random variable X. From the distribution function $F_X(\cdot)$ one may obtain the probability mass function $p_X(\cdot)$ by

$$p_X(x) = F_X(x) - \lim_{a \to x-} F_X(a). \tag{2.5}$$

A random variable X is *discrete* if the sum of the probability mass function over the points at which it is positive (there are at most a countably infinite number) is equal to 1; in symbols, X is discrete if

$$\sum_{\substack{\text{over all points } x \text{ such} \\ \text{that } p_X(x) > 0}} p_X(x) = 1. \tag{2.6}$$

In other words, a random variable X is discrete when one distributes a unit mass over the infinite line in accordance with the probability function $P_X[\cdot]$ if one does so by attaching a positive mass $p_X(x)$ to each of a finite or a countably infinite number of points.

If a random variable X is discrete, it suffices to know its probability mass function $p_X(\cdot)$ in order to know its probability function $P_X[\cdot]$, for we have the following formula expressing $P_X[\cdot]$ in terms of $p_X(\cdot)$. *If X is discrete, then for any Borel set B of real numbers*

$$P_X[B] = P[X \text{ is in } B] = \sum_{\substack{\text{over all points } x \text{ in } B \\ \text{such that } p_X(x) > 0}} p_X(x). \tag{2.7}$$

Thus, for a discrete random variable X, to evaluate the probability $P_X[B]$ that the random variable X will have an observed value lying in B, one has only to list the probability mass points of X which lie in B. One then adds the probability masses attached to these probability mass points to obtain $P_X[B]$.

The distribution function of a discrete random variable X is given in terms of its probability mass function by

$$F_X(x) = \sum_{\substack{\text{over all points } x' \leq x \\ \text{such that } p_X(x') > 0}} p_X(x'). \tag{2.8}$$

The distribution function $F_X(\cdot)$ of a discrete random variable X is what might be called a piecewise constant or "step" function, as diagrammed in Fig. 3A of Chapter 4. It consists of a series of horizontal lines over the intervals between probability mass points; at a probability mass point x, the graph of $F_X(\cdot)$ jumps upward by an amount $p_X(x)$.

▶ **Example 2B.** A random variable X has a *binomial distribution* with parameters n and p if it is a discrete random variable whose probability mass function $p_X(\cdot)$ is given by, for any real number x,

$$\begin{aligned} p_X(x) &= \binom{n}{x} p^x (1-p)^{n-x} \qquad \text{if } x = 0, 1, \cdots, n \\ &= 0 \qquad \text{otherwise.} \end{aligned} \tag{2.9}$$

Thus for a random variable X, which has a binomial distribution with parameters $n = 6$ and $p = \frac{1}{3}$,

$$P[1 < X \leq 2] = \binom{6}{2}\left(\frac{1}{3}\right)^2\left(\frac{2}{3}\right)^4 = 0.3292$$

$$P[1 \leq X \leq 2] = \binom{6}{1}\left(\frac{1}{3}\right)\left(\frac{2}{3}\right)^5 + \binom{6}{2}\left(\frac{1}{3}\right)^2\left(\frac{2}{3}\right)^4 = 0.5926. \quad ◀$$

▶ **Example 2C. Identically distributed random variables.** Some insight into the notion of identically distributed random variables may be gained by considering the following simple example of two random variables that are distinct as functions and yet are identically distributed. Suppose one is tossing a fair die; consider the random variables X and Y, defined as follows:

Value of X, if outcome of die is		*Value of Y, if outcome of die is*	
2	1, 2, 3	2	4, 5, 6
1	4, 5	1	2, 3
0	6	0	1

It is clear that both X and Y are discrete random variables, whose probability mass functions agree for all x; indeed, $p_X(2) = p_Y(2) = \frac{1}{2}$, $p_X(1) = p_Y(1) = \frac{1}{3}$, $p_X(0) = p_Y(0) = \frac{1}{6}$, $p_X(x) = p_Y(x) = 0$ for $x \neq 0, 1$, or 2. Consequently, the probability functions $P_X[B]$ and $P_Y[B]$ agree for all sets B. ◀

If a random variable X is continuous, there exists a nonnegative function $f_X(\cdot)$, called the probability density function of the random variable X, which has the following property: for any Borel set B of real numbers

$$P_X[B] = P[X \text{ is in } B] = \int_B f_X(x)\,dx. \tag{2.10}$$

In words, for a continuous random variable X, once the probability density function $f_X(\cdot)$ is known, the value $P_X[B]$ of the probability function at any Borel set B may be obtained by integrating the probability density function $f_X(\cdot)$ over the set B.

The distribution function $F_X(\cdot)$ of a continuous random variable is given in terms of its probability density function by

$$F_X(x) = \int_{-\infty}^{x} f_X(x')\,dx'. \tag{2.11}$$

In turn, the probability density function of a continuous random variable can be obtained from its distribution function by differentiation:

$$f_X(x) = \frac{d}{dx} F_X(x) \tag{2.12}$$

at all points x at which the derivative on the right-hand side of (2.12) exists.

▶ **Example 2D.** A random variable X is said to be *normally distributed* if it is continuous and if constants m and σ exist, where $-\infty < m < \infty$ and $\sigma > 0$, such that the probability density function $f_X(\cdot)$ is given by, for any real number x,

$$f_X(x) = \frac{1}{\sigma\sqrt{2\pi}} e^{-\frac{1}{2}\left(\frac{x-m}{\sigma}\right)^2}. \tag{2.13}$$

Then for any real numbers a and b

$$(2.14)\quad P[a \le X \le b] = \int_a^b f_X(x)\,dx = \Phi\left(\frac{b-m}{\sigma}\right) - \Phi\left(\frac{a-m}{\sigma}\right).$$

For a random variable X, which is normally distributed with parameters $m = 2$ and $\sigma = 2$,

$$P[1 \le X \le 2] = P[1 < X < 2] = \Phi\left(\frac{2-2}{2}\right) - \Phi\left(\frac{1-2}{2}\right) = 0.1915. \blacktriangleleft$$

We conclude this section by making explicit mention of our conventions concerning the use of the letters p, f, and F, and the subscripts $X, Y, \ldots$. We shall always use $p(\cdot)$ to denote a probability mass function and then add as a subscript the random variable (which could be denoted by X, Y, Z, U, V, W, etc.) of which it is the probability mass function. Thus, $p_U(\cdot)$ denotes the probability mass function of the random variable U, whereas $p_U(u)$ denotes the value of $p_U(\cdot)$ at the point u. Similarly, we write $f_X(\cdot)$, $f_Y(\cdot), f_Z(\cdot), f_U(\cdot), f_V(\cdot), f_W(\cdot)$ to denote the probability density function, respectively, of X, Y, Z, U, V, W. Similarly, we write $F_X(\cdot)$, $F_Y(\cdot)$, $F_Z(\cdot)$, $F_U(\cdot)$, $F_V(\cdot)$, $F_W(\cdot)$ to denote the distribution function, respectively, of X, Y, Z, U, V, W.

EXERCISES

In exercises 2.1 to 2.8 describe the probability law of the random variable given.

2.1. The number of aces in a hand of 13 cards drawn without replacement from a bridge deck.

2.2. The sum of numbers on 2 balls drawn with replacement (without replacement) from an urn containing 6 balls, numbered 1 to 6.

2.3. The maximum of the numbers on 2 balls drawn with replacement (without replacement) from an urn containing 6 balls, numbered 1 to 6.

2.4. The number of white balls drawn in a sample of size 2 drawn with replacement (without replacement) from an urn containing 6 balls, of which 4 are white.

2.5. The second digit in the decimal expansion of a number chosen on the unit interval in accordance with a uniform probability law.

2.6. The number of times a fair coin is tossed until heads appears (i) for the first time, (ii) for the second time, (iii) the third time.

2.7. The number of cards drawn without replacement from a deck of 52 cards until (i) a spade appears, (ii) an ace appears.

2.8. The number of balls in the first urn if 10 distinguishable balls are distributed in 4 urns in such a manner that each ball is equally likely to be placed in any urn.

In exercises 2.9 to 2.16 find $P[1 \leq X \leq 2]$ for the random variable X described.

2.9. X is normally distributed with parameters $m = 1$ and $\sigma = 1$.

2.10. X is Poisson distributed with parameter $\lambda = 1$.

2.11. X obeys a binomial probability law with parameters $n = 10$ and $p = 0.1$.

2.12. X obeys an exponential probability law with parameter $\lambda = 1$.

2.13. X obeys a geometric probability law with parameter $p = \frac{1}{3}$.

2.14. X obeys a hypergeometric probability law with parameters $N = 100$, $p = 0.1$, $n = 10$.

2.15. X is uniformly distributed over the interval $\frac{1}{2}$ to $\frac{3}{2}$.

2.16. X is Cauchy distributed with parameters $\alpha = 1$ and $\beta = 1$.

3. AN EXAMPLE, TREATED FROM THE POINT OF VIEW OF NUMERICAL *n*-TUPLE VALUED RANDOM PHENOMENA

In the next two sections we discuss an example that illustrates the need to introduce various concepts concerning random variables, which will, in turn, be presented in the course of the discussion. We begin in this section by discussing the example in terms of the notion of a numerical valued random phenomenon in order to show the similarities and differences between this notion and that of a random variable.

Let us consider a commuter who is in the habit of taking a train to the city; the time of departure from the station is given in the railroad timetable as 7:55 A.M. However, the commuter notices that the actual time of departure is a random phenomenon, varying between 7:55 and 8 A.M. Let us assume that the probability law of the random phenomenon is specified by a probability density function $f_1(\cdot)$; further, let us assume

$$
\begin{aligned}
(3.1) \qquad f_1(x_1) &= \tfrac{2}{25}(5 - x_1) \qquad \text{for } 0 \leq x_1 \leq 5 \\
&= 0 \qquad \text{otherwise.}
\end{aligned}
$$

in which x_1 represents the number of minutes after 7:55 A.M. that the train departs.

Let us suppose next that the time it takes the commuter to travel from his home to the station is a numerical valued random phenomenon, varying between 25 and 30 minutes. Then, if the commuter leaves his home at 7:30 A.M. every day, his time of arrival at the station is a random phenomenon, varying between 7:55 and 8 A.M. Let us suppose that the probability law of this random phenomenon is specified by a probability density function $f_2(\cdot)$; further, let us assume that $f_2(\cdot)$ is of the same functional form as $f_1(\cdot)$, so that

$$(3.2) \qquad \begin{aligned} f_2(x_2) &= \tfrac{2}{25}(5 - x_2) \qquad \text{for } 0 \le x_2 \le 5 \\ &= 0 \qquad \text{otherwise,} \end{aligned}$$

in which x_2 represents the number of minutes after 7:55 A.M. that the commuter arrives at the station.

The question now naturally arises: will the commuter catch the 7:55 A.M. train? Of course, this question cannot be answered by us; but perhaps we can answer the question: what is the probability that the commuter will catch the 7:55 A.M. train?

Before any attempt can be made to answer this question, we must express mathematically as a set on a sample description space the random event described verbally as the event that the commuter catches the train. Further, to compute the probability of the event, a probability function on the sample description space must be defined.

As our sample description space S, we take the space of 2-tuples (x_1, x_2) of real numbers, where x_1 represents the time (in minutes after 7:55 A.M.) at which the train departs from the station, and x_2 denotes the time (in minutes after 7:55 A.M.) at which the commuter arrives at the station. The event A that the man catches the train is then given as a set of sample descriptions by $A = \{(x_1, x_2)\colon\ x_1 > x_2\}$, since to catch the train his arrival time x_2 must be less than the train's departure time x_1. The event A is diagrammed in Fig. 3A.

We define next a probability function $P[\cdot]$ on the events in S. To do this, we use the considerations of section 7, Chapter 4, concerning numerical 2-tuple valued random phenomena. In particular, let us suppose that the probability function $P[\cdot]$ is specified by a 2-dimensional probability density function $f(.\,,.)$. From a knowledge of $f(.\,,.)$ we may compute the probability $P[A]$ that the commuter will catch his train by the formula

$$(3.3) \qquad \begin{aligned} P[A] &= \iint_A f(x_1, x_2)\, dx_1\, dx_2 \\ &= \int_{-\infty}^{\infty} dx_1 \int_{-\infty}^{x_1} dx_2 f(x_1, x_2) \\ &= \int_{-\infty}^{\infty} dx_2 \int_{x_2}^{\infty} dx_1 f(x_1, x_2) \end{aligned}$$

in which the second and third equations follow by the usual rules of calculus for evaluating double integrals (or integrals over the plane) by means of iterated (or repeated) single integrals.

We next determine whether the function $f(.\,,.)$ is specified by our having specified the probability density functions $f_1(\cdot)$ and $f_2(\cdot)$ by (3.1) and (3.2).

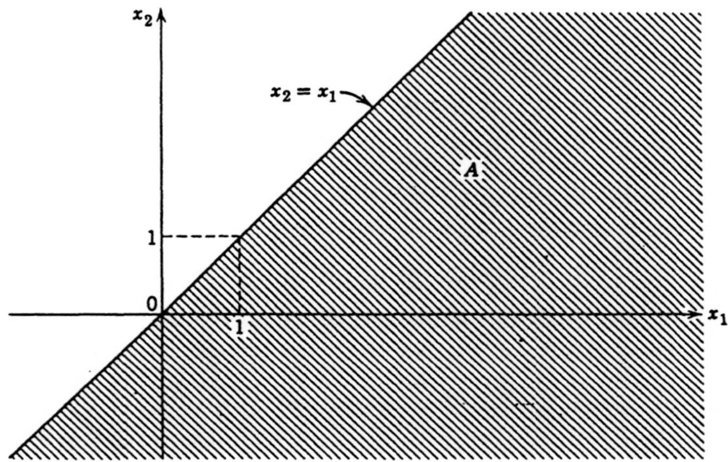

Fig. 3A. The event A that the man catches the train represented as a set of points in the (x_1, x_2)-plane.

More generally, we consider the question: *what relationship exists between the individual probability density functions $f_1(\cdot)$ and $f_2(\cdot)$ and the joint probability density function $f(.\,,.)$?* We show first that *from a knowledge of $f(.\,,.)$ one may obtain a knowledge of $f_1(\cdot)$ and $f_2(\cdot)$ by the formulas, for all real numbers x_1 and x_2,*

$$f_1(x_1) = \int_{-\infty}^{\infty} f(x_1, x_2)\, dx_2 \tag{3.4}$$

$$f_2(x_2) = \int_{-\infty}^{\infty} f(x_1, x_2)\, dx_1.$$

Conversely, we show by a general example that *from a knowledge of $f_1(\cdot)$ and $f_2(\cdot)$ one cannot obtain a knowledge of $f(.\,,.)$, since $f(.\,,.)$ is not uniquely determined by $f_1(\cdot)$ and $f_2(\cdot)$; more precisely, we show that to given probability density functions $f_1(\cdot)$ and $f_2(\cdot)$ there exists an infinity of functions $f(.\,,.)$ that satisfy (3.4) with respect to $f_1(\cdot)$ and $f_2(\cdot)$.*

To prove (3.4), let $F_1(\cdot)$ and $F_2(\cdot)$ be the distribution functions of the first and second random phenomena under consideration; in the example discussed, $F_1(\cdot)$ is the distribution function of the departure time of the train from the station, and $F_2(\cdot)$ is the distribution function of the arrival time of the man at the station. We may obtain expressions for $F_1(\cdot)$ and $F_2(\cdot)$ in terms of $f(.\,,.)$, for $F_1(x)$ is equal to the probability, according to the probability function $P[\cdot]$, of the set $\{(x_1', x_2')\colon\ x_1' \le x_1, -\infty < x_2' < \infty\}$, and similarly $F_2(x_2) = P[\{(x_1', x_2')\colon\ -\infty < x_1' < \infty, x_2' \le x_2\}]$. Consequently,

$$\begin{aligned} F_1(x_1) &= \int_{-\infty}^{x_1} dx_1' \int_{-\infty}^{\infty} dx_2' f(x_1', x_2') \\ F_2(x_2) &= \int_{-\infty}^{x_2} dx_2' \int_{-\infty}^{\infty} dx_1' f(x_1', x_2'). \end{aligned} \tag{3.5}$$

We next use the fact that

$$f_1(x_1) = \frac{d}{dx_1} F_1(x_1), \qquad f_2(x_2) = \frac{d}{dx_2} F_2(x_2). \tag{3.6}$$

By differentiation of (3.5), in view of (3.6), we obtain (3.4).

Conversely, given any two probability density functions $f_1(\cdot)$ and $f_2(\cdot)$, let us show how one may find many probability density functions $f(.\,,.)$ to satisfy (3.4). Let A be a positive number. Choose a finite nonempty interval a_1 to b_1 such that $f_1(x_1) \ge A$ for $a_1 \le x_1 \le b_1$. Similarly, choose a finite nonempty interval a_2 to b_2 such that $f_2(x_2) \ge A$ for $a_2 \le x_2 \le b_2$. Define a function of two variables $h(.\,,.)$ by

$$\begin{aligned} h(x_1, x_2) &= A^2 \sin\left[\frac{\pi}{b_1 - a_1}\left(x_1 - \frac{a_1 + b_1}{2}\right)\right] \\ &\qquad \sin\left[\frac{\pi}{b_2 - a_2}\left(x_2 - \frac{a_2 + b_2}{2}\right)\right] \\ &\qquad \text{if both } a_1 \le x_1 \le b_1, \quad a_2 \le x_2 \le b_2 \\ &= 0 \qquad \text{otherwise.} \end{aligned} \tag{3.7}$$

Clearly, by construction, for all x_1 and x_2

$$|h(x_1, x_2)| \le f_1(x_1) f_2(x_2), \tag{3.8}$$

$$\int_{-\infty}^{\infty} h(x_1, x_2)\, dx_1 = \int_{-\infty}^{\infty} h(x_1, x_2)\, dx_2 = \int_{-\infty}^{\infty}\int_{-\infty}^{\infty} h(x_1, x_2)\, dx_1\, dx_2 = 0.$$

Define the function $f(.\,,.)$ for any real numbers x_1 and x_2 by

$$f(x_1, x_2) = f_1(x_1) f_2(x_2) + h(x_1, x_2). \tag{3.9}$$

It may be verified, in view of (3.8), that $f(.\,,.)$ is a probability density function satisfying (3.4).

We now return to the question of how to determine $f(.\,,.)$. *There is one (and, in general, only one) circumstance in which the individual probability density functions $f_1(\cdot)$ and $f_2(\cdot)$ determine the joint probability density function $f(.\,,.)$, namely, when the respective random phenomena, whose probability density functions are $f_1(\cdot)$ and $f_2(\cdot)$, are independent.*

We define two random phenomena as independent, letting $P_1[\cdot]$ and $P_2[\cdot]$ denote their respective probability functions and $P[\cdot]$ their joint probability function, if it holds that for all real numbers a_1, b_1, a_2, and b_2

$$(3.10)\quad P[\{(x_1, x_2)\colon\ a_1 < x_1 \le b_1, a_2 < x_2 \le b_2\}]$$
$$= P_1[\{x_1\colon\ a_1 < x_1 \le b\}]P_2[\{x_2\colon\ a_2 < x_2 \le b_2\}].$$

Equivalently, two random phenomena are independent, letting $F_1(\cdot)$ and $F_2(\cdot)$ denote their respective distribution functions and $F(.\,,.)$ their joint distribution function, if it holds that for all real numbers x_1 and x_2

$$(3.11)\qquad F(x_1, x_2) = F_1(x_1)F_2(x_2).$$

Equivalently, two continuous random phenomena are independent, letting $f_1(\cdot)$ and $f_2(\cdot)$ denote their respective probability density functions and $f(.\,,.)$ their joint probability density, if it holds that for all real numbers x_1 and x_2

$$(3.12)\qquad f(x_1, x_2) = f_1(x_1)f_2(x_2).$$

Equivalently, two discrete random phenomena are independent, letting $p_1(\cdot)$ and $p_2(\cdot)$ denote their respective probability mass functions and $p(.\,,.)$ their joint probability mass function if it holds that for all real numbers x_1 and x_2

$$(3.13)\qquad p(x_1, x_2) = p_1(x_1)p_2(x_2).$$

The equivalence of the foregoing statements concerning independence may be shown more or less with ease by using the relationships developed in Chapter 4; indications of the proofs are contained in section 6.

Independence may also be defined in terms of the notion of *an event depending on a phenomenon,* which is analogous to the notion of an event depending on a trial developed in section 2 of Chapter 3. *An event A is said to depend on a random phenomenon if a knowledge of the outcome of the phenomenon suffices to determine whether or not the event A has occurred.* We then define two random phenomena as independent if, for any two events A_1 and A_2, depending, respectively, on the first and second phenomenon, the probability of the intersection of A_1 and A_2 is equal to the product of their probabilities:

$$P[A_1A_2] = P[A_1]P[A_2]. \tag{3.14}$$

As shown in section 2 of Chapter 3, two random phenomena are independent if and only if a knowledge of the outcome of one of the phenomena does not affect the probability of any event depending upon the other phenomenon.

Let us now return to the problem of the commuter catching his train, and let us assume that the commuter's arrival time and the train's departure time are independent random phenomena. Then (3.12) holds, and from (3.3)

$$\begin{aligned} P[A] &= \int_{-\infty}^{\infty} dx_1 f_1(x_1) \int_{-\infty}^{x_1} dx_2 f_2(x_2) \\ &= \int_{-\infty}^{\infty} dx_2 f_2(x_2) \int_{x_2}^{\infty} dx_1 f_1(x_1). \end{aligned} \tag{3.15}$$

Since $f_1(\cdot)$ and $f_2(\cdot)$ are specified by (5.1) and (5.2), respectively, the probability $P[A]$ that the commuter will catch his train can now be computed by evaluating the integrals in (3.15). However, in the present example there is a very special feature present that makes it possible to evaluate $P[A]$ without any laborious calculation.

The reader may have noticed that the probability density functions $f_1(\cdot)$ and $f_2(\cdot)$ have the same functional form. If we define $f(\cdot)$ by $f(x) = \frac{2}{25}(5-x)$ or 0, depending on whether $0 \leq x \leq 5$ or otherwise, we find that $f_1(x) = f_2(x) = f(x)$ for all real numbers x. In terms of $f(\cdot)$, we may write (3.15), making the change of variable $x_1' = x_2$ and $x_2' = x_1$ in the second integral,

$$\begin{aligned} P[A] &= \int_{-\infty}^{\infty} dx_1 f(x_1) \int_{-\infty}^{x_1} dx_2 f(x_2) \\ &= \int_{-\infty}^{\infty} dx_1' f(x_1') \int_{x_1'}^{\infty} dx_2' f(x_2'). \end{aligned} \tag{3.16}$$

By adding the two integrals in (3.16), it follows that

$$2P[A] = \int_{-\infty}^{\infty} dx_1 f(x_1) \int_{-\infty}^{\infty} dx_2 f(x_2) = 1.$$

We conclude that the probability $P[A]$ that the man will catch his train is equal to $\frac{1}{2}$.

EXERCISES

3.1. Consider the example in the text. Let the probability law of the train's departure time be given by (3.1). However, assume that the man's arrival time at the railroad station is uniformly distributed over the interval 7:55 to 8 A.M. Assume that the man's arrival time is independent of the train's departure time. Find the probability of the event that the man will catch the train.

3.2. Consider the example in the text. Assume that the train's departure time and the man's arrival time are independent random phenomena, each uniformly distributed over the interval 7:55 to 8 A.M. Find the probability of the event that the man will catch the train.

4. THE SAME EXAMPLE TREATED FROM THE POINT OF VIEW OF RANDOM VARIABLES

We now treat the example considered in the foregoing section in terms of random variables. We shall see that the notion of a random variable does not replace the idea of a numerical valued random phenomenon but rather extends it.

We let X_1 and X_2 denote, respectively, the departure time of the train and the arrival time of the commuter at the station. In order, with complete rigor, to regard X_1 and X_2 as random variables, we must state the probability space on which they are defined as functions. Let us first consider X_1. We may define X_1 as the identity function (so that $X_1(x_1) = x_1$, for all x_1 in R_1) on a real line R_1, on which a probability distribution (that is, a distribution of probability mass) has been placed in accordance with the probability density function $f_1(\cdot)$ given by (3.1). Or we may define X_1 as a function on a space S of 2-tuples (x_1, x_2) of real numbers, on which a probability distribution has been placed in accordance with the probability density function $f(.,.)$ given by (3.12); in this case we define $X_1(s) = X_1((x_1, x_2)) = x_1$. Similarly, we may regard X_2 as either the identity function on a real line R_2, on which a probability distribution has been placed in accordance with the probability density function $f_2(\cdot)$ given by

(3.2), or as the function with values $X_2((x_1, x_2)) = x_2$, defined on the probability space S. In order to consider X_1 and X_2 in the same context, they must be defined on the same probability space. Consequently, we regard X_1 and X_2 as being defined on S.

It should be noted that no matter how X_1 and X_2 are defined as functions the individual probability laws of X_1 and X_2 are specified by the probability density functions $f_{X_1}(\cdot)$ and $f_{X_2}(\cdot)$, with values at any real number x,

$$(4.1) \qquad f_{X_1}(x) = f_{X_2}(x) = \tfrac{2}{25}(5 - x) \qquad \text{for } 0 \leq x \leq 5$$
$$= 0 \qquad \text{otherwise.}$$

Consequently, the random variables X_1 and X_2 are identically distributed.

We now turn our attention to the problem of computing the probability that the man will catch the train. In the previous section we reduced this problem to one involving the computation of the probability of a certain event (set) on a probability space. In this section we reduce the problem to one involving the computation of the distribution function of a random variable; by so doing, we not only solve the problem given but also a number of related problems.

Let $Y = X_1 - X_2$ denote the difference between the train's departure time X_1 and the man's arrival time X_2. It is clear that the man catches the train if and only if $Y > 0$. Therefore, the probability that the man will catch the train is equal to $P[Y > 0]$. In order for $P[Y > 0]$ to be a meaningful expression, it is necessary that Y be a random variable, which is to say that Y is a function on some probability space. This will be the case if and only if the random variables X_1 and X_2 are defined as functions on the same probability space. Consequently, we must regard X_1 and X_2 as functions on the probability space S, defined in the second paragraph of this section. Then Y is a function on the probability space S, and $P[Y > 0]$ is meaningful. Indeed, we may compute the distribution function $F_Y(\cdot)$ of Y, defined for any real number y by

$$(4.2) \qquad F_Y(y) = P[Y \leq y] = P[\{s: \quad Y(s) \leq y\}].$$

Then $P[Y > 0] = 1 - F_Y(0)$.

To compute the distribution function $F_Y(\cdot)$ of Y, there are two methods available. In one method we use the fact that we know the probability space S on which Y is defined as a function and use (4.2). A second method is to use only the fact that Y is defined as a function of the random variables X_1 and X_2. The second method requires the introduction of the notion of the joint probability law of the random variables X_1 and X_2 and is discussed in the next section. We conclude this section by obtaining $F_Y(\cdot)$ by means of the first method.

As a function on the probability space S, Y is given, at each 2-tuple (x_1, x_2), by $Y((x_1, x_2)) = x_1 - x_2$. Consequently, by (4.2), for any real number y,

$$F_Y(y) = P[\{(x_1, x_2):\ x_1 - x_2 \leq y\}] \tag{4.3}$$

$$= \iint\limits_{\{(x_1,x_2):x_1-x_2\leq y\}} f(x_1, x_2)\,dx_1\,dx_2$$

$$= \int_{-\infty}^{\infty} dx_1 \int_{x_1-y}^{\infty} dx_2 f(x_1, x_2).$$

From (4.3) we obtain an expression for the probability density function $f_Y(\cdot)$ of the random variable Y. In the second integration in (4.3), make the change of variable $x_2' = -x_2 + x_1$. Then

$$F_Y(y) = \int_{-\infty}^{\infty} dx_1 \int_{-\infty}^{y} dx_2' f(x_1, x_1 - x_2').$$

By interchanging the order of integration, we have

$$F_Y(y) = \int_{-\infty}^{y} dx_2' \int_{-\infty}^{\infty} dx_1 f(x_1, x_1 - x_2'). \tag{4.4}$$

By differentiating the expression in (4.4) with respect to y, we obtain the integrand of the integration with respect to x_2', with x_2' replaced by y; thus

$$f_Y(y) = \frac{d}{dy} F_Y(y) = \int_{-\infty}^{\infty} dx_1 f(x_1, x_1 - y). \tag{4.5}$$

Equation (4.5) constitutes *a general expression for the probability density function of the random variable Y defined on a space S of 2-tuples (x_1, x_2) by $Y((x_1, x_2)) = x_1 - x_2$, where a probability function has been specified on S by the probability density function $f(. , .)$.*

To illustrate the use of (4.5), let us consider again the probability density functions introduced in connection with the problem of the commuter catching the train. The probability density function $f(. , .)$ is given by (3.12) in terms of the functions $f_1(\cdot)$ and $f_2(\cdot)$, given by (3.1) and (3.2), respectively.

In the case of independent phenomena, (4.5) becomes

$$f_Y(y) = \int_{-\infty}^{\infty} dx f_1(x) f_2(x - y) = \int_{-\infty}^{\infty} dx f_1(x + y) f_2(x). \tag{4.6}$$

If further, as is the case here, the two random phenomena [with respective probability density functions $f_1(\cdot)$ and $f_2(\cdot)$] are identically distributed, so that, for all real numbers x, $f_1(x) = f_2(x) = f(x)$, for some function $f(\cdot)$, then the probability density function $f_Y(\cdot)$ is an *even* function; that is, $f_Y(-y) = f_Y(y)$ for all y. It then suffices to evaluate $f_Y(y)$ for $y \geq 0$. One obtains, by using (3.1), (3.2), and (4.6),

$$(4.7) \qquad f_Y(y) = \int_0^5 dx \frac{2}{25}(5 - x)f(x + y)$$

$$= \int_0^{5-y} dx \left(\frac{2}{25}\right)^2 (5 - x)(5 - (x + y)) \qquad \text{if } 0 \leq y \leq 5$$

$$= 0 \qquad \text{if } y \geq 5.$$

Therefore,

$$(4.8) \qquad f_Y(y) = \frac{4|y|^3 - 300|y| + 1000}{6(5)^4} \qquad \text{if } |y| \leq 5$$

$$= 0 \qquad \text{otherwise.}$$

Consequently $$P[Y > 0] = \int_0^\infty f_Y(y)\, dy = \frac{1}{2}.$$

EXERCISES

4.1. Consider the random variable Y defined in the text. Find the probability density function of Y under the assumptions made in exercise 3.1.

4.2. Consider the random variable Y defined in the text. Find its probability density function under the assumptions made in exercise 3.2.

5. JOINTLY DISTRIBUTED RANDOM VARIABLES

Two random variables, X_1 and X_2, are said to be jointly distributed if they are defined as functions on the same probability space. It is then possible to make joint probability statements about X_1 and X_2 (that is, probability statements about the simultaneous behavior of the two random variables). In this section we introduce the notions used to describe the *joint probability law* of jointly distributed random variables.

The joint probability function, denoted by $P_{X_1,X_2}[\cdot]$ of two jointly distributed random variables, is defined for every Borel set B of 2-tuples of real numbers by

$$(5.1) \qquad P_{X_1,X_2}[B] = P[\{s \text{ in } S\colon \ (X_1(s), X_2(s)) \text{ is in } B\}],$$

in which S denotes the sample description space on which the random variables X_1 and X_2 are defined and $P[\cdot]$ denotes the probability function defined on S. In words, $P_{X_1,X_2}[B]$ represents the probability that the 2-tuple (X_1, X_2) of observed values of the random variables will lie in the set B. For brevity, we usually write

$$P_{X_1,X_2}[B] = P[(X_1, X_2) \text{ is in } B], \tag{5.2}$$

instead of (5.1). However, *it should be kept constantly in mind that the right-hand side of (5.2) is without mathematical content of its own; rather, it is an intuitively meaningful concise way of writing the right-hand side of (5.1).*

It is useful to think of the joint probability function $P_{X_1,X_2}[\cdot]$ of two jointly distributed random variables X_1 and X_2 as representing the distribution of a unit amount of probability mass over a 2-dimensional plane on which rectangular coordinates have been marked off, as in Fig. 7A of Chapter 4, so that to any point in the plane there corresponds a 2-tuple (x_1', x_2') of real numbers representing it. For any Borel set B of 2-tuples $P_{X_1,X_2}[B]$ represents the amount of probability mass distributed over the set B.

We are particularly interested in knowing the value $P_{X_1,X_2}[B]$ for sets B, which are combinatorial product sets in the plane. A set B is called a combinatorial product set if it is of the form $B = \{(x_1, x_2)\colon \ x_1 \text{ is in } B_1 \text{ and } x_2 \text{ is in } B_2\}$ for some Borel sets B_1 and B_2 of real numbers. If B is of this form, we then write, for brevity, $P_{X_1,X_2}[B] = P[X_1 \text{ is in } B_1, X_2 \text{ is in } B_2]$.

In order to know the joint probability function $P_{X_1,X_2}[B]$ for all Borel sets B of 2-tuples, it suffices to know it for all infinite rectangle sets B_{x_1,x_2}, where, for any two real numbers x_1 and x_2, we define the "infinite rectangle" set

$$B_{x_1,x_2} = \{(x_1', x_2')\colon \ x_1' \leq x_1, x_2' \leq x_2\} \tag{5.3}$$

as the set consisting of all 2-tuples (x_1', x_2') whose first component x_1' is less than the specified real number x_1 and whose second component x_2' is less than the specified real number x_2. To specify the joint probability function of X_1 and X_2, it suffices to specify the *joint distribution function* $F_{X_1,X_2}(\cdot\,,\cdot)$ of the random variables X_1 and X_2, defined for all real numbers x_1 and x_2 by the equation

$$F_{X_1,X_2}(x_1, x_2) = P[X_1 \leq x_1, X_2 \leq x_2] = P_{X_1,X_2}[B_{x_1,x_2}]. \tag{5.4}$$

In words, $F_{X_1,X_2}(x_1, x_2)$ represents the probability that the simultaneous observation (X_1, X_2) will have the property that $X_1 \leq x_1$ and $X_2 \leq x_2$.

In terms of the probability mass distributed over the plane of Fig. 7A of Chapter 4, $F_{X_1,X_2}(x_1, x_2)$ represents the amount of mass in the "infinite rectangle" B_{x_1,x_2}.

The reader should verify for himself the following important formula [compare (7.3) of Chapter 4]: for any real numbers a_1, a_2, b_1, and b_2, such that $a_1 \leq b_1$, $a_2 \leq b_2$, the probability $P[a_1 < X_1 \leq b_1, a_2 < X_2 \leq b_2]$ that the simultaneous observation (X_1, X_2) will be such that $a_1 < X_1 \leq b_1$ and $a_2 < X_2 \leq b_2$ may be given in terms of $F_{X_1,X_2}(\cdot\,,\cdot)$ by

$$(5.5)\quad P[a_1 < X_1 \leq b_1, a_2 < X_2 \leq b_2] = F_{X_1,X_2}(b_1, b_2) + F_{X_1,X_2}(a_1, a_2) - F_{X_1,X_2}(a_1, b_2) - F_{X_1,X_2}(b_1, a_2).$$

It is important to note that from a knowledge of the joint distribution function $F_{X_1,X_2}(\cdot\,,\cdot)$ of two jointly distributed random variables one may obtain the distribution functions $F_{X_1}(\cdot)$ and $F_{X_2}(\cdot)$ of each of the random variables X_1 and X_2. We have the formula for any real number x_1:

$$(5.6)\quad \begin{aligned} F_X(x_1) &= P[X_1 \leq x_1] = P[X_1 \leq x_1, X_2 < \infty] \\ &= \lim_{x_2 \to \infty} F_{X_1,X_2}(x_1, x_2) = F_{X_1,X_2}(x_1, \infty). \end{aligned}$$

Similarly, for any real number x_2

$$(5.7)\quad F_{X_2}(x_2) = \lim_{x_1 \to \infty} F_{X_1,X_2}(x_1, x_2) = F_{X_1,X_2}(\infty, x_2).$$

In terms of the probability mass distributed over the plane by the joint distribution function $F_{X_1,X_2}(\cdot\,,\cdot)$, the quantity $F_{X_1}(x_1)$ is equal to the amount of mass in the half-plane that consists of all 2-tuples (x_1', x_2') that are to the left of, or on, the line with equation $x_1' = x_1$.

The function $F_{X_1}(\cdot)$ is called the *marginal distribution function* of the random variable X_1 corresponding to the joint distribution function $F_{X_1,X_2}(\cdot\,,\cdot)$. Similarly, $F_{X_2}(\cdot)$ is called the marginal distribution function of X_2 corresponding to the joint distribution function $F_{X_1,X_2}(\cdot\,,\cdot)$.

We next define the *joint probability mass* function of two random variables X_1 and X_2, denoted by $p_{X_1,X_2}(\cdot\,,\cdot)$, as a function of 2 variables, with value, for any real numbers x_1 and x_2.

$$(5.8)\quad \begin{aligned} p_{X_1,X_2}(x_1, x_2) &= P\,[X_1 = x_1, X_2 = x_2] \\ &= P_{X_1,X_2}[\{(x_1', x_2'){:}\quad x_1' = x_1, x_2' = x_2\}]. \end{aligned}$$

It may be shown that there is only a finite or countably infinite number of 2-tuples (x_1, x_2) at which $p_{X_1,X_2}(x_1, x_2) > 0$. The jointly distributed random variables X_1 and X_2 are said to be *jointly discrete* if the sum of the joint probability mass function over the points (x_1, x_2) where $p_{X_1,X_2}(x_1, x_2)$ is positive is equal to 1. If the random variables X_1 and X_2 are jointly discrete, then they are individually discrete, with individual probability mass functions, for any real numbers x_1 and x_2.

$$p_{X_1}(x_1) = \sum_{\substack{\text{over all } x_2 \text{ such that} \\ p_{X_1,X_2}(x_1,x_2)>0}} p_{X_1,X_2}(x_1, x_2)$$

(5.9)

$$p_{X_2}(x_2) = \sum_{\substack{\text{over all } x_1 \text{ such that} \\ p_{X_1,X_2}(x_1,x_2)>0}} p_{X_1,X_2}(x_1, x_2).$$

Two jointly distributed random variables, X_1 and X_2, are said to be *jointly continuous* if they are specified by a joint probability density function.

Two jointly distributed random variables, X_1 and X_2, are said to be specified by a *joint probability density function* if there is a nonnegative Borel function $f_{X_1,X_2}(\cdot\,,\cdot)$, called the joint probability density of X_1 and X_2, such that for any Borel set B of 2-tuples of real numbers the probability $P[(X_1, X_2) \text{ is in } B]$ may be obtained by integrating $f_{X_1,X_2}(\cdot\,,\cdot)$ over B; in symbols,

$$(5.10)\quad P_{X_1,X_2}[B] = P[(X_1, X_2) \text{ is in } B] = \iint_B f_{X_1,X_2}(x_1', x_2')\, dx_1'\, dx_2'.$$

By letting $B = B_{x_1,x_2}$ in (5.10), it follows that the joint distribution function for any real numbers x_1 and x_2 may be given by

$$(5.11)\qquad F_{X_1,X_2}(x_1, x_2) = \int_{-\infty}^{x_1} dx_1' \int_{-\infty}^{x_2} dx_2' f_{X_1,X_2}(x_1', x_2').$$

Next, for any real numbers a_1, b_1, a_2, b_2, such that $a_1 \le b_1$, $a_2 \le b_2$, one may verify that

$$(5.12)\quad P[a_1 < X_1 \le b_1, a_2 < X_2 \le b_2] = \int_{a_1}^{b_1} dx_1' \int_{a_2}^{b_2} dx_2' f_{X_1,X_2}(x_1', x_2').$$

The joint probability density function may be obtained from the joint distribution function by routine differentiation, since

$$(5.13)\qquad f_{X_1,X_2}(x_1, x_2) = \frac{\partial^2}{\partial x_1\, \partial x_2} F_{X_1,X_2}(x_1, x_2)$$

at all 2-tuples (x_1, x_2), where the partial derivatives on the right-hand side of (5.13) are well defined.

If the random variables X_1 and X_2 are jointly continuous, then they are individually continuous, with individual probability density functions for any real numbers x_1 and x_2 given by

$$(5.14) \qquad \begin{aligned} f_{X_1}(x_1) &= \int_{-\infty}^{\infty} f_{X_1,X_2}(x_1, x_2)\, dx_2 \\ f_{X_2}(x_2) &= \int_{-\infty}^{\infty} f_{X_1,X_2}(x_1, x_2)\, dx_1. \end{aligned}$$

The reader should compare (5.14) with (3.4).

To prove (5.14), one uses the fact that by (5.6), (5.7), and (5.11),

$$F_{X_1}(x_1) = \int_{-\infty}^{x_1} dx_1' \int_{-\infty}^{\infty} dx_2' f_{X_1,X_2}(x_1', x_2')$$

$$F_{X_2}(x_2) = \int_{-\infty}^{x_2} dx_2' \int_{-\infty}^{\infty} dx_1' f_{X_1,X_2}(x_1', x_2').$$

The foregoing notions extend at once to the case of n random variables. We list here the most important notations used in discussing n jointly distributed random variables $X_1, X_2, \ldots, X_n$. The joint probability function for any Borel set B of n-tuples is given by

$$(5.15) \qquad P_{X_1,X_2,\cdots,X_n}[B] = P[(X_1, X_2, \cdots, X_n) \text{ is in } B].$$

The joint distribution function for any real numbers $x_1, x_2, \ldots, x_n$ is given by

$$(5.16) \qquad F_{X_1,X_2,\cdots,X_n}(x_1, x_2, \cdots, x_n) = P[X_1 \leq x_1, X_2 \leq x_2, \cdots, X_n \leq x_n].$$

The *joint probability density* function (if the derivative below exists) is given by

$$(5.17) \qquad f_{X_1,X_2,\cdots,X_n}(x_1, x_2, \cdots, x_n) = \frac{\partial^n}{\partial x_1\, \partial x_2 \cdots \partial x_n} F_{X_1,X_2,\cdots,X_n}(x_1, x_2, \cdots, x_n).$$

The *joint probability mass* function is given by

$$(5.18) \qquad p_{X_1,X_2,\cdots,X_n}(x_1, x_2, \cdots, x_n) = P[X_1 = x_1, X_2 = x_2, \cdots, X_n = x_n].$$

A *discrete joint probability law* is specified by its probability mass function: for any Borel set B of n-tuples

$$(5.19) \qquad P_{X_1,X_2,\cdots,X_n}[B] = \sum_{\substack{\text{over all } (x_1,x_2,\cdots,x_n) \text{ in } B \text{ such that} \\ p_{X_1,X_2,\cdots,X_n}(x_1,x_2,\cdots,x_n) > 0}} p_{X_1,X_2,\cdots,X_n}(x_1, x_2, \cdots, x_n).$$

A *continuous joint probability law* is specified by its probability density function: for any Borel set B of n-tuples

$$(5.20)\quad P_{X_1,X_2,\cdots,X_n}[B] = \iint_B \cdots \int f_{X_1,X_2,\cdots,X_n}(x_1, x_2, \cdots, x_n)\, dx_1\, dx_2 \cdots dx_n.$$

The *individual* (*or marginal*) *probability law* of each of the random variables $X_1, X_2, \ldots, X_n$ may be obtained from the joint probability law. In the continuous case, for any $k = 1, 2, \ldots, n$ and any fixed number x_k^0,

$$(5.21)\quad f_{X_k}(x_k^0) = \int_{-\infty}^{\infty} dx_1 \cdots \int_{-\infty}^{\infty} dx_{k-1} \int_{-\infty}^{\infty} dx_{k+1} \cdots \int_{-\infty}^{\infty} dx_n \; f_{X_1,X_2,\cdots,X_n}(x_1, \cdots, x_k^0, x_{k+1}, \cdots x_n).$$

An analogous formula may be written in the discrete case for $p_{X_k}(x_k^0)$.

▶ **Example 5A. Jointly discrete random variables.** Consider a sample of size 2 drawn with replacement (without replacement) from an urn containing two white, one black, and two red balls. Let the random variables X_1 and X_2 be defined as follows; for $k = 1, 2$, $X_k = 1$ or 0, depending on whether the ball drawn on the kth draw is white or nonwhite. (i) Describe the joint probability law of (X_1, X_2). (ii) Describe the individual (or marginal) probability laws of X_1 and X_2.

Solution: The random variables X_1 and X_2 are clearly jointly discrete. Consequently, to describe their joint probability law, it suffices to state their joint probability mass function $p_{X_1,X_2}(x_1, x_2)$. Similarly, to describe their individual probability laws, it suffices to describe their individual probability mass functions $p_{X_1}(x_1)$ and $p_{X_2}(x_2)$. These functions are conveniently presented in the following tables:

Sampling with replacement

	$p_{X_1,X_2}(x_1, x_2)$		
$x_2 \backslash x_1$	0	1	$p_{X_2}(x_2)$
0	$\frac{3}{5}\,\frac{3}{5}$	$\frac{2}{5}\,\frac{3}{5}$	$\frac{3}{5}$
1	$\frac{3}{5}\,\frac{2}{5}$	$\frac{2}{5}\,\frac{2}{5}$	$\frac{2}{5}$
$p_{X_1}(x_1)$	$\frac{3}{5}$	$\frac{2}{5}$	

Sampling without replacement

	$p_{X_1,X_2}(x_1, x_2)$		
$x_2 \backslash x_1$	0	1	$p_{X_2}(x_2)$
0	$\frac{3}{5}\,\frac{2}{4}$	$\frac{2}{5}\,\frac{3}{4}$	$\frac{3}{5}$
1	$\frac{3}{5}\,\frac{2}{4}$	$\frac{2}{5}\,\frac{1}{4}$	$\frac{2}{5}$
$p_{X_1}(x_1)$	$\frac{3}{5}$	$\frac{2}{5}$	

◀

▶ **Example 5B. Jointly continuous random variables.** Suppose that at two points in a room (or on a city street or in the ocean) one measures the intensity of sound caused by general background noise. Let X_1 and X_2 be random variables representing the intensity of sound at the two points. Suppose that the joint probability law of the sound intensities, X_1 and X_2, is continuous, with the joint probability density function given by

$$\begin{aligned} f_{X_1,X_2}(x_1, x_2) &= x_1 x_2 \exp\left[-\tfrac{1}{2}(x_1^2 + x_2^2)\right] \qquad \text{if } x_1 > 0, \qquad x_2 > 0 \\ &= 0 \qquad \text{otherwise.} \end{aligned}$$

Find the individual probability density functions of X_1 and X_2. Further, find $P[X_1 \leq 1, X_2 \leq 1]$ and $P[X_1 + X_2 \leq 1]$.

Solution: By (5.14), the individual probability density functions are given by

$$f_{X_1}(x_1) = \int_0^\infty x_1 x_2 \exp\left[-\tfrac{1}{2}(x_1^2 + x_2^2)\right] dx_2 = x_1 \exp\left(-\tfrac{1}{2}x_1^2\right)$$

$$f_{X_2}(x_2) = \int_0^\infty x_1 x_2 \exp\left[-\tfrac{1}{2}(x_1^2 + x_2^2)\right] dx_1 = x_2 \exp\left(-\tfrac{1}{2}x_2^2\right).$$

Note that the random variables X_1 and X_2 are identically distributed. Next, the probability that each sound intensity is less than or equal to 1 is given by

$$\begin{aligned} P[X_1 \leq 1, X_2 \leq 1] &= \int_{-\infty}^1 \int_{-\infty}^1 f_{X_1,X_2}(x_1, x_2)\, dx_1\, dx_2 \\ &= \left(\int_0^1 x_1 e^{-\frac{1}{2}x_1^2}\, dx_1\right)\left(\int_0^1 x_2 e^{-\frac{1}{2}x_2^2}\, dx_2\right) = 0.1548. \end{aligned}$$

The probability that the sum of the sound intensities is less than 1 is given by

$$\begin{aligned} P[X_1 + X_2 \leq 1] &= \iint_{\{(x_1,x_2):\ x_1 + x_2 \leq 1\}} f_{X_1,X_2}(x_1, x_2)\, dx_1\, dx_2 \\ &= \int_0^1 dx_1\, x_1 e^{-\frac{1}{2}x_1^2} \int_0^{1-x_1} dx_2 x_2 e^{-\frac{1}{2}x_2^2} = 0.2433. \end{aligned}$$ ◀

▶ **Example 5C. The maximum noise intensity.** Suppose that at five points in the ocean one measures the intensity of sound caused by general background noise (the so-called ambient noise). Let X_1, X_2, X_3, X_4, and X_5 be random variables representing the intensity of sound at the various points. Suppose that their joint probability law is continuous, with joint probability density function given by

$$
\begin{aligned}
f_{X_1,X_2,X_3,X_4,X_5}(x_1, x_2, x_3, x_4, x_5) &= x_1x_2x_3x_4x_5 \\
&\qquad \times \exp\left[-\tfrac{1}{2}(x_1^2 + x_2^2 + x_3^2 + x_4^2 + x_5^2)\right] \\
&\qquad \text{if } 0 \le x_1, x_2, x_3, x_4, x_5 \\
&= 0 \qquad \text{otherwise.}
\end{aligned}
$$

Define Y as the maximum intensity; in symbols, $Y = \text{maximum}$ $(X_1, X_2, X_3, X_4, X_5)$. For any positive number y the probability that Y is less than or equal to y is given by

$$
\begin{aligned}
P[Y \le y] &= P[X_1 \le y, X_2 \le y, \cdots, X_5 \le y] \\
&= \int_{-\infty}^{y} dx_1 \int_{-\infty}^{y} dx_2 \cdots \int_{-\infty}^{y} dx_5 f_{X_1,X_2,X_3,X_4,X_5}(x_1, x_2, \cdots, x_5) \\
&= \left(\int_0^y xe^{-\frac{1}{2}x^2}\, dx\right)^5 = (1 - e^{-\frac{1}{2}y^2})^5.
\end{aligned}
$$

◀

THEORETICAL EXERCISE

5.1. Multivariate distributions with given marginal distributions. Let $f_1(\cdot)$ and $f_2(\cdot)$ be two probability density functions. An infinity of joint probability densities $f(.\,,.)$ exist, of which $f_1(\cdot)$ and $f_2(\cdot)$ are the marginal probability density functions [that is, such that (3.4) holds]. One method of constructing $f(.\,,.)$ is given by (3.9); verify this assertion. Show that another method of constructing a joint probability density function $f(.\,,.)$, with given marginal probability density functions $f_1(\cdot)$ and $f_2(\cdot)$, is by defining for a given constant a, such that $|a| \le 1$,

$$
(5.22) \qquad f(x_1, x_2) = f_1(x_1)f_2(x_2)\{1 + a[2F_1(x_1) - 1][2F_2(x_2) - 1]\}
$$

in which $F_1(\cdot)$ and $F_2(\cdot)$ are the distribution functions corresponding to $f_1(\cdot)$ and $f_2(\cdot)$, respectively. Show that the distribution function $F(.\,,.)$ corresponding to $f(.\,,.)$ is given by

$$
(5.23) \qquad F(x_1, x_2) = F_1(x_1)F_2(x_2)\{1 + a[1 - F_1(x_1)][1 - F_2(x_2)]\}
$$

Equations (5.22) and (5.23) are due to E. J. Gumbel, "Distributions à plusieurs variables dont les marges sont données," *C. R. Acad. Sci. Paris*, Vol. 246 (1958), pp. 2717–2720.

EXERCISES

In exercises 5.1 to 5.3 consider a sample of size 3 drawn with replacement (without replacement) from an urn containing (i) 1 white and 2 black balls,

(ii) 1 white, 1 black, and 1 red ball. For $k = 1, 2, 3$ let $X_k = 1$ or 0 depending on whether the ball drawn on the kth draw is white or nonwhite.

5.1. Describe the joint probability law of (X_1, X_2, X_3).

5.2. Describe the individual (marginal) probability laws of X_1, X_2, X_3.

5.3. Describe the individual probability laws of the random variables Y_1, Y_2, and Y_3, in which $Y_1 = X_1 + X_2 + X_3$, $Y_2 = \text{maximum } (X_1, X_2, X_3)$. and $Y_3 = \text{minimum } (X_1, X_2, X_3)$.

In exercises 5.4 to 5.6 consider 2 random variables, X_1 and X_2, with joint probability law specified by the joint probability density function

$$(a)\quad f_{X_1,X_2}(x_1, x_2) = \tfrac{1}{4} \quad \text{if } 0 \leq x_1 \leq 2 \quad \text{and} \quad 0 \leq x_2 \leq 2$$
$$= 0 \quad \text{otherwise.}$$

$$(b)\quad f_{X_1,X_2}(x_1, x_2) = e^{-(x_1+x_2)} \quad \text{if } x_1 \geq 0 \quad \text{and} \quad x_2 \geq 0$$
$$= 0 \quad \text{otherwise.}$$

5.4. Find (i) $P[X_1 \leq 1, X_2 \leq 1]$, (ii) $P[X_1 + X_2 \leq 1]$, (iii) $P[X_1 + X_2 > 2]$.

5.5. Find (i) $P[X_1 < 2X_2]$, (ii) $P[X_1 > 1]$, (iii) $P[X_1 = X_2]$.

5.6. Find (i) $P[X_2 > 1 \mid X_1 \leq 1]$, (ii) $P[X_1 > X_2 \mid X_2 > 1]$.

In exercises 5.7 to 5.10 consider 2 random variables, X_1 and X_2, with the joint probability law specified by the probability mass function $p_{X_1,X_2}(\cdot\,,\cdot)$ given for all x_1 and x_2 at which it is positive by (*a*) Table 5A, (*b*) Table 5B, in which for brevity we write h for $\frac{1}{60}$.

TABLE 5A

	$p_{X_1,X_2}(x_1, x_2)$			
x_2 \ x_1	0	1	2	$p_{X_2}(x_2)$
0	h	$2h$	$3h$	$6h$
1	$2h$	$4h$	$6h$	$12h$
2	$3h$	$6h$	$9h$	$18h$
3	$4h$	$8h$	$12h$	$24h$
$p_{X_1}(x_1)$	$10h$	$20h$	$30h$	

5.7. Show that the individual probability mass functions of X_1 and X_2 may be obtained by summing the respective columns and rows as indicated. Are X_1 and X_2 (i) jointly discrete, (ii) individually discrete?

5.8. Find (i) $P[X_1 \leq 1, X_2 \leq 1]$, (ii) $P[X_1 + X_2 \leq 1]$, (iii) $P[X_1 + X_2 > 2]$.

TABLE 5B

x_2 \ x_1	$p_{X_1,X_2}(x_1, x_2)$: 0	1	2	$p_{X_2}(x_2)$
0	h	$4h$	$9h$	$14h$
1	$2h$	$6h$	$12h$	$20h$
2	$3h$	$8h$	$3h$	$14h$
3	$4h$	$2h$	$6h$	$12h$
$p_{X_1}(x_1)$	$10h$	$20h$	$30h$	

5.9. Find (i) $P[X_1 < 2X_2]$, (ii) $P[X_1 > 1]$, (iii) $P[X_1 = X_2]$.

5.10. Find (i) $P[X_1 \geq X_2 \mid X_2 > 1]$, (ii) $P[X_1^2 + X_2^2 \leq 1]$.

6. INDEPENDENT RANDOM VARIABLES

In section 2 of Chapter 3 we defined the notion of a series of independent trials. In this section we define the notion of independent random variables. This notion plays the same role in the theory of jointly distributed random variables that the notion of independent trials plays in the theory of sample description spaces consisting of n trials. We consider first the case of two jointly distributed random variables.

Let X_1 and X_2 be jointly distributed random variables, with individual distribution functions $F_{X_1}(\cdot)$ and $F_{X_2}(\cdot)$, respectively, and joint distribution function $F_{X_1,X_2}(\cdot\,,\cdot)$. We say that the random variables X_1 and X_2 are independent if for any two Borel sets of real numbers B_1 and B_2 the events [X_1 is in B_1] and [X_2 is in B_2] are independent; that is,

$$(6.1) \quad P[X_1 \text{ is in } B_1 \text{ and } X_2 \text{ is in } B_2] = P[X_1 \text{ is in } B_1]P[X_2 \text{ is in } B_2].$$

The foregoing definition may be expressed equivalently: the random variables X_1 and X_2 are independent if for any event A_1, depending only on the random variable X_1, and any event A_2, depending only on the random variable X_2, $P[A_1A_2] = P[A_1]P[A_2]$, so that the events A_1 and A_2 are independent.

It may be shown that if (6.1) holds for sets B_1 and B_2, which are infinitely extended intervals of the form $B_1 = \{x_1'\colon\ x_1' \leq x_1\}$ and $B_2 = \{x_2'\colon\ x_2' \leq x_2\}$, for any real numbers x_1 and x_2, then (6.1) holds for any Borel sets B_1 and B_2 of real numbers. We therefore have the following equivalent formulation of the notion of the independence of two jointly distributed random variables X_1 and X_2.

Two jointly distributed random variables, X_1 and X_2 are independent if their joint distribution function $F_{X_1,X_2}(\cdot\,,\cdot)$ may be written as the product of their individual distribution functions $F_{X_1}(\cdot)$ and $F_{X_2}(\cdot)$ in the sense that, for any real numbers x_1 and x_2,

$$F_{X_1,X_2}(x_1, x_2) = F_{X_1}(x_1)F_{X_2}(x_2). \tag{6.2}$$

Similarly, two jointly continuous random variables, X_1 and X_2 are independent if their joint probability density function $f_{X_1,X_2}(\cdot\,,\cdot)$ may be written as the product of their individual probability density functions $f_{X_1}(\cdot)$ and $f_{X_2}(\cdot)$ in the sense that, for any real numbers x_1 and x_2,

$$f_{X_1,X_2}(x_1, x_2) = f_{X_1}(x_1)f_{X_2}(x_2). \tag{6.3}$$

Equation (6.3) follows from (6.2) by differentiating both sides of (6.2) first with respect to x_1 and then with respect to x_2. Equation (6.2) follows from (6.3) by integrating both sides of (6.3).

Similarly, two jointly discrete random variables, X_1 and X_2 are independent if their joint probability mass function $p_{X_1,X_2}(\cdot\,,\cdot)$ may be written as the product of their individual probability mass functions $p_{X_1}(\cdot)$ and $p_{X_2}(\cdot)$ in the sense that, for all real numbers x_1 and x_2,

$$p_{X_1,X_2}(x_1, x_2) = p_{X_1}(x_1)p_{X_2}(x_2). \tag{6.4}$$

Two random variables X_1 and X_2, which do not satisfy any of the foregoing relations, are said to be dependent or nonindependent.

▶ **Example 6A. Independent and dependent random variables.** In example 5A the random variables X_1 and X_2 are independent in the case of sampling with replacement but are dependent in the case of sampling without replacement. In either case, the random variables X_1 and X_2 are identically distributed. In example 5B the random variables X_1 and X_2 are independent and identically distributed. It may be seen from the definitions given at the end of the section that the random variables $X_1, X_2, \ldots, X_5$ considered in example 5C are independent and identically distributed. ◀

Independent random variables have the following exceedingly important property:

THEOREM 6A. *Let the random variables Y_1 and Y_2 be obtained from the random variables X_1 and X_2 by some functional transformation, so that $Y_1 = g_1(X_1)$ and $Y_2 = g_2(X_2)$ for some Borel functions $g_1(\cdot)$ and $g_2(\cdot)$ of a real variable. Independence of the random variables X_1 and X_2 implies independence of the random variables Y_1 and Y_2.*

This assertion is proved as follows. First, for any set B_1 of real numbers, write $g_1^{-1}(B_1) = \{\text{real numbers } x\colon \quad g_1(x) \text{ is in } B_1\}$. It is clear that the event that Y_1 is in B_1 occurs if and only if the event that X_1 is in $g_1^{-1}(B_1)$ occurs. Similarly, for any set B_2 the events that Y_2 is in B_2 and X_2 is in $g_2^{-1}(B_2)$ occur, or fail to occur, together. Consequently, by (6.1)

$$
\begin{aligned}
(6.5) \quad P[Y_1 \text{ is in } B_1,\ Y_2 \text{ is in } B_2] &= P[X_1 \text{ is in } g_1^{-1}(B_1),\ X_2 \text{ is in } g_2^{-1}(B_2)] \\
&= P[X_1 \text{ is in } g_1^{-1}(B_1)]P[X_2 \text{ is in } g_2^{-1}(B_2)] \\
&= P[g_1(X_1) \text{ is in } B_1]P[g_2(X_2) \text{ is in } B_2] \\
&= P[Y_1 \text{ is in } B_1]P[Y_2 \text{ is in } B_2],
\end{aligned}
$$

and the proof of theorem 6A is concluded.

▶ **Example 6B.** Sound intensity is often measured in decibels. A reference level of intensity I_0 is adopted. Then a sound of intensity X is reported as having Y decibels:

$$Y = 10 \log_{10} \frac{X}{I_0}.$$

Now if X_1 and X_2 are the sound intensities at two different points on a city street, let Y_1 and Y_2 be the corresponding sound intensities measured in decibels. If the original sound intensities X_1 and X_2 are independent random variables, then from theorem 6A it follows that Y_1 and Y_2 are independent random variables. ◀

The foregoing notions extend at once to several jointly distributed random variables. We define n jointly distributed random variables $X_1, X_2, \ldots, X_n$ as independent if any one of the following equivalent conditions holds: (i) for any n Borel sets $B_1, B_2, \ldots, B_n$ of real numbers

$$
\begin{aligned}
(6.6) \quad & P[X_1 \text{ is in } B_1,\ X_2 \text{ is in } B_2, \cdots, X_n \text{ is in } B_n] \\
&\qquad = P[X_1 \text{ is in } B_1]P[X_2 \text{ is in } B_2] \cdots P[X_n \text{ is in } B_n],
\end{aligned}
$$

(ii) for any real numbers $x_1, x_2, \ldots, x_n$

$$(6.7) \quad F_{X_1,X_2,\cdots,X_n}(x_1, x_2, \cdots, x_n) = F_{X_1}(x_1)F_{X_2}(x_2) \cdots F_{X_n}(x_n);$$

(iii) if the random variables are jointly continuous, then for any real numbers $x_1, x_2, \ldots, x_n$

$$(6.8) \quad f_{X_1,X_2,\cdots,X_n}(x_1, x_2, \cdots, x_n) = f_{X_1}(x_1)f_{X_2}(x_2) \cdots f_{X_n}(x_n);$$

(iv) if the random variables are jointly discrete, then for any real numbers $x_1, x_2, \ldots, x_n$

$$(6.9) \quad p_{X_1,X_2,\cdots,X_n}(x_1, x_2, \cdots, x_n) = p_{X_1}(x_1)p_{X_2}(x_2) \cdots p_{X_n}(x_n).$$

THEORETICAL EXERCISES

6.1. Give an example of 3 random variables, X_1, X_2, X_3, which are independent when taken two at a time but not independent when taken together. *Hint:* Let A_1, A_2, A_3 be events that have the properties asserted; see example 1C of Chapter 3. Define $X_i = 1$ or 0, depending on whether the event A_i has or has not occurred.

6.2. Give an example of two random variables, X_1 and X_2, which are not independent, but such that X_1^2 and X_2^2 are independent. Does such an example prove that the converse of theorem 6A is false?

6.3. Factorization rule for the probability density function of independent random variables. Show that n jointly continuous random variables $X_1, X_2, \ldots, X_n$ are independent if and only if their joint probability density function for all real numbers $x_1, x_2, \ldots, x_n$ may be written

$$f_{X_1,X_2,\cdots,X_n}(x_1, x_2, \cdots, x_n) = h_1(x_1)h_2(x_2)\cdots h_n(x_n)$$

in terms of some Borel functions $h_1(\cdot)$, $h_2(\cdot)$, ..., and $h_n(\cdot)$.

EXERCISES

6.1. The output of a certain electronic apparatus is measured at 5 different times. Let $X_1, X_2, \ldots, X_5$ be the observations obtained. Assume that $X_1, X_2, \ldots, X_5$ are independent random variables, each Rayleigh distributed with parameter $\alpha = 2$. Find the probability that maximum $(X_1, X_2, X_3, X_4, X_5) > 4$. (Recall that $f_{X_i}(x) = \frac{x}{4} e^{-x^2/8}$ for $x > 0$ and is equal to 0 elsewhere.)

6.2. Suppose 10 identical radar sets have a failure law following the exponential distribution. The sets operate independently of one another and have a failure rate of $\lambda = 1$ set/10^3 hours. What length of time will all 10 radar sets operate satisfactorily with a probability of 0.99?

6.3. Let X and Y be jointly continuous random variables, with a probability density function

$$f_{X,Y}(x, y) = \frac{1}{2\pi} \exp\left[-\tfrac{1}{2}(x^2 + y^2)\right].$$

(i) Are X and Y independent random variables?

(ii) Are X and Y identically distributed random variables?

(iii) Are X and Y normally distributed random variables?

(iv) Find $P[X^2 + Y^2 \leq 4]$. *Hint:* Use polar coordinates.

(v) Are X^2 and Y^2 independent random variables? *Hint:* Use theorem 6A.

(vi) Find $P[X^2 \leq 2]$, $P[Y^2 \leq 2]$.

(vii) Find the individual probability density functions of X^2 and Y^2. [Use (8.8).]

(viii) Find the joint probability density function of X^2 and Y^2. [Use (6.3).]

(ix) Would you expect that $P[X^2Y^2 \le 4] \ge P[X^2 \le 2]P[Y^2 \le 2]$?

(x) Would you expect that $P[X^2Y^2 \le 4] = P[X^2 \le 2]P[Y^2 \le 2]$?

6.4. Let X_1, X_2, and X_3 be independent random variables, each uniformly distributed on the interval 0 to 1. Determine the number a such that

(i) P[at least one of the numbers X_1, X_2, X_3 is greater than a] $= 0.9$.

(ii) P[at least 2 of the numbers X_1, X_2, X_3 is greater than a] $= 0.9$.

Hint: To obtain a numerical answer, use the table of binomial probabilities.

6.5. Consider two events A and B such that $P[A] = \frac{1}{4}$, $P[B \mid A] = \frac{1}{2}$, and $P[A \mid B] = \frac{1}{4}$. Let the random variables X and Y be defined as $X = 1$ or 0, depending on whether the event A has or has not occurred, and $Y = 1$ or 0, depending on whether the event B has or has not occurred. State whether each of the following statements, is true or false:

(i) The random variables X and Y are independent;

(ii) $P[X^2 + Y^2 = 1] = \frac{1}{4}$;

(iii) $P[XY = X^2Y^2] = 1$;

(iv) The random variable X is uniformly distributed on the interval 0 to 1;

(v) The random variables X and Y are identically distributed.

6.6. Show that the two random variables X_1 and X_2 considered in exercise 5.7 are independent if their joint probability mass function is given by Table 5A, and are dependent if their joint probability mass function is given by Table 5B.

In exercises 6.7 to 6.9 let X_1 and X_2 be independent random variables, uniformly distributed over the interval 0 to 1.

6.7. Find (i) $P[X_1 + X_2 < 0.5]$, (ii) $P[X_1 - X_2 < 0.5]$.

6.8. Find (i) $P[X_1X_2 < 0.5]$, (ii) $P[X_1/X_2 < 0.5]$, (iii) $P[X_1^2 < 0.5]$.

6.9. Find (i) $P[X_1^2 + X_2^2 < 0.5]$, (ii) $P[e^{-X_1} < 0.5]$, (iii) $P[\cos \pi X_2 < 0.5]$.

7. RANDOM SAMPLES, RANDOMLY CHOSEN POINTS (GEOMETRICAL PROBABILITY), AND RANDOM DIVISION OF AN INTERVAL

The concepts now assembled enable us to explain some of the major meanings assigned to the word "random" in the mathematical theory of probability.

One meaning arises in connection with the notion of a *random sample of a random variable*. Let us consider a random variable X, of which it is possible to make repeated measurements, denoted by $X_1, X_2, \ldots, X_n, \ldots$. For example, $X_1, X_2, \ldots, X_n$ may be the lifetimes of each of n electric light bulbs, or they may be the numbers on balls drawn (with or without replacement) from an urn containing balls numbered 1 to 100, and so on. The set of n measurements $X_1, X_2, \ldots, X_n$ is spoken of as a *sample* of size n of the random variable X, by which is meant that each of the measurements X_k, for $k = 1, 2, \ldots, n$, is a random variable whose distribution function $F_{X_k}(\cdot)$ is equal, as a function of x, to the distribution function $F_X(\cdot)$ of the random variable X. If, further, the random variables $X_1, X_2, \ldots, X_n$ are independent, then we say that $X_1, X_2, \ldots, X_n$ is a *random sample* (or an independent sample) of size n of the random variable X. Thus the adjective "random," when used to describe a sample of a random variable, indicates that the members of the sample are independent identically distributed random variables.

▶ **Example 7A.** Suppose that the life in hours of electronic tubes of a certain type is known to be approximately normally distributed with parameters $m = 160$ and $\sigma = 20$. What is the probability that a random sample of four tubes will contain no tube with a lifetime of less than 180 hours?

Solution: Let X_1, X_2, X_3, and X_4 denote the respective lifetimes of the four tubes in the sample. The assumption that the tubes constitute a random sample of a random variable normally distributed with parameters $m = 160$ and $\sigma = 20$ is to be interpreted as assuming that the random variables X_1, X_2, X_3, and X_4 are independent, with individual probability density functions, for $k = 1, \ldots, 4$,

$$f_{X_k}(x) = \frac{1}{20\sqrt{2\pi}} \exp\left[-\frac{1}{2}\left(\frac{x - 160}{20}\right)^2\right]. \tag{7.1}$$

The probability that each tube in the sample has a lifetime greater than, or equal to, 180 hours, is given by

$$\begin{aligned} &P[X_1 \geq 180, X_2 \geq 180, X_3 \geq 180, X_4 \geq 180] \\ &\qquad = P[X_1 \geq 180]P[X_2 \geq 180]P[X_3 \geq 180]P[X_4 \geq 180] = (0.159)^4, \end{aligned}$$

since $P[X_k \geq 180] = 1 - \Phi\left(\dfrac{180 - 160}{20}\right) = 1 - \Phi(1) = 0.1587.$ ◀

A second meaning of the word "random" arises when it is used to describe a sample drawn from a finite population. A sample, each of whose components is drawn from a finite population, is said to be a random sample if at each draw all candidates available for selection have an equal probability of being selected. The word "random" was used in this sense throughout Chapter 2.

▶ **Example 7B.** As in example 7A, consider electronic tubes of a certain type whose lifetimes are normally distributed with parameters $m = 160$

and $\sigma = 20$. Let a random sample of four tubes be put into a box. Choose a tube at random from the box. What is the probability that the tube selected will have a lifetime greater than 180 hours?

Solution: For $k = 0, 1, \ldots, 4$ let A_k be the event that the box contains k tubes with a lifetime greater than 180 hours. Since the tube lifetimes are independent random variables with probability density functions given by (7.1), it follows that

$$P[A_k] = \binom{4}{k}(0.1587)^k(0.8413)^{4-k}. \tag{7.2}$$

Let B be the event that the tube selected from the box has a lifetime greater than 180 hours. The assumption that the tube is selected at random is to be interpreted as assuming that

$$P[B \mid A_k] = \frac{k}{4}, \qquad k = 0, 1, \cdots, 4. \tag{7.3}$$

The probability of the event B is then given by

$$P[B] = \sum_{k=0}^{4} P[B \mid A_k]P[A_k] = \sum_{k=1}^{4} \frac{k}{4} \binom{4}{k} p^k q^{4-k}, \tag{7.4}$$

where we have let $p = 0.1587$, $q = 0.8413$. Then

$$P[B] = p\sum_{k=1}^{4} \binom{3}{k-1} p^{k-1}q^{3-(k-1)} = p, \tag{7.5}$$

so that the probability that a tube selected at random from a random sample will have a lifetime greater than 180 hours is the same as the probability that any tube of the type under consideration will have a lifetime greater than 180 hours. A similar result was obtained in example 4D of Chapter 3. A theorem generalizing and unifying these results is given in theoretical exercise 4.1 of Chapter 4. ◀

The word random has a third meaning, which is frequently encountered. The phrase "a point randomly chosen from the interval a to b" is used for brevity to describe a random variable obeying a uniform probability law over the interval a to b, whereas the phrase "n points chosen randomly from the interval a to b" is used for brevity to describe n independent random variables obeying uniform probability laws over the interval a to b. Problems involving randomly chosen points have long been discussed by probabilists under the heading of "geometrical probabilities." In modern terminology problems involving geometrical probabilities may be formulated as problems involving independent random variables, each obeying a uniform probability law.

▶ **Example 7C.** Two points are selected randomly on a line of length a so as to be on opposite sides of the mid-point of the line. Find the probability that the distance between them is less than $\frac{1}{3}a$.

Solution: Introduce a coordinate system on the line so that its left-hand endpoint is 0 and its right-hand endpoint is a. Let X_1 be the coordinate of the point selected randomly in the interval 0 to $\frac{1}{2}a$, and let X_2 be the coordinate of the point selected randomly in the interval $\frac{1}{2}a$ to a. We assume that the random variables X_1 and X_2 are independent and that each obeys a uniform probability law over its interval. The joint probability density function of X_1 and X_2 is then

$$(7.6) \qquad \begin{aligned} f_{X_1,X_2}(x_1, x_2) &= \frac{4}{a^2} \qquad \text{for } 0 < x_1 < \frac{a}{2}, \quad \frac{a}{2} < x_2 < a \\ &= 0 \qquad \text{otherwise.} \end{aligned}$$

The event B that the distance between the two points selected is less than $\frac{1}{3}a$ is then the event $[X_2 - X_1 < \frac{1}{3}a]$. The probability of this event is the probability attached to the cross-hatched area in Fig. 7A. However, this probability can be represented as the ratio of the area of the cross-hatched triangle and the area of the shaded rectangle; thus

$$(7.7) \qquad P\left[X_2 - X_1 < \frac{1}{3}a\right] = \frac{1/2[(1/3)a]^2}{[(1/2)a]^2} = \frac{2}{9}. \qquad ◀$$

▶ **Example 7D.** Consider again the random variables X_1 and X_2 defined in example 7C. Find the probability that the three line segments (from 0 to X_1, from X_1 to X_2, and from X_2 to a) could be made to form the three sides of a triangle.

Solution: In order that the three-line segments mentioned can form a triangle, it is necessary and sufficient that the following inequalities be fulfilled (why?):

$$(7.8) \qquad \begin{aligned} &X_1 < (X_2 - X_1) + (a - X_2) &&\text{or} \quad 2X_1 < a \\ &(a - X_2) < (X_2 - X_1) + X_1 &&\text{or} \quad a < 2X_2 \\ &(X_2 - X_1) < X_1 + (a - X_2) &&\text{or} \quad 2X_2 < a + 2X_1. \end{aligned}$$

The probability of these inequalities being fulfilled is the probability of the cross-hatched area in Fig. 7B, which is clearly $\frac{1}{2}$. It might be noted that if each of the two points, with coordinates X_1 and X_2, are chosen randomly on the interval 0 to a, then the probability is only $\frac{1}{4}$ that the three line segments determined by the two points could be made to form the three sides of a triangle. ◀

Problems involving geometrical probability have played a major role in the development of the modern conception of probability. In the nineteenth century the Laplacean definition of probability was widely accepted. It was thought that probability problems could be given unique solutions by means of finding the proper framework of "equally likely" descriptions. To contradict this point of view, examples were constructed that admitted of several equally plausible, but incompatible, solutions. We now discuss an example similar to one given by Joseph Bertrand in his treatise *Calcul des probabilités*, Paris, 1889, p. 4, and later called by Poincaré, "Bertrand's paradox." It was pointed out to the author by one of his students that this example should serve as a warning to all persons who adopt practical

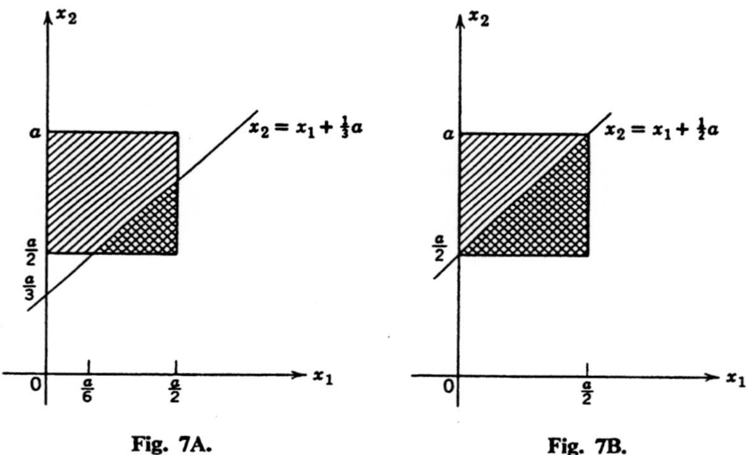

Fig. 7A. **Fig. 7B.**

policies on the basis of theoretical solutions, without first establishing that the assumptions underlying the solutions are in accord with the experimentally observed facts.

▶ **Example 7E. Bertrand's paradox.** Let a chord be chosen randomly in a circle of radius r. What is the probability that the length X of the chord will be less than the radius r?

Solution: It is not clear what is meant by a randomly chosen chord. In order to give meaning to this phrase, we shall reformulate the problem as one involving randomly chosen points. We shall state two methods for randomly choosing points to determine a chord. In this manner we obtain two distinct answers for the probability $P[X < r]$ that the length X of a randomly chosen chord will be less than the radius r.

One method is as follows: let Y_1 and Y_2 be points chosen randomly in the interval 0 to 2π and the interval 0 to r, respectively. Draw a chord by letting Y_1 be the angle made by the chord with a fixed reference line and by letting Y_2 be the (perpendicular) distance of the mid-point of the chord from the center of the circle (see Fig. 7C). A second method of randomly choosing a chord is as follows: let Z_1 and Z_2 be points chosen randomly in the interval 0 to 2π and the interval 0 to $\pi/2$, respectively. Draw a chord by letting Z_1 and Z_2 be the angles indicated in Fig. 7D. The reader may be able to think of other methods of choosing points to determine a chord. Six different solutions of Bertrand's paradox are given in Czuber, *Wahrscheinlichkeitsrechnung*, B. G. Teubner, Leipzig, 1908, pp. 106–109.

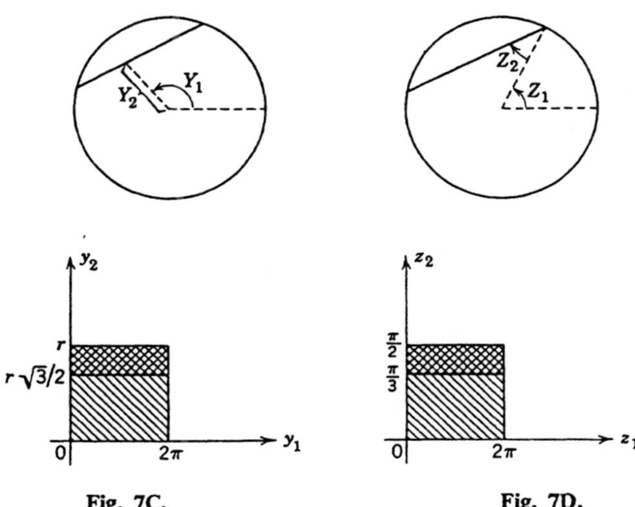

Fig. 7C. **Fig. 7D.**

The length X of the chord may be expressed in terms of the random variables Y_1, Y_2, Z_1, and Z_2:

$$X = 2\sqrt{r^2 - Y_2^2} \qquad \text{or} \qquad X = 2r\cos Z_2. \tag{7.9}$$

Consequently $P[X < r] = P[Y_2 > r\frac{1}{2}\sqrt{3}]$, or $P[X < r] = P[\cos Z_2 < \frac{1}{2}] = P[Z_2 > (\pi/3)]$. In both cases the required probability is equal to the ratio of the areas of the cross-hatched regions in Figs. 7C and 7D to the areas of the corresponding shaded regions. The first solution yields the answer

$$P[X < r] = \frac{2\pi r(1 - \sqrt{3}/2)}{2\pi r} = 1 - \frac{1}{2}\sqrt{3} \doteq 0.134, \tag{7.10}$$

whereas the second solution yields the answer

$$P[X < r] = \frac{[(\pi/2) - \pi/3]2\pi}{(\pi/2)2\pi} = \frac{1}{3} \doteq 0.333 \tag{7.11}$$

for the probability that the length of the chord chosen will be less than the radius of the circle.

It should be noted that random experiments could be performed in such a way that either (7.10) or (7.11) would be the correct probability in the sense of the frequency definition of probability. If a disk of diameter $2r$ were cut out of cardboard and thrown at random on a table ruled with parallel lines a distance $2r$ apart, then one and only one of these lines would cross the disk. All distances from the center would be equally likely, and (7.10) would represent the probability that the chord drawn by the line across the disk would have a length less than r. On the other hand, if the disk were held by a pivot through a point on its edge, which point lay upon a certain straight line, and spun randomly about this point, then (7.11) would represent the probability that the chord drawn by the line across the disk would have a length less than r. ◀

The following example has many important extensions and practical applications.

▶ **Example 7F. The probability of an uncrowded road.** Along a straight road, L miles long, are n distinguishable persons, distributed at random. Show that the probability that no two persons will be less than a distance d miles apart is equal to, for d such that $(n-1)d \leq L$,

$$\left(1 - (n-1)\frac{d}{L}\right)^n. \tag{7.12}$$

Solution: For $j = 1, 2, \ldots, n$ let X_j denote the position of the jth person. We assume that $X_1, X_2, \ldots, X_n$ are independent random variables, each uniformly distributed over the interval 0 to L. Their joint probability density function is then given by

$$\begin{aligned} f_{X_1, X_2, \cdots, X_n}(x_1, x_2, \cdots, x_n) &= \frac{1}{L^n}, \qquad 0 \leq x_1, x_2, \cdots, x_n \leq L \\ &= 0 \qquad \text{otherwise}. \end{aligned} \tag{7.13}$$

Next, for each permutation, or ordered n-tuple chosen without replacement, $(i_1, i_2, \ldots, i_n)$ of the integers 1 to n, define

$$I(i_1, i_2, \cdots, i_n) = \{(x_1, x_2, \cdots, x_n): \ x_{i_1} < x_{i_2} < \cdots < x_{i_n}\}. \tag{7.14}$$

Thus $I(i_1, i_2, \ldots, i_n)$ is a zone of points in n-dimensional Euclidean space. There are $n!$ such zones that are mutually exclusive. The union of all zones does not include all the points in n-dimensional space, since an n-tuple $(x_1, x_2, \ldots, x_n)$ that contains two equal components does not lie in any zone. However, we are able to ignore the set of points not included in any of the zones, since this set has probability zero under a continuous probability law. Now the event B that no two persons are less than a distance d apart may be represented as the set of n-tuples $(x_1, x_2, \ldots, x_n)$ for which the distance $|x_i - x_j|$ between any two components is greater than d. To find the probability of B, we must first find the probability of the intersection of B and each zone $I(i_1, i_2, \ldots, i_n)$. We may represent this intersection as follows:

$$(7.15)\quad BI(i_1, i_2, \cdots, i_n) = \{(x_1, x_2, \cdots, x_n):\ 0 < x_{i_1} < L - (n-1)d,\ x_{i_1} + d < x_{i_2} < L - (n-2)d,\ x_{i_2} + d < x_{i_3} < L - (n-3)d, \cdots, x_{i_{n-1}} + d < x_{i_n} < L\}.$$

Consequently,

$$(7.16)\quad P_{X_1, X_2, \cdots, X_n}[BI(i_1, i_2, \cdots, i_n)]$$

$$= \frac{1}{L^n} \int_0^{L-(n-1)d} dx_{i_1} \int_{x_{i_1}+d}^{L-(n-2)d} dx_{i_2} \cdots \int_{x_{i_{n-1}}+d}^{L} dx_{i_n}$$

$$= \int_0^{1-(n-1)d'} du_1 \int_{u_1+d'}^{1-(n-2)d'} du_2 \cdots \int_{u_{n-2}+d'}^{1-d'} du_{n-1} \int_{u_{n-1}+d'}^{1} du_n,$$

in which we have made the change of variables $u_1 = x_{i_1}/L, \ldots, u_n = x_{i_n}/L$, and have set $d' = d/L$. The last written integral is readily evaluated and is seen to be equal to, for $k = 2, \ldots, n-1$,

$$(7.17)\quad \int_0^{1-(n-1)d'} du_1 \int_{u_1+d'}^{1-(n-2)d'} du_2 \cdots \int_{u_{k-1}+d'}^{1-(n-k)d'} du_k \times \frac{1}{(n-k)!}(1-(n-k)d' - u_k)^{n-k}.$$

The probability of B is equal to the product of $n!$ and the probability of the intersection of B and any zone $I(i_1, i_2, \ldots, i_n)$. The proof of (7.12) is now complete. ◀

In a similar manner one may solve the following problem.

▶ **Example 7G. Packing cylinders randomly on a rod.** Consider a horizontal rod of length L on which n equal cylinders, each of length c, are distributed at random. The probability that no two cylinders will be less than d apart is equal to, for d such that $L \geq nc + (n-1)d$,

$$(7.18)\quad \left(1 - \frac{(n-1)d}{L - nc}\right)^n. \qquad ◀$$

The foregoing considerations, together with (6.2) of Chapter 2, establish an extremely useful result.

The Random Division of an Interval or a Circle. Suppose that a straight line of length L is divided into n subintervals by $(n - 1)$ points chosen at random on the line or that a circle of circumference L is divided into n subintervals by n points chosen at random on the circle. Then the probability P_k that exactly k of the subintervals will exceed d in length is given by

$$P_k = \binom{n}{k} \sum_{j=k}^{[L/d]} (-1)^{j-k} \binom{n-k}{n-j} \left(1 - \frac{jd}{L}\right)^{n-1}. \tag{7.19}$$

It may clarify the meaning of (7.19) to express it in terms of random variables. Let $X_1, X_2, \ldots, X_{n-1}$ be the coordinates of the $n - 1$ points chosen randomly on the line (a similar discussion may be given for the circle.) Then $X_1, X_2, \ldots, X_{n-1}$ are independent random variables, each uniformly distributed on the interval 0 to L. Define new random variables $Y_1, Y_2, \ldots, Y_{n-1}$: Y_1 is equal to the minimum of $X_1, X_2, \ldots, X_{n-1}$; Y_2 is equal to the second smallest number among $X_1, X_2, \ldots, X_{n-1}$; and, so on, up to Y_{n-1}, which is equal to the maximum of $X_1, X_2, \ldots, X_{n-1}$. The random variables $Y_1, Y_2, \ldots, Y_{n-1}$ thus constitute a reordering of the random variables $X_1, X_2, \ldots, X_{n-1}$, according to increasing magnitude. For this reason, the random variables $Y_1, Y_1, \ldots, Y_{n-1}$ are called the *order statistics* corresponding to $X_1, X_2, \ldots, X_{n-1}$. The random variable Y_k, for $k = 1, 2, \ldots, n - 1$, is usually spoken of as the kth smallest value among $X_1, X_2, \ldots, X_{n-1}$.

The lengths $W_1, W_2, \ldots, W_n$ of the n successive subintervals into which the $(n - 1)$ randomly chosen points divide the line may now be expressed:

$$W_1 = Y_1, \quad W_2 = Y_2 - Y_1, \cdots, W_j = Y_j - Y_{j-1}, \cdots, \quad W_n = L - Y_{n-1}. \tag{7.20}$$

The probability P_k is the probability that exactly k of the n events $[W_1 > d], [W_2 > d], \ldots, [W_n > d]$ will occur. To prove (7.19), one needs only to verify that for any integer j the probability that j specified subintervals will exceed d in length is equal to

$$\left(1 - \frac{jd}{L}\right)^{n-1} \quad \text{if } 0 \leq j \leq L/d. \tag{7.21}$$

References to the large variety of problems to which (7.19) is applicable may be found in two papers: J. O. Irwin, "A Unified Derivation of Some Well-known Frequency Distributions of Interest in Biometry and Statistics," *Journal of the Royal Statistical Society A*, Vol. 118 (1955), pp. 389–398, and L. Takacs; "On a general probability theorem and its application in the theory of stochastic processes," *Proceedings of the Cambridge Philosophical Society*, Vol. 54 (1958), pp. 219–224.

THEORETICAL EXERCISES

7.1. **Buffon's Needle Problem.** A smooth table is ruled with equidistant parallel lines at distance D apart. A needle of length $L < D$ is dropped on the table. Show that the probability that it will cross one of the lines is $(2L)/(\pi D)$. For an account of some experiments made in connection with the Buffon Needle Problem see J. V. Uspensky, *Introduction to Mathematical Probability*, McGraw-Hill, New York, 1937, pp. 112–113.

7.2. A straight line of unit length is divided into n subintervals by $n - 1$ points chosen at random. For $r = 1, 2, \ldots, n - 1$, show that the probability that none of r specified subintervals will be less than d in length is equal to

$$(1 - rd)^{n-1} \tag{7.22}$$

Hence, using (6.3) of Chapter 2, conclude that the probability that at least 1 of the n subintervals will exceed d in length is equal to

$$n(1 - d)^{n-1} - \binom{n}{2}(1 - 2d)^{n-1} + \cdots (-1)^{r-1}\binom{n}{r}(1 - rd)^{n-1} + \cdots, \tag{7.23}$$

the series continuing as long as the terms $(1 - rd)^{n-1}$, $r = 1, 2, \ldots$ are positive.

EXERCISES

7.1. A young man and a young lady plan to meet between 5 and 6 P.M., each agreeing not to wait more than 10 minutes for the other. Find the probability that they will meet if they arrive independently at random times between 5 and 6 P.M.

7.2. Consider light bulbs produced by a machine for which it is known that the life X in hours of a light bulb produced by the machine is a random variable with probability density function

$$\begin{aligned} f_X(x) &= \frac{1}{1000} e^{-(x/1000)} \qquad \text{for } x > 0 \\ &= 0 \qquad \text{otherwise.} \end{aligned}$$

Consider a box containing 100 such bulbs, selected randomly from the output of the machine.

(i) What is the probability that a bulb selected randomly from the box will have a lifetime greater than 1020 hours?

(ii) What is the probability that a sample of 5 bulbs selected randomly from the box will contain (*a*) at least 1 bulb, (*b*) 4 or more bulbs with a lifetime greater than 1020 hours?

(iii) Find approximately the probability that the box will contain between 30 and 40 bulbs, inclusive, with a lifetime greater than 1020 hours.

7.3. Six soldiers take up random positions on a road 2 miles long. What is the probability that the distance between any two soldiers will be more than (i) $\frac{1}{2}$, (ii) $\frac{1}{3}$, (iii) $\frac{1}{4}$ of a mile?

7.4. **Another version of Bertrand's paradox.** Let a chord be drawn at random in a given circle. What is the probability that the length of the chord will be greater than the side of the equilateral triangle inscribed in that circle?

7.5. A point is chosen randomly on each of 2 adjacent sides of a square. Find the probability that the area of the triangle formed by the sides of the square and the line joining the 2 points will be (i) less than $\frac{1}{8}$ of the area of the square, (ii) greater than $\frac{1}{2}$ of the area of the square.

7.6. Three points are chosen randomly on the circumference of a circle. What is the probability that there will be a semicircle in which all will lie?

7.7. A line is divided into 3 subintervals by choosing 2 points randomly on the line. Find the probability that the 3-line segments thus formed could be made to form the sides of a triangle.

7.8. Find the probability that the roots of the equation $x^2 + 2X_1x + X_2 = 0$ will be real if (i) X_1 and X_2 are randomly chosen between 0 and 1, (ii) X_1 is randomly chosen between 0 and 1, and X_2 is randomly chosen between -1 and 1.

7.9. In the interval 0 to 1, n points are chosen randomly. Find (i) the probability that the point lying farthest to the right will be to the right of the number 0.6, (ii) the probability that the point lying farthest to the left will be to the left of the number 0.6, (iii) the probability that the point lying next farthest to the left will be to the right of the number 0.6.

7.10. A straight line of unit length is divided into 10 subintervals by 9 points chosen at random. For any (i) number $d > \frac{1}{2}$, (ii) number $d > \frac{1}{3}$ find the probability that none of the subintervals will exceed d in length.

8. THE PROBABILITY LAW OF A FUNCTION OF A RANDOM VARIABLE

In this section we develop formulas for the probability law of a random variable Y, which arises as a function of another random variable X, so that for some Borel function $g(\cdot)$

$$Y = g(X). \tag{8.1}$$

To find the probability law of Y, it is best in general first to find its distribution function $F_Y(\cdot)$, from which one may obtain the probability density function $f_Y(\cdot)$ or the probability mass function $p_Y(\cdot)$ in cases in which these functions exist. From (2.2) we obtain the following formula for the value $F_Y(y)$ at the real number y of the distribution function $F_Y(\cdot)$:

$$F_Y(y) = P_X[\{x: \quad g(x) \leq y\}] \qquad \text{if } Y = g(X). \tag{8.2}$$

Of great importance is the special case of a linear function $g(x) = ax + b$, in which a and b are given real numbers so that $a > 0$ and $-\infty < b < \infty$. The distribution function of the random variable $Y = aX + b$ is given by

$$F_{aX+b}(y) = P[aX + b \leq y] = P\left[X \leq \frac{y - b}{a}\right] = F_X\left(\frac{y - b}{a}\right). \tag{8.3}$$

If X is continuous, so is $Y = aX + b$, with a probability density function for any real number y given by

$$f_{aX+b}(y) = \frac{1}{a} f_X\left(\frac{y - b}{a}\right). \tag{8.4}$$

If X is discrete, so is $Y = aX + b$, with a probability mass function for any real number y given by

$$p_{aX+b}(y) = p_X\left(\frac{y - b}{a}\right). \tag{8.5}$$

Next, let us consider $g(x) = x^2$. Then $Y = X^2$. For $y < 0$, $\{x: \quad x^2 \leq y\}$ is the empty set of real numbers. Consequently,

$$F_{X^2}(y) = 0 \qquad \text{for } y < 0. \tag{8.6}$$

For $y \geq 0$

$$\begin{aligned} F_{X^2}(y) &= P[X^2 \leq y] = P[-\sqrt{y} \leq X \leq \sqrt{y}] \\ &= F_X(\sqrt{y}) - F_X(-\sqrt{y}) + p_X(-\sqrt{y}). \end{aligned} \tag{8.7}$$

One sees from (8.7) that if X possesses a probability density function $f_X(\cdot)$ then the distribution function $F_{X^2}(\cdot)$ of X^2 may be expressed as an integral; this is the necessary and sufficient condition that X^2 possess a probability density function $f_{X^2}(\cdot)$. To evaluate the value of $f_{X^2}(y)$ at a real number y, we differentiate (8.7) and (8.6) with respect to y. We obtain

$$\begin{aligned} f_{X^2}(y) &= [f_X(\sqrt{y}) + f_X(-\sqrt{y})] \frac{1}{2\sqrt{y}} \qquad \text{for } y > 0 \\ &= 0 \qquad \text{for } y < 0. \end{aligned} \tag{8.8}$$

It may help the reader to recall the so-called chain rule for differentiation of a function of a function, required to obtain (8.8), if we point out that

$$(8.9)\quad \frac{d}{dy}F_X(\sqrt{y}) = \lim_{h\to 0}\frac{F_X(\sqrt{y+h}) - F_X(\sqrt{y})}{h}$$

$$= \lim_{h\to 0}\frac{F_X(\sqrt{y+h}) - F_X(\sqrt{y})}{\sqrt{y+h}-\sqrt{y}}\lim_{h\to 0}\frac{\sqrt{y+h}-\sqrt{y}}{h}$$

$$= F_X'(\sqrt{y})\frac{d}{dy}\sqrt{y}$$

If X is discrete, it then follows from (8.7) that X^2 is discrete, since the distribution function $F_{X^2}(\cdot)$ may be expressed entirely as a sum. The probability mass function of X^2 for any real number y is given by

$$(8.10)\quad \begin{aligned} p_{X^2}(y) &= p_X(\sqrt{y}) + p_X(-\sqrt{y}) \qquad \text{for } y \geq 0 \\ &= 0 \qquad \text{for } y < 0. \end{aligned}$$

► **Example 8A. The random sine wave.** Let

$$(8.11)\qquad X = A\sin\theta,$$

in which the amplitude A is a known positive constant and the phase θ is a random variable uniformly distributed on the interval $-\pi/2$ to $\pi/2$. The distribution function $F_X(\cdot)$ for $|x| < A$ is given by

$$\begin{aligned} F_X(x) &= P[A\sin\theta \leq x] = P[\sin\theta \leq x/A] \\ &= P[\theta \leq \sin^{-1}(x/A)] = F_\theta\left(\sin^{-1}\frac{x}{A}\right) \\ &= \frac{1}{\pi}\left[\sin^{-1}\left(\frac{x}{A}\right) + \frac{\pi}{2}\right]. \end{aligned}$$

Consequently, the probability density function is given by

$$(8.12)\quad \begin{aligned} f_X(x) &= \frac{1}{\pi A}\left(1 - \left(\frac{x}{A}\right)^2\right)^{-1/2}, \qquad |x| \leq A \\ &= 0 \qquad \text{otherwise.} \end{aligned}$$

Random variables of the form of (8.11) arise in the theory of ballistics. If a projectile is fired at an angle α to the earth, with a velocity of magnitude v, then the point at which the projectile returns to the earth is at a distance R from the point at which it was fired; R is given by the equation $R = (v^2/g)\sin 2\alpha$, in which g is the gravitational constant, equal to

980 cm/sec² or 32.2 ft/sec². If the firing angle α is a random variable with a known probability law, then the range R of the projectile is also a random variable with a known probability law.

A random variable similar to the one given in (8.11) was encountered in the discussion of Bertrand's paradox in section 7; namely, the random variable $X = 2r \cos Z$, in which Z is uniformly distributed over the interval 0 to $\pi/2$. ◀

▶ **Example 8B. The positive part of a random variable.** Given any real number x, we define the symbols x^+ and x^- as follows:

$$x^+ = x \quad \text{if } x \geq 0, \qquad x^- = 0 \quad \text{if } x \geq 0 \tag{8.13}$$
$$= 0 \quad \text{if } x < 0. \qquad = -x \quad \text{if } x < 0.$$

Then $x = x^+ - x^-$ and $|x| = x^+ + x^-$. Given a random variable X, let $Y = X^+$. We call Y the positive part of X. The distribution function of the positive part of X is given by

$$F_{X^+}(y) = 0 \quad \text{if } y < 0 \tag{8.14}$$
$$= F_X(0) \quad \text{if } y = 0$$
$$= F_X(y) \quad \text{if } y > 0.$$

Thus, if X is normally distributed with parameters $m = 0$ and $\sigma = 1$,

$$F_{X^+}(y) = 0 \quad \text{if } y < 0 \tag{8.15}$$
$$= \Phi(0) = \tfrac{1}{2} \quad \text{if } y = 0$$
$$= \Phi(y) \quad \text{if } y > 0.$$

The positive part X^+ of a normally distributed random variable is neither continuous nor discrete but has a distribution function of mixed type. ◀

The Calculus of Probability Density Functions. Let X be a continuous random variable, and let $Y = g(X)$. Unless some conditions are imposed on the function $g(\cdot)$, it is not necessarily true that Y is continuous. For example, $Y = X^+$ is not continuous if X has a positive probability of being negative. We now state some conditions on the function $g(\cdot)$ under which $g(X)$ is a continuous random variable if X is a continuous random variable. At the same time, we give formulas for the probability density function of $g(X)$ in terms of the probability density function of X and the derivatives of $g(\cdot)$.

We first consider the case in which the function $g(\cdot)$ is differentiable at every real number x and, further, either $g'(x) > 0$ for all x or $g'(x) < 0$ for all x. We may then prove the following facts (see R. Courant,

Differential and Integral Calculus, Interscience, New York, 1937, pp. 144–145): (i) as x goes from $-\infty$ to ∞, $g(x)$ is either monotone increasing or monotone decreasing; (ii) the limits

$$\text{(8.16)}\qquad \begin{aligned} \alpha' &= \lim_{x\to-\infty} g(x), \qquad & \beta' &= \lim_{x\to\infty} g(x) \\ \alpha &= \min(\alpha', \beta'), \qquad & \beta &= \max(\alpha', \beta') \end{aligned}$$

exist (although they may be infinite); (iii) for every value of y such that $\alpha < y < \beta$ there exists exactly one value of x such that $y = g(x)$ [this value of x is denoted by $g^{-1}(y)$]; (iv) the inverse function $x = g^{-1}(y)$ is differentiable and its derivative is given by

$$\text{(8.17)}\qquad \frac{dx}{dy} = \frac{d}{dy} g^{-1}(y) = \left(\frac{d}{dx} g(x)\bigg|_{x=g^{-1}(y)}\right)^{-1} = \frac{1}{dy/dx}.$$

For example, let $g(x) = \tan^{-1} x$. Then $g'(x) = 1/(1 + x^2)$ is positive for all x. Here $\alpha = -\pi/2$ and $\beta = \pi/2$. The inverse function is $\tan y$, defined for $|y| \le \pi/2$. The derivative of the inverse function is given by $dx/dy = \sec^2 y$. One sees that $(dy/dx)^{-1} = 1 + (\tan y)^2$ is equal to dx/dy, as asserted by (8.17). We may now state the following theorem:

If $y = g(x)$ is differentiable for all x, and either $g'(x) > 0$ for all x or $g'(x) < 0$ for all x, and if X is a continuous random variable, then $Y = g(X)$ is a continuous random variable with probability density function given by

$$\text{(8.18)}\qquad \begin{aligned} f_Y(y) &= f_X[g^{-1}(y)]\left|\frac{d}{dy} g^{-1}(y)\right| \qquad && \text{if } \alpha < y < \beta \\ &= 0 && \text{otherwise,} \end{aligned}$$

in which α and β are defined by (8.16).

To illustrate the use of (8.18), let us note the formula: if X is a continuous random variable, then

$$\text{(8.19)}\qquad \begin{aligned} f_{\tan^{-1}X}(y) &= f_X(\tan y)\sec^2 y \qquad && \text{for } |y| < \frac{\pi}{2} \\ &= 0 && \text{otherwise.} \end{aligned}$$

To prove (8.18), we distinguish two cases; the case in which the function $y = g(x)$ is monotone increasing and that in which it is monotone decreasing. In the first case the distribution function of Y for $\alpha < y < \beta$ may be written

$$\text{(8.20)}\qquad F_Y(y) = P[g(X) \le y] = P[X \le g^{-1}(y)] = F_X[g^{-1}(y)].$$

In the second case, for $\alpha < y < \beta$,

$$\text{(8.20')}\qquad F_Y(y) = P[g(X) \le y] = P[X \ge g^{-1}(y)] = 1 - F_X[g^{-1}(y)].$$

If (8.20) is differentiated with respect to y, (8.18) is obtained. We leave it to the reader to consider the case in which $y < \alpha$ or $y > \beta$.

One may extend (8.18) to the case in which the derivative $g'(x)$ is continuous and vanishes at only a finite number of points. We leave the proof of the following assertion to the reader.

Let $y = g(x)$ be differentiable for all x and assume that the derivative $g'(x)$ is continuous and nonzero at all but a finite number of values of x. Then, to every real number y, (i) there is a positive integer $m(y)$ and points $x_1(y), x_2(y), \ldots, x_m(y)$ such that, for $k = 1, 2, \ldots, m(y)$,

$$g[x_k(y)] = y, \qquad g'[x_k(y)] \neq 0, \tag{8.21}$$

or (ii) there is no value of x such that $g(x) = y$ and $g'(x) \neq 0$; in this case we write $m(y) = 0$. If X is a continuous random variable, then $Y = g(X)$ is a continuous random variable with a probability density function given by

$$\begin{aligned} f_Y(y) &= \sum_{k=1}^{m(y)} f_X[x_k(y)]|g'[x_k(y)]|^{-1} \qquad \text{if } m(y) > 0 \\ &= 0 \qquad \text{if } m(y) = 0. \end{aligned} \tag{8.22}$$

We obtain as an immediate consequence of (8.22): if X is a continuous random variable, then

$$\begin{aligned} f_{|X|}(y) &= f_X(y) + f_X(-y) \qquad \text{for } y > 0 \\ &= 0 \qquad \text{for } y < 0; \end{aligned} \tag{8.23}$$

$$\begin{aligned} f_{\sqrt{|X|}}(y) &= 2y(f_X(y^2) + f_X(-y^2)) \qquad \text{for } y > 0 \\ &= 0 \qquad \text{for } y < 0. \end{aligned} \tag{8.24}$$

Equations (8.23) and (8.24) may also be obtained directly, by using the same technique with which (8.8) was derived.

The Probability Integral Transformation. It is a somewhat surprising fact, of great usefulness both in theory and in practice, that to obtain a random sample of a random variable X it suffices to obtain a random sample of a random variable U, which is uniformly distributed over the interval 0 to 1. This follows from the fact that the distribution function $F_X(\cdot)$ of the random variable X is a nondecreasing function. Consequently, an inverse function $F_X^{-1}(\cdot)$ may be defined for values of y between 0 and 1: $F_X^{-1}(y)$ is equal to the smallest value of x satisfying the condition that $F_X(x) \geq y$.

▶ **Example 8C.** If X is normally distributed with parameters m and σ, then $F_X(x) = \Phi[(x - m)/\sigma]$ and $F_X^{-1}(y) = m + \sigma\Phi^{-1}(y)$, in which $\Phi^{-1}(y)$ denotes the value of x satisfying the equation $\Phi(\Phi^{-1}(y)) = y$. ◀

In terms of the inverse function $F_X^{-1}(y)$ to the distribution function $F_X(\cdot)$ of the random variable X, we may state the following theorem, the proof of which we leave as an exercise for the reader.

THEOREM 8A. Let $U_1, U_2, \ldots, U_n$ be independent random variables, each uniformly distributed over the interval 0 to 1. The random variables defined by

$$(8.25) \quad X_1 = F_X^{-1}(U_1), \qquad X_2 = F_X^{-1}(U_2), \cdots, X_n = F_X^{-1}(U_n)$$

are then a random sample of the random variable X. Conversely, if $X_1, X_2, \ldots, X_n$ are a random sample of the random variable X and if the distribution function $F_X(\cdot)$ is continuous, then the random variables

$$(8.26) \qquad U_1 = F_X(X_1), \qquad U_2 = F_X(X_2), \cdots, U_n = F_X(X_n)$$

are a random sample of the random variable $U = F_X(X)$, which is uniformly distributed on the interval 0 to 1.

The transformation of a random variable X into a uniformly distributed random variable $U = F_X(X)$ is called the *probability integral transformation.* It plays an important role in the modern theory of goodness-of-fit tests for distribution functions; see T. W. Anderson and D. Darling, "Asymptotic theory of certain goodness of fit criteria based on stochastic processes," *Annals of Mathematical Statistics*, Vol. 23 (1952), pp. 195–212.

EXERCISES

8.1. Let X have a χ^2 distribution with parameters n and σ. Show that $Y = \sqrt{X/n}$ has a χ distribution with parameters n and σ.

8.2. The temperature T of a certain object, recorded in degrees Fahrenheit, obeys a normal probability law with mean 98.6 and variance 2. The temperature θ measured in degrees centigrade is related to T by $\theta = \frac{5}{9}(T - 32)$. Describe the probability law of θ.

8.3. The magnitude v of the velocity of a molecule with mass m in a gas at absolute temperature T is a random variable, which, according to the kinetic theory of gas, possesses the Maxwell distribution with parameter $\alpha = (2kT/m)^{1/2}$ in which k is Boltzmann's constant. Find and sketch the probability density function of the kinetic energy $E = \frac{1}{2}mv^2$ of a molecule. Describe in words the probability law of E.

8.4. A hardware store discovers that the number X of electric toasters it sells in a week obeys a Poisson probability law with mean 10. The profit on each toaster sold is \$2. If at the beginning of the week 10 toasters are in stock, the profit Y from sale of toasters during the week is $Y = 2$ minimum $(X, 10)$. Describe the probability law of Y.

8.5. Find the probability density function of $X = \cos \theta$, in which θ is uniformly distributed on $-\pi$ to π.

8.6. Find the probability density function of the random variable $X = A \sin \omega t$, in which A and ω are known constants and t is a random variable uniformly distributed on the interval $-T$ to T, in which (i) T is a constant such that $0 \leq \omega T \leq \pi/2$, (ii) $T = n(2\pi/\omega)$ for some integer $n \geq 2$.

8.7. Find the probability density function of $Y = e^X$, in which X is normally distributed with parameters m and σ. The random variable Y is said to have a lognormal distribution with parameters m and σ. (The importance and usefulness of the lognormal distribution is discussed by J. Aitchison and J. A. C. Brown, *The Lognormal Distribution*, Cambridge University Press, 1957.)

In exercises 8.8 to 8.11 let X be uniformly distributed on (a) the interval 0 to 1, (b) the interval -1 to 1. Find and sketch the probability density function of the functions given.

8.8. (i) X^2, (ii) $\sqrt{|X|}$.

8.9. (i) e^X, (ii) $-\log_e |X|$.

8.10. (i) $\cos \pi X$, (ii) $\tan \pi X$.

8.11. (i) $2X + 1$, (ii) $2X^2 + 1$.

In exercises 8.12 to 8.15 let X be normally distributed with parameters $m = 0$ and $\sigma = 1$. Find and sketch the probability density functions of the functions given.

8.12. (i) X^2, (ii) e^X.

8.13. (i) $|X|^{1/2}$, (ii) $|X|^{1/3}$.

8.14. (i) $2X + 1$, (ii) $2X^2 + 1$.

8.15. (i) $\sin \pi X$, (ii) $\tan^{-1} X$.

8.16. At time $t = 0$, a particle is located at the point $x = 0$ on an x-axis. At a time T randomly selected from the interval 0 to 1, the particle is suddenly given a velocity v in the positive x-direction. For any time $t > 0$ let $X(t)$ denote the position of the particle at time t. Then $X(t) = 0$, if $t < T$, and $X(t) = v(t - T)$, if $t \geq T$. Find and sketch the distribution function of the random variable $X(t)$ for any given time $t > 0$.

In exercises 8.17 to 8.20 suppose that the amplitude $X(t)$ at a time t of the signal emitted by a certain random signal generator is known to be a random variable (a) uniformly distributed over the interval -1 to 1, (b) normally distributed with parameters $m = 0$ and $\sigma > 0$, (c) Rayleigh distributed with parameter σ.

8.17. The waveform $X(t)$ is passed through a squaring circuit; the output $Y(t)$ of the squaring circuit at time t is assumed to be given by $Y(t) = X^2(t)$. Find and sketch the probability density function of $Y(t)$ for any time $t > 0$.

8.18. The waveform $X(t)$ is passed through a rectifier, giving as its output $Y(t) = |X(t)|$. Describe the probability law of $Y(t)$ for any time $t > 0$.

8.19. The waveform $X(t)$ is passed through a half-wave rectifier, giving as its output $Y(t) = X^+(t)$, the positive part of $X(t)$. Describe the probability law of $Y(t)$ for any $t > 0$.

8.20. The waveform $X(t)$ is passed through a clipper, giving as its output $Y(t) = g[X(t)]$, where $g(x) = 1$ or 0, depending on whether $x > 0$ or $x < 0$. Find and sketch the probability mass function of $Y(t)$ for any $t > 0$.

8.21. Prove that the function given in (8.12) is a probability density function. Does the fact that the function is unbounded cause any difficulty?

9. THE PROBABILITY LAW OF A FUNCTION OF RANDOM VARIABLES

In this section we develop formulas for the probability law of a random variable Y, which arises as a function $Y = g(X_1, X_2, \ldots, X_n)$ of n jointly distributed random variables $X_1, X_2, \ldots, X_n$. All of the formulas developed in this section are consequences of the following basic theorem.

THEOREM 9A. Let $X_1, X_2, \ldots, X_n$ be n jointly distributed random variables, with joint probability law $P_{X_1, X_2, \ldots, X_n}[\cdot]$. Let $Y = g(X_1, X_2, \ldots, X_n)$. Then, for any real number y

$$
\begin{aligned}
(9.1)\quad F_Y(y) &= P[Y \leq y] \\
&= P_{X_1, X_2, \cdots, X_n}[\{(x_1, x_2, \cdots, x_n)\colon \; g(x_1, x_2, \cdots, x_n) \leq y\}].
\end{aligned}
$$

The proof of theorem 9A is immediate, since the event that $Y \leq y$ is logically equivalent to the event that $g(X_1, \ldots, X_n) \leq y$, which is the event that the observed values of the random variables $X_1, X_2, \ldots, X_n$ lie in the set of n-tuples $\{(x_1, \ldots, x_n)\colon \; g(x_1, x_2, \ldots, x_n) \leq y\}$.

We are especially interested in the case in which the random variables $X_1, X_2, \ldots, X_n$ are jointly continuous, with joint probability density $f_{X_1, X_2, \ldots, X_n}(\cdot, \cdot, \ldots, \cdot)$. Then (9.1) may be written

$$
(9.2)\quad F_Y(y) = \iint\cdots\int_{\{(x_1, x_2, \cdots, x_n): g(x_1, x_2, \cdots, x_n) \leq y\}} f_{X_1, X_2, \cdots, X_n}(x_1, x_2, \cdots, x_n)\, dx_1\, dx_2 \cdots dx_n.
$$

To begin with, let us obtain the probability law of the sum of two jointly continuous random variables X_1 and X_2, with a joint probability density function $f_{X_1,X_2}(.\,,.)$. Let $Y = X_1 + X_2$. Then

$$(9.3)\quad F_Y(y) = P[X_1 + X_2 \le y] = P_{X_1,X_2}[\{(x_1, x_2): \; x_1 + x_2 \le y\}]$$

$$= \iint_{\{(x_1,x_2):\; x_1+x_2 \le y\}} f_{X_1,X_2}(x_1, x_2)\, dx_1\, dx_2$$

$$= \int_{-\infty}^{\infty} dx_1 \int_{-\infty}^{y-x_1} dx_2 f_{X_1,X_2}(x_1, x_2)$$

$$= \int_{-\infty}^{\infty} dx_1 \int_{-\infty}^{y} dx_2' f_{X_1,X_2}(x_1, x_2' - x_1).$$

By differentiation of the last equation in (9.3), we obtain the *formula for the probability density function* of $X_1 + X_2$: for any real number y

$$(9.4)\quad f_{X_1+X_2}(y) = \int_{-\infty}^{\infty} dx_1 f_{X_1,X_2}(x_1, y - x_1) = \int_{-\infty}^{\infty} dx_2 f_{X_1,X_2}(y - x_2, x_2).$$

If the random variables X_1 and X_2 are independent, then for any real number y

$$(9.5)\quad f_{X_1+X_2}(y) = \int_{-\infty}^{\infty} dx\, f_{X_1}(x) f_{X_2}(y - x) = \int_{-\infty}^{\infty} dx\, f_{X_1}(y - x) f_{X_2}(x).$$

The mathematical operation involved in (9.5) arises in many parts of mathematics. Consequently, it has been given a name. Consider three functions $f_1(\cdot)$, $f_2(\cdot)$, and $f_3(\cdot)$, which are such that for every real number y

$$(9.6)\qquad f_3(y) = \int_{-\infty}^{\infty} f_1(x) f_2(y - x)\, dx;$$

the function $f_3(\cdot)$ is then said to be the *convolution* of the functions $f_1(\cdot)$ and $f_2(\cdot)$, and in symbols we write $f_3(\cdot) = f_1(\cdot) * f_2(\cdot)$.

In terms of the notion of convolution, we may express (9.5) as follows. *The probability density function $f_{X_1+X_2}(\cdot)$ of the sum of two independent continuous random variables is the convolution of the probability density functions $f_{X_1}(\cdot)$ and $f_{X_2}(\cdot)$ of the random variables.*

One can prove similarly that if the random variables X_1 and X_2 are jointly discrete then the probability mass function of their sum, $X_1 + X_2$, for any real number y is given by

$$(9.7)\qquad p_{X_1+X_2}(y) = \sum_{\substack{\text{over all } x \text{ such that} \\ p_{X_1,X_2}(x,y-x)>0}} p_{X_1,X_2}(x, y - x)$$

$$= \sum_{\substack{\text{over all } x \text{ such that} \\ p_{X_1,X_2}(y-x,x)>0}} p_{X_1,X_2}(y - x, x)$$

In the same way that we proved (9.4) we may prove the formulas for the probability density function of the difference, product, and quotient of two jointly continuous random variables:

$$(9.8) \qquad f_{X-Y}(y) = \int_{-\infty}^{\infty} dx\, f_{X,Y}(y + x, x) = \int_{-\infty}^{\infty} dx\, f_{X,Y}(x, x - y).$$

$$(9.9) \qquad f_{XY}(y) = \int_{-\infty}^{\infty} dx \frac{1}{|x|} f_{X,Y}\left(\frac{y}{x}, x\right) = \int_{-\infty}^{\infty} dx \frac{1}{|x|} f_{X,Y}\left(x, \frac{y}{x}\right).$$

$$(9.10) \qquad f_{X/Y}(y) = \int_{-\infty}^{\infty} dx\, |x| f_{X,Y}(yx, x).$$

We next consider the function of two variables given by $g(x_1, x_2) = \sqrt{x_1^2 + x_2^2}$ and obtain the probability law of $Y = \sqrt{X_1^2 + X_2^2}$. Suppose one is taking a walk in a plane; starting at the origin, one takes a step of magnitude X_1 in one direction and then in a perpendicular direction one takes a step of magnitude X_2. One will then be at a distance Y from the origin given by $Y = \sqrt{X_1^2 + X_2^2}$. Similarly, suppose one is shooting at a target; let X_1 and X_2 denote the coordinates of the shot, taken along perpendicular axes, the center of which is the target. Then $Y = \sqrt{X_1^2 + X_2^2}$ is the distance from the target to the point hit by the shot.

The distribution function of $Y = \sqrt{X_1^2 + X_2^2}$ clearly satisfies $F_Y(y) = 0$ for $y < 0$, and for $y \geq 0$

$$(9.11) \qquad F_Y(y) = \iint_{\{(x_1,x_2):\, x_1^2+x_2^2 \leq y^2\}} f_{X_1,X_2}(x_1, x_2)\, dx_1\, dx_2.$$

We express the double integral in (9.11) by means of polar coordinates. We have, letting $x_1 = r \cos \theta$, $x_2 = r \sin \theta$,

$$(9.12) \qquad F_Y(y) = \int_0^{2\pi} d\theta \int_0^y r\, dr\, f_{X_1,X_2}(r \cos \theta, r \sin \theta).$$

If X_1 and X_2 are jointly continuous, then Y is continuous, with a probability density function obtained by differentiating (9.12) with respect to y. Consequently,

$$(9.13) \qquad f_{\sqrt{X_1^2+X_2^2}}(y) = y \int_0^{2\pi} d\theta\, f_{X_1,X_2}(y \cos \theta, y \sin \theta) \qquad \text{for } y > 0$$
$$= 0 \qquad \text{for } y < 0,$$

$$(9.14) \qquad f_{X_1^2+X_2^2}(y) = \frac{1}{2} \int_0^{2\pi} d\theta\, f_{X_1,X_2}(\sqrt{y} \cos \theta, \sqrt{y} \sin \theta) \qquad \text{for } y > 0$$
$$= 0 \qquad \text{for } y < 0,$$

where (9.14) follows from (9.13) and (8.8).

The formulas given in this section provide tools for the solution of a great many problems of theoretical and applied probability theory, as examples 9A to 9F indicate. In particular, the important problem of finding the probability distribution of the sum of two independent random variables can be treated by using (9.5) and (9.7). One may prove results such as the following:

THEOREM 9B. Let X_1 and X_2 be independent random variables.

(i) If X_1 is normally distributed with parameters m_1 and σ_1 and X_2 is normally distributed with parameters m_2 and σ_2, then $X_1 + X_2$ is normally distributed with parameters $m = m_1 + m_2$ and $\sigma = \sqrt{\sigma_1^2 + \sigma_2^2}$.

(ii) If X_1 obeys a binomial probability law with parameters n_1 and p and X_2 obeys a binomial probability law with parameters n_2 and p, then $X_1 + X_2$ obeys a binomial probability law with parameters $n_1 + n_2$ and p.

(iii) If X_1 is Poisson distributed with parameter λ_1 and X_2 is Poisson distributed with parameter λ_2, then $X_1 + X_2$ is Poisson distributed with parameter $\lambda = \lambda_1 + \lambda_2$.

(iv) If X_1 obeys a Cauchy probability law with parameters a_1 and b_1 and X_2 obeys a Cauchy probability law with parameters a_2 and b_2, then $X_1 + X_2$ obeys a Cauchy probability law with parameters $a_1 + a_2$ and $b_1 + b_2$.

(v) If X_1 obeys a gamma probability law with parameters r_1 and λ and X_2 obeys a gamma probability law with parameters r_2 and λ, then $X_1 + X_2$ obeys a gamma probability law with parameters $r_1 + r_2$ and λ.

A proof of part (i) of theorem 9B is given in example 9A. The other parts of theorem 9B are left to the reader as exercises. A proof of theorem 9B from another point of view is given in section 4 of Chapter 9.

▶ **Example 9A.** Let X_1 and X_2 be independent random variables; X_1 is normally distributed with parameters m_1 and σ_1, whereas X_2 is normally distributed with parameters m_2 and σ_2. Show that their sum $X_1 + X_2$ is normally distributed, with parameters m and σ satisfying the relations

$$(9.15) \qquad m = m_1 + m_2, \qquad \sigma^2 = \sigma_1^2 + \sigma_2^2.$$

Solution: By (9.5),

$$f_{X_1+X_2}(y) = \frac{1}{2\pi\sigma_1\sigma_2}\int_{-\infty}^{\infty} dx \exp\left[-\tfrac{1}{2}\left(\frac{x - m_1}{\sigma_1}\right)^2\right] \exp\left[-\tfrac{1}{2}\left(\frac{y - x - m_2}{\sigma_2}\right)^2\right]$$

By (6.9) of Chapter 4, it follows that

$$(9.16)\quad f_{X_1+X_2}(y) = \frac{1}{\sqrt{2\pi}\sigma} \exp\left[-\tfrac{1}{2}\left(\frac{y-m}{\sigma}\right)^2\right]$$

$$\times \left\{\frac{1}{\sqrt{2\pi}\sigma^*}\int_{-\infty}^{\infty} dx \exp\left[-\tfrac{1}{2}\left(\frac{x-m^*}{\sigma^*}\right)^2\right]\right\},$$

where

$$m^* = \frac{m_1\sigma_2^2 + (y - m_2)\sigma_1^2}{\sigma_1^2 + \sigma_2^2}, \qquad \sigma^{*2} = \frac{\sigma_1^2\sigma_2^2}{\sigma_1^2 + \sigma_2^2}.$$

However, the expression in braces in equation (9.16) is equal to 1. Therefore, it follows that $X_1 + X_2$ is normally distributed with parameters m and σ, given by (9.15). ◀

▶ **Example 9B. The assembly of parts.** It is often the case that a dimension of an assembled article is the sum of the dimensions of several parts. An electrical resistance may be the sum of several electrical resistances. The weight or thickness of the article may be the sum of the weights or thicknesses of individual parts. The probability law of the individual dimensions may be known; what is of interest is the probability law of the dimension of the assembled article. An answer to this question may be obtained from (9.5) and (9.7) if the individual dimensions are independent random variables. For example, let us consider two 10-ohm resistors assembled in series. Suppose that, in fact, the resistances of the resistors are independent random variables, each obeying a normal probability law with mean 10 ohms and standard deviation 0.5 ohms. The unit, consisting of the two resistors assembled in series, has resistance equal to the sum of the individual resistances; therefore, the resistance of the unit obeys a normal probability law with mean 20 ohms and standard deviation $\{(0.5)^2 + (0.5)^2\}^{1/2} = 0.707$ ohms. Now suppose one wishes to measure the resistance of the unit, using an ohmmeter whose error of measurement is a random variable obeying a normal probability law with mean 0 and standard deviation 0.5 ohms. The measured resistance of the unit is a random variable obeying a normal probability law with mean 20 ohms and standard deviation $\sqrt{(0.707)^2 + (0.5)^2} = 0.866$ ohms. ◀

▶ **Example 9C.** Let X_1 and X_2 be independent random variables, each normally distributed with parameters $m = 0$ and $\sigma > 0$. Then

$$f_{X_1,X_2}(y\cos\theta, y\sin\theta) = \frac{1}{2\pi\sigma^2} e^{-\frac{1}{2}\left(\frac{y}{\sigma}\right)^2}.$$

Consequently, for $y > 0$

$$f_{\sqrt{X_1^2+X_2^2}}(y) = \frac{y}{\sigma^2} e^{-\frac{1}{2}\left(\frac{y}{\sigma}\right)^2}. \tag{9.17}$$

$$f_{X_1^2+X_2^2}(y) = \frac{1}{2\sigma^2} e^{-\frac{1}{2\sigma^2}y} \tag{9.18}$$

In words, $\sqrt{X_1^2 + X_2^2}$ has a Rayleigh distribution with parameter σ, whereas $X_1^2 + X_2^2$ has a χ^2 distribution with parameters $n = 2$ and σ. ◀

▶ **Example 9D. The probability distribution of the envelope of narrow-band noise.** A family of random variables $X(t)$, defined for $t > 0$, is said to represent a narrow-band noise voltage [see S. O. Rice, "Mathematical Analysis of Random Noise," *Bell System. Tech. Jour.*, Vol. 24 (1945), p. 81] if $X(t)$ is represented in the form

$$X(t) = X_c(t) \cos \omega t + X_s(t) \sin \omega t, \tag{9.19}$$

in which ω is a known frequency, whereas $X_c(t)$ and $X_s(t)$ are independent normally distributed random variables with means 0 and equal variances σ^2. The envelope of $X(t)$ is then defined as

$$R(t) = [X_c^2(t) + X_s^2(t)]^{1/2}. \tag{9.20}$$

In view of example 9C, it is seen that the *envelope* $R(t)$ *has a Rayleigh distribution* with parameter $\alpha = \sigma$. ◀

▶ **Example 9E.** Let U and V be independent random variables, such that U is normally distributed with mean 0 and variance σ^2 and V has a χ distribution with parameters n and σ. Show that the quotient $T = U/V$ has Student's distribution with parameter n.

Solution: By (9.10), the probability density function of T for any real number is given by

$$\begin{aligned} f_T(y) &= \int_0^\infty dx\, x f_U(yx) f_V(x) \\ &= \frac{K}{\sigma^{n+1}} \int_0^\infty dx\, x \exp\left[-\tfrac{1}{2}\left(\frac{yx}{\sigma}\right)^2\right] x^{n-1} \exp\left[-\frac{n}{2}\left(\frac{x}{\sigma}\right)^2\right] \end{aligned}$$

where

$$K = \frac{2(n/2)^{n/2}}{\Gamma(n/2)\sqrt{2\pi}}.$$

By making the change of variable $u = x\sqrt{(y^2 + n)}/\sigma$, it follows that

$$\begin{aligned} f_T(y) &= K(y^2 + n)^{-(n+1)/2} \int_0^\infty du\, u^n e^{-\frac{1}{2}u^2} \\ &= K(y^2 + n)^{-(n+1)/2} 2^{(n-1)/2} \Gamma\left(\frac{n+1}{2}\right), \end{aligned}$$

from which one may immediately deduce that the probability density function of T is given by (4.15) of Chapter 4. ◀

▶ **Example 9F. Distribution of the range.** A ship is shelling a target on an enemy shore line, firing n independent shots, all of which may be assumed to fall on a straight line and to be distributed according to the distribution function $F(x)$ with probability density function $f(x)$. Define the range (or span) R of the attack as the interval between the location of the extreme shells. Find the probability density function of R.

Solution: Let $X_1, X_2, \ldots, X_n$ be independent random variables representing the coordinates locating the position of the n shots. The range R may be written $R = V - U$, in which $V =$ maximum $(X_1, X_2, \ldots, X_n)$ and $U =$ minimum $(X_1, X_2, \ldots, X_n)$. The joint distribution function $F_{U,V}(u, v)$ is found as follows. If $u \geq v$, then $F_{U,V}(u, v)$ is the probability that simultaneously $X_1 \leq v, \ldots, X_n \leq v$; consequently,

$$(9.21) \qquad F_{U,V}(u, v) = [F(v)]^n \qquad \text{if } u \geq v,$$

since $P[X_k \leq v] = F(v)$ for $k = 1, 2, \ldots, n$. If $u < v$, then $F_{U,V}(u, v)$ is the probability that simultaneously $X_1 \leq v, \ldots, X_n \leq v$ but not simultaneously $u < X_1 \leq v, \ldots, u < X_n \leq v$; consequently,

$$(9.22) \qquad F_{U,V}(u, v) = [F(v)]^n - [F(v) - F(u)]^n \qquad \text{if } u < v.$$

The joint probability density of U and V is then obtained by differentiation. It is given by

$$\begin{aligned}(9.23) \quad f_{U,V}(u, v) &= 0 \qquad \text{if } u > v \\ &= n(n-1)[F(v) - F(u)]^{n-2} f(u) f(v), \qquad \text{if } u < v.\end{aligned}$$

From (9.8) and (9.23) it follows that the *probability density function of the range* R of n independent continuous random variables, whose individual distribution functions are all equal to $F(x)$ and whose individual probability density functions are all equal to $f(x)$, is given by

$$\begin{aligned}(9.24) \quad f_R(x) &= \int_{-\infty}^{\infty} dv\, f_{U,V}(v - x, v) \\ &= 0 \qquad \text{for } x < 0 \\ &= n(n-1)\int_{-\infty}^{\infty} [F(v) - F(v - x)]^{n-2} f(v - x) f(v)\, dv, \\ &\qquad\qquad \text{for } x > 0.\end{aligned}$$

The distribution function of R is then given by

$$\begin{aligned}(9.25) \qquad F_R(x) &= 0 \qquad \text{if } x < 0 \\ &= n\int_{-\infty}^{\infty} [F(v) - F(v - x)]^{n-1} f(v)\, dv, \qquad \text{if } x \geq 0.\end{aligned}$$

Equations (9.24) and (9.25) can be explicitly evaluated only in a few cases, such as that in which each random variable $X_1, X_2, \ldots, X_n$ is uniformly distributed on the interval 0 to 1. Then from (9.24) it follows that

$$(9.26) \qquad \begin{aligned} f_R(x) &= n(n-1)x^{n-2}(1-x) \qquad 0 \le x \le 1 \\ &= 0 \qquad \text{elsewhere.} \end{aligned}$$

◀

A Geometrical Method for Finding the Probability Law of a Function of Several Random Variables. Consider n jointly continuous random variables $X_1, X_2, \ldots, X_n$, and the random variable $Y = g(X_1, X_2, \ldots, X_n)$. Suppose that the joint probability density function of $X_1, X_2, \ldots, X_n$ has the property that it is constant on the surface in n-dimensional space obtained by setting $g(x_1, \ldots, x_n)$ equal to a constant; more precisely, suppose that there is a function of a real variable, denoted by $f_g(\cdot)$, such that

$$(9.27) \qquad f_{X_1, X_2, \cdots, X_n}(x_1, \cdots, x_n) = f_g(y), \qquad \text{if } g(x_1, x_2, \cdots, x_n) = y.$$

If (9.27) holds and $g(\cdot)$ is a continuous function, we obtain a simple formula for the probability density function $f_Y(\cdot)$ of the random variable $Y = g(X_1, X_2, \ldots, X_n)$; for any real number y

$$(9.28) \qquad f_Y(y) = f_g(y) \frac{dV_g(y)}{dy},$$

in which $V_g(y)$ represents the volume within the surface in n-dimensional space with equation $g(x_1, x_2, \ldots, x_n) = y$; in symbols,

$$(9.29) \qquad V_g(y) = \iint\cdots\int_{\{(x_1, x_2, \cdots, x_n): g(x_1, \cdots, x_n) \le y\}} dx_1 dx_2 \cdots dx_n.$$

We sketch a proof of (9.28). Let $B(y; h) = \{(x_1, x_2, \ldots, x_n)\colon\ y < g(x_1, \ldots, x_n) \le y + h\}$. Then, by the law of the mean for integrals,

$$\begin{aligned} F_Y(y+h) - F_Y(y) &= \iint\cdots\int_{B(y;h)} f_{X_1, \cdots, X_n}(x_1, \cdots, x_n)\, dx_1 \cdots dx_n \\ &= f_{X_1, \cdots, X_n}(x_1', \cdots, x_n')[V_g(y+h) - V_g(y)] \end{aligned}$$

for some point $(x_1', \ldots, x_n')$ in the set $B(y; h)$. Now, as h tends to 0, $f_{X_1, \cdots, X_n}(x_1', \ldots, x_n')$ tends to $f_g(y)$, assuming $f_g(\cdot)$ is a continuous function, and $[V_g(y+h) - V_g(y)]/h$ tends to $dV_g(y)/dy$. From these facts, one immediately obtains (9.28).

We illustrate the use of (9.28) by obtaining a basic formula, which generalizes example 9C.

▶ **Example 9G.** Let $X_1, X_2, \ldots, X_n$ be independent random variables, each normally distributed with mean 0 and variance 1. Let $Y = \sqrt{X_1^2 + X_2^2 + \ldots + X_n^2}$. Show that

$$f_Y(y) = \frac{y^{n-1}e^{-\frac{1}{2}y^2}}{\int_0^\infty y^{n-1}e^{-\frac{1}{2}y^2}\,dy}, \qquad \text{for } y > 0, \tag{9.30}$$

$$= 0, \qquad \text{for } y < 0,$$

where $\int_0^\infty y^{n-1}e^{-\frac{1}{2}y^2}\,dy = 2^{(n-2)/2}\Gamma(n/2)$. In words, Y has a χ distribution with parameters n and $\sigma = \sqrt{n}$.

Solution: Define $g(x_1, \ldots, x_n) = \sqrt{x_1^2 + \ldots + x_n^2}$ and $f_g(y) = (2\pi)^{-n/2}e^{-\frac{1}{2}y^2}$. Then (9.27) holds. Now $V_g(y)$ is the volume within a sphere in n-dimensional space of radius y. Clearly, $V_g(y) = 0$ for $y < 0$, and for $y \geq 0$

$$V_g(y) = y^n \iint\cdots\int_{\{(x_1,\cdots,x_n):x_1^2+\cdots+x_n^2\leq 1\}} dx_1 \cdots dx_n,$$

so that $V_g(y) = Ky^n$ for some constant K. Then $dV_g(y)/dy = nKy^{n-1}$. By (9.28), $f_Y(y) = 0$ for $y < 0$, and for $y \geq 0$ $f_Y(y) = K'y^{n-1}e^{-\frac{1}{2}y^2}$, for some constant K'. To obtain K', use the normalization condition $\int_{-\infty}^{\infty} f_Y(y)\,dy = 1$. The proof of (9.30) is complete. ◀

▶ **Example 9H. The energy of an ideal gas is χ^2 distributed.** Consider an ideal gas composed of N particles of respective masses $m_1, m_2, \ldots, m_N$. Let $v_x^{(i)}, v_y^{(i)}, v_z^{(i)}$ denote the velocity components at a given time instant of the ith particle. Assume that the total energy E of the gas is given by its kinetic energy

$$E = \sum_{i=1}^{N} \frac{m_i}{2}\{(v_x^{(i)})^2 + (v_y^{(i)})^2 + (v_z^{(i)})^2\}.$$

Assume that the joint probability density function of the $3N$-velocities $(v_x^{(1)}, v_y^{(1)}, v_z^{(1)}, v_x^{(2)}, v_y^{(2)}, v_z^{(2)}, \ldots, v_x^{(N)}, v_y^{(N)}, v_z^{(N)})$ is proportional to $e^{-E/kT}$, in which k is Boltzmann's constant and T is the absolute temperature of the gas; in statistical mechanics one says that the state of the gas has as its probability law Gibb's canonical distribution. The energy E of the gas is a random variable whose probability density function may be derived by the geometrical method. For $x > 0$

$$f_E(x) = K_1 e^{-x/kT}\frac{dV_E(x)}{dx}$$

for some constant K_1, in which $V_E(x)$ is the volume within the ellipsoid in $3N$-dimensional space consisting of all $3N$-tuples of velocities whose kinetic energy $E \le x$. One may show that

$$V_E(x) = K_2 x^{3N/2}, \qquad \frac{dV_E(x)}{dx} = K_2 \tfrac{3}{2} N x^{(3N/2)-1}$$

for some constant K_2 in the same way that $V_g(y)$ is shown in example 9G to be proportional to y^n. Consequently, for $x > 0$

$$f_E(x) = \frac{x^{(3N/2)-1} e^{-x/kT}}{\int_0^\infty x^{(3N/2)-1} e^{-x/kT}\, dx}.$$

In words, E has a χ^2 distribution with parameters $n = 3N$ and $\sigma^2 = kT/2$. ◀

We leave it for the reader to verify the validity of the next example.

▶ **Example 9I. The joint normal distribution.** Consider two jointly normally distributed random variables X_1 and X_2; that is, X_1 and X_2 have a joint probability density function

$$f_{X_1,X_2}(x_1, x_2) = \frac{1}{2\pi\sigma_1\sigma_2\sqrt{1-\rho^2}} e^{-Q(x_1,x_2)} \tag{9.31}$$

for some constants $\sigma_1 > 0$, $\sigma_2 > 0$, $-1 < \rho < 1$, $-\infty < m_1 < \infty$, $-\infty < m_2 < \infty$, in which the function $Q(.\,,.)$ for any two real numbers x_1 and x_2 is defined by

$$Q(x_1, x_2) = \frac{1}{2(1-\rho^2)}\left[\left(\frac{x_1 - m_1}{\sigma_1}\right)^2 - 2\rho\left(\frac{x_1 - m_1}{\sigma_1}\right)\left(\frac{x_2 - m_2}{\sigma_2}\right) + \left(\frac{x_2 - m_2}{\sigma_2}\right)^2\right].$$

The curve $Q(x_1, x_2) =$ constant is an ellipse. Let $Y = Q(X_1, X_2)$. Then $P[Y > y] = e^{-y}$ for $y > 0$. ◀

THEORETICAL EXERCISES

Various probability laws (or, equivalently, probability distributions), which are of importance in statistics, arise as the probability laws of various functions of normally distributed random variables.

9.1. The χ^2 distribution. Show that if $X_1, X_2, \ldots, X_n$ are independent random variables, each normally distributed with parameters $m = 0$ and $\sigma > 0$, and if $Z = X_1^2 + X_2^2 + \ldots + X_n^2$, then Z has a χ^2 distribution with parameters n and σ^2.

9.2. The χ distribution. Show that if $X_1, X_2, \ldots, X_n$ are independent random variables, each normally distributed with parameters $m = 0$ and $\sigma > 0$, then

$$V = \sqrt{\frac{1}{n}\sum_{k=1}^{n} X_k^2}$$

has a χ distribution with parameters n and σ.

9.3. Student's distribution. Show that if $X_0, X_1, \ldots, X_n$ are $(n + 1)$ independent random variables, each normally distributed with parameters $m = 0$ and $\sigma > 0$, then the random variable

$$X_0 \Big/ \sqrt{\frac{1}{n}\sum_{k=1}^{n} X_k^2}$$

has as its probability law Student's distribution with parameter n (which, it should be noted, is independent of σ)!

9.4. The F distribution. Show that if Z_1 and Z_2 are independent random variables, χ^2 distributed with n_1 and n_2 degrees of freedom, respectively, then the quotient $n_2 Z_1/n_1 Z_2$ obeys the F distribution with parameters n_1 and n_2. Consequently, conclude that if $X_1, \ldots, X_m, X_{m+1}, \ldots, X_{m+n}$ are $(m + n)$ independent random variables, each normally distributed with parameters $m = 0$ and $\sigma > 0$, then the random variable

$$\frac{(1/m)\sum_{k=1}^{m} X_k^2}{(1/n)\sum_{k=1}^{n} X_{m+k}^2}$$

has as its probability law the F distribution with parameters m and n. In statistics the parameters m and n are spoken of as "degrees of freedom."

9.5. Show that if X_1 has a binomial distribution with parameters n_1 and p, if X_2 has a binomial distribution with parameters n_2 and p, and X_1 and X_2 are independent, then $X_1 + X_2$ has a binomial distribution with parameters $n_1 + n_2$ and p.

9.6. Show that if X_1 has a Poisson distribution with parameter λ_1, if X_2 has a Poisson distribution with parameter λ_2, and X_1 and X_2 are independent, then $X_1 + X_2$ is Poisson distributed with parameter $\lambda_1 + \lambda_2$.

9.7. Show that if X_1 and X_2 are independently and uniformly distributed over the interval a to b, then

$$\begin{aligned} (9.32) \qquad f_{X_1+X_2}(y) &= 0 && \text{for } y < 2a \text{ or } y > 2b \\ &= \frac{y - 2a}{(b-a)^2} && \text{for } 2a < y < a + b \\ &= \frac{2b - y}{(b-a)^2} && \text{for } a + b < y < 2b. \end{aligned}$$

9.8. Prove the validity of the assertion made in example 9I. Identify the probability law of Y. Find the probability law of $Z = 2(1 - \rho^2)Y$.

9.9. Let X_1 and X_2 have a joint probability density function given by (9.31). Show that the sum $X_1 + X_2$ is normally distributed, with parameters $m = m_1 + m_2$ and $\sigma^2 = \sigma_1^2 + 2\rho\sigma_1\sigma_2 + \sigma_2^2$.

9.10. Let X_1 and X_2 have a joint probability density function given by equation (9.31), with $m_1 = m_2 = 0$. Show that

$$f_{X_1/X_2}(y) = \frac{\sigma_1\sigma_2\sqrt{1-\rho^2}}{\pi(\sigma_2^2y^2 - 2\rho\sigma_1\sigma_2 y + \sigma_1^2)}. \tag{9.33}$$

If X_1 and X_2 are independent, then the quotient X_1/X_2 has a Cauchy distribution.

9.11. Use the proof of example 9G to prove that the volume $V_n(r)$ of an n-dimensional sphere of radius r is given by

$$V_n(r) = \iint\cdots\int_{\{(x_1,\cdots,x_n):x_1^2+\cdots+x_n^2\le r^2\}} dx_1\,dx_2\cdots dx_n = \frac{\pi^{n/2}}{\Gamma[(n/2)+1]}r^n. \tag{9.34}$$

Prove that the surface area of the sphere is given by $dV_n(r)/dr$.

9.12. Prove that it is impossible for two independent identically distributed random variables, X_1 and X_2, each taking the values 1 to 6, to have the property that $P[X_1 + X_2 = k] = \frac{1}{11}$ for $k = 2, 3, \ldots, 12$. Consequently, conclude that it is impossible to weight a pair of dice so that the probability of occurrence of every sum from 2 to 12 will be the same.

9.13. Prove that if two independent identically distributed random variables, X_1 and X_2, each taking the values 1 to 6, have the property that their sum will satisfy $P[X_1 + X_2 = k] = P[X_1 + X_2 = 14 - k] = (k-1)/36$ for $k = 2$, $3, 4, 5, 6$, and $P[X_1 + X_2 = 7] = \frac{6}{36}$ then $P[X_1 = k] = P[X_2 = k] = \frac{1}{6}$ for $k = 1, 2, \ldots, 6$.

EXERCISES

9.1. Suppose that the load on an airplane wing is a random variable X obeying a normal probability law with mean 1000 and variance 14,400, whereas the load Y that the wing can withstand is a random variable obeying a normal probability law with mean 1260 and variance 2500. Assuming that X and Y are independent, find the probability that $X < Y$ (that the load encountered by the wing is less than the load the wing can withstand).

In exercises 9.2 to 9.4 let X_1 and X_2 be independently and uniformly distributed over the intervals 0 to 1.

9.2. Find and sketch the probability density function of (i) $X_1 + X_2$, (ii) $X_1 - X_2$, (iii) $|X_1 - X_2|$.

9.3. (i) Maximum (X_1, X_2), (ii) minimum (X_1, X_2).

9.4. (i) X_1X_2, (ii) X_1/X_2.

In exercises 9.5 to 9.7 let X_1 and X_2 be independent random variables, each normally distributed with parameters $m = 0$ and $\sigma > 0$.

9.5. Find and sketch the probability density function of (i) $X_1 + X_2$, (ii) $X_1 - X_2$, (iii) $|X_1 - X_2|$, (iv) $(X_1 + X_2)/2$, (v) $(X_1 - X_2)/2$.

9.6. (i) $X_1^2 + X_2^2$, (ii) $(X_1^2 + X_2^2)/2$.

9.7. (i) X_1/X_2, (ii) $X_1/|X_2|$.

9.8. Let X_1, X_2, X_3, and X_4 be independent random variables, each normally distributed with parameters $m = 0$ and $\sigma^2 = 1$. Find and sketch the probability density functions of (i) $X_3/\sqrt{(X_1^2 + X_2^2)/2}$, (ii) $2X_3^2/(X_1^2 + X_2^2)$, (iii) $3X_4^2/(X_1^2 + X_2^2 + X_3^2)$, (iv) $(X_1^2 + X_2^2)/(X_3^2 + X_4^2)$.

9.9. Let X_1, X_2, and X_3 be independent random variables, each exponentially distributed with parameter $\lambda = \frac{1}{2}$. Find the probability density function of (i) $X_1 + X_2 + X_3$, (ii) minimum (X_1, X_2, X_3), (iii) maximum (X_1, X_2, X_3), (iv) X_1/X_2.

9.10. Find and sketch the probability density function of $\theta = \tan^{-1}(Y/X)$ if X and Y are independent random variables, each normally distributed with mean 0 and variance σ^2.

9.11. The envelope of a narrow-band noise is sampled periodically, the samples being sufficiently far apart to assure independence. In this way n independent random variables $X_1, X_2, \ldots, X_n$ are observed, each of which is Rayleigh distributed with parameter σ. Let $Y = \text{maximum}\,(X_1, X_2, \ldots, X_n)$ be the largest value in the sample. Find the probability density function of Y.

9.12. Let $v = (v_x^2 + v_y^2 + v_z^2)^{1/2}$ be the magnitude of the velocity of a particle whose velocity components v_x, v_y, v_z are independent random variables, each normally distributed with mean 0 and variance kT/M; k is Boltzmann's constant, T is the absolute temperature of the medium in which the particle is immersed, and M is the mass of the particle. Describe the probability law of v.

9.13. Let $X_1, X_2, \ldots, X_n$ be independent random variables, uniformly distributed over the interval 0 to 1. Describe the probability law of $-2 \log (X_1 X_2 \ldots X_n)$. Using this result, describe a procedure for forming a random sample of a random variable with a χ^2 distribution with $2n$ degrees of freedom.

9.14. Let X and Y be independent random variables, each exponentially distributed with parameter λ. Find the probability density function of $Z = X/(X + Y)$.

9.15. Show that if $X_1, X_2, \ldots, X_n$ are independent identically distributed random variables, whose minimum $Y = \text{minimum}\,(X_1, X_2, \ldots, X_n)$ obeys an exponential probability law with parameter λ, then each of the random variables $X_1, \ldots, X_n$ obeys an exponential probability law with parameter (λ/n). If you prefer to solve the problem for the special case that $n = 2$, this will suffice. *Hint:* Y obeys an exponential probability law with parameter λ if and only if $F_Y(y) = 1 - e^{-\lambda y}$ or 0, depending on whether $y \geq 0$ or $y < 0$.

9.16. Let $X_1, X_2, \ldots, X_n$ be independent random variables (i) uniformly distributed on the interval -1 to 1, (ii) exponentially distributed with mean 2. Find the distribution of the range $R = \text{maximum}(X_1, X_2, \ldots, X_n) - \text{minimum}(X_1, X_2, \ldots, X_n)$.

9.17. Find the probability that in a random sample of size n of a random variable uniformly distributed on the interval 0 to 1 the range will exceed 0.8.

9.18. Determine how large a random sample one must take of a random variable uniformly distributed on the interval 0 to 1 in order that the probability will be more than 0.95 that the range will exceed 0.90.

9.19. The random variable X represents the amplitude of a sine wave; Y represents the amplitude of a cosine wave. Both are independently and uniformly distributed over the interval 0 to 1.

(i) Let the random variable R represent the amplitude of their resultant; that is, $R^2 = X^2 + Y^2$. Find and sketch the probability density function of R.

(ii) Let the random variable θ represent the phase angle of the resultant; that is, $\theta = \tan^{-1}(Y/X)$. Find and sketch the probability density function of θ.

9.20. The noise output of a quadratic detector in a radio receiver can be represented as $X^2 + Y^2$, where X and Y are independently and normally distributed with parameters $m = 0$ and $\sigma > 0$. If, in addition to noise, there is a signal present, the output is represented by $(X + a)^2 + (Y + b)^2$, where a and b are given constants. Find the probability density function of the output of the detector, assuming that (i) noise alone is present, (ii) both signal and noise are present.

9.21. Consider 3 jointly distributed random variables X, Y, and Z with a joint probability density function

$$f_{X,Y,Z}(x, y, z) = \frac{6}{(1 + x + y + z)^4} \qquad \text{for } x > 0, \quad y > 0, \quad z > 0$$

$$= 0 \qquad \text{otherwise.}$$

Find the probability density function of the sum $X + Y + Z$.

10. THE JOINT PROBABILITY LAW OF FUNCTIONS OF RANDOM VARIABLES

In section 9 we treated in some detail the problem of obtaining the individual probability law of a function of random variables. It is natural to consider next the problem of obtaining the joint probability law of several random variables which arise as functions. In principle, this problem is no different from those previously considered. However, the details are more complicated. Consequently, in this section, we content ourselves with stating an often-used formula for the joint probability density function of n random variables $Y_1, Y_2, \ldots, Y_n$, which arise as functions of n jointly continuous random variables $X_1, X_2, \ldots, X_n$:

$$(10.1) \quad Y_1 = g_1(X_1, X_2, \cdots, X_n),\ Y_2 = g_2(X_1, X_2, \cdots, X_n), \cdots, \quad Y_n = g_n(X_1, X_2, \cdots, X_n).$$

We consider only the case in which the functions $g_1(x_1, x_2, \ldots, x_n)$, $g_2(x_1, x_2, \ldots, x_n)$, $g_n(x_1, x_2, \ldots, x_n)$ have continuous first partial derivatives at all points $(x_1, x_2, \ldots, x_n)$ and are such that the Jacobian

$$(10.2) \qquad J(x_1, x_2, \cdots, x_n) = \begin{vmatrix} \dfrac{\partial g_1}{\partial x_1} & \dfrac{\partial g_1}{\partial x_2} & \cdots & \dfrac{\partial g_1}{\partial x_n} \\ \dfrac{\partial g_2}{\partial x_1} & \dfrac{\partial g_2}{\partial x_2} & \cdots & \dfrac{\partial g_2}{\partial x_n} \\ \cdot & \cdot & \cdot & \cdot \\ \dfrac{\partial g_n}{\partial x_1} & \dfrac{\partial g_n}{\partial x_2} & \cdots & \dfrac{\partial g_n}{\partial x_n} \end{vmatrix} \neq 0$$

at all points $(x_1, x_2, \ldots, x_n)$. Let C be the set of points $(y_1, y_2, \ldots, y_n)$ such that the n equations

$$(10.3) \quad y_1 = g_1(x_1, x_2, \cdots, x_n), \qquad y_2 = g_2(x_1, x_2, \cdots, x_n), \cdots, \quad y_n = g_n(x_1, x_2, \cdots, x_n)$$

possess at least one solution $(x_1, x_2, \ldots, x_n)$. The set of equations in (10.3) then possesses exactly one solution, which we denote by

$$(10.4) \quad x_1 = g^{-1}(y_1, y_2, \cdots, y_n), \qquad x_2 = g^{-1}(y_1, y_2, \cdots, y_n), \cdots, \quad x_n = g^{-1}(y_1, y_2, \cdots, y_n).$$

If $X_1, X_2, \ldots, X_n$ are jointly continuous random variables, whose joint probability density function is continuous at all but a finite number of points in the $(x_1, x_2, \ldots, x_n)$ space, then the random variables $Y_1, Y_2, \ldots, Y_n$ defined by (10.1) are jointly continuous with a joint probability density function given by

$$(10.5) \quad f_{Y_1, Y_2, \cdots, Y_n}(y_1, y_2, \cdots, y_n) = f_{X_1, X_2, \cdots, X_n}(x_1, x_2, \cdots, x_n)|J(x_1, x_2, \cdots, x_n)|^{-1},$$

if $(y_1, y_2, \ldots, y_n)$ belongs to C, and $x_1, x_2, \ldots, x_n$ are given by (10.4); for $(y_1, y_2, \ldots, y_n)$ not belonging to C

$$f_{Y_1,Y_2,\cdots,Y_n}(y_1, y_2, \cdots, y_n) = 0. \tag{10.5'}$$

It should be noted that (10.5) extends (8.18). We leave it to the reader to formulate a similar extension of (8.22).

We omit the proof that the random variables $Y_1, Y_2, \ldots, Y_n$ are jointly continuous and possess a joint probability density. We sketch a proof of the formula given by (10.5) for the joint probability density function. One may show that for any real numbers $u_1, u_2, \ldots, u_n$

$$\begin{aligned} f_{Y_1,Y_2,\cdots,Y_n}(u_1, u_2, \cdots, u_n) &= \lim_{h_1\to 0, h_2\to 0, \cdots, h_n\to 0} \frac{1}{h_1 h_2 \cdots h_n} P[u_1 \le Y_1 \le u_1 + h_1, \\ &\qquad u_2 \le Y_2 \le u_2 + h_2, \cdots, u_n \le Y_n \le u_n + h_n]. \end{aligned} \tag{10.6}$$

The probability on the right-hand side of (10.6) is equal to

$$\begin{aligned} P[u_1 \le g_1(X_1, X_2, \cdots, X_n) \le u_1 + h_1, \cdots, \\ u_n \le g_n(X_1, X_2, \cdots, X_n) \le u_n + h_n] \\ = \iint_{D_n}\cdots\int f_{X_1,X_2,\cdots,X_n}(x_1, x_2, \cdots, x_n)\, dx_1\, dx_2 \cdots dx_n, \end{aligned} \tag{10.7}$$

in which

$$\begin{aligned} D_n = \{(x_1, x_2, \cdots, x_n)\colon\ u_1 \le g_1(x_1, x_2, \cdots, x_n) \le u_1 + h_1, \cdots, \\ u_n \le g_n(x_1, x_2, \cdots, x_n) \le u_n + h_n\}. \end{aligned}$$

Now, if $(u_1, u_2, \ldots, u_n)$ does not belong to C, then for sufficiently small values of $h_1, h_2, \ldots, h_n$ there are no points $(x_1, x_2, \ldots, x_n)$ in D_n and the probability in (10.7) is 0. From the fact that the quantities in (10.6), whose limit is being taken, are 0 for sufficiently small values of $h_1, h_2, \ldots, h_n$, it follows that $f_{Y_1,Y_2,\ldots,Y_n}(u_1, u_2, \ldots, u_n) = 0$ for $(u_1, u_2, \ldots, u_n)$ not in C. Thus (10.5′) is proved. To prove (10.5), we use the celebrated formula for change of variables in multiple integrals (see R. Courant, *Differential and Integral Calculus*, Interscience, New York, 1937, Vol II, p. 253, or T. Apostol, *Mathematical Analysis*, Addison-Wesley, Reading, Massachusetts, 1957, p. 271) to transform the integral on the right-hand side of (10.7) to the integral

$$\begin{aligned} \int_{u_1}^{u_1+h_1} dy_1 \int_{u_2}^{u_2+h_2} dy_2 \cdots \int_{u_n}^{u_n+h_n} dy_n \\ f_{X_1,X_2,\cdots,X_n}(x_1, x_2, \cdots, x_n)|J(x_1, x_2, \cdots, x_n)|^{-1}. \end{aligned} \tag{10.8}$$

Replacing the probability on the right-hand side of (10.6) by the integral in (10.8) and then taking the limits indicated in (10.6), we finally obtain (10.5).

▶ **Example 10A.** Let X_1 and X_2 be jointly continuous random variables. Let $U_1 = X_1 + X_2$, $U_2 = X_1 - X_2$. For any real numbers u_1 and u_2 show that

$$f_{U_1,U_2}(u_1, u_2) = \frac{1}{2} f_{X_1,X_2}\left(\frac{u_1 + u_2}{2}, \frac{u_1 - u_2}{2}\right). \tag{10.9}$$

Solution: Let $g_1(x_1, x_2) = x_1 + x_2$ and $g_2(x_1, x_2) = x_1 - x_2$. The equations $u_1 = x_1 + x_2$ and $u_2 = x_1 - x_2$ clearly have as their solution $x_1 = (u_1 + u_2)/2$ and $x_2 = (u_1 - u_2)/2$. The Jacobian J is given by

$$J = \begin{vmatrix} \frac{\partial g_1}{\partial x_1} & \frac{\partial g_1}{\partial x_2} \\ \frac{\partial g_2}{\partial x_1} & \frac{\partial g_2}{\partial x_2} \end{vmatrix} = \begin{vmatrix} 1 & 1 \\ 1 & -1 \end{vmatrix} = -2.$$

In view of these facts, (10.9) is an immediate consequence of (10.5). ◀

In exactly the same way one may establish the following result:

▶ **Example 10B.** Let X_1 and X_2 be jointly continuous random variables. Let

$$R = \sqrt{X_1^2 + X_2^2}, \qquad \theta = \tan^{-1}(X_2/X_1). \tag{10.10}$$

Then for any real numbers r and α, such that $r \geq 0$ and $0 \leq \alpha \leq 2\pi$,

$$f_{R,\theta}(r, \alpha) = r f_{X_1,X_2}(r \cos \alpha, r \sin \alpha). \tag{10.11}$$

It should be noted that we immediately obtain from (10.11) the formula for $f_R(r)$ given by (9.13), since

$$f_R(r) = \int_0^{2\pi} d\alpha f_{R,\theta}(r, \alpha). \tag{10.12}$$ ◀

▶ **Example 10C. Rotation of axes.** Let X_1 and X_2 be jointly distributed random variables. Let

$$Y_1 = X_1 \cos \alpha + X_2 \sin \alpha, \qquad Y_2 = -X_1 \sin \alpha + X_2 \cos \alpha \tag{10.13}$$

for some angle α in the interval $0 \leq \alpha \leq 2\pi$. Then

$$f_{Y_1,Y_2}(y_1, y_2) = f_{X_1,X_2}(y_1 \cos \alpha - y_2 \sin \alpha, y_1 \sin \alpha + y_2 \cos \alpha). \tag{10.14}$$

To illustrate the use of (10.14), consider two jointly normally distributed random variables with a joint probability density function given by (9.31), with $m_1 = m_2 = 0$. Then

$$f_{Y_1,Y_2}(y_1, y_2) = \frac{1}{2\pi\sigma_1\sigma_2\sqrt{1-\rho^2}} \times \exp\left[\frac{-1}{2(1-\rho^2)}(Ay_1^2 - 2By_1y_2 + Cy_2^2)\right], \tag{10.15}$$

where

$$A = \frac{\cos^2\alpha}{\sigma_1^2} - 2\rho\frac{\cos\alpha\sin\alpha}{\sigma_1\sigma_2} + \frac{\sin^2\alpha}{\sigma_2^2}$$

$$B = \frac{\cos\alpha\sin\alpha}{\sigma_1^2} - \rho\frac{\sin^2\alpha - \cos^2\alpha}{\sigma_1\sigma_2} - \frac{\cos\alpha\sin\alpha}{\sigma_2^2} \tag{10.16}$$

$$C = \frac{\sin^2\alpha}{\sigma_1^2} + 2\rho\frac{\cos\alpha\sin\alpha}{\sigma_1\sigma_2} + \frac{\cos^2\alpha}{\sigma_2^2}.$$

From (10.15) one sees that two random variables Y_1 and Y_2, obtained by a rotation of axes from jointly normally distributed random variables, X_1 and X_2, are jointly normally distributed. Further, if the angle of rotation α is chosen so that

$$\tan 2\alpha = \frac{2\rho\sigma_1\sigma_2}{\sigma_1^2 - \sigma_2^2}, \tag{10.17}$$

then $B = 0$, and Y_1 and Y_2 are independent normally distributed. Thus *by a suitable rotation of axes, two jointly normally distributed random variables may be transformed into two independent normally distributed random variables.* ◀

THEORETICAL EXERCISES

10.1. Let X_1 and X_2 be independent random variables, each exponentially distributed with parameter λ. Show that the random variables $X_1 + X_2$ and X_1/X_2 are independent.

10.2. Let X_1 and X_2 be independent random variables, each normally distributed with parameters $m = 0$ and $\sigma > 0$. Show that $X_1^2 + X_2^2$ and X_1/X_2 are independent.

10.3. Let X_1 and X_2 be independent random variables, χ^2 distributed with n_1 and n_2 degrees of freedom, respectively. Show that $X_1 + X_2$ and X_1/X_2 are independent.

10.4. Let X_1, X_2, and X_3 be independent identically normally distributed random variables. Let $\bar{X} = (X_1 + X_2 + X_3)/3$ and $S = (X_1 - \bar{X})^2 + (X_2 - \bar{X})^2 + (X_3 - \bar{X})^2$. Show that $\bar{X}$ and S are independent.

10.5. Generation of a random sample of a normally distributed random variable. Let U_1, U_2 be independent random variables, each uniformly distributed on the interval 0 to 1. Show that the random variables

$$X_1 = (-2 \log_e U_1)^{1/2} \cos 2\pi U_2$$
$$X_2 = (-2 \log_e U_1)^{1/2} \sin 2\pi U_2$$

are independent random variables, each normally distributed with mean 0 and variance 1. (For a discussion of this result, see G. E. P. Box and Mervin E. Muller, "A note on the generation of random normal deviates," *Annals of Mathematics Statistics*, Vol. 29 (1958), pp. 610–611.)

EXERCISES

10.1. Let X_1 and X_2 be independent random variables, each exponentially distributed with parameter $\lambda = \frac{1}{2}$. Find the joint probability density function of Y_1 and Y_2, in which (i) $Y_1 = X_1 + X_2$, $Y_2 = X_1 - X_2$, (ii) $Y_1 = \text{maximum}(X_1, X_2)$, $Y_2 = \text{minimum}(X_1, X_2)$.

10.2. Let X_1 and X_2 have joint probability density function given by

$$f_{X_1, X_2}(x_1, x_2) = \frac{1}{\pi} \qquad \text{if } x_1^2 + x_2^2 \le 1$$
$$= 0 \qquad \text{otherwise.}$$

Find the joint probability density function of (R, θ), in which $R = \sqrt{X_1^2 + X_2^2}$ and $\theta = \tan^{-1} X_2/X_1$. Show that, and explain why, R^2 is uniformly distributed but R is not.

10.3. Let X and Y be independent random variables, each uniformly distributed over the interval 0 to 1. Find the individual and joint probability density functions of the random variables R and θ, in which $R = \sqrt{X^2 + Y^2}$ and $\theta = \tan^{-1} Y/X$.

10.4. Two voltages $X(t)$ and $Y(t)$ are independently and normally distributed with parameters $m = 0$ and $\sigma = 1$. These are combined to give two new voltages, $U(t) = X(t) + Y(t)$ and $V(t) = X(t) - Y(t)$. Find the joint probability density function of $U(t)$ and $V(t)$. Are $U(t)$ and $V(t)$ independent? Find $P[U(t) > 0, V(t) < 0]$.

11. CONDITIONAL PROBABILITY OF AN EVENT GIVEN A RANDOM VARIABLE. CONDITIONAL DISTRIBUTIONS

In this section we introduce a notion that is basic to the theory of random processes, the notion of *the conditional probability of a random event A, given a random variable X*. This notion forms the basis of the mathematical treatment of jointly distributed random variables that are not independent and, consequently, are dependent.

Given two events, A and B, on the same probability space, the conditional probability $P[A \mid B]$ of the event A, given the event B, has been defined:

$$(11.1) \qquad P[A \mid B] = \frac{P[AB]}{P[B]} \qquad \text{if } P[B] > 0$$

$$= \text{undefined if } P[B] = 0.$$

Now suppose we are given an event A and a random variable X, both defined on the same probability space. We wish to define, for any real number x, the conditional probability of the event A, given the event that the observed value of X is equal to x, denoted in symbols by $P[A \mid X = x]$. Now if $P[X = x] > 0$, we may define this conditional probability by (11.1). However, for any random variable X, $P[X = x] = 0$ for all (except, at most, a countable number of) values of x. Consequently, the conditional probability $P[A \mid X = x]$ of the event A, given that $X = x$, must be regarded as being undefined insofar as (11.1) is concerned.

The meaning that one intuitively assigns to $P[A \mid X = x]$ is that it represents the probability that A has occurred, knowing that X was observed as equal to x. Therefore, it seems natural to define

$$(11.2) \qquad P[A \mid X = x] = \lim_{h \to 0} P[A \mid x - h < X < x + h]$$

if the conditioning events $[x - h < X < x + h]$ have positive probability for every $h > 0$. However, we have to be very careful how we define the limit in (11.2). As stated, (11.2) is essentially false, in the sense that the limit does not exist in general. However, we can define a limiting operation, similar to (11.2) in spirit, although different in detail, that in advanced probability theory is shown always to exist.

Given a real number x, define $H_n(x)$ as that interval, of length $1/2^n$, starting at a multiple of $1/2^n$, that contains x; in symbols,

$$(11.3) \qquad H_n(x) = \left\{x': \quad \frac{[x \cdot 2^n]}{2^n} \leq x' < \frac{[x \cdot 2^n] + 1}{2^n}\right\}.$$

Then we define the conditional probability of the event A, given that the random variable X has an observed value equal to x, by

$$(11.4) \qquad P[A \mid X = x] = \lim_{n \to \infty} P[A \mid X \text{ is in } H_n(x)].$$

It may be proved that the conditional probability $P[A \mid X = x]$, defined by (11.4), has the following properties.

First, the convergence set C of points x on the real line at which the limit in (11.4) exists has probability one, according to the probability function of the random variable X; that is, $P_X[C] = 1$. For practical purposes this suffices, since we expect that all observed values of X lie in the set C, and we wish to define $P[A \mid X = x]$ only at points x that could actually arise as observed values of X.

Second, from a knowledge of $P[A \mid X = x]$ one may obtain $P[A]$ by the following formulas:

$$P[A] = \begin{cases} \displaystyle\int_{-\infty}^{\infty} P[A \mid X = x]\, dF_X(x) \\ \displaystyle\int_{-\infty}^{\infty} P[A \mid X = x] f_X(x)\, dx \\ \displaystyle\sum_{\substack{\text{over all } x \text{ such} \\ \text{that } p_X(x) > 0}} P[A \mid X = x] p_X(x), \end{cases} \tag{11.5}$$

in which the last two equations hold if X is respectively continuous or discrete. More generally, for every Borel set B of real numbers, the probability of the intersection of the event A and the event $\{X \text{ is in } B\}$ that the observed value of X is in B is given by

$$P[A\{X \text{ is in } B\}] = \int_B P[A \mid X = x]\, dF_X(x). \tag{11.6}$$

Indeed, in advanced studies of probability theory the conditional probability $P[A \mid X = x]$ is defined not constructively by (11.4) but descriptively, as the unique (almost everywhere) function of x satisfying (11.6) for every Borel set B of real numbers. This characterization of $P[A \mid X = x]$ is used to prove (11.15).

▶ **Example 11A.** A young man and a young lady plan to meet between 5:00 and 6:00 P.M., each agreeing not to wait more than ten minutes for the other. Assume that they arrive independently at random times between 5:00 and 6:00 P.M. Find the conditional probability that the young man and the young lady will meet, given that the young man arrives at 5:30 P.M.

Solution: Let X be the man's arrival time (in minutes after 5:00 P.M.) and let Y be the lady's arrival time (in minutes after 5:00 P.M.). If the man arrives at a time x, there will be a meeting if and only if the lady's arrival time Y satisfies $|Y - x| \leq 10$ or $-10 + x \leq Y \leq x + 10$. Let A denote the event that the man and lady meet. Then, for any x between 0 and 60

$$\begin{aligned} P[A \mid X = x] &= P[-10 \leq Y - X \leq 10 \mid X = x] \\ &= P[-10 + x \leq Y \leq x + 10 \mid X = x] \\ &= P[-10 + x \leq Y \leq x + 10], \end{aligned} \tag{11.7}$$

in which we have used (11.9) and (11.11). Next, using the fact that Y is uniformly distributed between 0 and 60, we obtain (as graphed in Fig. 11A)

$$P[A \mid X = x] = \frac{10 + x}{60} \quad \text{if } 0 \leq x \leq 10 \tag{11.8}$$

$$= \tfrac{1}{3} \quad \text{if } 10 \leq x \leq 50$$

$$= \frac{70 - x}{60} \quad \text{if } 50 \leq x \leq 60$$

$$= \text{undefined if } x < 0 \text{ or } x > 60.$$

Consequently, $P[A \mid X = 30] = \frac{1}{3}$, so that the conditional probability that

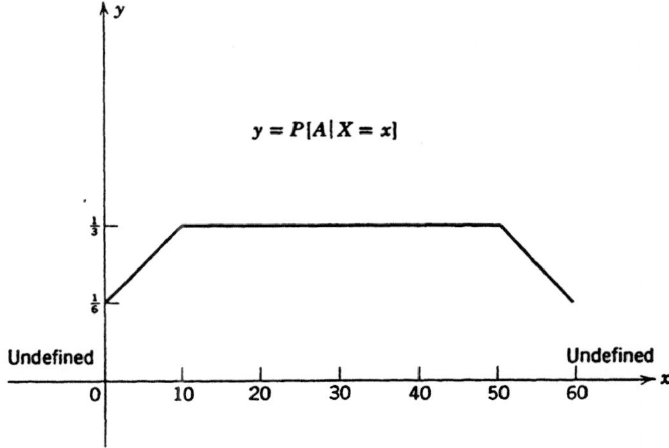

Fig. 11A. The conditional probability $P[A \mid X = x]$, graphed as a function of x.

the young man and the young lady will meet, given that the young man arrives at 5:30 P.M., is $\frac{1}{3}$. Further, by applying (11.5), we determine that $P[A] = \frac{11}{36}$. ◀

In (11.7) we performed certain manipulations that arise frequently when one is dealing with conditional probabilities. We now justify these manipulations.

Consider two jointly distributed random variables X and Y. Let $g(x, y)$ be a Borel function of two variables. Let z be a fixed real number. Let $A = [g(X, Y) \leq z]$ be the event that the random variable $g(X, Y)$ has an observed value less than or equal to z. Next, let x be a fixed real number, and let $A(x) = [g(x, Y) \leq z]$ be the event that the random variable $g(x, Y)$, which is a function only of Y, has an observed value less than or equal to z. It appears formally reasonable that

$$(11.9) \qquad P[g(X, Y) \le z \mid X = x] = P[g(x, Y) \le z \mid X = x].$$

In words, a statement involving the random variable X, conditioned by the hypothesis that the value of X is a given number x, has the same conditional probability given $X = x$, as the corresponding statement obtained by replacing the random variable X by its observed value. The proof of (11.9) is omitted, since it is beyond the scope of this book.

It may help to comprehend (11.9) if we state it in terms of the events $A = [g(X, Y) \le z]$ and $A(x) = [g(x, Y) \le z]$. Equation (11.9) asserts that the functions of u,

$$(11.10) \qquad P[A|X = u] \quad \text{and} \quad P[A(x)|X = u]$$

have the same value at $u = x$.

Another important formula is the following. If the random variables X and Y are independent, then

$$(11.11) \qquad P[g(x, Y) \le z \mid X = x] = P[g(x, Y) \le z],$$

since it holds that

$$(11.12) \quad P[A \mid X = x] = P[A] \qquad \text{if the event } A \text{ is independent of } X.$$

We thus obtain the basic fact that *if the random variables X and Y are independent*

$$(11.13) \quad P[g(X, Y) \le z | X = x] = P[g(x, Y) \le z \mid X = x] = P[g(x, Y) \le z].$$

We next define the notion of the *conditional distribution function* of one random variable Y given another random variable X, denoted $F_{Y|X}(.|.)$. For any real numbers x and y, it is defined by

$$(11.14) \qquad F_{Y|X}(y \mid x) = P[Y \le y \mid X = x].$$

The conditional distribution function $F_{Y|X}(.|.)$ has the basic property that for any real numbers x and y the joint distribution function $F_{X,Y}(x, y)$ may be expressed in terms of $F_{Y|X}(y \mid x)$ by

$$(11.15) \qquad F_{X,Y}(x, y) = \int_{-\infty}^{x} F_{Y|X}(y \mid x')\, dF_X(x').$$

To prove (11.15), let X and Y be two jointly distributed random variables. For two given real numbers x and y define $A = [Y \le y]$. Then (11.15) may be written

$$(11.16) \qquad P[X \le x, Y \le y] = \int_{-\infty}^{x} P[A \mid X = x']\, dF_X(x').$$

If in (11.6) $B = \{x'\colon \ x' \le x\}$, (11.16) is obtained.

Now suppose that the random variables X and Y are jointly continuous. We may then define the *conditional probability density function* of the random variable Y, given the random variable X, denoted by $f_{Y|X}(y \mid x)$. It is defined for any real numbers x and y by

$$f_{Y|X}(y \mid x) = \frac{\partial}{\partial y} F_{Y|X}(y \mid x). \tag{11.17}$$

We now prove the basic formula: *if* $f_X(x) > 0$, *then*

$$f_{Y|X}(y \mid x) = \frac{f_{X,Y}(x, y)}{f_X(x)}. \tag{11.18}$$

To prove (11.18), we differentiate (11.15) with respect to x (first replacing $dF_X(x')$ by $f_X(x')\,dx'$). Then

$$\frac{\partial}{\partial x} F_{X,Y}(x, y) = F_{Y|X}(y \mid x) f_X(x). \tag{11.19}$$

Now differentiating (11.19) with respect to y, we obtain

$$f_{X,Y}(x, y) = f_{Y|X}(y \mid x) f_X(x) \tag{11.20}$$

from which (11.18) follows immediately.

▶ **Example 11B.** Let X_1 and X_2 be jointly normally distributed random variables whose probability density function is given by (9.31). Then the conditional probability density of X_1, given X_2, is equal to

$$f_{X_1|X_2}(x \mid y) = \frac{1}{\sqrt{2\pi}\sigma_1\sqrt{1-\rho^2}} \times \exp\left\{-\frac{1}{2(1-\rho^2)\sigma_1^2}\left[x - m_1 - \rho\frac{\sigma_1}{\sigma_2}(y - m_2)\right]^2\right\} \tag{11.21}$$

In words, the conditional probability law of the random variable X_1, given X_2, is the normal probability law with parameters $m = m_1 + \rho(\sigma_1/\sigma_2)(x_2 - m_2)$ and $\sigma = \sigma_1\sqrt{1-\rho^2}$. To prove (11.21), one need only verify that it is equal to the quotient $f_{X_1,X_2}(x, y)/f_{X_2}(y)$. Similarly, one may establish the following result. ◀

▶ **Example 11C.** Let X and Y be jointly distributed random variables. Let

$$R = \sqrt{X^2 + Y^2}, \qquad \theta = \tan^{-1}(Y/X). \tag{11.22}$$

Then, for $r > 0$

$$f_{\theta|R}(\theta \mid r) = \frac{f_{X,Y}(r\cos\theta, r\sin\theta)}{\int_0^{2\pi} d\theta\, f_{X,Y}(r\cos\theta, r\sin\theta)}. \tag{11.23}$$

◀

In the foregoing examples we have considered the problem of obtaining $f_{X|Y}(x \mid y)$, knowing $f_{X,Y}(x, y)$. We next consider the converse problem of obtaining the individual probability law of X from a knowledge of the conditional probability law of X, given Y, and of the individual probability law of Y.

▶ **Example 11D.** Consider the decay of particles in a cloud chamber (or, similarly the breakdown of equipment or the occurrence of accidents). Assume that the time X of any particular particle to decay is a random variable obeying an exponential probability law with parameter y. However, it is not assumed that the value of y is the same for all particles. Rather, it is assumed that there are particles of different types (or equipment of different types or individuals of different accident proneness). More specifically, it is assumed that for a particle randomly selected from the cloud chamber the parameter y is a particular value of a random variable Y obeying a gamma probability law with a probability density function,

$$f_Y(y) = \frac{\beta^\alpha}{\Gamma(\alpha)} y^{\alpha-1}e^{-\beta y}, \qquad \text{for } y > 0, \tag{11.24}$$

in which the parameters α and β are positive constants characterizing the experimental conditions under which the particles are observed.

The assumption that the time X of a particle to decay obeys an exponential law is now expressed as an assumption on the conditional probability law of X given Y:

$$f_{X|Y}(x \mid y) = ye^{-xy} \qquad \text{for } x > 0. \tag{11.25}$$

We find the individual probability law of the time X (of a particle selected at random to decay) as follows; for $x > 0$

$$\begin{aligned} f_X(x) &= \int_{-\infty}^{\infty} f_{X,Y}(x, y)\, dy = \int_{-\infty}^{\infty} f_{X|Y}(x \mid y) f_Y(y)\, dy \\ &= \int_0^\infty ye^{-xy} \frac{\beta^\alpha}{\Gamma(\alpha)} y^{\alpha-1}e^{-\beta y}\, dy \\ &= \frac{\alpha\beta^\alpha}{(\beta + x)^{\alpha+1}}. \end{aligned} \tag{11.26}$$

The reader interested in further study of the foregoing model, as well as a number of other interesting topics, should consult J. Neyman, "The Problem of Inductive Inference," *Communications on Pure and Applied Mathematics*, Vol. 8 (1955), pp. 13–46. ◀

The foregoing notions may be extended to several random variables. In particular, let us consider n random variables $X_1, X_2, \ldots, X_n$ and a random variable U, all of which are jointly distributed. By suitably adapting the foregoing considerations, we may define a function

$$(11.27) \qquad F_{X_1, X_2, \cdots, X_n|U}(x_1, x_2, \cdots, x_n \mid u),$$

called the conditional distribution function of the random variables $X_1, X_2, \ldots, X_n$, given the random variable U, which may be shown to satisfy, for all real numbers $x_1, x_2, \ldots, x_n$ and u,

$$(11.28) \quad F_{X_1, \cdots, X_n, U}(x_1, \cdots, x_n, u) = \int_{-\infty}^{u} F_{X_1, \cdots, X_n|U}(x_1, \cdots, x_n \mid u')\, dF_U(u').$$

THEORETICAL EXERCISES

11.1. Let T be a random variable, and let t be a fixed number. Define the random variable U by $U = T - t$ and the event A by $A = [T > t]$. Evaluate $P[A \mid U = x]$ and $P[U > x \mid A]$ in terms of the distribution function of T. Explain the difference in meaning between these concepts.

11.2. If X and Y are independent Poisson random variables, show that the conditional distribution of X, given $X + Y$, is binomial.

11.3. Given jointly distributed random variables, X_1 and X_2, prove that, for any x_2 and almost all x_1, $F_{X_2|X_1}(x_2 \mid x_1) = F_{X_2}(x_2)$ if and only if X_1 and X_2 are independent.

11.4. Prove that for any jointly distributed random variables X_1 and X_2

$$\int_{-\infty}^{\infty} f_{X_1|X_2}(x_1 \mid x_2)\, dx_1 = 1, \qquad \int_{-\infty}^{\infty} f_{X_2|X_1}(x_2 \mid x_1)\, dx_2 = 1.$$

For contrast evaluate

$$\int_{-\infty}^{\infty} f_{X_1|X_2}(x_1 \mid x_2)\, dx_2, \qquad \int_{-\infty}^{\infty} f_{X_2|X_1}(x_2 \mid x_1)\, dx_1.$$

EXERCISES

In exercises 11.1 to 11.3 let X and Y be independent random variables. Let $Z = Y - X$. Let $A = [|Y - X| \leq 1]$. Find (i) $P[A \mid X = 1]$, (ii) $F_{Z|X}(0 \mid 1)$, (iii) $f_{Z|X}(0 \mid 1)$, (iv) $P[Z \leq 0 \mid A]$.

11.1. If X and Y are each uniformly distributed over the interval 0 to 2.

11.2. If X and Y are each normally distributed with parameters $m = 0$ and $\sigma = 2$.

11.3. If X and Y are each exponentially distributed with parameter $\lambda = 1$.

In exercises 11.4 to 11.6 let X and Y be independent random variables. Let $U = X + Y$ and $V = Y - X$. Let $A = [|V| \leq 1]$. Find (i) $P[A \mid U = 1]$, (ii) $F_{V|U}(0 \mid 1)$, (iii) $f_{V|U}(0 \mid 1)$, (iv) $P[U \geq 0 \mid A]$, (v) $f_{V|U}(v|u)$.

11.4. If X and Y are each uniformly distributed over the interval 0 to 2.

11.5. If X and Y are each normally distributed with parameters $m = 0$ and $\sigma = 2$.

11.6. If X and Y are each exponentially distributed with parameter $\lambda = 1$.

11.7. Let X_1 and X_2 be jointly normally distributed random variables (representing the observed amplitudes of a noise voltage recorded a known time interval apart). Assume that their joint probability density function is given by (9.31) with (i) $m_1 = m_2 = 0$, $\sigma_1 = \sigma_2 = 1$, $\rho = 0.5$, (ii) $m_1 = 1$, $m_2 = 2$, $\sigma_1 = 1$, $\sigma_2 = 4$, $\rho = 0.5$. Find $P[X_2 > 1 \mid X_1 = 1]$.

11.8. Let X_1 and X_2 be jointly normally distributed random variables, representing the daily sales (in thousands of units) of a certain product in a certain store on two successive days. Assume that the joint probability density function of X_1 and X_2 is given by (9.31), with $m_1 = m_2 = 3$, $\sigma_1 = \sigma_2 = 1$, $\rho = 0.8$. Find K so that (i) $P[X_2 > K] = 0.05$, (ii) $P[X_2 > K \mid X_1 = 2] = 0.05$, (iii) $P[X_2 > K \mid X_1 = 1] = 0.05$. Suppose the store desires to have on hand on a given day enough units of the product so that with probability 0.95 it can supply all demands for the product on the day. How large should its inventory be on a given morning if (iv) yesterday's sales were 2000 units, (v) yesterday's sales are not known.

CHAPTER 8

Expectation of a Random Variable

In dealing with random variables, it is as important to know their means and variances as it is to know their probability laws. In this chapter we define the notion of the expectation of a random variable and describe the significant role that this notion plays in probability theory.

1. EXPECTATION, MEAN, AND VARIANCE OF A RANDOM VARIABLE

Given the random variable X, we define the expectation of the random variable, denoted by $E[X]$, as the mean of the probability law of X; in symbols,

$$E[X] = \begin{cases} \int_{-\infty}^{\infty} x \, dF_X(x) \\ \int_{-\infty}^{\infty} x f_X(x) \, dx \\ \sum\limits_{\substack{\text{over all } x \text{ such} \\ \text{that } p_X(x) > 0}} x \, p_X(x), \end{cases} \tag{1.1}$$

depending on whether X is specified by its distribution function $F_X(\cdot)$, its probability density function $f_X(\cdot)$, or its probability mass function $p_X(\cdot)$.

Given a random variable Y, which arises as a Borel function of a random variable X so that

$$Y = g(X) \tag{1.2}$$

for some Borel function $g(\cdot)$, the expectation $E[g(X)]$, in view of (1.1), is given by

$$E[g(X)] = \int_{-\infty}^{\infty} y \, dF_{g(X)}(y). \tag{1.3}$$

On the other hand, given the Borel function $g(\cdot)$ and the random variable X, we can form the expectation of $g(x)$ with respect to the probability law of X, denoted by $E_X[g(x)]$ and defined by

$$E_X[g(x)] = \begin{cases} \displaystyle\int_{-\infty}^{\infty} g(x) \, dF_X(x) \\ \displaystyle\int_{-\infty}^{\infty} g(x) f_X(x) \, dx \\ \displaystyle\sum_{\substack{\text{over all } x \text{ such} \\ \text{that } p_X(x) > 0}} g(x) p_X(x), \end{cases} \tag{1.4}$$

depending on whether X is specified by its distribution function $F_X(\cdot)$, its probability density function $f_X(\cdot)$, or its probability mass function $p_X(\cdot)$.

It is a striking fact, of great importance in probability theory, *that for any random variable X and Borel function $g(\cdot)$*

$$E[g(X)] = E_X[g(x)] \tag{1.5}$$

if either of these expectations exists. In words, (1.5) says that the expectation of the random variable $g(X)$ is equal to the expectation of the function $g(\cdot)$ with respect to the random variable X.

The validity of (1.5) is a direct consequence of the fact that the integrals used to define expectations are required to be absolutely convergent.* Some idea of the proof of (1.5), in the case that $g(\cdot)$ is continuous, can be gained. Partition the y-axis in Fig. 1A into subintervals by points $y_0 < y_1 < \ldots < y_n$. Then approximately

$$\begin{aligned} E_{g(X)}[y] &= \int_{-\infty}^{\infty} y \, dF_{g(X)}(y) \\ &\doteq \sum_{j=1}^{n} y_j [F_{g(X)}(y_j) - F_{g(X)}(y_{j-1})] \\ &\doteq \sum_{j=1}^{n} y_j P_X[\{x\colon \ y_{j-1} < g(x) \le y_j\}]. \end{aligned} \tag{1.6}$$

* At the end of the section we give an example that shows that (1.5) does not hold if the integrals used to define expectations are not required to converge absolutely.

To each point y_j on the y-axis, there is a number of points $x_j^{(1)}, x_j^{(2)}, \ldots,$ at which $g(x)$ is equal to y. Form the set of all such points on the x-axis that correspond to the points $y_1, \ldots, y_n$. Arrange these points in increasing order, $x_0 < x_1 < \ldots < x_m$. These points divide the x-axis into subintervals. Further, it is clear upon reflection that the last sum in (1.6) is equal to

$$\sum_{k=1}^{m} g(x_k)P_X[\{x: \quad x_{k-1} < x \le x_k\}] \doteq E_X[g(x)], \tag{1.7}$$

which completes our intuitive proof of (1.5). A rigorous proof of (1.5) cannot be attempted here, since a more careful treatment of the integration process does not lie within the scope of this book.

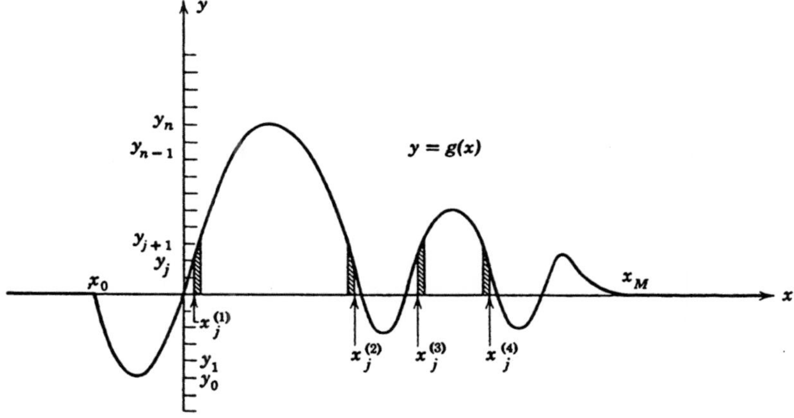

Fig. 1A. With the aid of this graph of a possible function $g(\cdot)$, one can see that (1.5) holds.

Given a random variable X and a function $g(\cdot)$, we thus find two distinct notions, represented by $E[g(X)]$ and $E_X[g(x)]$, which nevertheless, are always numerically equal. It has become customary always to use the notation $E[g(X)]$, since this notation is the most convenient for technical manipulation. However, the reader should be aware that although we write $E[g(X)]$ the concept in which we are really very often interested is $E_X[g(x)]$, the expectation of the function $g(x)$ with respect to the random variable X. Thus, for example, the nth moment of a random variable X (for any integer n) is often defined as $E[X^n]$, the expectation of the nth power of X. From the point of view of the intuitive meaning of the nth moment, however, it should be defined as the expectation $E_X[x^n]$ of the function $g(x) = x^n$ with respect to the probability law of the random variable X. We shall define the *moments of a random variable* in terms of the notation of the expectation of a random variable. However, it should be borne in mind that we could define as well the moments of a random variable as the corresponding moments of the probability law of the random variable.

Given a random variable X, we denote its *mean* by $E[X]$, its *mean square* by $E[X^2]$, its *square mean* by $E^2[X]$, its *n*th *moment about the point c* by $E[(X - c)^n]$, and its *n*th *central moment* (that is, *n*th moment about its mean) by $E[(X - E[X])^n]$. In particular, the *variance* of a random variable, denoted by Var $[X]$, is defined as its second central moment, so that

$$\text{Var}[X] = E[(X - E[X])^2] = E[X^2] - E^2[X]. \tag{1.8}$$

The *standard deviation* of a random variable, denoted by $\sigma[X]$, is defined as the positive square root of its variance, so that

$$\sigma[X] = \sqrt{\text{Var}[X]}, \qquad \sigma^2[X] = \text{Var}[X]. \tag{1.9}$$

The *moment generating function* of a random variable, denoted by $\psi_X(\cdot)$, is defined for every real number t by

$$\psi_X(t) = E[e^{tX}]. \tag{1.10}$$

It is shown in section 5 that if $X_1, X_2, \ldots, X_n$ constitute a random sample of the random variable X then the arithmetic mean $(X_1 + X_2 + \ldots + X_n)/n$ is, for large n, approximately equal to the mean $E[X]$. This fact has led early writers on probability theory to call $E[X]$ the *expected value* of the random variable X; this terminology, however, is somewhat misleading, for if $E[X]$ is the expected value of any random variable it is the expected value of the arithmetic mean of a random sample of the random variable.

▶ **Example 1A. The mean duration of the game of "odd man out."** The game of "odd man out" was described in example 3D of Chapter 3. On each independent play of the game, N players independently toss fair coins. The game concludes when there is an odd man; that is, the game concludes the first time that exactly one of the coins falls heads or exactly one of the coins falls tails. Let X be the number of plays required to conclude the game; more briefly, X is called the duration of the game. Find the mean and standard deviation of X.

Solution: It has been shown that the random variable X obeys a geometric probability law with parameter $p = N/2^{N-1}$. The mean of X is then equal to the mean of the geometric probability law, so that $E[X] = 1/p$. Similarly, $\sigma^2[X] = q/p^2$. Thus, if $N = 5$, $E[X] = 2^4/5 = 3.2$, $\sigma^2[X] = (11/16)/(5/16)^2 = (11)(16)/25$, and $\sigma[X] = 4\sqrt{11}/5 = 2.65$. The mean duration $E[X]$ has the following interpretation; if $X_1, X_2, \ldots, X_n$ are the durations of n independent games of "odd man out," then the average duration $(X_1 + X_2 + \ldots + X_n)/n$ of the n games is approximately equal to $E[X]$ if the number n of games is large. Note that in a game with five players the mean duration $E[X](= 3.2)$ is not equal to an integer. Consequently, one will never observe a game whose duration is equal to the mean duration; nevertheless, the arithmetic mean of a large number of observed durations can be expected to be equal to the mean duration. ◀

To find the mean and variance of the random variable X, in the foregoing example we found the mean and variance of the probability law of X. If a random variable Y can be represented as a Borel function $Y = g_1(X)$ of a random variable X, one can find the mean and variance of Y without actually finding the probability law of Y. To do this, we make use of an extension of (1.5).

Let X and Y be random variables such that $Y = g_1(X)$ for some Borel function $g_1(\cdot)$. Then for any Borel function $g(\cdot)$

$$E[g(Y)] = E[g(g_1(X))] \tag{1.11}$$

in the sense that if either of these expectations exists then so does the other, and the two are equal.

To prove (1.11) we must prove that

$$\int_{-\infty}^{\infty} g(y)\, dF_{g_1(X)}(y) = \int_{-\infty}^{\infty} g(g_1(x))\, dF_X(x). \tag{1.12}$$

The proof of (1.12) is beyond the scope of this book.

To illustrate the meaning of (1.11), we write it for the case in which the random variable X is continuous and $g_1(x) = x^2$. Using the formula for the probability density function of $Y = X^2$, given by (8.8) of Chapter 7, we have for any continuous function $g(\cdot)$

$$\begin{aligned} E[g(Y)] &= \int_0^{\infty} g(y) f_Y(y)\, dy \\ &= \int_0^{\infty} g(y) \frac{1}{2\sqrt{y}} [f_X(\sqrt{y}) + f_X(-\sqrt{y})]\, dy, \end{aligned} \tag{1.13}$$

whereas

$$E[g(g_1(X))] = E[g(X^2)] = \int_{-\infty}^{\infty} g(x^2) f_X(x)\, dx. \tag{1.14}$$

One may verify directly that the integrals on the right-hand sides of (1.13) and (1.14) are equal, as asserted by (1.11).

As one immediate consequence of (1.11), we have the following formula for *the variance of a random variable* $g(X)$, *which arises as a function of another random variable:*

$$\text{Var}\,[g(X)] = E[g^2(X)] - E^2[g(X)]. \tag{1.15}$$

▶ **Example 1B. The square of a normal random variable.** Let X be a normally distributed random variable with mean 0 and variance σ^2. Let $Y = X^2$. Then the mean and variance of Y are given by $E[Y] = E[X^2] = \sigma^2$, $\text{Var}\,[Y] = E[X^4] - E^2[X^2] = 3\sigma^4 - \sigma^4 = 2\sigma^4$. ◀

If a random variable X is known to be normally distributed with mean m and variance σ^2, then for brevity one often writes X is $N(m, \sigma^2)$.

▶ **Example 1C. The logarithmic normal distribution.** A random variable X is said to have a logarithmic normal distribution if its logarithm $\log X$ is normally distributed. One may find the mean and variance of X by finding the mean and variance of $X = e^Y$, in which Y is $N(m, \sigma^2)$. Now $E[X] = E[e^Y]$ is the value at $t = 1$ of the moment-generating function $\psi_Y(t)$ of Y. Similarly $E[X^2] = E[e^{2Y}] = \psi_Y(2)$. Since $\psi_Y(t) = \exp(mt + \frac{1}{2}\sigma^2 t^2)$, it follows that $E[X] = \exp(m + \frac{1}{2}\sigma^2)$ and $\text{Var}\,[X] = E[X^2] - E^2[X] = \exp(2m + 2\sigma^2) - \exp(2m + \sigma^2)$. ◀

Example 1D shows how the mean (or the expectation) of a random variable is interpreted.

▶ **Example 1D. Disadvantageous or unfair bets.** Roulette is played by spinning a ball on a circular wheel, which has been divided into thirty-seven arcs of equal length, bearing numbers from 0 to 36.* Let X denote the number of the arc on which the ball comes to rest. Assume each arc is equally likely to occur, so that the probability mass function of X is given by $p_X(x) = 1/37$ for $x = 0, 1, \ldots, 36$. Suppose that one is given even odds on a bet that the observed value of X is an odd number; that is, on a 1 dollar bet one is paid 2 dollars (including one's stake) if X is odd, and one is paid nothing (so that one loses one's stake) if X is not odd. How much can one expect to win at roulette by consistently betting on an odd outcome?

Solution: Define a random variable Y as equal to the amount won by betting 1 dollar on an odd outcome at a play of the game of roulette. Then $Y = 1$ if X is odd and $Y = -1$ if X is not odd. Consequently,

* The roulette table described is the one traditionally in use in most European casinos. The roulette tables in many American casinos have wheels that are divided into 38 arcs, bearing numbers 00, 0, 1, . . . , 36.

$P[Y = 1] = \frac{18}{37}$ and $P[Y = -1] = \frac{19}{37}$. The mean $E[Y]$ of the random variable Y is then given by

$$(1.16) \qquad E[Y] = 1 \cdot p_Y(1) + (-1) \cdot p_Y(-1) = -\tfrac{1}{37} = -0.027.$$

The amount one can expect to win at roulette by betting on an odd outcome may be regarded as equal to the mean $E[Y]$ in the following sense. Let $Y_1, Y_2, \ldots, Y_n, \ldots$ be one's winnings in a succession of plays of roulette at which one has bet on an odd outcome. It is shown in section 5 that the average winnings $(Y_1 + Y_2 + \ldots + Y_n)/n$ in n plays tends, as the number of plays becomes infinite, to $E[Y]$. The fact that $E[Y]$ is equal to a negative number implies that betting on an odd outcome at roulette is *disadvantageous* (or unfair) for the bettor, since after a long series of plays he can expect to have lost money at a rate of 2.7 cents per dollar bet. Many games of chance are disadvantageous for the bettor in the sense that the mean winnings is negative. However, the mean (or expected) winnings describe just one aspect of what will occur in a long series of plays. For a gambler who is interested only in a modest increase in his fortune it is more important to know the probability that as a result of a series of bets on an odd outcome in roulette the size of his 1000-dollar fortune will increase to 1200 dollars before it decreases to zero. A home owner insures his home against destruction by fire, even though he is making a disadvantageous bet (in the sense that his expected money winnings are negative) because he is more concerned with making equal to zero the probability of a large loss. ◀

Most random variables encountered in applications of probability theory have finite means and variances. However, random variables without finite means have long been encountered by physicists in connection with problems of return to equilibrium. The following example illustrates a random variable of this type that has infinite mean.

▶ **Example 1E. On long leads in fair games.** Consider two players engaged in a friendly game of matching pennies with fair coins. The game is played as follows. One player tosses a coin, while the other player guesses the outcome, winning one cent if he guesses correctly and losing one cent if he guesses incorrectly. The two friends agree to stop playing the moment neither is winning. Let N be the duration of the game; that is, N is equal to the number of times coins are tossed before the players are even. Find $E[N]$, the mean duration of the game.

Solution: It is clear that the game of matching pennies with fair coins is not disadvantageous to either player in the sense that if Y is the winnings of a given player on any play of the game then $E[Y] = 0$. From this fact one may be led to the conclusion that the total winnings S_n of a given player in n plays will be equal to 0 in half the plays, over a very large number of plays. However, no such inference can be made. Indeed, consider the random variable N, which represents the first trial N at which $S_N = 0$. We now show that $E[N] = \infty$; in words, the mean duration of the game of matching pennies is infinite. Note that this does not imply that the duration N is infinite; it may be shown that there is probability one that in a finite number of plays the fortunes of the two players will equalize. To compute $E[N]$, we must compute its probability law. The duration N of the game cannot be equal to an odd integer, since the fortunes will equalize if and only if each player has won on exactly half the tosses. We omit the computation of the probability that $N = n$, for n an even integer, and quote here the result (see W. Feller, *An Introduction to Probability Theory and its Applications*, second edition, Wiley, New York, 1957, p. 75):

$$P[N = 2m] = \frac{1}{2m}\binom{2m-2}{m-1}2^{-2(m-1)}. \tag{1.17}$$

The mean duration of the game is then given by

$$E[N] = \sum_{m=1}^{\infty} (2m)P[N = 2m]. \tag{1.18}$$

It may be shown, using Stirling's formula, that

$$\binom{2n}{n}2^{-2n} \sim \frac{1}{(n\pi)^{1/2}}, \tag{1.19}$$

the sign $\sim$ indicating that the ratio of the two sides in (1.19) tends to 1 as n tends to infinity. Consequently, $(2m)P[N = 2m] \geq K/\sqrt{m}$ for some constant K. Therefore, the infinite series in (1.18) diverges, and $E[N] = \infty$. ◀

To conclude this section, let us justify the fact that the integrals defining expectations are required to be absolutely convergent by showing, by example, that if the expectation of a continuous random variable X is defined by

$$E[X] = \lim_{a\to\infty} \int_{-a}^{a} xf_X(x)\,dx \tag{1.20}$$

then it is not necessarily true that for any constant c

$$E[X + c] = E[X] + c.$$

Let X be a random variable whose probability density function is an even function, that is, $f_X(-x) = f_X(x)$. Then, under the definition given by (1.20), the mean $E[X]$ exists and equals 0, since $\int_{-a}^{a} xf_X(x)\,dx = 0$ for every a. Now

$$E[X + c] = \lim_{a\to\infty} \int_{-a}^{a} yf_X(y - c)\,dy. \tag{1.21}$$

Assuming $c > 0$, and letting $u = y - c$, we may write

$$\begin{aligned}\int_{-a}^{a} yf_X(y - c)\,dy &= \int_{-a-c}^{a-c} (u + c)f_X(u)\,du \\ &= \int_{-a-c}^{a+c} uf_X(u)\,du - \int_{a-c}^{a+c} uf_X(u)\,du + c\int_{-a-c}^{a-c} f_X(u)\,du.\end{aligned}$$

The first of these integrals vanishes, and the last tends to 1 as a tends to ∞. Consequently, to prove that if $E[X]$ is defined by (1.20) one can find a random variable X and a constant c such that $E[X + c] \neq E[X] + c$, it suffices to prove that one can find an even probability density function $f(\cdot)$ and a constant $c > 0$ such that

$$\text{it is not so that } \lim_{a\to\infty} \int_{a-c}^{a+c} uf(u)\,du = 0. \tag{1.22}$$

An example of a continuous even probability density function satisfying (1.22) is the following. Letting $A = 3/\pi^2$, define

$$\begin{aligned} f(x) &= A\frac{1}{|k|}(1 - |k - x|) \qquad \text{if } k = \pm 1, \pm 2^2, \pm 3^2, \cdots \text{ is such that } |k - x| \le 1 \\ &= 0 \qquad \text{elsewhere.}\end{aligned} \tag{1.23}$$

In words, $f(x)$ vanishes, except for points x, which lie within a distance 1 from a point that in absolute value is a perfect square $1, 2^2, 3^2, 4^2, \ldots$. That $f(\cdot)$ is a probability density function follows from the fact that

$$\int_{-\infty}^{\infty} f(x)\,dx = 2A\sum_{k=1}^{\infty}\frac{1}{k^2} = 2A\frac{\pi^2}{6} = 1.$$

That (1.22) holds for $c > 1$ follows from the fact that for $k = 2^2, 3^2, \ldots$

$$\int_{k-1}^{k+1} uf(u)\,du \ge (k - 1)\int_{k-1}^{k+1} f(u)\,du = \frac{(k-1)}{k}A \ge \frac{A}{2}.$$

THEORETICAL EXERCISES

1.1. **The mean and variance of a linear function of a random variable.** Let X be a random variable with finite mean and variance. Let a and b be real numbers. Show that

$$\begin{aligned} E[aX + b] &= aE[X] + b, \qquad & \operatorname{Var}[aX + b] &= |a|^2 \operatorname{Var}[X], \\ \sigma[aX + b] &= |a|\sigma[X], \qquad & \varphi_{aX+b}(t) &= e^{ibt}\varphi_X(at).\end{aligned} \tag{1.24}$$

1.2. **Chebyshev's inequality for random variables.** Let X be a random variable with finite mean and variance. Show that for any $h > 0$ and any $\epsilon > 0$

$$(1.25)\quad \begin{aligned} &P[|X - E[X]| \le h\sigma[X]] \ge 1 - \frac{1}{h^2}, \qquad P[|X - E[X]| > h\sigma[X]] \le \frac{1}{h^2} \\ &P[|X - E[X]| \le \epsilon] \ge 1 - \frac{\sigma^2[X]}{\epsilon^2}, \qquad P[|X - E[X]| > \epsilon] \le \frac{\sigma^2[X]}{\epsilon^2}. \end{aligned}$$

Hint: $P[|X - E[X]| \le h\sigma[X]] = F_X(E[X] + h\sigma[X]) - F_X(E[X] - h\sigma[X])$ if $F_X(\cdot)$ is continuous at these points.

1.3. **Continuation of example 1E.** Using (1.17), show that $P[N < \infty] = 1$.

EXERCISES

1.1. Consider a gambler who is to win 1 dollar if a 6 appears when a fair die is tossed; otherwise he wins nothing. Find the mean and variance of his winnings.

1.2. Suppose that 0.008 is the probability of death within a year of a man aged 35. Find the mean and variance of the number of deaths within a year among 20,000 men of this age.

1.3. Consider a man who buys a lottery ticket in a lottery that sells 100 tickets and that gives 4 prizes of 200 dollars, 10 prizes of 100 dollars, and 20 prizes of 10 dollars. How much should the man be willing to pay for a ticket in this lottery?

1.4. Would you pay 1 dollar to buy a ticket in a lottery that sells 1,000,000 tickets and gives 1 prize of 100,000 dollars, 10 prizes of 10,000 dollars, and 100 prizes of 1000 dollars?

1.5. Nine dimes and a silver dollar are in a red purse, and 10 dimes are in a black purse. Five coins are selected without replacement from the red purse and placed in the black purse. Then 5 coins are selected without replacement from the black purse and placed in the red purse. The amount of money in the red purse at the end of this experiment is a random variable. What is its mean and variance?

1.6. **St. Petersburg problem (or paradox?).** How much would you be willing to pay to play the following game of chance. A fair coin is tossed by the player until heads appears. If heads appears on the first toss, the bank pays the player 1 dollar. If heads appears for the first time on the second throw the bank pays the player 2 dollars. If heads appears for the first time on the third throw the player receives $4 = 2^2$ dollars. In general, if heads appears for the first time on the nth throw, the player receives 2^{n-1} dollars. The amount of money the player will win in this game is a random variable; find its mean. Would you be willing to pay this amount to play the game? (For a discussion of this problem and why it is sometimes called a paradox see T. C. Fry, *Probability and Its Engineering Uses*, Van Nostrand, New York, 1928, pp. 194–199.)

1.7. The output of a certain manufacturer (it may be radio tubes, textiles, canned goods, etc.) is graded into 5 grades, labeled A^5, A^4, A^3, A^2, and A (in decreasing order of quality). The manufacturer's profit, denoted by X, on an item depends on the grade of the item, as indicated in the table. The grade of an item is random; however, the proportions of the manufacturer's output in the various grades is known and is given in the table below. Find the mean and variance of X, in which X denotes the manufacturer's profit on an item selected randomly from his production.

Grade of an Item	Profit on an Item of This Grade	Probability that an Item Is of This Grade
A^5	\$1.00	$\frac{5}{16}$
A^4	0.80	$\frac{1}{4}$
A^3	0.60	$\frac{1}{4}$
A^2	0.00	$\frac{1}{16}$
A	−0.60	$\frac{1}{8}$

1.8. Consider a person who commutes to the city from a suburb by train. He is accustomed to leaving his home between 7:30 and 8:00 A.M. The drive to the railroad station takes between 20 and 30 minutes. Assume that the departure time and length of trip are independent random variables, each uniformly distributed over their respective intervals. There are 3 trains that he can take, which leave the station and arrive in the city precisely on time. The first train leaves at 8:05 A.M. and arrives at 8:40 A.M., the second leaves at 8:25 A.M. and arrives at 8:55 A.M., the third leaves at 9:00 A.M. and arrives at 9:43 A.M.

(i) Find the mean and variance of his time of arrival in the city.

(ii) Find the mean and variance of his time of arrival under the assumption that he leaves his home between 7:30 and 7:55 A.M.

1.9. Two athletic teams play a series of games; the first team to win 4 games is the winner. Suppose that one of the teams is stronger than the other and has probability p [equal to (i) 0.5, (ii) $\frac{2}{3}$] of winning each game, independent of the outcomes of any other game. Assume that a game cannot end in a tie. Find the mean and variance of the number of games required to conclude the series. (Use exercise 3.26 of Chapter 3.)

1.10. Consider an experiment that consists of N players independently tossing fair coins. Let A be the event that there is an "odd" man (that is, either exactly one of the coins falls heads or exactly one of the coins falls tails). For $r = 1, 2, \ldots$ let X_r be the number of times the experiment is repeated until the event occurs for the rth time.

(i) Find the mean and variance of X_r.

(ii) Evaluate $E[X_r]$ and Var $[X_r]$ for $N = 3, 4, 5$ and $r = 1, 2, 3$.

1.11. Let an urn contain 5 balls, numbered 1 to 5. Let a sample of size 3 be drawn with replacement (without replacement) from the urn and let X be the largest number in the sample. Find the mean and variance of X.

1.12. Let X be $N(m, \sigma^2)$. Find the mean and variance of (i) $|X|$, (ii) $|X - c|$ where (*a*) c is a given constant, (*b*) $\sigma = m = c = 1$, (*c*) $\sigma = m = 1, c = 2$.

1.13. Let X and Y be independent random variables, each $N(0, 1)$. Find the mean and variance of $\sqrt{X^2 + Y^2}$.

1.14. Find the mean and variance of a random variable X that obeys the probability law of Laplace, specified by the probability density function, for some constants α and $\beta > 0$:

$$f(x) = \frac{1}{2\beta} \exp\left(-\frac{|x - \alpha|}{\beta}\right) \qquad -\infty < x < \infty.$$

1.15. The velocity v of a molecule with mass m in a gas at absolute temperature T is a random variable obeying the Maxwell Boltzmann law:

$$\begin{aligned} f_v(x) &= \frac{4}{\sqrt{\pi}} \beta^{3/2} x^2 e^{-\beta x^2} \qquad && x > 0 \\ &= 0 && x < 0, \end{aligned}$$

in which $\beta = m/(2kT)$, k = Boltzmann's constant. Find the mean and variance of (i) the velocity of a molecule, (ii) the kinetic energy $E = \frac{1}{2}mv^2$ of a molecule.

2. EXPECTATIONS OF JOINTLY DISTRIBUTED RANDOM VARIABLES

Consider two jointly distributed random variables X_1 and X_2. The expectation $E_{X_1,X_2}[g(x_1, x_2)]$ of a function $g(x_1, x_2)$ of two real variables is defined as follows:

If the random variables X_1 and X_2 are jointly continuous, with joint probability density function $f_{X_1,X_2}(x_1, x_2)$, then

$$(2.1) \qquad E_{X_1,X_2}[g(x_1, x_2)] = \int_{-\infty}^{\infty} \int_{-\infty}^{\infty} g(x_1, x_2) f_{X_1,X_2}(x_1, x_2)\, dx_1\, dx_2.$$

If the random variables X_1 and X_2 are jointly discrete, with joint probability mass function $p_{X_1,X_2}(x_1, x_2)$, then

$$(2.2) \qquad E_{X_1,X_2}[g(x_1, x_2)] = \sum_{\substack{\text{over all } (x_1,x_2) \text{ such} \\ \text{that } p_{X_1,X_2}(x_1,x_2) > 0}} g(x_1, x_2) p_{X_1,X_2}(x_1, x_2).$$

If the random variables X_1 and X_2 have joint distribution function $F_{X_1,X_2}(x_1, x_2)$, then

$$(2.3) \qquad E_{X_1,X_2}[g(x_1, x_2)] = \int_{-\infty}^{\infty} \int_{-\infty}^{\infty} g(x_1, x_2)\, dF_{X_1,X_2}(x_1, x_2),$$

where the two-dimensional Stieltjes integral may be defined in a manner similar to that in which the one-dimensional Stieltjes integral was defined in section 6 of Chapter 5.

On the other hand, $g(X_1, X_2)$ is a random variable, with expectation

$$(2.4) \qquad E[g(X_1, X_2)] = \begin{cases} \displaystyle\int_{-\infty}^{\infty} y \, dF_{g(X_1,X_2)}(y) \\ \displaystyle\int_{-\infty}^{\infty} y f_{g(X_1,X_2)}(y) \, dy \\ \displaystyle\sum_{\substack{\text{over all points } y \\ \text{where } p_{g(X_1,X_2)}(y) > 0}} y p_{g(X_1,X_2)}(y), \end{cases}$$

depending on whether the probability law of $g(X_1, X_2)$ is specified by its distribution function, probability density function, or probability mass function.

It is a basic fact of probability theory that for any jointly distributed random variables X_1 and X_2 and any Borel function $g(x_1, x_2)$

$$(2.5) \qquad E[g(X_1, X_2)] = E_{X_1,X_2}[g(x_1, x_2)],$$

in the sense that if either of the expectations in (2.5) exists then so does the other, and the two are equal. A rigorous proof of (2.5) is beyond the scope of this book.

In view of (2.5) we have two ways of computing the expectation of a function of jointly distributed random variables. Equation (2.5) generalizes (1.5). Similarly, (1.11) may also be generalized.

Let X_1, X_2, and Y be random variables such that $Y = g_1(X_1, X_2)$ for some Borel function $g_1(x_1, x_2)$. Then for any Borel function $g(\cdot)$

$$(2.6) \qquad E[g(Y)] = E[g(g_1(X_1, X_2))].$$

The most important property possessed by the operation of expectation of a random variable is its *linearity property*: if X_1 and X_2 are jointly distributed random variables with finite expectations $E[X_1]$ and $E[X_2]$, then the sum $X_1 + X_2$ has a finite expectation given by

$$(2.7) \qquad E[X_1 + X_2] = E[X_1] + E[X_2].$$

Let us sketch a proof of (2.7) in the case that X_1 and X_2 are jointly continuous. The reader may gain some idea of how (2.7) is proved in general by consulting the proof of (6.22) in Chapter 2.

From (2.5) it follows that

$$(2.7') \qquad E[X_1 + X_2] = \int_{-\infty}^{\infty} \int_{-\infty}^{\infty} (x_1 + x_2) f_{X_1,X_2}(x_1, x_2) \, dx_1 \, dx_2.$$

Now

$$(2.7'') \qquad \begin{aligned} \int_{-\infty}^{\infty} dx_1 \, x_1 \int_{-\infty}^{\infty} dx_2 \, f_{X_1,X_2}(x_1, x_2) &= \int_{-\infty}^{\infty} dx_1 \, x_1 f_{X_1}(x_1) = E[X_1] \\ \int_{-\infty}^{\infty} dx_2 \, x_2 \int_{-\infty}^{\infty} dx_1 \, f_{X_1,X_2}(x_1, x_2) &= \int_{-\infty}^{\infty} dx_2 \, x_2 f_{X_2}(x_2) = E[X_2]. \end{aligned}$$

The integral on the right-hand side of (2.7') is equal to the sum of the integrals on the left-hand sides of (2.7"). The proof of (2.7) is now complete.

The moments and moment-generating function of jointly distributed random variables are defined by a direct generalization of the definitions given for a single random variable. For any two nonnegative integers n_1 and n_2 we define

$$\alpha_{n_1,n_2} = E[X_1^{n_1}X_2^{n_2}] \tag{2.8}$$

as a moment of the jointly distributed random variables X_1 and X_2. The sum $n_1 + n_2$ is called the order of the moment. For the moments of orders 1 and 2 we have the following names; $\alpha_{1,0}$ and $\alpha_{0,1}$ are, respectively, the means of X_1 and X_2, whereas $\alpha_{2,0}$ and $\alpha_{0,2}$ are, respectively, the mean squares of X_1 and X_2. The moment $\alpha_{11} = E[X_1X_2]$ is called the *product moment*.

We next define the central moments of the random variables X_1 and X_2. For any two nonnegative integers, n_1 and n_2, we define

$$\mu_{n_1,n_2} = E[(X_1 - E[X_1])^{n_1}(X_2 - E[X_2])^{n_2}] \tag{2.8'}$$

as a central moment of order $n_1 + n_2$. We are again particularly interested in the central moments of orders 1 and 2. The central moments $\mu_{1,0}$ and $\mu_{0,1}$ of order 1 both vanish, whereas $\mu_{2,0}$ and $\mu_{0,2}$ are, respectively, the variances of X_1 and X_2. The central moment $\mu_{1,1}$ is called the *covariance* of the random variables X_1 and X_2 and is written Cov $[X_1, X_2]$; in symbols,

$$\text{Cov}\,[X_1, X_2] = \mu_{1,1} = E[(X_1 - E[X_1])(X_2 - E[X_2])]. \tag{2.9}$$

We leave it to the reader to prove that *the covariance is equal to the product moment, minus the product of the means;* in symbols,

$$\text{Cov}\,[X_1, X_2] = E[X_1X_2] - E[X_1]E[X_2]. \tag{2.10}$$

The covariance derives its importance from the role it plays in the *basic formula for the variance of the sum of two random variables:*

$$\text{Var}\,[X_1 + X_2] = \text{Var}\,[X_1] + \text{Var}\,[X_2] + 2\,\text{Cov}\,[X_1, X_2] \tag{2.11}$$

To prove (2.11), we write

$$\begin{aligned}\text{Var}\,[X_1 + X_2] &= E[(X_1 + X_2)^2] - E^2[X_1 + X_2]\\ &= E[X_1^2] - E^2[X_1] + E[X_2^2] - E^2[X_2]\\ &\qquad + 2(E[X_1X_2] - E[X_1]E[X_2]),\end{aligned}$$

from which (2.11) follows by (1.8) and (2.10).

The joint moment-generating function is defined for any two real numbers, t_1 and t_2, by

$$\psi_{X_1,X_2}(t_1, t_2) = E[e^{(t_1X_1+t_2X_2)}].$$

The moments can be read off from the power-series expansion of the moment-generating function, since formally

$$\psi_{X_1,X_2}(t_1, t_2) = \sum_{n_1=0}^{\infty} \sum_{n_2=0}^{\infty} \frac{t_1^{n_1}}{n_1!} \frac{t_2^{n_2}}{n_2!} E[X_1^{n_1}X_2^{n_2}]. \tag{2.12}$$

In particular, the means, variances, and covariance of X_1 and X_2 may be expressed in terms of the derivatives of the moment-generating function:

$$E[X_1] = \frac{\partial}{\partial t_1} \psi_{X_1,X_2}(0, 0), \qquad E[X_2] = \frac{\partial}{\partial t_2} \psi_{X_1,X_2}(0, 0). \tag{2.13}$$

$$E[X_1^2] = \frac{\partial^2}{\partial t_1^2} \psi_{X_1,X_2}(0, 0), \qquad E[X_2^2] = \frac{\partial^2}{\partial t_2^2} \psi_{X_1,X_2}(0, 0). \tag{2.14}$$

$$E[X_1X_2] = \frac{\partial^2}{\partial t_1\, \partial t_2} \psi_{X_1,X_2}(0, 0). \tag{2.15}$$

$$\text{Var}\,[X_1] = \frac{\partial^2}{\partial t_1^2} \psi_{X_1-m_1,X_2-m_2}(0, 0), \qquad \text{Var}\,[X_2] = \frac{\partial^2}{\partial t_2^2} \psi_{X_1-m_1,X_2-m_2}(0, 0). \tag{2.16}$$

$$\text{Cov}\,[X_1, X_2] = \frac{\partial^2}{\partial t_1\, \partial t_2} \psi_{X_1-m_1,X_2-m_2}(0, 0), \tag{2.17}$$

in which $m_1 = E[X_1]$, $m_2 = E[X_2]$.

▶ **Example 2A. The joint moment-generating function and covariance of jointly normal random variables.** Let X_1 and X_2 be jointly normally distributed random variables with a joint probability density function

$$f_{X_1,X_2}(x_1, x_2) = \frac{1}{2\pi\sigma_1\sigma_2\sqrt{1-\rho^2}} \exp\left\{ -\frac{1}{2(1-\rho^2)} \left[\left(\frac{x_1 - m_1}{\sigma_1}\right)^2 - 2\rho\left(\frac{x_1 - m_1}{\sigma_1}\right)\left(\frac{x_2 - m_2}{\sigma_2}\right) + \left(\frac{x_2 - m_2}{\sigma_2}\right)^2 \right] \right\}. \tag{2.18}$$

The joint moment-generating function is given by

$$\psi_{X_1,X_2}(t_1, t_2) = \int_{-\infty}^{\infty}\int_{-\infty}^{\infty} e^{(t_1x_1+t_2x_2)} f_{X_1,X_2}(x_1, x_2)\, dx_1\, dx_2. \tag{2.19}$$

To evaluate the integral in (2.19), let us note that since

$$u_1^2 - 2\rho u_1 u_2 + u_2^2 = (1 - \rho^2)u_1^2 + (u_2 - \rho u_1)^2$$

we may write

$$(2.20) \quad f_{X_1,X_2}(x_1, x_2) = \frac{1}{\sigma_1}\phi\left(\frac{x_1 - m_1}{\sigma_1}\right)\frac{1}{\sigma_2\sqrt{1-\rho^2}} \times \phi\left(\frac{x_2 - m_2 - (\sigma_2/\sigma_1)\rho(x_1 - m_1)}{\sigma_2\sqrt{1-\rho^2}}\right),$$

in which $\phi(u) = \frac{1}{\sqrt{2\pi}} e^{-\frac{1}{2}u^2}$ is the normal density function. Using our knowledge of the moment-generating function of a normal law, we may perform the integration with respect to the variable x_2 in the integral in (2.19). We thus determine that $\psi_{X_1,X_2}(t_1, t_2)$ is equal to

$$(2.21) \quad \int_{-\infty}^{\infty} dx_1 \frac{1}{\sigma_1}\phi\left(\frac{x_1 - m_1}{\sigma_1}\right) \exp(t_1 x_1) \exp\left\{t_2\left[m_2 + \frac{\sigma_2}{\sigma_1}\rho(x_1 - m_1)\right]\right\} \times \exp\left[\tfrac{1}{2}t_2^2\sigma_2^2(1-\rho^2)\right]$$

$$= \exp\left[\tfrac{1}{2}t_2^2\sigma_2^2(1-\rho^2) + t_2 m_2 - t_2\frac{\sigma_2}{\sigma_1}\rho m_1\right] \times \exp\left[m_1\left(t_1 + t_2\frac{\sigma_2}{\sigma_1}\rho\right) + \tfrac{1}{2}\sigma_1^2\left(t_1 + t_2\frac{\sigma_2}{\sigma_1}\rho\right)^2\right].$$

By combining terms in (2.21), we finally obtain that

$$(2.22) \quad \psi_{X_1,X_2}(t_1, t_2) = \exp\left[t_1 m_1 + t_2 m_2 + \tfrac{1}{2}(t_1^2\sigma_1^2 + 2\rho\sigma_1\sigma_2 t_1 t_2 + t_2^2\sigma_2^2)\right].$$

The covariance is given by

$$(2.23) \quad \operatorname{Cov}[X_1, X_2] = \frac{\partial}{\partial t_1\,\partial t_2} e^{-(t_1 m_1 + t_2 m_2)}\psi_{X_1,X_2}(t_1, t_2)\Big|_{t_1=0,t_2=0} = \rho\sigma_1\sigma_2.$$

Thus, *if two random variables are jointly normally distributed, their joint probability law is completely determined from a knowledge of their first and second moments, since* $m_1 = E[X_1]$, $m_2 = E[X_2]$, $\sigma_1^2 = \operatorname{Var}[X_1]$, $\sigma_2^2 = \operatorname{Var}[X_2^2]$, $\rho\sigma_1\sigma_2 = \operatorname{Cov}[X_1, X_2]$. ◀

The foregoing notions may be extended to the case of n jointly distributed random variables, $X_1, X_2, \ldots, X_n$. For any Borel function $g(x_1, x_2, \ldots, x_n)$ of n real variables, the expectation $E[g(X_1, X_2, \ldots, X_n)]$ of the random variable $g(X_1, X_2, \ldots, X_n)$ may be expressed in terms of the joint probability law of $X_1, X_2, \ldots, X_n$.

If $X_1, X_2, \ldots, X_n$ are jointly continuous, with a joint probability density function $f_{X_1,X_2,\cdots,X_n}(x_1, x_2, \ldots, x_n)$, it may be shown that

$$(2.24)\quad E[g(X_1, X_2, \cdots, X_n)] = \int_{-\infty}^{\infty}\int_{-\infty}^{\infty}\cdots\int_{-\infty}^{\infty} g(x_1, x_2, \cdots, x_n) \times f_{X_1,X_2,\cdots,X_n}(x_1, x_2, \cdots, x_n)\, dx_1 dx_2 \cdots dx_n.$$

If $X_1, X_2, \ldots, X_n$ are jointly discrete, with a joint probability mass function $p_{X_1,X_2,\cdots,X_n}(x_1, x_2, \ldots, x_n)$, it may be shown that

$$(2.25)\quad E[g(X_1, X_2, \cdots, X_n)] = \sum_{\substack{\text{over all } (x_1,x_2,\cdots,x_n) \text{ such that} \\ p_{X_1,X_2,\cdots,X_n}(x_1,x_2,\cdots,x_n)>0}} g(x_1, x_2, \cdots, x_n) p_{X_1,X_2,\cdots,X_n}(x_1, x_2, \cdots, x_n)$$

The joint moment-generating function of n jointly distributed random variables is defined by

$$(2.26)\quad \psi_{X_1,X_2,\cdots,X_n}(t_1, t_2, \ldots, t_n) = E[e^{(t_1X_1+t_2X_2+\cdots+t_nX_n)}].$$

It may also be proved that if $X_1, X_2, \ldots, X_n$ and Y are random variables, such that $Y = g_1(X_1, X_2, \ldots, X_n)$ for some Borel function $g_1(x_1, x_2, \ldots, x_n)$ of n real variables, then for any Borel function $g(\cdot)$ of one real variable

$$(2.27)\quad E[g(Y)] = E[g(g_1(X_1, X_2, \cdots, X_n))].$$

THEORETICAL EXERCISES

2.1. Linearity property of the expectation operation. Let X_1 and X_2 be jointly discrete random variables with finite means. Show that (2.7) holds.

2.2. Let X_1 and X_2 be jointly distributed random variables whose joint moment-generating function has a logarithm given by

$$(2.28)\quad \log \psi_{X_1,X_2}(t_1, t_2) = \nu \int_{-\infty}^{\infty} du \int_{-\infty}^{\infty} dy\, f_Y(y)\, \{e^{y[t_1W_1(u)+t_2W_2(u)]} - 1\}$$

in which Y is a random variable with probability density function $f_Y(\cdot)$, $W_1(\cdot)$ and $W_2(\cdot)$ are known functions, and $\nu > 0$. Show that

$$E[X_1] = \nu E[Y]\int_{-\infty}^{\infty} W_1(u)\, du, \qquad E[X_2] = \nu E[Y]\int_{-\infty}^{\infty} W_2(u)\, du,$$

$$(2.29)\quad \operatorname{Var}[X_1] = \nu E[Y^2]\int_{-\infty}^{\infty} W_1^2(u)\, du,$$

$$\operatorname{Var}[X_2] = \nu E[Y^2]\int_{-\infty}^{\infty} W_2^2(u)\, du,$$

$$\operatorname{Cov}[X_1, X_2] = \nu E[Y^2]\int_{-\infty}^{\infty} W_1(u) W_2(u)\, du.$$

Moment-generating functions of the form of (2.28) play an important role in the mathematical theory of the phenomenon of shot noise in radio tubes.

2.3. The random telegraph signal. For $t > 0$ let $X(t) = U(-1)^{N(t)}$, where U is a discrete random variable such that $P[U = 1] = P[U = -1] = \frac{1}{2}$, $\{N(t), t > 0\}$ is a family of random variables such that $N(0) = 0$, and for any times $t_1 < t_2$, the random variables U, $N(t_1)$, and $N(t_2) - N(t_1)$ are independent. For any $t_1 < t_2$, suppose that $N(t_2) - N(t_1)$ obeys (i) a Poisson probability law with parameter $\lambda = \nu(t_2 - t_1)$, (ii) a binomial probability law with parameters p and $n = (t_2 - t_1)$. Show that $E[X(t)] = 0$ for any $t > 0$, and for any $t \geq 0$, $\tau \geq 0$

$$E[X(t)X(t + \tau)] = e^{-2\nu\tau} \quad \text{Poisson case,} \tag{2.30}$$

$$= (q - p)^{\tau} \quad \text{binomial case.}$$

Regarded as a random function of time, $X(t)$ is called a "random telegraph signal." *Note:* in the binomial case, t takes only integer values.

EXERCISES

2.1. An ordered sample of size 5 is drawn without replacement from an urn containing 8 white balls and 4 black balls. For $j = 1, 2, \ldots, 5$ let X_j be equal to 1 or 0, depending on whether the ball drawn on the jth draw is white or black. Find $E[X_2]$, $\sigma^2[X_2]$, Cov $[X_1, X_2]$, Cov $[X_2, X_3]$.

2.2. An urn contains 12 balls, of which 8 are white and 4 are black. A ball is drawn and its color noted. The ball drawn is then replaced; at the same time 2 balls of the same color as the ball drawn are added to the urn. The process is repeated until 5 balls have been drawn. For $j = 1, 2, \ldots, 5$ let X_j be equal to 1 or 0, depending on whether the ball drawn on the jth draw is white or black. Find $E[X_2]$, $\sigma^2[X_2]$, Cov $[X_1, X_2]$.

2.3. Let X_1 and X_2 be the coordinates of 2 points randomly chosen on the unit interval. Let $Y = |X_1 - X_2|$ be the distance between the points. Find the mean, variance, and third and fourth moments of Y.

2.4. Let X_1 and X_2 be independent normally identically distributed random variables, with mean m and variance σ^2. Find the mean of the random variable $Y = \max(X_1, X_2)$. *Hint:* for any real numbers x_1 and x_2 show and use the fact that $2 \max(x_1, x_2) = |x_1 - x_2| + x_1 + x_2$.

2.5. Let X_1 and X_2 be jointly normally distributed with mean 0, variance 1, and covariance ρ. Find $E[\max(X_1, X_2)]$.

2.6. Let X_1 and X_2 have a joint moment-generating function

$$\psi_{X_1,X_2}(t_1, t_2) = a(e^{t_1+t_2} + 1) + b(e^{t_1} + e^{t_2}),$$

in which a and b are positive constants such that $2a + 2b = 1$. Find $E[X_1]$, $E[X_2]$, Var $[X_1]$, Var $[X_2]$, Cov $[X_1, X_2]$.

2.7. Let X_1 and X_2 have a joint moment-generating function

$$\psi_{X_1,X_2}(t_1, t_2) = [a(e^{t_1+t_2} + 1) + b(e^{t_1} + e^{t_2})]^2,$$

in which a and b are positive constants such that $2a + 2b = 1$. Find $E[X_1]$, $E[X_2]$, Var $[X_1]$, Var $[X_2]$, Cov $[X_1, X_2]$.

2.8. Let X_1 and X_2 be jointly distributed random variables whose joint moment-generating function has a logarithm given by (2.28), with $\nu = 4$, Y uniformly distributed over the interval -1 to 1, and

$$\begin{aligned} W_1(u) &= e^{-(u-a_1)}, & u &\geq a_1, & W_2(u) &= e^{-(u-a_2)}, & u &\geq a_2 \\ &= 0, & u &< a_1, & &= 0, & u &< a_2. \end{aligned}$$

in which a_1, a_2 are given constants such that $0 < a_1 < a_2$. Find $E[X_1]$, $E[X_2]$, Var $[X_1]$, Var $[X_2]$, Cov $[X_1, X_2]$.

2.9. Do exercise 2.8 under the assumption that Y is $N(1, 2)$.

3. UNCORRELATED AND INDEPENDENT RANDOM VARIABLES

The notion of independence of two random variables, X_1 and X_2, is defined in section 6 of Chapter 7. In this section we show how the notion of independence may be formulated in terms of expectations. At the same time, by a modification of the condition for independence of random variables, we are led to the notion of uncorrelated random variables.

We begin by considering the properties of expectations of products of random variables. Let X_1 and X_2 be jointly distributed random variables. By the linearity properties of the operation of taking expectations, it follows that for any two functions, $g_1(\cdot\,,\cdot)$ and $g_2(\cdot\,,\cdot)$,

$$E[g_1(X_1, X_2) + g_2(X_1, X_2)] = E[g_1(X_1, X_2)] + E[g_2(X_1, X_2)] \tag{3.1}$$

if the expectations on the right side of (3.1) exist. However, it is *not* true that a similar relation holds for products; namely, it is *not* true in general that $E[g_1(X_1, X_2)g_2(X_1, X_2)] = E[g_1(X_1, X_2)]E[g_2(X_1, X_2)]$. There is one special circumstance in which a relation similar to the foregoing is valid, namely, if the random variables X_1 and X_2 are independent and if the functions are functions of one variable only. More precisely, we have the following theorem:

THEOREM 3A: *If the random variables X_1 and X_2 are independent, then for any two Borel functions $g_1(\cdot)$ and $g_2(\cdot)$ of one real variable the product moment of $g_1(X_1)$ and $g_2(X_2)$ is equal to the product of their means; in symbols,*

$$E[g_1(X_1)g_2(X_2)] = E[g_1(X_1)]E[g_2(X_2)] \tag{3.2}$$

if the expectations on the right side of (3.2) exist.

To prove equation (3.2), it suffices to prove it in the form

$$E[Y_1 Y_2] = E[Y_1]E[Y_2] \quad \text{if } Y_1 \text{ and } Y_2 \text{ are independent,} \tag{3.3}$$

since independence of X_1 and X_2 implies independence of $g(X_1)$ and $g(X_2)$. We write out the proof of (3.3) only for the case of jointly continuous random variables. We have

$$\begin{aligned} E[Y_1 Y_2] &= \int_{-\infty}^{\infty}\int_{-\infty}^{\infty} y_1 y_2 f_{Y_1,Y_2}(y_1, y_2)\, dy_1\, dy_2 \\ &= \int_{-\infty}^{\infty}\int_{-\infty}^{\infty} y_1 y_2 f_{Y_1}(y_1) f_{Y_2}(y_2)\, dy_1\, dy_2 \\ &= \int_{-\infty}^{\infty} y_1 f_{Y_1}(y_1)\, dy_1 \int_{-\infty}^{\infty} dy_2 f_{Y_2}(y_2)\, dy_2 = E[Y_1]E[Y_2]. \end{aligned}$$

Now suppose that we modify (3.2) and ask only that it hold for the functions $g_1(x) = x$ and $g_2(x) = x$, so that

$$E[X_1 X_2] = E[X_1]E[X_2]. \tag{3.4}$$

For reasons that are explained after (3.7), two random variables, X_1 and X_2, which satisfy (3.4), are said to be *uncorrelated.* From (2.10) it follows that X_1 and X_2 satisfy (3.4) and therefore are uncorrelated if and only if

$$\text{Cov}\,[X_1, X_2] = 0. \tag{3.5}$$

For uncorrelated random variables the formula given by (2.11) for the variance of the sum of two random variables becomes particularly elegant; the variance of the sum of two uncorrelated random variables is equal to the sum of their variances. *Indeed,*

$$\text{Var}\,[X_1 + X_2] = \text{Var}\,[X_1] + \text{Var}\,[X_2] \tag{3.6}$$

if and only if X_1 and X_2 are uncorrelated.

Two random variables that are independent are uncorrelated, for if (3.2) holds then, a fortiori, (3.4) holds. The converse is not true in general; an example of two uncorrelated random variables that are not independent is given in theoretical exercise 3.2. In the important special case in which X_1 and X_2 are jointly normally distributed, it follows that they are independent if they are uncorrelated (see theoretical exercise 3.3).

The *correlation coefficient* $\rho(X_1, X_2)$ of two jointly distributed random variables with finite positive variances is defined by

$$\rho(X_1, X_2) = \frac{\text{Cov}\,[X_1, X_2]}{\sigma[X_1]\sigma[X_2]}. \tag{3.7}$$

In view of (3.7) and (3.5), two random variables X_1 and X_2 are uncorrelated if and only if their correlation coefficient is zero.

The correlation coefficient provides a measure of how good a prediction of the value of one of the random variables can be formed on the basis of an observed value of the other. It is subsequently shown that

$$|\rho(X_1, X_2)| \leq 1. \tag{3.8}$$

Further $\rho(X_1, X_2) = 1$ if and only if

$$\frac{X_2 - E[X_2]}{\sigma[X_2]} = \frac{X_1 - E[X_1]}{\sigma[X_1]} \tag{3.9}$$

and $\rho(X_1, X_2) = -1$ if and only if

$$\frac{X_2 - E[X_2]}{\sigma[X_2]} = -\frac{X_1 - E[X_1]}{\sigma[X_1]}. \tag{3.10}$$

From (3.9) and (3.10) it follows that if the correlation coefficient equals 1 or -1 then there is perfect prediction; to a given value of one of the random variables there is one and only one value that the other random variable can assume. What is even more striking is that $\rho(X_1, X_2) = \pm 1$ if and only if X_1 and X_2 are linearly dependent.

That (3.8), (3.9), and (3.10) hold follows from the following important theorem.

THEOREM 3B. *For any two jointly distributed random variables, X_1 and X_2, with finite second moments*

$$E^2[X_1 X_2] = |E[X_1 X_2]|^2 \leq E[X_1^2]E[X_2^2]. \tag{3.11}$$

Further, equality holds in (3.11), that is, $E^2[X_1 X_2] = E[X_1^2]E[X_2^2]$ *if and only if, for some constant t, $X_2 = tX_1$, which means that the probability mass distributed over the (x_1, x_2)-plane by the joint probability law of the random variables is situated on the line $x_2 = tx_1$.*

Applied to the random variables $X_1 - E[X_1]$ and $X_2 - E[X_2]$, (3.11) states that

$$|\text{Cov}[X_1, X_2]|^2 \leq \text{Var}[X_1]\,\text{Var}[X_2], \qquad |\text{Cov}[X_1, X_2]| \leq \sigma[X_1]\sigma[X_2]. \tag{3.12}$$

We prove (3.11) as follows. Define, for any real number t, $h(t) = E[(tX_1 - X_2)^2] = t^2E[X_1^2] - 2tE[X_1X_2] + E[X_2^2]$. Clearly $h(t) \geq 0$ for all t. Consequently, the quadratic equation $h(t) = 0$ has either no solutions or one solution. The equation $h(t) = 0$ has no solutions if and only if $E^2[X_1X_2] - E[X_1^2]E[X_2^2] < 0$. It has exactly one solution if and only if $E^2[X_1X_2] = E[X_1^2]E[X_2^2]$. From these facts one may immediately infer (3.11) and the sentence following it.

The inequalities given by (3.11) and (3.12) are usually referred to as *Schwarz's* inequality or *Cauchy's* inequality.

Conditions for Independence. It is important to note the difference between two random variables being independent and being uncorrelated. They are uncorrelated if and only if (3.4) holds. It may be shown that they are independent if and only if (3.2) holds for all functions $g_1(\cdot)$ and $g_2(\cdot)$, for which the expectations in (3.2) exist. More generally, theorem 3C can be proved.

THEOREM 3C. Two jointly distributed random variables X_1 and X_2 are independent if and only if each of the following equivalent statements is true:

(i) *Criterion in terms of probability functions.* For any Borel sets B_1 and B_2 of real numbers, $P[X_1 \text{ is in } B_1, X_2 \text{ is in } B_2] = P[X_1 \text{ is in } B_1]P[X_2 \text{ is in} B_2]$.

(ii) *Criterion in terms of distribution functions.* For any two real numbers, x_1 and x_2, $F_{X_1,X_2}(x_1, x_2) = F_{X_1}(x_1)F_{X_2}(x_2)$.

(iii) *Criterion in terms of expectations.* For any two Borel functions, $g_1(\cdot)$ and $g_2(\cdot)$, $E[g_1(X_1)g_2(X_2)] = E[g_1(X_1)]E[g_2(X_2)]$ if the expectations involved exist.

(iv) *Criterion in terms of moment-generating functions* (if they exist). For any two real numbers, t_1 and t_2,

$$\psi_{X_1,X_2}(t_1, t_2) = E[e^{t_1X_1+t_2X_2}] = \psi_{X_1}(t_1)\psi_{X_2}(t_2). \tag{3.13}$$

THEORETICAL EXERCISES

3.1. *The standard deviation has the properties of the operation of taking the absolute value of a number*: show first that for any 2 real numbers, x and y, $|x + y| \le |x| + |y|$, $||x| - |y|| \le |x - y|$. *Hint:* Square both sides of the equations. Show next that for any 2 random variables, X and Y,

$$\sigma[X + Y] \le \sigma[X] + \sigma[Y], \quad |\sigma[X] - \sigma[Y]| \le \sigma[X - Y]. \tag{3.14}$$

Give an example to prove that the variance does not satisfy similar relationships.

3.2. Show that independent random variables are uncorrelated. Give an example to show that the converse is false. *Hint:* Let $X = \sin 2\pi U$, $Y = \cos 2\pi U$, in which U is uniformly distributed over the interval 0 to 1.

3.3. Prove that if X_1 and X_2 are jointly normally distributed random variables whose correlation coefficient vanishes then X_1 and X_2 are independent. *Hint:* Use example 2A.

3.4. Let α and β be the values of a and b which minimize

$$f(a, b) = E|X_2 - a - bX_1|^2.$$

Express α, β, and $f(\alpha, \beta)$ in terms of $\rho(X_1, X_2)$. The random variable $\alpha + \beta X_1$ is called the best linear predictor of X_2, given X_1 [see Section 7, in particular, (7.13) and (7.14)].

3.5. Prove that (3.9) and (3.10) hold under the conditions stated.

3.6. Let X_1 and X_2 be jointly distributed random variables possessing finite second moments. State conditions under which it is possible to find 2 *uncorrelated* random variables, Y_1 and Y_2, which are linear combinations of X_1 and X_2 (that is, $Y_1 = a_{11}X_1 + a_{12}X_2$ and $Y_2 = a_{21}X_1 + a_{22}X_2$ for some constants a_{11}, a_{12}, a_{21}, a_{22} and $\text{Cov}[Y_1, Y_2] = 0$).

3.7. Let X and Y be jointly normally distributed with mean 0, arbitrary variances, and correlation ρ. Show that

$$P[X \geq 0, Y \geq 0] = P[X \leq 0, Y \leq 0] = \frac{1}{4} + \frac{1}{2\pi}\sin^{-1}\rho.$$

$$P[X \leq 0, Y \geq 0] = P[X \geq 0, Y \leq 0] = \frac{1}{4} - \frac{1}{2\pi}\sin^{-1}\rho.$$

Hint: Consult H. Cramér, *Mathematical Methods of Statistics*, Princeton University Press, 1946, p. 290.

3.8. Suppose that n tickets bear arbitrary numbers $x_1, x_2, \ldots, x_n$, which are not all the same. Suppose further that 2 of the tickets are selected at random without replacement. Show that the correlation coefficient ρ between the numbers appearing on the 2 tickets is equal to $(-1)/(n-1)$.

3.9. In an urn containing N balls, a proportion p is white and $q = 1 - p$ are black. A ball is drawn and its color noted. The ball drawn is then replaced, and Nr balls are added of the same color as the ball drawn. The process is repeated until n balls have been drawn. For $j = 1, 2, \ldots, n$ let X_j be equal to 1 or 0, depending on whether the ball drawn on the jth draw is white or black. Show that the correlation coefficient between X_i and X_j is equal to $r/(1 + r)$. Note that the case $r = -1/N$ corresponds to sampling without replacement, and $r = 0$ corresponds to sampling with replacement.

EXERCISES

3.1. Consider 2 events A and B such that $P[A] = \frac{1}{4}$, $P[B \mid A] = \frac{1}{2}$, $P[A \mid B] = \frac{1}{4}$. Define random variables X and Y: $X = 1$ or 0, depending on whether the event A has or has not occurred, and $Y = 1$ or 0, depending on whether the event B has or has not occurred. Find $E[X]$, $E[Y]$, $\text{Var}[X]$, $\text{Var}[Y]$, $\rho(X, Y)$. Are X and Y independent?

3.2. Consider a sample of size 2 drawn with replacement (without replacement) from an urn containing 4 balls, numbered 1 to 4. Let X_1 be the smallest and X_2 be the largest among the numbers drawn in the sample. Find $\rho(X_1, X_2)$.

3.3. Two fair coins, each with faces numbered 1 and 2, are thrown independently. Let X denote the sum of the 2 numbers obtained, and let Y denote the maximum of the numbers obtained. Find the correlation coefficient between X and Y.

3.4. Let U, V, and W be uncorrelated random variables with equal variances. Let $X = U + V$, $Y = U + W$. Find the correlation coefficient between X and Y.

3.5. Let X_1 and X_2 be uncorrelated random variables. Find the correlation $\rho(Y_1, Y_2)$ between the random variables $Y_1 = X_1 + X_2$ and $Y_2 = X_1 - X_2$ in terms of the variances of X_1 and X_2.

3.6. Let X_1 and X_2 be uncorrelated normally distributed random variables. Find the correlation $\rho(Y_1, Y_2)$ between the random variables $Y_1 = X_1^2$ and $Y_2 = X_2^2$.

3.7. Consider the random variables whose joint moment-generating function is given in exercise 2.6. Find $\rho(X_1, X_2)$.

3.8. Consider the random variables whose joint moment-generating function is given in exercise 2.7. Find $\rho(X_1, X_2)$.

3.9. Consider the random variables whose joint moment-generating function is given in exercise 2.8. Find $\rho(X_1, X_2)$.

3.10. Consider the random variables whose joint moment-generating function is given in exercise 2.9. Find $\rho(X_1, X_2)$.

4. EXPECTATIONS OF SUMS OF RANDOM VARIABLES

Random variables, which arise as, or may be represented as, *sums* of other random variables, play an important role in probability theory. In this section we obtain formulas for the mean, mean square, variance, and moment-generating function of a sum of random variables.

Let $X_1, X_2, \ldots, X_n$ be n jointly distributed random variables. Using the linearity properties of the expecration operation, we immediately obtain the following formulas for the mean, mean square, and variance of the sum:

$$E\left[\sum_{k=1}^{n} X_k\right] = \sum_{k=1}^{n} E[X_k]; \tag{4.1}$$

$$E\left[\left(\sum_{k=1}^{n} X_k\right)^2\right] = \sum_{k=1}^{n} E[X_k^2] + 2\sum_{k=1}^{n} \sum_{j=k+1}^{n} E[X_k X_j]; \tag{4.2}$$

$$\operatorname{Var}\left[\sum_{k=1}^{n} X_k\right] = \sum_{k=1}^{n} \operatorname{Var}[X_k] + 2\sum_{k=1}^{n} \sum_{j=k+1}^{n} \operatorname{Cov}[X_k, X_j]. \tag{4.3}$$

Equations (4.2) and (4.3) follow from the facts

$$\left(\sum_{k=1}^{n} X_k\right)^2 = \sum_{k=1}^{n} \sum_{j=1}^{n} X_k X_j = \sum_{k=1}^{n} \left(\sum_{j=1}^{k-1} X_k X_j + X_k^2 + \sum_{j=k+1}^{n} X_k X_j\right), \tag{4.4}$$

$$\sum_{k=1}^{n} \sum_{j=1}^{k-1} X_k X_j = \sum_{j=1}^{n} \sum_{k=j+1}^{n} X_k X_j = \sum_{k=1}^{n} \sum_{j=k+1}^{n} X_k X_j. \tag{4.5}$$

Equation (4.3) simplifies considerably if the random variables $X_1, X_2, \ldots, X_n$ are *uncorrelated* (by which is meant that $\text{Cov}[X_k, X_j] = 0$ for every $k \neq j$). Then the variance of the sum of the random variables is equal to the sum of the variances of the random variables; in symbols,

$$(4.6)\quad \text{Var}\left[\sum_{k=1}^{n} X_k\right] = \sum_{k=1}^{n} \text{Var}[X_k] \qquad \text{if Cov}[X_k, X_j] = 0 \qquad \text{for } k \neq j.$$

If the random variables $X_1, X_2, \ldots, X_n$ are *independent*, then we may give a formula for the moment-generating function of their sum; for any real number t

$$(4.7)\qquad \psi_{X_1+X_2+\cdots+X_n}(t) = \psi_{X_1}(t)\psi_{X_2}(t)\cdots\psi_{X_n}(t).$$

In words, *the moment-generating function of the sum of independent random variables is equal to the product of their moment-generating functions.* The importance of the moment-generating function in probability theory derives as much from the fact that (4.7) holds as from the fact that the moment-generating function may be used to compute moments. The proof of (4.7) follows immediately, once we rewrite (4.7) explicitly in terms of expectations:

$$(4.7')\qquad E[e^{t(X_1+\cdots+X_n)}] = E[e^{tX_1}]\cdots E[e^{tX_n}].$$

Equations (4.1)–(4.3) are useful for finding the mean and variance of a random variable Y (without knowing the probability law of Y) if one can represent Y as a sum of random variables $X_1, X_2, \ldots, X_n$, the mean, variances, and covariances of which are known.

▶ **Example 4A. A binomial random variable as a sum.** The number of successes in n independent repeated Bernoulli trials with probability p of success at each trial is a random variable. Let us denote it by S_n. It has been shown that S_n obeys a binomial probability law with parameters n and p. Consequently,

$$(4.8)\qquad E[S_n] = np, \qquad \text{Var}[S_n] = npq, \qquad \psi_{S_n}(t) = (pe^t + q)^n.$$

We now show that (4.8) is an immediate consequence of (4.1), (4.6), and (4.7). Define random variables $X_1, X_2, \ldots, X_n$ by $X_k = 1$ or 0, depending on whether the outcome of the kth trial is a success or a failure. One may verify that (i) $S_n = X_1 + X_2 + \ldots + X_n$; (ii) $X_1, \ldots, X_n$ are independent random variables; (iii) for $k = 1, 2, \ldots, n$, X_k is a Bernoulli random variable, with mean $E[X_k] = p$, variance $\text{Var}[X_k] = pq$, and moment-generating function $\psi_{X_k}(t) = pe^t + q$. The desired conclusion may now be inferred. ◀

▶ **Example 4B. A hypergeometric random variable as a sum.** The number of white balls drawn in a sample of size n drawn without replacement from an urn containing N balls, of which $a = Np$ are white, is a random variable. Let us denote it by S_n. It has been shown that S_n obeys a hypergeometric probability law. Consequently,

$$E[S_n] = np, \qquad \text{Var}\,[S_n] = npq\frac{N-n}{N-1}. \tag{4.9}$$

We now show that (4.9) can be derived by means of (4.1) and (4.3), without knowing the probability law of S_n. Define random variables $X_1, X_2, \ldots, X_n$: $X_k = 1$ or 0, depending on whether a white ball is or is not drawn on the kth draw. Verify that (i) $S_n = X_1 + X_2 + \ldots + X_n$; (ii) for $k = 1, 2, \ldots, n$, X_k is a Bernoulli random variable, with mean $E[X_k] = p$ and $\text{Var}\,[X_k] = pq$. However, the random variables $X_1, \ldots, X_n$ are not independent, and we need to compute their product moments $E[X_jX_k]$ and covariances $\text{Cov}\,[X_j, X_k]$ for any $j \neq k$. Now, $E[X_jX_k] = P[X_j = 1, X_k = 1]$, so that $E[X_jX_k]$ is equal to the probability that the balls drawn on the jth and kth draws are both white, which is equal to $[a(a-1)]/[N(N-1)]$. Therefore,

$$\text{Cov}\,[X_j, X_k] = E[X_jX_k] - E[X_j]E[X_k] = \frac{a(a-1)}{N(N-1)} - p^2 = \frac{-pq}{N-1}.$$

Consequently,

$$\text{Var}\,[S_n] = npq + n(n-1)\left(\frac{-pq}{N-1}\right) = npq\left(1 - \frac{n-1}{N-1}\right).$$

The desired conclusions may now be inferred. ◀

▶ **Example 4C. The number of occupied urns as a sum.** If n distinguishable balls are distributed into M distinguishable urns in such a way that each ball is equally likely to go into any urn, what is the expected number of occupied urns?

Solution: For $k = 1, 2, \ldots, M$ let $X_k = 1$ or 0, depending on whether the kth urn is or is not occupied. Then $S = X_1 + X_2 + \ldots + X_M$ is the number of occupied urns, and $E[S]$ the expected number of occupied urns. The probability that a given urn will be occupied is equal to $1 - [1 - (1/M)]^n$. Therefore, $E[X_k] = 1 - [1 - (1/M)]^n$ and $E[S] = M\{1 - [1 - (1/M)]^n\}$. ◀

THEORETICAL EXERCISES

4.1. Waiting times in coupon collecting. Assume that each pack of cigarettes of a certain brand contains one of a set of N cards and that these cards are distributed among the packs at random (assume that the number of packs available is infinite). Let S_N be the minimum number of packs that must be purchased in order to obtain a complete set of N cards. Show that $E[S_N] = N\sum_{k=1}^{N}(1/k)$, which may be evaluated by using the formula (see H. Cramér, *Mathematical Methods of Statistics*, Princeton University Press, 1946, p. 125)

$$\sum_{k=1}^{N}\frac{1}{k} = 0.57722 + \log_e N + \frac{1}{2N} + R_N,$$

in which $0 < R_N < 1/8N^2$. Verify that $E[S_{52}] \doteq 236$ if $N = 52$. *Hint:* For $k = 0, 1, \ldots, N-1$ let X_k be the number of packs that must be purchased after k distinct cards have been collected in order to collect the $(k+1)$st distinct card. Show that $E[X_k] = N/(N-k)$ by using the fact that X_k has a geometric distribution.

4.2. Continuation of (4.1). For $r = 1, 2, \ldots, N$ let S_r be the minimum number of packs that must be purchased in order to obtain r different cards. Show that

$$E[S_r] = N\left(\frac{1}{N} + \frac{1}{N-1} + \frac{1}{N-2} + \cdots + \frac{1}{N-r+1}\right)$$

$$\operatorname{Var}[S_r] = N\left(\frac{1}{(N-1)^2} + \frac{2}{(N-2)^2} + \cdots + \frac{r-1}{(N-r+1)^2}\right).$$

Show that approximately (for large N)

$$E[S_r] \doteq N\log\frac{N}{N-r+1}.$$

Show further that the moment-generating function of S_r is given by

$$\psi_{S_r}(t) = \prod_{k=0}^{r-1}\frac{(N-k)e^t}{(N-ke^t)}.$$

4.3. Continuation of (4.1). For r preassigned cards let T_r be the minimum number of packs that must be purchased in order to obtain all r cards. Show that

$$E[T_r] = \sum_{k=1}^{r}\frac{N}{r-k+1}, \qquad \operatorname{Var}[T_r] = \sum_{k=1}^{r}\frac{N(N-r+k-1)}{(r-k+1)^2}.$$

4.4. The mean and variance of the number of matches. Let S_M be the number of matches obtained by distributing, 1 to an urn, M balls, numbered 1 to M, among M urns, numbered 1 to M. It was shown in theoretical exercise 3.3 of Chapter 5 that $E[S_M] = 1$ and $\operatorname{Var}[S_M] = 1$. Show this, using the fact that $S_M = X_1 + \ldots + X_M$, in which $X_k = 1$ or 0, depending on whether the kth urn does or does not contain ball number k. *Hint:* Show that $\operatorname{Cov}[X_j, X_k] = (M-1)/M^2$ or $1/M^2(M-1)$, depending on whether $j = k$ or $j \neq k$.

4.5. Show that if $X_1, \ldots, X_n$ are independent random variables with zero means and finite fourth moments, then the third and fourth moments of the sum $S_n = X_1 + \ldots + X_n$ are given by

$$E[S_n^3] = \sum_{k=1}^{n} E[X_k^3], \qquad E[S_n^4] = \sum_{k=1}^{n} E[X_k^4] + 6\sum_{k=1}^{n} E[X_k^2] \sum_{j=k+1}^{n} E[X_j^2].$$

If the random variables $X_1, \ldots, X_n$ are independent and identically distributed as a random variable X, then

$$E[S_n^3] = nE[X^3], \qquad E[S_n^4] = nE[X^4] + 3n(n-1)E^2[X^2].$$

4.6. Let $X_1, X_2, \ldots, X_n$ be a random sample of a random variable X. Define the sample mean $\bar{X}$ and the sample variance S^2 by

$$\bar{X} = \frac{1}{n}\sum_{k=1}^{n} X_k, \qquad S^2 = \frac{1}{n-1}\sum_{k=1}^{n}(X_k - \bar{X})^2.$$

(i) Show that $E[S^2] = \sigma^2$, $\text{Var}[S^2] = (\sigma^4/n)[(\mu_4/\sigma^4) - (n - 3/n - 1)]$, in which $\sigma^2 = \text{Var}[X]$, $\mu_4 = E[(X - E[X])^4]$. *Hint:* show that

$$\sum_{k=1}^{n}(X_k - E[X])^2 = \sum_{k=1}^{n}(X_k - \bar{X})^2 + n(\bar{X} - E[X])^2.$$

(ii) Show that $\rho(X_i - \bar{X}, X_j - \bar{X}) = \dfrac{-1}{n-1}$ for $i \neq j$.

EXERCISES

4.1. Let X_1, X_2, and X_3 be independent normally distributed random variables, each with mean 1 and variance 3. Find $P[X_1 + X_2 + X_3 > 0]$.

4.2. Consider a sequence of independent repeated Bernoulli trials in which the probability of success on any trial is $p = \frac{5}{16}$.

(i) Let S_n be the number of trials required to achieve the nth success. Find $E[S_n]$ and $\text{Var}[S_n]$. *Hint:* Write S_n as a sum, $S_n = X_1 + \ldots + X_n$, in which X_k is the number of trials between the $k - 1$st and kth successes. The random variables $X_1, \ldots, X_n$ are independent and identically distributed.

(ii) Let T_n be the number of failures encountered before the nth success is achieved. Find $E[T_n]$ and $\text{Var}[T_n]$.

4.3. A fair coin is tossed n times. Let T_n be the number of times in the n tosses that a tail is followed by a head. Show that $E[T_n] = (n-1)/4$ and $E[T_n^2] = (n-1)/4 + [(n-2)(n-3)]/16$. Find $\text{Var}[T_n]$.

4.4. A man with n keys wants to open his door. He tries the keys independently and at random. Let N_n be the number of trials required to open the door. Find $E[N_n]$ and $\text{Var}[N_n]$ if (i) unsuccessful keys are not eliminated from further selections, (ii) if they are. Assume that exactly one of the keys can open the door.

In exercises 4.5 and 4.6 consider an item of equipment that is composed by assembling in a straight line 4 components of lengths X_1, X_2, X_3, and X_4, respectively. Let $E[X_1] = 20$, $E[X_2] = 30$, $E[X_3] = 40$, $E[X_4] = 60$.

4.5. Assume $\text{Var}[X_j] = 4$ for $j = 1, \ldots, 4$.
(i) Find the mean and variance of the length $L = X_1 + X_2 + X_3 + X_4$ of the item if X_1, X_2, X_3, and X_4 are uncorrelated.
(ii) Find the mean and variance of L if $\rho(X_j, X_k) = 0.2$ for $1 \leq j < k \leq 4$.

4.6. Assume that $\sigma[X_j] = (0.1)E[X_j]$ for $j = 1, \ldots, 4$. Find the ratio $E[L]/\sigma[L]$, called the measurement signal-to-noise ratio of the length L (see section 6), for both cases considered in exercise 4.5.

5. THE LAW OF LARGE NUMBERS AND THE CENTRAL LIMIT THEOREM

In the applications of probability theory to real phenomena two results of the mathematical theory of probability play a conspicuous role. These results are known as the law of large numbers and central limit theorem. At this point in this book we have sufficient mathematical tools available to show how to apply these basic results. In Chapters 9 and 10 we develop the additional mathematical tools required to prove these theorems with a sufficient degree of generality.

A set of n observations $X_1, X_2, \ldots, X_n$ are said to constitute a random sample of a random variable X if $X_1, X_2, \ldots, X_n$ are independent random variables, identically distributed as X. Let

$$S_n = X_1 + X_2 + \cdots + X_n \tag{5.1}$$

be the sum of the observations. Their arithmetic mean

$$M_n = \frac{1}{n} S_n \tag{5.2}$$

is called the *sample mean*.

By (4.1), (4.6), and (4.7), we obtain the following expressions for the mean, variance, and moment-generating function of S_n and M_n, in terms of the mean, variance, and moment-generating function of X (assuming these exist):

$$E[S_n] = nE[X], \qquad \text{Var}[S_n] = n\,\text{Var}[X], \qquad \psi_{S_n}(t) = [\psi_X(t)]^n. \tag{5.3}$$

$$E[M_n] = E[X], \qquad \text{Var}[M_n] = \frac{1}{n}\text{Var}[X], \qquad \psi_{M_n}(t) = \left[\psi_X\left(\frac{t}{n}\right)\right]^n. \tag{5.4}$$

From (5.4) we obtain the striking fact that the variance of the sample mean $(1/n)S_n$ tends to 0 as the sample size n tends to infinity. Now, by Chebyshev's inequality, it follows that if a random variable has a small variance then it is approximately equal to its mean, in the sense that with probability close to 1 an observation of the random variable will yield an observed value approximately equal to the mean of the random variable; in particular, the probability is 0.99 that an observed value of the random variable is within 10 standard deviations of the mean of the random variable. We have thus established that the sample mean of a random sample $X_1, X_2, \ldots, X_n$ of a random variable, with a probability that can be made as close to 1 as desired by taking a large enough sample, is approximately equal to the ensemble mean $E[X]$. This fact, known as the *law of large numbers*, was first established by Bernoulli in 1713 (see section 5 of Chapter 5). The validity of the law of large numbers is the mathematical expression of the fact that increasingly accurate measurements of a quantity (such as the length of a rod) are obtained by averaging an increasingly large number of observations of the value of the quantity. A precise mathematical statement and proof of the law of large numbers is given in Chapter 10.

However, even more can be proved about the sample mean than that it tends to be equal to the mean. *One can approximately evaluate, for any interval about the mean, the probability that the sample mean will have an observed value in that interval*, since the *sample mean is approximately normally distributed.* More generally, it may be shown that *if S_n is the sum of independent identically distributed random variables $X_1, X_2, \ldots, X_n$, with finite means and variances then, for any real numbers $a < b$*

$$(5.5) \quad P[a \le S_n \le b] = P\left[\frac{a - E[S_n]}{\sigma[S_n]} \le \frac{S_n - E[S_n]}{\sigma[S_n]} \le \frac{b - E[S_n]}{\sigma[S_n]}\right]$$
$$\doteq \Phi\left(\frac{b - E[S_n]}{\sigma[S_n]}\right) - \Phi\left(\frac{a - E[S_n]}{\sigma[S_n]}\right).$$

In words, (5.5) may be expressed as follows: *the sum of a large number of independent identically distributed random variables with finite means and variances, normalized to have mean zero and variance* 1, *is approximately normally distributed.* Equation (5.5) represents a rough statement of one of the most important theorems of probability theory. In 1920 G. Polya gave this theorem the name "the central limit theorem of probability theory." This name continues to be used today, although a more apt description would be "the normal convergence theorem." The central limit theorem was first proved by De Moivre in 1733 for the case in which $X_1, X_2, \ldots, X_n$ are Bernoulli random variables, so that S_n is then a binomial random variable. A proof of (5.5) in this case (with a continuity correction) was given in section 2 of Chapter 6. The determination of the exact conditions for the validity of (5.5) constituted the outstanding problem of probability theory from its beginning until the decade of the 1930's. A precise mathematical statement and proof of the central limit theorem is given in Chapter 10.

It may be of interest to outline the basic idea of the proof of (5.5), even though the mathematical tools are not at hand to justify the statements made. To prove (5.5) it suffices to prove that the moment-generating function

$$\psi_n(t) = E[e^{t(S_n - E[S_n])/\sigma[S_n]}] = \left\{\psi_{X-E[X]}\left(\frac{t}{\sqrt{n}\sigma[X]}\right)\right\}^n \tag{5.6}$$

satisfies for t in a neighborhood of 0

$$\lim_{n\to\infty} \log \psi_n(t) = \frac{t^2}{2}, \tag{5.7}$$

in which $t^2/2$ is the logarithm of the moment-generating function of a random variable X, which is $N(0, 1)$. Now, expanding in Taylor series,

$$\psi_{X-E[X]}(u) = 1 + \tfrac{1}{2}\sigma^2[X]u^2 + A(u), \tag{5.8}$$

where the remainder $A(u)$ satisfies the condition $\lim_{u\to 0} A(u)/u^2 = 0$. Similarly, $\log(1 + v) = v + B(v)$ where $\lim_{v\to 0} B(v)/v = 0$. Consequently one may show that for values of u sufficiently close to 0

$$\log \psi_{X-E[X]}(u) = \tfrac{1}{2}\sigma^2[X^2]u^2 + C(u) \tag{5.9}$$

where

$$\lim_{u\to 0} \frac{C(u)}{u^2} = 0. \tag{5.10}$$

It then follows that

$$\log \psi_n(t) = n \log \psi_{X-E[X]}\left(\frac{t}{\sqrt{n}\sigma[X]}\right) = \frac{t^2}{2} + nC\left(\frac{t}{\sqrt{n}\sigma[X]}\right) \tag{5.11}$$

where

$$\lim_{n\to\infty} nC\left(\frac{t}{\sqrt{n}\sigma[X]}\right) = 0. \tag{5.12}$$

From (5.11) and (5.12) one obtains (5.7). Our heuristic outline of the proof of (5.5) is now complete.

Given any random variable X with finite mean and variance, we define its *standardization*, denoted by X^*, as the random variable

$$X^* = \frac{X - E[X]}{\sigma[X]}. \tag{5.13}$$

The standardization X^* is a dimensionless random variable, with mean $E[X^*] = 0$ and variance $\sigma^2[X^*] = 1$.

The central limit theorem of probability theory can now be formulated: *The standardization* $(S_n)^*$ *of the sum* S_n *of a large number* n *of independent and identically distributed random variables is approximately normally distributed.* In Chapter 10 it is shown that this result may be considerably extended to include cases in which S_n is the sum of dependent nonidentically distributed random variables.

▶ **Example 5A. Reliability.** Evaluation of the reliability of rockets is a problem of obvious importance in the space age. By the reliability of a rocket one means the probability p that an attempted launching of the rocket will be successful. Suppose that rockets of a certain type have, by many tests, been established as 90% reliable. Suppose that a modification of the rocket design is being considered. Which of the following sets of evidence throws more doubt on the hypothesis that the modified rocket is only 90% reliable: (i) of 100 modified rockets tested, 96 performed satisfactorily, (ii) of 64 modified rockets tested, 62 (equal to 96.9%) performed satisfactorily.

Solution: Let S_1 be the number of rockets in the group of 100 which performed satisfactorily, and let S_2 be the number of rockets in the group of 64 which performed satisfactorily. If p is the reliability of a rocket, then S_1 and S_2 have standardizations (since S_1 and S_2 have binomial distributions):

$$(S_1)^* = \frac{S_1 - 100p}{10\sqrt{pq}}, \qquad (S_2)^* = \frac{S_2 - 64p}{8\sqrt{pq}}.$$

If $p = 0.9$, $S_1 = 96$, and $S_2 = 62$, then $(S_1)^* = 2$ and $(S_2)^* = 1\frac{5}{6}$. If $(S_1)^*$ is $N(0, 1)$, the probability of observing a value of $(S_1)^*$ greater than or equal to 2 is 0.023. If $(S_2)^*$ is $N(0, 1)$, the probability of observing a value of $(S_2)^*$ greater than or equal to 1.83 is 0.034. Consequently, scoring 96 successes in 100 tries is better evidence than scoring 62 successes in 64 tries for the hypothesis that the modified rocket has a higher reliability than the original rocket. ◀

▶ **Example 5B. Brownian motion and random walk.** A particle (of diameter 10^{-4} centimeter, say) immersed in a liquid or gas exhibits ceaseless irregular motions that are discernible under the microscope. The motion of such a particle is called Brownian, after the English botanist Robert Brown, who noticed the phenomenon in 1827. The same phenomenon is also exhibited in striking fashion by smoke particles suspended in air. The explanation of the phenomenon of Brownian motion was one of the major successes of statistical mechanics and kinetic theory. In 1905 Einstein showed that the Brownian motion could be explained by assuming that the particles are subject to the continual bombardment of the molecules of the surrounding medium. The theoretical results of Einstein were soon confirmed by the exact experimental work of Perrin. To appreciate the importance of these events, the reader should be aware that in the years around 1900 atoms and molecules were far from being accepted as they are today—there were still physicists who did not believe in them. After Einstein's work this was possible no longer (see Max Born, *Natural Philosophy of Cause and Chance*, Oxford, 1949, p.63). If we let S_t denote the displacement after t minutes of a particle in Brownian motion from its starting point, Einstein showed that S_t has probability density function

$$f_{S_t}(x) = \left(\frac{1}{4\pi Dt}\right)^{1/2} e^{-x^2/4Dt} \tag{5.14}$$

in which D is a constant, called the *diffusion coefficient*, which depends on the absolute temperature and friction coefficient of the surrounding medium. In words, S_t is normally distributed with mean 0 and variance

$$E[S_t^2] = 2Dt. \tag{5.15}$$

The result given by (5.15) is especially important; it states that the mean square displacement $E[S_t^2]$ of a particle in Brownian motion is proportional to the time t. A model for Brownian motion is provided by a particle undergoing a random walk. Let $X_1, X_2, \ldots, X_n$ be independent random variables, identically distributed as a random variable X, which has mean $E[X] = 0$ and finite variance $E[X^2]$. The sum $S_n = X_1 + X_2 + \ldots + X_n$ represents the displacement from its starting position of a point (or particle) performing a random walk on a straight line by taking at the kth step a displacement X_k. After n steps, the total displacement S_n has a mean and mean square given by

$$E[S_n] = 0, \qquad E[S_n^2] = nE[X^2]. \tag{5.16}$$

Thus the mean-square displacement of a particle undergoing a random walk is proportional to the number of steps n. Since S_n is approximately normally distributed in the sense that (5.5) holds, it might be thought that the probability density function of S_n is approximately given by

$$f_{S_n}(x) = \left(\frac{1}{2\pi Bn}\right)^{1/2} e^{-x^2/2Bn}, \tag{5.17}$$

in which $B = E[X^2]$. However, (5.17) represents a stronger conclusion than (5.5). Equation (5.17) is a normal convergence theorem for probability density functions, whereas (5.5) is a normal convergence theorem for distribution functions; (5.17) implies (5.5), but the converse is not true. It may be shown that a sufficient condition for the validity of (5.17) is that the random variable X possesses a square integrable probability density function. From the fact that S_n is approximately normally distributed in the sense that (5.5) holds it follows that it is very improbable that a value of S_n will be observed more than 3 or 4 standard deviations from its mean. Consequently, in a random walk in which the individual steps have mean 0 it is very unlikely after n steps that the distance from the origin will be greater than $4\sigma[X]\sqrt{n}$. ◀

EXERCISES

5.1. Which of the following sets of evidence throws more doubt on the hypothesis that new born babies are as likely to be boys as girls: (i) of 10,000 new born babies, 5100 are male; (ii) of 1000 new born babies, 510 are male.

5.2. The game of roulette is described in example 1D. Find the probability that the total amount of money lost by a gambling house on 100,000 bets made by the public on an odd outcome at roulette will be negative.

5.3. As an estimate of the unknown mean $E[X]$ of a random variable, it is customary to take the sample mean $\bar{X} = (X_1 + X_2 + \ldots + X_n)/n$ of a random sample $X_1, X_2, \ldots, X_n$ of the random variable X. How large a sample should one observe if there is to be a probability of at least 0.95 that the sample mean $\bar{X}$ will not differ from the true mean $E[X]$ by more than 25% of the standard deviation $\sigma[X]$?

5.4. A man plays a game in which his probability of winning or losing a dollar is $\frac{1}{2}$. Let S_n be the man's fortune (that is, the amount he has won or lost) after n independent plays of the game.

(i) Find $E[S_n]$ and Var $[S_n]$. *Hint:* Write $S_n = X_1 + \ldots + X_n$, in which X_i is the change in the man's fortune on the ith play of the game.

(ii) Find *approximately* the probability that after 10,000 plays of the game the change in the man's fortune will be between -50 and 50 dollars.

5.5. Consider a game of chance in which one may win 10 dollars or lose 1, 2, 3, or 4 dollars; each possibility has probability 0.20. How many times can this game be played if there is to be a probability of at least 95% that in the final outcome the average gain or loss per game will be between -2 and $+2$?

5.6. A certain gambler's daily income (in dollars) is a random variable X uniformly distributed over the interval -3 to 3.

(i) Find approximately the probability that after 100 days of independent play he will have won more than 200 dollars.

(ii) Find the quantity A that the probability is greater than 95% that the gambler's winnings (which may be negative) in 100 independent days of play will be greater than A.

(iii) Determine the number of days the gambler can play in order to have a probability greater than 95% that his total winnings on these days will be less than 180 dollars in absolute value.

5.7. Add 100 real numbers, each of which is rounded off to the nearest integer. Assume that each rounding-off error is a random variable uniformly distributed between $-\frac{1}{2}$ and $\frac{1}{2}$ and that the 100 rounding-off errors are independent. Find approximately the probability that the error in the sum will be between -3 and 3. Find the quantity A that the probability is approximately 99% that the error in the sum will be less than A in absolute value.

5.8. If each strand in a rope has a breaking strength, with mean 20 pounds and standard deviation 2 pounds, and the breaking strength of a rope is the sum of the (independent) breaking strengths of all the strands, what is the probability that a rope made up of 64 strands will support a weight of (i) 1280 pounds, (ii) 1240 pounds.

5.9. A delivery truck carries loaded cartons of items. If the weight of each carton is a random variable, with mean 50 pounds and standard deviation 5 pounds, how many cartons can the truck carry so that the probability of the total load exceeding 1 ton will be less than 5%? State any assumptions made.

5.10. Consider light bulbs, produced by a machine, whose life X in hours is a random variable obeying an exponential probability law with a mean lifetime of 1000 hours.

(i) Find approximately the probability that a sample of 100 bulbs selected at random from the output of the machine will contain between 30 and 40 bulbs with a lifetime greater than 1020 hours.

(ii) Find approximately the probability that the sum of the lifetimes of 100 bulbs selected randomly from the output of the machine will be less than 110,000 hours.

5.11. The apparatus known as Galton's quincunx is described in exercise 2.10 of Chapter 6. Assume that in passing from one row to the next the change X in the abscissa of a ball is a random variable, with the following probability law: $P[X = \frac{1}{2}] = P[X = -\frac{1}{2}] = \frac{1}{2} - \eta$, $P[X = \frac{3}{2}] = P[X = -\frac{3}{2}] = \eta$, in which η is an unknown constant. In an experiment performed with a quincunx consisting of 100 rows, it was found that 80% of the balls inserted into the apparatus passed through the 21 central openings of the last row (that is, the openings with abscissas 0, ± 1, $\pm 2, \ldots, \pm 10$). Determine the value of η consistent with this result.

5.12. A man invests a total of N dollars in a group of n securities, whose rates of return (interest rates) are independent random variables $X_1, X_2, \ldots, X_n$, respectively, with means $i_1, i_2, \ldots, i_n$ and variances $\sigma_1^2, \sigma_2^2, \ldots, \sigma_n^2$, respectively. If the man invests N_j dollars in the jth security, then his return in dollars on this particular portfolio is a random variable R given by $R = N_1X_1 + N_2X_2 + \ldots + N_nX_n$. Let the standard deviation $\sigma[R]$ of R be used as a measure of the risk involved in selecting a given portfolio of securities. In particular, let us consider the problem of distributing investments of 5500 dollars between two securities, one of which has a rate of return X_1, with mean 6% and standard deviation 1%, whereas the other has a rate of return X_2 with mean 15% and standard deviation 10%. (i) If it is desired to hold the risk to a minimum, what amounts N_1 and N_2 should be invested in the respective securities? What is the mean and variance of the return from this portfolio?

(ii) What is the amount of risk that must be taken in order to achieve a portfolio whose mean return is equal to 400 dollars?

(iii) By means of Chebyshev's inequality, find an interval, symmetric about 400 dollars, that, with probability greater than 75%, will contain the return R from the portfolio with a mean return $E[R] = 400$ dollars. Would you be justified in assuming that the return R is approximately normally distributed?

6. THE MEASUREMENT SIGNAL-TO-NOISE RATIO OF A RANDOM VARIABLE

A question of great importance in science and engineering is the following: under what conditions can an observed value of a random variable X be identified with its mean $E[X]$? We have seen in section 5 that if X is the arithmetic mean of a very large number of independent identically distributed random variables then for any preassigned distance ϵ an observed value of X will, with high probability, be within ϵ of $E[X]$. In this section we discuss some conditions under which an observed value of a random variable may be identified with its mean.

If X has finite mean $E[X]$ and variance $\sigma^2[X]$, then the condition that an observed value of X is, with high probability, within a preassigned distance ϵ from its mean may be obtained from Chebyshev's inequality: for any $\epsilon > 0$

$$P[|X - E[X]| \leq \epsilon] \geq 1 - \frac{\sigma^2[X]}{\epsilon^2}. \tag{6.1}$$

From (6.1) one obtains these conclusions:

$$\begin{aligned} P[|X - E[X]| \leq \epsilon] &\geq 95\% \qquad \text{if } \epsilon \geq 4.5\sigma[X], \\ &\geq 99\% \qquad \text{if } \epsilon \geq 10\sigma[X]. \end{aligned} \tag{6.2}$$

If X may be assumed to be approximately normally distributed, then

$$P[|X - E[X]| \leq \epsilon] = \Phi(\epsilon/\sigma[X]) - \Phi(-\epsilon/\sigma[X]). \tag{6.3}$$

From (6.3) one obtains these conclusions:

$$\begin{aligned} P[|X - E[X]| \leq \epsilon] &\geq 95\% \qquad \text{if } \epsilon \geq 1.96\sigma[X], \\ &\geq 99\% \qquad \text{if } \epsilon > 2.58\sigma[X]. \end{aligned} \tag{6.4}$$

As a measure of how close the observed value of X will be to its mean $E[X]$, one often uses not the absolute deviation $|X - E[X]|$ but the *relative deviation*

$$\frac{|X - E[X]|}{|E[X]|} = \left|1 - \frac{X}{E[X]}\right|, \tag{6.5}$$

assuming that $E[X] \neq 0$.

Chebyshev's inequality may be reformulated in terms of the relative deviation: for any $\delta > 0$

$$P\left[\left|\frac{X - E[X]}{E[X]}\right| \leq \delta\right] \geq 1 - \frac{1}{\delta^2}\frac{\sigma^2[X]}{E^2[X]}. \tag{6.6}$$

From (6.6) one obtains these conclusions:

$$\begin{aligned} P\left[\left|\frac{X - E[X]}{E[X]}\right| \leq \delta\right] &\geq 95\% \qquad \text{if } \delta \geq 4.5\,\frac{\sigma[X]}{|E[X]|}, \\ &\geq 99\% \qquad \text{if } \delta \geq 10\,\frac{\sigma[X]}{|E[X]|}. \end{aligned} \tag{6.7}$$

Similarly, if X is approximately normally distributed,

$$\begin{aligned} P\left[\left|\frac{X - E[X]}{E[X]}\right| \leq \delta\right] &\geq 95\% \qquad \text{if } \delta \geq 1.96\,\frac{\sigma[X]}{|E[X]|}, \\ &\geq 99\% \qquad \text{if } \delta \geq 2.58\,\frac{\sigma[X]}{|E[X]|}. \end{aligned} \tag{6.8}$$

From the foregoing inequalities we obtain this basic conclusion for a random variable X with nonzero mean and finite variance.

In order that the percentage error of X as an estimate of $E[X]$ may with high probability be small, it is sufficient that the ratio

$$\frac{|E[X]|}{\sigma[X]} \tag{6.9}$$

be large. The quantity in (6.9) is called the *measurement signal-to-noise ratio* of the random variable X.*

How large must the measurement signal-to-noise ratio of a random variable X be in order that its observed value X be a good estimate of its mean? By (6.7) and (6.8), various answers to this question can be obtained.

* The measurement signal-to-noise ratio of a random variable is the reciprocal of the *coefficient of variation* of the random variable. (For a definition of the latter, see M. G. Kendall and A. Stuart, *The Advanced Theory of Statistics*, Griffin, London, 1958, p. 47.)

For example, if it is desired that

$$(6.10)\qquad P\left[\left|\frac{X - E[X]}{E[X]}\right| \le 10\%\right] \ge 95\%,$$

then the measurement signal-to-noise ratio must satisfy approximately

$$(6.11)\qquad \begin{aligned} &\frac{|E[X]|}{\sigma[X]} \ge 45 \quad \text{if Chebyshev's inequality applies,} \\ &\frac{|E[X]|}{\sigma[X]} \ge 20 \quad \text{if the normal approximation applies.} \end{aligned}$$

The measurement signal-to-noise ratio of various random variables is given in Table 6A. One sees that for most of the random variables given the measurement signal-to-noise ratio is proportional to the *square root* of some parameter. For example, suppose the number of particles

TABLE 6A

MEASUREMENT SIGNAL-TO-NOISE RATIO OF RANDOM VARIABLES OBEYING VARIOUS PROBABILITY LAWS

Probability Law of X	$E[X]$	$\sigma^2[X]$	$\left(\frac{E[X]}{\sigma[X]}\right)^2$
Poisson, with parameter $\lambda > 0$	λ	λ	λ
Binomial, with parameters n and p	np	$np(1-p)$	$\frac{np}{1-p}$
Geometric, with parameter p	$\frac{1}{p}$	$\frac{q}{p^2}$	$\frac{1}{q}$
Uniform over the interval a to b	$\frac{a+b}{2}$	$\frac{1}{12}(b-a)^2$	$3\left(\frac{b+a}{b-a}\right)^2$
Normal, with parameters m and σ	m	σ^2	$\left(\frac{m}{\sigma}\right)^2$
Exponential, with parameter λ	$\frac{1}{\lambda}$	$\frac{1}{\lambda^2}$	1
χ^2, with n degrees of freedom	n	$2n$	$\frac{n}{2}$
F with n_1, n_2 degrees of freedom	$\frac{n_2}{n_2-2}$ if $n_2 > 2$	$\frac{2n_2^2(n_1+n_2-2)}{n_1(n_2-2)^2(n_2-4)}$ if $n_2 > 4$	$\frac{n_1(n_2-4)}{2(n_1+n_2-2)}$ if $n_2 > 4$

emitted by a radioactive source during a certain time interval is being counted. The number of particles emitted obeys a Poisson probability law with some parameter λ whose value is unknown. If the true value of λ is known to be very large, then the observed number X of emitted particles is a good estimate of λ, since the measurement signal-to-noise ratio of X is $\sqrt{\lambda}$.

It is shown in Chapter 10 that many of the random variables in Table 6A are approximately normally distributed in cases in which their measurement signal-to-noise ratio is very large.

▶ **Example 6A. The density of an ideal gas.** An ideal gas can be regarded as a collection of n molecules distributed randomly in a volume V. The density of the gas in a subvolume v, contained in V, is a random variable d given by $d = Nm/v$, in which m is the mass of one gas molecule and N is the number of molecules in the volume v. Since it is assumed that each of the n molecules has an independent probability v/V of being in the subvolume v, the number N of molecules in v obeys a binomial probability law with mean $E[N] = nv/V$ and variance $\sigma^2[N] = npq$, in which we have let $p = v/V$ and $q = 1 - p$. The density then has mean $E[d] = nm/V$. In speaking of the density of gas in the volume v, the physicist usually has in mind the mean density. The question naturally arises: under what circumstances is the relative deviation $(d - E[d])/E[d]$ of the true density d from the mean density $E[d]$ within a preassigned percentage error δ? More specifically, what values must n, m, v, and V have in order that

$$P\left[\left|\frac{d - E[d]}{E[d]}\right| \le \delta\right] \ge 1 - \eta, \tag{6.12}$$

in which δ and η are preassigned positive quantities. By Chebyshev's inequality

$$P\left[\left|\frac{d - E[d]}{E[d]}\right| \le \delta\right] \ge 1 - \frac{\sigma^2[d]}{\delta^2 E^2[d]} = 1 - \frac{q}{\delta^2 np}. \tag{6.13}$$

Consequently, if the quantities n, m, v, and V are such that

$$\frac{1 - (v/V)}{n(v/V)} \le \delta^2\eta, \tag{6.14}$$

then (6.12) holds. Because of the enormous size of n (which is of the order 10^{20} per cm^3), one would expect (6.14) to be satisfied for $\eta = \delta = 10^{-5}$, say, as long as (v/V) is not too small. In this case it makes sense to speak of the density of gas in v, even though the number of molecules in v is not fixed but fluctuates. However, if v/V is very small, the fluctuations become sufficiently pronounced, and the ordinary notion of density, which identifies density with mean density, loses its meaning. The "density fluctuations" in small volumes can actually be detected experimentally inasmuch as they cause scattering of sufficiently short wavelengths. ◀

▶ **Example 6B. The law of $\sqrt{n}$.** The physicist Erwin Schrödinger has pointed out in the following statement (*What is Life*, Cambridge University Press, 1945, p.16), "... the degree of inaccuracy to be expected in any physical law, the so-called $\sqrt{n}$ law. The laws of physics and physical chemistry are inaccurate within a probable relative error of the order of $1/\sqrt{n}$, where n is the number of molecules that cooperate to bring about that law." From the law of $\sqrt{n}$ Schrödinger draws the conclusion that in order for the laws of physics and chemistry to be sufficient to explain the laws governing the behavior of living organisms it is necessary that the biologically relevant processes of such an organism involve the cooperation of a very large number of atoms, for only in this case do the laws of physics become exact laws. Since one can show that there are "incredibly small groups of atoms, much too small to display exact statistical laws, which play a dominating role in the very orderly and lawful events within a living organism", Schrödinger conjectures that it may not be possible to interpret life by the ordinary laws of physics, based on the "statistical mechanism which produces order from disorder." We state here a mathematical formulation of the law of $\sqrt{n}$. If $X_1, X_2, \ldots, X_n$ are independent random variables identically distributed as a random variable X, then the sum $S_n = X_1 + X_2 + \ldots + X_n$ and the sample mean $M_n = S_n/n$ have measurement signal-to-noise ratios given by

$$\frac{E[S_n]}{\sigma[S_n]} = \frac{E[M_n]}{\sigma[M_n]} = \sqrt{n}\,\frac{E[X]}{\sigma[X]}.$$

In words, the sum or average of n repeated independent measurements of a random variable X has a measurement signal-to-noise ratio of the order of $\sqrt{n}$. ◀

▶ **Example 6C. Can the energy of an ideal gas be both constant and a χ^2 distributed random variable?** In example 9H of Chapter 7 it is shown that if the state of an ideal gas is a random phenomenon whose probability law is given by Gibbs's canonical distribution then the energy E of the gas is a random variable possessing a χ^2 distribution with $3N$ degrees of freedom, in which N is the number of particles comprising the gas. Does this mean that if a gas has constant energy its state as a point in the space of all possible velocities cannot be regarded as obeying Gibbs's canonical distribution? The answer to this question is no. From a practical point of view there is no contradiction in regarding the energy E of the gas as being both a constant and a random variable with a χ^2 distribution if the number of degrees of freedom is very large, for then the measurement signal-to-noise ratio of E (which, from Table 6A, is equal to $(3N/2)^{1/2}$) is also very large. ◀

The terminology "signal-to-noise ratio" originated in communications theory. The mean $E[X]$ of a random variable X is regarded as a signal that one is attempting to receive (say, at a radio receiver). However, X is actually received. The difference between the desired value $E[X]$ and the received value X is called *noise*. The less noise present, the better one is able to receive the signal accurately. As a measure of signal strength to noise strength, one takes the signal-to-noise ratio defined by (6.9). The higher the signal-to-noise ratio, the more accurate the observed value X as an estimate of the desired value $E[X]$.

Any time a scientist makes a measurement he is attempting to obtain a signal in the presence of noise or, equivalently, to estimate the mean of a random variable. The skill of the experimental scientist lies in being able to conduct experiments that have a high measurement signal-to-noise ratio. However, there are experimental situations in which this may not be possible. For example, there is an inherent limit on how small one can make the variance of measurements taken with electronic devices. This limit arises from the noise or spontaneous current fluctuations present in such devices (see example 3D of Chapter 6). To measure weak signals in the presence of noise (that is, to measure the mean of a random variable with a small measurement signal-to-noise ratio) one should have a good knowledge of the modern theories of statistical inference.

On the one hand, the scientist and engineer should know statistics in order to interpret best the statistical significance of the data he has obtained. On the other hand, a knowledge of statistics will help the scientist or engineer to solve the basic problem confronting him in taking measurements: given a parameter θ, which he wishes to measure, to find random variables $X_1, X_2, \ldots, X_n$, whose observed values can be used to form estimates of θ that are best according to some criteria.

Measurement signal-to-noise ratios play a basic role in the evaluation of modern electronic apparatus. The reader interested in such questions may consult J. J. Freeman, *Principles of Noise*, Wiley, New York, 1958, Chapters 7 and 9.

EXERCISES

6.1. A random variable X has an unknown mean and known variance 4. How large a random sample should one take if the probability is to be at least 0.95 that the sample mean will not differ from the true mean $E[X]$ by (i) more than 0.1, (ii) more than 10% of the standard deviation of X, (iii) more than 10% of the true mean of X, if the true mean of X is known to be greater than 10.

6.2. Let $X_1, X_2, \ldots, X_n$ be independent normally distributed random variables with known mean 0 and unknown common variance σ^2. Define

$$S_n = \frac{1}{n}(X_1^2 + X_2^2 + \cdots + X_n^2).$$

Since $E[S_n] = \sigma^2$, S_n might be used as an estimate of σ^2. How large should n be in order to have a measurement signal-to-noise ratio of S_n greater than 20? If the measurement signal-to-noise ratio of S_n is greater than 20, how good is S_n as an estimate of σ^2?

6.3. Consider a gas composed of molecules (with mass of the order of 10^{-24} grams and at room temperature) whose velocities obey the Maxwell-Boltzmann law (see exercise 1.15). Show that one may assume that all the molecules move with the same velocity, which may be taken as either the mean velocity, the root mean square velocity, or the most probable velocity.

7. CONDITIONAL EXPECTATION. BEST LINEAR PREDICTION

An important tool in the study of the relationships that exist between two jointly distributed random variables, X and Y, is provided by the notion of conditional expectation. In section 11 of Chapter 7 the notion of the conditional distribution function $F_{Y|X}(\cdot \mid x)$ of the random variable Y, given the random variable X, is defined. We now define the *conditional mean* of Y, given X, by

$$(7.1) \qquad E[Y \mid X = x] = \begin{cases} \displaystyle\int_{-\infty}^{\infty} y \, d_y F_{Y|X}(y \mid x) \\ \displaystyle\int_{-\infty}^{\infty} y f_{Y|X}(y \mid x) \, dy \\ \displaystyle\sum_{\substack{\text{over all } y \text{ such that} \\ p_{Y|X}(y|x) > 0}} y \, p_{Y|X}(y \mid x); \end{cases}$$

the last two equations hold, respectively, in the cases in which $F_{Y|X}(\cdot \mid x)$ is continuous or discrete. From a knowledge of the conditional mean of Y, given X, the value of the mean $E[Y]$ may be obtained:

$$(7.2) \qquad E[Y] = \begin{cases} \displaystyle\int_{-\infty}^{\infty} E[Y \mid X = x] \, dF_X(x) \\ \displaystyle\int_{-\infty}^{\infty} E[Y \mid X = x] f_X(x) \, dx \\ \displaystyle\sum_{\substack{\text{over all } x \text{ such that} \\ p_X(x) > 0}} E[Y \mid X = x] p_X(x) \end{cases}$$

▶ **Example 7A. Sampling from an urn of random composition.** Let a random sample of size n be drawn without replacement from an urn containing N balls. Suppose that the number X of white balls in the urn is a random variable. Let Y be the number of white balls contained in the sample. The conditional distribution of Y, given X, is discrete, with probability mass function for $x = 0, 1, \ldots, N$ and $y = 0, 1, \ldots, x$ given by

$$p_{Y|X}(y \mid x) = P[Y = y \mid X = x] = \frac{\binom{x}{y}\binom{N-x}{n-y}}{\binom{N}{n}}, \tag{7.3}$$

since the conditional probability law of Y, given X, is hypergeometric. The conditional mean of Y, given X, can be readily obtained from a knowledge of the mean of a hypergeometric random variable;

$$E[Y \mid X = x] = n\frac{x}{N}. \tag{7.4}$$

The mean number of white balls in the sample drawn is then equal to

$$E[Y] = \sum_{x=0}^{N} E[Y \mid X = x]p_X(x) = \frac{n}{N}\sum_{x=0}^{N} xp_X(x) = \frac{n}{N}E[X]. \tag{7.5}$$

Now $E[X]/N$ is the mean proportion of white balls in the urn. Consequently (7.5) is analogous to the formulas for the mean of a binomial or hypergeometric random variable. Note that the probability law of Y is hypergeometric if X is hypergeometric and Y is binomial if X is binomial. (See theoretical exercise 4.1 of Chapter 4.) ◀

▶ **Example 7B. The conditional mean of jointly normal random variables.** Two random variables, X_1 and X_2, are jointly normally distributed if they possess a joint probability density function given by (2.18). Then

$$f_{X_2|X_1}(x_2 \mid x_1) = \frac{1}{\sigma_2\sqrt{1-\rho^2}}\phi\left(\frac{x_2 - m_2 - (\sigma_2/\sigma_1)\rho(x_1 - m_1)}{\sigma_2\sqrt{1-\rho^2}}\right). \tag{7.6}$$

Consequently, the conditional mean of X_2, given X_1, is given by

$$E[X_2 \mid X_1 = x_1] = m_2 + \frac{\sigma_2}{\sigma_1}\rho(x_1 - m_1) = \alpha_1 + \beta_1 x_1, \tag{7.7}$$

in which we define the constants α_1 and β_1 by

$$\alpha_1 = m_2 - \frac{\sigma_2}{\sigma_1}\rho m_1, \qquad \beta_1 = \frac{\sigma_2}{\sigma_1}\rho. \tag{7.8}$$

Similarly,

$$(7.9)\quad E[X_1 \mid X_2 = x_2] = \alpha_2 + \beta_2 x_2; \qquad \alpha_2 = m_1 - \frac{\sigma_1}{\sigma_2}\rho m_2, \qquad \beta_2 = \frac{\sigma_1}{\sigma_2}\rho.$$

From (7.7) it is seen that the conditional mean of a random variable X_2, given the value x_1 of a random variable X_1 with which X_2 is jointly normally distributed, is a *linear function* of x_1. Except in the case in which the two random variables, X_1 and X_2, are jointly normally distributed, it is generally to be expected that $E[X_2 \mid X_1 = x_1]$ is a *nonlinear* function of x_1. ◀

The conditional mean of one random variable, given another random variable, represents one possible answer to the problem of *prediction*. Suppose that a prospective father of height x_1 wishes to predict the height of his unborn son. If the height of the son is regarded as a random variable X_2 and the height x_1 of the father is regarded as an observed value of a random variable X_1, then as the prediction of the son's height we take the conditional mean $E[X_2 \mid X_1 = x_1]$. The justification of this procedure is that the conditional mean $E[X_2 \mid X_1 = x_1]$ may be shown to have the property that

$$(7.10)\quad E[(X_2 - E[X_2 \mid X_1 = x_1])^2]$$
$$= \int_{-\infty}^{\infty}\int_{-\infty}^{\infty} (x_2 - E[X_2 \mid X_1 = x_1])^2 f_{X_1,X_2}(x_1, x_2)\,dx_1\,dx_2$$
$$\leq E[(X_2 - g(X_1))^2] = \int_{-\infty}^{\infty}\int_{-\infty}^{\infty} [x_2 - g(x_1)]^2 f_{X_1,X_2}(x_1, x_2)\,dx_1\,dx_2$$

for any function $g(x_1)$ for which the last written integral exists. In words, (7.10) is interpreted to mean that if X_2 is to be predicted by a function $g(X_1)$ of the random variable X_1 then the conditional mean $E[X_2 \mid X_1 = x_1]$ has the smallest mean square error among all possible predictors $g(X_1)$.

From (7.7) it is seen that in the case in which the random variables are jointly normally distributed the problem of computing the conditional mean $E[X_2 \mid X_1 = x_1]$ may be reduced to that of computing the constants α_1 and β_1, for which one requires a knowledge only of the means, variances, and correlation coefficient of X_1 and X_2. If these moments are not known, they must be estimated from observed data. The part of statistics concerned with the estimation of the parameters α_1 and β_1 is called *regression analysis*.

It may happen that the joint probability law of the random variables X_1 and X_2 is unknown or is known but is such that the calculation of the conditional mean $E[X_2 \mid X_1 = x_1]$ is intractable. Suppose, however, that one knows the means, variances (assumed to be positive), and correlation coefficient of X_1 and X_2. Then the prediction problem may be solved by forming the *best linear predictor of X_2, given X_1*, denoted by $E^*[X_2 \mid X_1 = x_1]$. The best linear predictor of X_2, given X_1, is defined as that linear function $a + bX_1$ of the random variable X_1, that minimizes the mean square error of prediction $E[(X_2 - (a + bX_1))^2]$ involved in the use of $a + bX_1$ as a predictor of X_2. Now

$$
\begin{aligned}
-\frac{\partial}{\partial a} E[(X_2 - (a + bX_1))^2] &= 2E[X_2 - (a + bX_1)] \\
-\frac{\partial}{\partial b} E[(X_2 - (a + bX_1))^2] &= 2E[(X_2 - (a + bX_1))X_1].
\end{aligned} \tag{7.11}
$$

Solving for the values of a and b, denoted by α and β, at which these derivatives are equal to 0, one sees that α and β satisfy the equations

$$
\begin{aligned}
\alpha + \beta E[X_1] &= E[X_2] \\
\alpha E[X_1] + \beta E[X_1^2] &= E[X_1 X_2].
\end{aligned} \tag{7.12}
$$

Therefore, $E^*[X_2 \mid X_1 = x_1] = \alpha + \beta x_1$, in which

$$
\alpha = E[X_2] - \beta E[X_1], \qquad \beta = \frac{\operatorname{Cov}[X_1, X_2]}{\operatorname{Var}[X_1]} = \frac{\sigma[X_2]}{\sigma[X_1]} \rho(X_1, X_2). \tag{7.13}
$$

Comparing (7.7) and (7.13), one sees that the best linear predictor $E^*[X_2 \mid X_1 = x_1]$ coincides with the best predictor, or conditional mean, $E[X_2 \mid X_1 = x_1]$, in the case in which the random variables X_1 and X_2 are jointly normally distributed.

We can readily compute the mean square error of prediction achieved with the use of the best linear predictor. We have

(7.14)

$$
\begin{aligned}
E[(X_2 - E^*[X_2 \mid X_1 = x_1])^2] &= E[\{(X_2 - E[X_2]) - \beta(X_1 - E[X_1])\}^2] \\
&= \operatorname{Var}[X_2] + \beta^2 \operatorname{Var}[X_1] - 2\beta \operatorname{Cov}[X_2, X_1] \\
&= \operatorname{Var}[X_2] - \frac{\operatorname{Cov}^2[X_1, X_2]}{\operatorname{Var}[X_1]} \\
&= \operatorname{Var}[X_2]\{1 - \rho^2(X_1, X_2)\}.
\end{aligned}
$$

From (7.14) one obtains the important conclusion that *the closer the correlation between two random variables is to 1, the smaller the mean square error of prediction involved in predicting the value of one of the random variables from the value of the other.*

The Phenomenon of "Spurious" Correlation. Given three random variables U, V, and W, let X and Y be defined by

$$(7.15) \qquad X = U + W, \qquad Y = V + W \qquad \text{or} \qquad X = \frac{U}{W}, \qquad Y = \frac{V}{W},$$

(or in some similar way) as functions of U, V, and W. The reader should be careful not to infer the existence of a correlation between U and V from the existence of a correlation between X and Y.

▶ **Example 7C. Do storks bring babies?** Let W be the number of women of child-bearing age in a certain geographical area, U, the number of storks in the area, and V, the number of babies born in the area during a specified period of time. The random variables X and Y, defined by

$$(7.16) \qquad X = \frac{U}{W}, \qquad Y = \frac{V}{W},$$

then represent, respectively, the number of storks per woman and the number of babies born per woman in the area. If the correlation coefficient $\rho(X, Y)$ between X and Y is close to 1, does that not prove that storks bring babies? Indeed, even if it is proved only that the correlation coefficient $\rho(X, Y)$ is positive, would that not prove that the presence of storks in an area has a beneficial influence on the birth rate there? The reader interested in a discussion of these delightful questions would be well advised to consult J. Neyman, *Lectures and Conferences on Mathematical Statistics and Probability*, Washington, D.C., 1952, pp. 143–154. ◀

THEORETICAL EXERCISES

In the following exercises let X_1, X_2, and Y be jointly distributed random variables whose first and second moments are assumed known and whose variances are positive.

7.1. The *best linear predictor*, denoted by $E^*[Y \mid X_1, X_2]$, of Y, given X_1 and X_2, is defined as the linear function $a + b_1X_1 + b_2X_2$, which minimizes $E[(Y - (a + b_1X_1 + b_2X_2))^2]$. Show that

$$E^*[Y \mid X_1, X_2] = E[Y] + \beta_1(X_1 - E[X_1]) + \beta_2(X_2 - E[X_2])$$

where

$$\beta_1 = \operatorname{Cov}[Y, X_1]\Sigma_{11} + \operatorname{Cov}[Y, X_2]\Sigma_{12}$$
$$\beta_2 = \operatorname{Cov}[Y, X_1]\Sigma_{21} + \operatorname{Cov}[Y, X_2]\Sigma_{22},$$

in which we define

$$\Sigma_{11} = \operatorname{Var}[X_2]/\Delta, \qquad \Sigma_{22} = \operatorname{Var}[X_1]/\Delta, \qquad \Sigma_{12} = \Sigma_{21} = -\operatorname{Cov}[X_1, X_2]/\Delta.$$
$$\Delta = \operatorname{Var}[X_1]\operatorname{Var}[X_2][1 - \rho^2(X_1, X_2)].$$

7.2. The *residual* of Y with respect to X_1 and X_2, denoted by $\eta[Y \mid X_1, X_2]$, is defined by

$$\eta[Y \mid X_1, X_2] = Y - E^*[Y \mid X_1, X_2].$$

Show that $\eta[Y \mid X_1, X_2]$ is uncorrelated with X_1 and X_2. Consequently, conclude that the mean square error of prediction, called the *residual variance* of Y, given X_1 and X_2, is given by

$$E[\eta^2[Y \mid X_1, X_2]] = \text{Var}[Y] - \text{Var}[E^*[Y \mid X_1, X_2]].$$

Next show that the variance of the predictor is given by

$$\begin{aligned}\text{Var}[E^*[Y \mid X_1, X_2]] &= \beta_1^2 \text{Var}[X_1] + \beta_2^2 \text{Var}[X_2] + 2\beta_1\beta_2 \text{Cov}[X_1, X_2] \\ &= \Sigma_{11} \text{Cov}^2[Y, X_1] + \Sigma_{22} \text{Cov}^2[Y, X_2] \\ &\qquad + 2\Sigma_{12} \text{Cov}[Y, X_1] \text{Cov}[Y, X_2].\end{aligned}$$

The positive quantity $R[Y \mid X_1, X_2]$, defined by

$$R^2[Y \mid X_1, X_2] = \frac{\text{Var}[E^*[Y \mid X_1, X_2]]}{\text{Var}[Y]} = \rho^2(Y, E^*[Y \mid X_1, X_2]),$$

is called the *multiple correlation coefficient* between Y and the random vector (X_1, X_2). To understand the meaning of the multiple correlation coefficient, express in terms of it the residual variance of Y, given X_1 and X_2.

7.3. The *partial correlation coefficient* of X_1 and X_2 with respect to Y is defined by

$$\rho[X_1, X_2 \mid Y] = \rho(\eta[X_1 \mid Y], \eta[X_2 \mid Y]),$$

in which $\eta[X_i \mid Y] = X_i - E^*[X_i \mid Y]$ for $i = 1, 2$. Show that

$$\rho[X_1, X_2 \mid Y] = \frac{\rho(X_1, X_2) - \rho(X_1, Y)\rho(X_2, Y)}{\sqrt{(1 - \rho^2(X_1, Y))(1 - \rho^2(X_2, Y))}}.$$

7.4. (Continuation of example 7A). Show that

$$\text{Var}[Y] = n\frac{E[X]}{N}\left(1 - \frac{E[X]}{N}\right)\frac{N-n}{N-1} + \frac{n-1}{N-1}\frac{n}{N}\text{Var}[X]. \tag{7.17}$$

EXERCISES

7.1. Let X_1, X_2, X_3 be jointly distributed random variables with zero means, unit variances, and covariances $\text{Cov}[X_1, X_2] = 0.80$, $\text{Cov}[X_1, X_3] = -0.40$, $\text{Cov}[X_2, X_3] = -0.60$. Find (i) the best linear predictor of X_1, given X_2, (ii) the best linear predictor of X_3, given X_2, (iii) the partial correlation between X_1 and X_3, given X_2, (iv) the best linear predictor of X_1, given X_2 and X_3, (v) the residual variance of X_1, given X_2 and X_3, (vi) the residual variance of X_1, given X_2.

7.2. Find the conditional mean of Y, given X, if X and Y are jointly continuous random variables with a joint probability density function $f_{X,Y}(x, y)$ vanishing except for $x > 0$, $y > 0$, and in the case in which $x > 0$, $y > 0$ given by

(i) $$\frac{4}{5}(x + 3y)e^{-x-2y},$$

(ii) $$\frac{y}{(1 + x)^4} e^{-y/(1+x)},$$

(iii) $$\frac{9}{2}\frac{1 + x + y}{(1 + x)^4(1 + y)^4}.$$

7.3. Let $X = \cos 2\pi U$, $Y = \sin 2\pi U$, in which U is uniformly distributed on 0 to 1. Show that for $|x| \leq 1$

$$E^*[Y \mid X = x] = 0, \qquad E[Y \mid X = x] = \sqrt{1 - x^2}.$$

Find the mean square error of prediction achieved by the use of (i) the best linear predictor, (ii) the best predictor.

7.4. Let U, V, and W be uncorrelated random variables with equal variances. Let $X = U \pm W$, $Y = V \pm W$. Show that

$$\rho(X, W) = \rho(Y, W) = 1/\sqrt{2}, \qquad \rho(X, Y) = 0.5.$$

CHAPTER 9

Sums of Independent Random Variables

Chapters 9 and 10 are much less elementary in character than the first eight chapters of this book. They constitute an introduction to the limit theorems of probability theory and to the role of characteristic functions in probability theory. These chapters seek to provide a careful and rigorous derivation of the law of large numbers and the central limit theorem.

In this chapter we treat the problem of finding the probability law of a random variable that arises as the sum of independent random variables. A major tool in this study is the characteristic function of a random variable, introduced in section 2. In section 3 it is shown that the probability law of a random variable can be determined from its characteristic function. Section 4 discusses some consequences of the basic result that the characteristic function of a sum of independent random variables is the product of the characteristic functions of the individual random variables. Section 5 gives the proofs of the inversion formulas stated in section 3.

1. THE PROBLEM OF ADDITION OF INDEPENDENT RANDOM VARIABLES

A large number of the problems which arise in applications of probability theory may be regarded as special cases of the following general problem, which we call the *problem of addition of independent random variables; find, either exactly or approximately, the probability law of a random variable that arises as the sum of n independent random variables* $X_1, X_2, \ldots,$ X_n, *whose joint probability law is known.* The fundamental role played by this problem in probability theory is best described by a quotation from an article by Harald Cramér, "Problems in Probability Theory," *Annals of Mathematical Statistics*, Volume 18 (1947), p. 169.

During the early development of the theory of probability, the majority of problems considered were connected with gambling. The gain of a player in a certain game may be regarded as a random variable, and his total gain in a sequence of repetitions of the game is the sum of a number of independent variables, each of which represents the gain in a single performance of the game. Accordingly, a great amount of work was devoted to the study of the probability distributions of such sums. A little later, problems of a similar type appeared in connection with the theory of errors of observation, when the total error was considered as the sum of a certain number of partial errors due to mutually independent causes. At first, only particular cases were considered; but gradually general types of problems began to arise, and in the classical work of Laplace several results are given concerning the general problem to study the distribution of a sum

$$S_n = X_1 + X_2 + \cdots + X_n$$

of independent variables, when the distributions of the X_j are given. This problem may be regarded as the very starting point of a large number of those investigations by which the modern Theory of Probability was created. The efforts to prove certain statements of Laplace, and to extend his results further in various directions, have largely contributed to the introduction of rigorous foundations of the subject, and to the development of the analytical methods.

In this chapter we discuss the methods and notions by which a precise formulation and solution is given to the problem of addition of independent random variables. To begin with, in this section we discuss the two most important ways in which this problem can arise, namely in the analysis of *sample averages* and in the analysis of *random walks.*

Sample Averages. We have defined a *sample* of size n of a random variable X as a set of n jointly distributed random variables $X_1, X_2, \ldots, X_n$, whose individual probability laws coincide, for $k = 1, 2, \ldots, n$, with the probability law of X; in particular, the distribution function $F_{X_k}(\cdot)$ of X_k coincides with the distribution function $F_X(\cdot)$ of X. We have defined the sample as a *random sample* if the random variables $X_1, X_2, \ldots, X_n$ are independent.

Given a sample $X_1, X_2, \ldots, X_n$ of size n of the random variable X and any Borel function $g(\cdot)$ of a real variable, we define the *sample average* of $g(\cdot)$, denoted by $M_n[g(x)]$, as the arithmetic mean of the values $g(X_1), g(X_2),$ $\ldots, g(X_n)$ of the function at the members of the sample; in symbols,

$$M_n[g(x)] = \frac{1}{n} \sum_{k=1}^{n} g(X_k). \tag{1.1}$$

Of special importance are the sample mean m_n, defined by

$$m_n = M_n[x] = \frac{1}{n}\sum_{k=1}^{n} X_k, \tag{1.2}$$

and the *sample variance* S_n^2, defined by

$$\begin{aligned} S_n^2 &= M_n[(x - m_n)^2] = M_n[x^2] - M_n^2[x] \\ &= \frac{1}{n}\sum_{k=1}^{n} (X_k - m_n)^2 = \frac{1}{n}\sum_{k=1}^{n} X_k^2 - \left(\frac{1}{n}\sum_{k=1}^{n} X_k\right)^2. \end{aligned} \tag{1.3}$$

For a given function $g(\cdot)$ *the sample average* $M_n[g(x)]$ *is a random variable*, for it is a function of the random variables $X_1, X_2, \ldots, X_n$. The value of $M_n[g(x)]$ will, in general, be different when it is computed on the basis of two different samples of size n. The sample average $M_n[g(x)]$, like any other random variable, has a mean $E[M_n[g(x)]]$, a variance Var $[M_n[g(x)]]$, a distribution function $F_{M_n[g(x)]}(\cdot)$, a moment-generating function $\psi_{M_n[g(x)]}(\cdot)$, and, depending on whether it is a continuous or a discrete random variable, a probability density function $f_{M_n[g(x)]}(\cdot)$ or a probability mass function $p_{M_n[g(x)]}(\cdot)$. Our aim in this chapter and the next is to develop techniques for computing these quantities, both exactly and approximately, and especially to study their behavior for large sample sizes. The reader who goes on to the study of mathematical statistics will find that these techniques provide the framework for many of the concepts of statistics.

To study sample averages $M_n[g(x)]$ with respect to a random sample, it suffices to consider the sum $\sum_{k=1}^{n} Y_k$ of independent random variables $Y_1, \ldots, Y_n$, since the random variables $Y_1 = g(X_1), \ldots, Y_n = g(X_n)$ are independent if the random variables $X_1, \ldots, X_n$ are. Thus it is seen that *the study of sample averages has been reduced to the study of sums of independent random variables*.

Random Walk. Consider a particle that at a certain time is located at the point 0 on a certain straight line. Suppose that it then suffers displacements along the straight line in the form of a series of steps, denoted by $X_1, X_2, \ldots, X_n$, in which, for any integer k, X_k represents the displacement suffered by the particle at the kth step. The size X_k of the kth step is assumed to be a random variable with a known probability law. The particle can thus be imagined as executing a *random walk* along the line, its position (denoted by S_n) after n steps being the sum of the n steps $X_1, X_2, \ldots, X_n$; in symbols, $S_n = X_1 + X_2 + \ldots + X_n$. Clearly, S_n is a random variable, and the problem of finding the probability law of S_n naturally arises; in other words, one wishes to know, for any integer n and any interval a to b, the probability $P[a \leq S_n \leq b]$ that after n steps the particle will lie between a and b, inclusive.

The problem of random walks can be generalized to two or more dimensions. Suppose that the particle at each stage suffers a displacement in an (x, y) plane, and let X_k and Y_k denote, respectively, the change in the x- and y-coordinates of the particle at the kth step. The position of the particle after n steps is given by the random 2-tuple (S_n, T_n), in which $S_n = X_1 + X_2 + \ldots + X_n$ and $T_n = Y_1 + Y_2 + \ldots + Y_n$. We now have the problem of determining the joint probability law of the random variables S_n and T_n.

The problem of random walks occurs in many branches of physics, especially in its 2-dimensional form. The eminent mathematical statistician, Karl Pearson, was the first to formulate explicitly the problem of the 2-dimensional random walk. After Pearson had formulated this problem in 1905, the renowned physicist, Lord Rayleigh, pointed out that the problem of random walks was formally "the same as that of the composition of n isoperiodic vibrations of unit amplitude and of phases distributed at random," which he had considered as early as 1880 (for this quotation and a history of the problem of random walks, see p. 87 of S. Chandrasekhar, "Stochastic Problems in Physics and Astronomy," *Reviews of Modern Physics*, Volume 15 (1943), pp. 1–89). Almost all scattering problems in physics are instances of the problem of random walks.

▶ **Example 1A. A physical example of random walk.** Consider the amplitude and phase of a radar signal that has been reflected by a cloud. Each of the water drops in the cloud reflects a signal with a different amplitude and phase. The return signal received by the radar system is the resultant of all the signals reflected by each of the water drops in the cloud; thus one sees that formally the amplitude and phase of the signal returned by the cloud to the radar system is the sum of a (large) number of (presumably independent) random variables. ◀

In the study of sums of independent random variables a basic role is played by the notion of the characteristic function of a random variable. This notion is introduced in section 2.

2. THE CHARACTERISTIC FUNCTION OF A RANDOM VARIABLE

It has been pointed out that the probability law of a random variable X may be specified in a variety of ways. To begin with, either its probability function $P_X[\cdot]$ or its distribution function $F_X(\cdot)$ may be stated. Further, if the probability law is known to be continuous or discrete, then it may be specified by stating either its probability density function $f_X(\cdot)$ or its probability mass function $p_X(\cdot)$. We now describe yet another function, denoted by $\phi_X(\cdot)$ called the *characteristic function of the random variable X*, which has the property that a knowledge of $\phi_X(\cdot)$ serves to specify the probability law of the random variable X. Further, we shall see that the characteristic function has properties which render it particularly useful for the study of a sum of independent random variables.

To begin our introduction of the characteristic function, let us note the following fact about the probability function $P_X[\cdot]$ and the distribution function $F_X(\cdot)$ of a random variable X. Both functions can be regarded as the value of the expectation (with respect to the probability law of X) of various Borel functions $g(\cdot)$. Thus, for every Borel set B of real numbers

$$P_X[B] = E_X[I_B(x)] = E[I_B(X)], \tag{2.1}$$

in which $I_B(\cdot)$ is a function of a real variable, called the *indicator* function of the set B, with value $I_B(x)$ at any point x given by

$$\begin{aligned} I_B(x) &= 1 \quad \text{if } x \text{ belongs to } B, \\ &= 0 \quad \text{if } x \text{ does not belong to } B. \end{aligned} \tag{2.2}$$

On the other hand, for every real number y

$$F_X(y) = E_X[I_y(x)] = E[I_y(X)], \tag{2.3}$$

in which $I_y(\cdot)$ is a function of a real variable, defined by

$$\begin{aligned} I_y(x) &= 1 \quad \text{if } x \leq y \\ &= 0 \quad \text{if } x > y. \end{aligned} \tag{2.4}$$

We thus see that if one knows the expectation $E_X[g(x)]$ of every bounded Borel function $g(\cdot)$, with respect to the probability law of the random variable X, one will know by (2.1) and (2.3) the probability function and distribution function of X. Conversely, a knowledge of the probability function or of the distribution function of X yields a knowledge of $E[g(X)]$ for every function $g(\cdot)$ for which the expectation exists. Consequently, stating the expectation functional $E_X[\cdot]$ of a random variable [which is a function whose argument is a function $g(\cdot)$] constitutes another equivalent way of specifying the probability law of a random variable.

The question arises: is there any other family of functions on the real line in addition to those of the form of (2.2) and (2.4) such that a knowledge of the expectations of these functions with respect to the probability law of a random variable X would suffice to specify the probability law? We now show that the *complex exponential functions* provide such a family.

We define the expectation, with respect to a random variable X, of a function $g(\cdot)$, which takes values that are complex numbers, by

$$E[g(X)] = E[\text{Re } g(X)] + iE[\text{Im } g(X)] \tag{2.5}$$

in which the symbols Re and Im, respectively, are abbreviations of the phrases "real part of" and "imaginary part of." Note that

$$g(x) = \text{Re } g(x) + i \text{ Im } g(x).$$

It may be shown that under these definitions all the usual properties of the operation of taking expectations continue to hold for complex-valued functions whose expectations exist. We define $E[g(X)]$ as *existing* if $E[|g(X)|]$ is finite. If this is the case, it then follows that

$$|E[g(X)]| \leq E[|g(X)|], \tag{2.6}$$

or, more explicitly,

$$\{E^2[\text{Re } g(X)] + E^2[\text{Im } g(X)]\}^{1/2} \leq E[\{[\text{Re } g(X)]^2 + [\text{Im } g(X)]^2\}^{1/2}]. \tag{2.7}$$

The validity of (2.7) is proved in theoretical exercise 2.2. In words, (2.6) states that the modulus of the expectation of a complex-valued function is less than or equal to the expectation of the modulus of the function.

The notions are now at hand to define the *characteristic function* $\phi_X(\cdot)$ *of a random variable* X. We define $\phi_X(\cdot)$ as a function of a real variable u, whose value is the expectation of the complex exponential function e^{iux} with respect to the probability law of X; in symbols,

$$\phi_X(u) = E[e^{iuX}] = \int_{-\infty}^{\infty} e^{iux}\, dF_X(x). \tag{2.8}$$

The quantity e^{iux} for any real numbers x and u is defined by

$$e^{iux} = \cos ux + i \sin ux, \tag{2.9}$$

in which i is the imaginary unit, defined by $i = \sqrt{-1}$ or $i^2 = -1$. Since $|e^{iux}|^2 = (\cos ux)^2 + (\sin ux)^2 = 1$, it follows that, for any random variable X, $E[|e^{iuX}|] = E[1] = 1$. Consequently, the characteristic function always exists.

The characteristic function of a random variable has all the properties of the moment-generating function of a random variable. All the moments of the random variable X *that exist* may be obtained from a knowledge of the characteristic function by the formula

$$E[X^k] = \frac{1}{i^k} \frac{d^k}{du^k} \phi_X(0). \tag{2.10}$$

To prove (2.10), one must employ the techniques discussed in section 5.

More generally, *from a knowledge of the characteristic function of a random variable one may obtain a knowledge of its distribution function, its probability density function (if it exists), its probability mass function,* and many other expectations. These facts are established in section 3.

The importance of characteristic functions in probability theory derives from the fact that they have the following basic property. Consider any two random variables X and Y. If the characteristic functions are approximately equal [that is, $\phi_X(u) \doteq \phi_Y(u)$ for every real number u], then their probability laws are approximately equal over intervals (that is, for any finite numbers a and b, $P[a \leq X \leq b] \doteq P[a \leq Y \leq b]$) or, equivalently, their distribution functions are approximately equal [that is, $F_X(a) \doteq F_Y(a)$ for all real numbers a]. A precise formulation and proof of this assertion is given in Chapter 10.

Characteristic functions represent the ideal tool for the study of the problem of addition of independent random variables, since the sum $X_1 + X_2$ of two independent random variables X_1 and X_2 has as its characteristic function the product of the characteristic functions of X_1 and X_2; in symbols, for every real number u

$$\phi_{X_1+X_2}(u) = \phi_{X_1}(u)\phi_{X_2}(u) \tag{2.11}$$

if X_1 and X_2 are independent. It is natural to inquire whether there is some other function that enjoys properties similar to those of the characteristic function. The answer appears to be in the negative. In his paper "An essential property of the Fourier transforms of distribution functions," *Proceedings of the American Mathematical Society*, Vol. 3 (1952), pp. 508–510, E. Lukacs has proved the following theorem. Let $K(x, u)$ be a complex valued function of two real variables x and u, which is a bounded Borel function of x. Define for any random variable X

$$\phi_X(u) = E[K(X, u)].$$

In order that the function $\phi_X(u)$ satisfy (2.11) and the uniqueness condition

$$\phi_{X_1}(u) = \phi_{X_2}(u) \text{ for all } u \qquad \text{if and only if } F_{X_1}(x) = F_{X_2}(x) \text{ for all } x, \tag{2.12}$$

it is necessary and sufficient that $K(x, u)$ have the form

$$K(x, u) = e^{iuA(x)},$$

in which $A(x)$ is a suitable real valued function.

► **Example 2A.** If X is $N(0, 1)$, then its characteristic function $\phi_X(u)$ is given by

$$\phi_X(u) = e^{-\frac{1}{2}u^2}. \tag{2.13}$$

To prove (2.13), we make use of the Taylor series expansion of the exponential function:

$$(2.14)\quad \begin{aligned}\phi_X(u) &= \frac{1}{\sqrt{2\pi}}\int_{-\infty}^{\infty} e^{iux}\, e^{-\frac{1}{2}x^2}\, dx = \frac{1}{\sqrt{2\pi}}\int_{-\infty}^{\infty} \sum_{n=0}^{\infty} \frac{(iux)^n}{n!} e^{-\frac{1}{2}x^2}\, dx \\ &= \sum_{n=0}^{\infty} \frac{(iu)^n}{n!} \frac{1}{\sqrt{2\pi}} \int_{-\infty}^{\infty} x^n e^{-\frac{1}{2}x^2}\, dx \\ &= \sum_{m=0}^{\infty} \frac{(iu)^{2m}}{(2m)!}\frac{(2m)!}{2^m m!} = \sum_{m=0}^{\infty}\left(-\frac{1}{2}u^2\right)^m \frac{1}{m!} = e^{-\frac{1}{2}u^2}.\end{aligned}$$

The interchange of the order of summation and integration in (2.14) may be justified by the fact that the infinite series is dominated by the integrable function $\exp(|ux| - \frac{1}{2}x^2)$. ◀

▶ **Example 2B.** If X is $N(m, \sigma^2)$, then its characteristic function $\phi_X(u)$ is given by

$$(2.15)\qquad \phi_X(u) = \exp(imu - \tfrac{1}{2}\sigma^2u^2).$$

To prove (2.15), define $Y = (X - m)/\sigma$. Then Y is $N(0, 1)$, and $\phi_Y(u) = e^{-\frac{1}{2}u^2}$. Since X may be written as a linear combination, $X = \sigma Y + m$, the validity of (2.15) follows from the general formula

$$(2.16)\qquad \phi_X(u) = e^{ibu}\phi_Y(au) \qquad \text{if } X = aY + b.$$ ◀

▶ **Example 2C.** If X is Poisson distributed with mean $E[X] = \lambda$, then its characteristic function $\phi_X(u)$ is given by

$$(2.17)\qquad \phi_X(u) = e^{\lambda(e^{iu}-1)}.$$

To prove (2.17), we write

$$(2.18)\quad \begin{aligned}\phi_X(u) &= \sum_{k=0}^{\infty} e^{iuk} p_X(k) = \sum_{k=0}^{\infty} e^{iuk}\frac{\lambda^k}{k!}e^{-\lambda} \\ &= e^{-\lambda}\sum_{k=0}^{\infty}\frac{(\lambda e^{iu})^k}{k!} = e^{-\lambda}e^{\lambda e^{iu}}.\end{aligned}$$ ◀

▶ **Example 2D.** Consider a random variable X with a probability density function, for some positive constant a,

$$(2.19)\qquad f_X(x) = \frac{a}{2}e^{-a|x|}, \qquad -\infty < x < \infty,$$

which is called *Laplace's distribution*. The characteristic function $\phi_X(u)$ is given by

$$(2.20)\qquad \phi_X(u) = \frac{a^2}{a^2 + u^2}.$$

To prove (2.20), we note that since $f_X(x)$ is an even function of x we may write

$$\phi_X(u) = 2\int_0^\infty \cos ux\, f_X(x)\, dx = a\int_0^\infty e^{-ax}\cos ux\, dx \tag{2.21}$$

$$= a\,\frac{e^{-ax}(u\sin ux - a\cos ux)}{a^2+u^2}\bigg|_0^\infty = \frac{a^2}{a^2+u^2}. \quad ◀$$

THEORETICAL EXERCISES

2.1. Cumulants and the log-characteristic function. The logarithm (to the base e) of the characteristic function of a random variable X is often easy to differentiate. Its nth derivative may be used to form the nth *cumulant* of X, written $K_n[X]$, which is defined by

$$K_n[X] = \frac{1}{i^n}\frac{d^n}{du^n}\log\phi_X(u)\bigg|_{u=0} \tag{2.22}$$

If the nth absolute moment $E[|X|^n]$ exists, then both $\phi_X(\cdot)$ and $\log\phi_X(\cdot)$ are differentiable n times and may be expanded in terms of their first n derivatives; in particular,

$$\log\phi_X(u) = K_1[X](iu) + K_2[X]\frac{(iu)^2}{2!} + \cdots + K_n[X]\frac{(iu)^n}{n!} + R_n(u), \tag{2.23}$$

in which the remainder $R_n(u)$ is such that $|u|^nR_n(u)$ tends to 0 as $|u|$ tends to 0. From a knowledge of the cumulants of a probability law one may obtain a knowledge both of its moments and its central moments. Show by evaluating the derivatives at $t = 0$ of $e^{K(t)}$, in which $K(t) = \log\phi_X(t)$, that

$$\begin{aligned} E[X] &= K_1 \\ E[X^2] &= K_2 + K_1^2 \\ E[X^3] &= K_3 + 3K_2K_1 + K_1^3 \\ E[X^4] &= K_4 + 4K_3K_1 + 3K_2^2 + 3K_2K_1^2 + K_1^4 \end{aligned} \tag{2.24}$$

Show, by evaluating the derivatives of $e^{K_m(t)}$, in which $K_m(t) = \log\phi_X(t) - itm$ and $m = E[X]$, that

$$\begin{aligned} E[(X-m)^2] &= K_2 \\ E[(X-m)^3] &= K_3 \\ E[(X-m)^4] &= K_4 + 3K_2^2. \end{aligned} \tag{2.25}$$

2.2. The square root of sum of squares inequality. Prove that (2.7) holds by showing that for any 2 random variables, X and Y,

$$\sqrt{E^2[X] + E^2[Y]} \le E[\sqrt{X^2+Y^2}]. \tag{2.26}$$

Hint: Show, and use the fact, that $\sqrt{x^2+y^2} - \sqrt{x_0^2+y_0^2} \ge [(x-x_0)x_0 + (y-y_0)y_0]/\sqrt{x_0^2+y_0^2}$ for real x, y, x_0, y_0 with $x_0y_0 \ne 0$.

EXERCISE

2.1. Compute the characteristic function of a random variable X that has as its probability law (i) the binomial distribution with mean 3 and standard deviation $\frac{3}{2}$, (ii) the Poisson distribution with mean 3, (iii) the geometric distribution with parameter $p = \frac{1}{4}$, (iv) the normal distribution with mean 3 and standard deviation $\frac{3}{2}$, (v) the gamma distribution with parameters $r = 2$ and $\lambda = 3$.

3. THE CHARACTERISTIC FUNCTION OF A RANDOM VARIABLE SPECIFIES ITS PROBABILITY LAW

In this section we give various *inversion* formulas for the distribution function, probability mass function, and probability density function of a random variable in terms of its characteristic function. As a consequence of these formulas, it follows that *to describe the probability law of a random variable it suffices to specify its characteristic function.*

We first prove a theorem that gives in terms of characteristic functions an explicit formula for $E[g(X)]$ for a fairly large class of functions $g(\cdot)$.

THEOREM 3A. Let $g(\cdot)$ be a *bounded* Borel function of a real variable that at every point x possesses a limit from the right $g(x + 0)$ and a limit from the left $g(x - 0)$. Let

$$g^*(x) = \frac{g(x + 0) + g(x - 0)}{2} \tag{3.1}$$

be the arithmetic mean of these limits. Assume further that $g(x)$ is absolutely integrable; that is,

$$\int_{-\infty}^{\infty} |g(x)|\, dx < \infty. \tag{3.2}$$

Define $\gamma(\cdot)$ as the Fourier integral (or transform) of $g(\cdot)$; that is, for every real number u

$$\gamma(u) = \frac{1}{2\pi}\int_{-\infty}^{\infty} e^{-iux} g(x)\, dx. \tag{3.3}$$

Then, for any random variable X the expectation $E[g^*(X)]$ may be expressed in terms of the characteristic function $\phi_X(\cdot)$:

$$E[g^*(X)] = \int_{-\infty}^{\infty} g^*(x)\, dF_X(x) = \lim_{U\to\infty} \int_{-U}^{U}\left(1 - \frac{|u|}{U}\right)\gamma(u)\phi_X(u)\, du. \tag{3.4}$$

The proof of this important theorem is given in section 5. In this section we discuss its consequences.

If the product $\gamma(u)\phi_X(u)$ is absolutely integrable, that is,

$$\int_{-\infty}^{\infty} |\gamma(u)\phi_X(u)| \, du < \infty, \tag{3.5}$$

then (3.4) may be written

$$E[g^*(X)] = \int_{-\infty}^{\infty} \gamma(u)\phi_X(u) \, du. \tag{3.6}$$

Without imposing the condition (3.5), it is incorrect to write (3.6). Indeed, in order even to write (3.6) the integral on the right-hand side of (3.6) must exist; this is equivalent to (3.5) being true.

We next take for $g(\cdot)$ a function defined as follows, for some finite numbers a and b (with $a < b$):

$$\begin{aligned} g(x) &= 1 && \text{if } a < x < b \\ &= \tfrac{1}{2} && \text{if } x = a \text{ or } x = b \\ &= 0 && \text{if } x < a \text{ or } x > b. \end{aligned} \tag{3.7}$$

The function $g(\cdot)$ defined by (3.7) fulfills the hypotheses of theorem 3A; it is bounded, absolutely integrable, and possesses right-hand and left-hand limits at any point x. Further, for every x, $g^*(x) = g(x)$. Now, if a and b are points at which the distribution function $F_X(\cdot)$ is continuous, then

$$\int_{-\infty}^{\infty} g(x) \, dF_X(x) = F_X(b) - F_X(a). \tag{3.8}$$

Further,

$$\gamma(u) = \frac{1}{2\pi} \frac{e^{-iub} - e^{-iua}}{-iu}. \tag{3.9}$$

Consequently, with this choice of function $g(\cdot)$, theorem 3A yields *an expression for the distribution function of a random variable in terms of its characteristic function.*

THEOREM 3B. If a and b, where $a < b$, are finite real numbers at which the distribution function $F_X(\cdot)$ is continuous, then

$$F_X(b) - F_X(a) = \lim_{U \to \infty} \frac{1}{2\pi} \int_{-U}^{U} \left(1 - \frac{|u|}{U}\right) \frac{e^{-iub} - e^{-iua}}{-iu} \phi_X(u) \, du. \tag{3.10}$$

Equation (3.10) constitutes an *inversion formula*, whereby, with a knowledge of the characteristic function $\phi_X(\cdot)$, a knowledge of the distribution function $F_X(\cdot)$ may be obtained.

An explicit inversion formula for $F_X(x)$ in terms of $\phi_X(\cdot)$ may be written in various ways. Since $\lim_{a\to-\infty} F_X(a) = 0$, we determine from (3.10) that at any point x where $F_X(\cdot)$ is continuous

$$(3.11)\quad F_X(x) = \lim_{a\to-\infty} \lim_{U\to\infty} \frac{1}{2\pi}\int_{-U}^{U}\left(1 - \frac{|u|}{U}\right)\frac{e^{-iux} - e^{-iua}}{-iu}\phi_X(u)\,du.$$

The limit is taken over the set of points a, which are continuity points of $F_X(\cdot)$.

A more useful inversion formula, the proof of which is given in section 5, is the following: at any point x, where $F_X(\cdot)$ is continuous,

$$(3.12)\qquad F_X(x) = \frac{1}{2} - \frac{1}{\pi}\int_0^\infty \frac{\operatorname{Im}[e^{-iux}\phi_X(u)]}{u}\,du.$$

The integral is an improper Riemann integral, defined as

$$\lim_{U\to\infty}\int_{1/U}^{U}\frac{\operatorname{Im}[e^{-iux}\phi_X(u)]}{u}\,du.$$

Equations (3.11) and (3.12) lead immediately to the *uniqueness theorem*, which states that there is a one-to-one correspondence between distribution functions and characteristic functions; two characteristic functions that are equal at all points (or equal at all except a countable number of points) are the characteristic functions of the same distribution function, and two distribution functions that are equal at all except a countable number of points give rise to the same characteristic function.

We may express the probability mass function $p_X(\cdot)$ of the random variable X in terms of its characteristic function; for any real number x

$$(3.13)\qquad p_X(x) = P[X = x] = F_X(x + 0) - F_X(x - 0)$$
$$= \lim_{U\to\infty}\frac{1}{2U}\int_{-U}^{U} e^{-iux}\phi_X(u)\,du.$$

The proof of (3.13) is given in section 5.

It is possible to give a criterion in terms of characteristic functions that a random variable X has an absolutely continuous probability law.* *If the characteristic function $\phi_X(\cdot)$ is absolutely integrable, that is,*

$$(3.14)\qquad \int_{-\infty}^{\infty}|\phi_X(u)|\,du < \infty,$$

* In this section we use the terminology "an absolutely continuous probability law" for what has previously been called in this book "a continuous probability law". This is to call the reader's attention to the fact that in advanced probability theory it is customary to use the expression "absolutely continuous" rather than "continuous." A continuous probability law is then defined as one corresponding to a continuous distribution function.

then the random variable X obeys the absolutely continuous probability law specified by the probability density function $f_X(\cdot)$ for any real number x given by

$$f_X(x) = \frac{1}{2\pi}\int_{-\infty}^{\infty} e^{-iux}\phi_X(u)\,du. \tag{3.15}$$

One expresses (3.15) in words by saying that $f_X(\cdot)$ is the Fourier transform, or Fourier integral, of $\phi_X(\cdot)$.

The proof of (3.15) follows immediately from the fact that at any continuity points x and a of $F_X(\cdot)$

$$F_X(x) - F_X(a) = \frac{1}{2\pi}\int_{-\infty}^{\infty} \frac{e^{-iux} - e^{-iua}}{-iu}\phi_X(u)\,du. \tag{3.16}$$

Equation (3.16) follows from (3.6) in the same way that (3.10) followed from (3.4). It may be proved from (3.16) that (*i*) $F_X(\cdot)$ is continuous at every point x, (ii) $f_X(x) = (d/dx)F_X(x)$ exists at every real number x and is given by (3.15), (iii) for any numbers a and b, $F_X(b) - F_X(a) = \int_a^b f_X(x)\,dx$. From these facts it follows that $F_X(\cdot)$ is specified by $f_X(\cdot)$ and that $f_X(x)$ is given by (3.15).

The inversion formula (3.15) provides a powerful method of calculating Fourier transforms and characteristic functions. Thus, for example, from a knowledge that

$$\left(\frac{\sin (u/2)}{u/2}\right)^2 = \int_{-\infty}^{\infty} e^{iux}f(x)\,dx, \tag{3.17}$$

where $f(\cdot)$ is defined by

$$\begin{aligned} f(x) &= 1 - |x| \qquad \text{for } |x| \le 1 \\ &= 0 \qquad \text{otherwise,} \end{aligned} \tag{3.18}$$

it follows by (3.15) that

$$\int_{-\infty}^{\infty} e^{-iux}\frac{1}{2\pi}\left(\frac{\sin (x/2)}{x/2}\right)^2 dx = f(u). \tag{3.19}$$

Similarly, from

$$\frac{1}{1 + u^2} = \int_{-\infty}^{\infty} e^{iux}\frac{1}{2}e^{-|x|}\,dx \tag{3.20}$$

it follows that

$$e^{-|u|} = \int_{-\infty}^{\infty} e^{-iux}\frac{1}{\pi}\frac{1}{1 + x^2}\,dx. \tag{3.21}$$

We note finally the following important formulas concerning *sums of independent random variables, convolution of distribution functions, and products of characteristic functions.* Let X_1 and X_2 be two independent random variables, with respective distribution functions $F_{X_1}(\cdot)$ and $F_{X_2}(\cdot)$ and respective characteristic functions $\phi_{X_1}(\cdot)$ and $\phi_{X_2}(\cdot)$. It may be proved (see section 9 of Chapter 7) that the distribution function of the sum $X + Y$ for any real number z is given by

$$F_{X_1+X_2}(z) = \int_{-\infty}^{\infty} F_{X_1}(z - x)\, dF_{X_2}(x). \tag{3.22}$$

On the other hand, it is clear that the characteristic function of the sum for any real number u is given by

$$\phi_{X_1+X_2}(u) = \phi_{X_1}(u)\phi_{X_2}(u), \tag{3.23}$$

since, by independence of X_1 and X_2, $E[e^{iu(X_1+X_2)}] = E[e^{iuX_1}]E[e^{iuX_2}]$. The distribution function $F_{X_1+X_2}(\cdot)$, given by (3.22), is said to be the convolution of the distribution functions $F_{X_1}(\cdot)$ and $F_{X_2}(\cdot)$; in symbols, one writes $F_{X_1+X_2} = F_{X_1} * F_{X_2}$.

EXERCISES

3.1. Verify (3.17), (3.19), (3.20), and (3.21).

3.2. Prove that if $f_1(\cdot)$ and $f_2(\cdot)$ are probability density functions, whose corresponding characteristic functions $\phi_1(\cdot)$ and $\phi_2(\cdot)$ are absolutely integrable, then

$$\int_{-\infty}^{\infty} f_1(y - x) f_2(x)\, dx = \frac{1}{2\pi}\int_{-\infty}^{\infty} e^{-iuy}\phi_1(u)\phi_2(u)\, du. \tag{3.24}$$

3.3. Use (3.15), (3.17), and (3.24) to prove that

$$\frac{1}{2\pi}\int_{-\infty}^{\infty} e^{-iuy}\left(\frac{\sin (u/2)}{u/2}\right)^4 du = \int_{-\infty}^{\infty} f(y - x) f(x)\, dx. \tag{3.25}$$

Evaluate the integral on the right-hand side of (3.25).

3.4. Let X be uniformly distributed over the interval 0 to π. Let $Y = A \cos X$. Show directly that the probability density function of Y for any real number y is given by

$$\begin{aligned} f_Y(y) &= \frac{1}{\pi\sqrt{A^2 - y^2}} \qquad \text{for } |y| < A \\ &= 0 \qquad \text{otherwise.} \end{aligned} \tag{3.26}$$

The characteristic function of Y may be written

$$\phi_Y(u) = \frac{1}{\pi}\int_0^{\pi} e^{iuA\cos\theta}\, du = J_0(Au), \tag{3.27}$$

in which $J_0(\cdot)$ is the Bessel function of order 0, defined for our purposes by the integral in (3.27). Is it true or false that

$$\frac{1}{2\pi}\int_{-\infty}^{\infty} e^{-iuy} J_0(Au)\, du = \frac{1}{\pi\sqrt{A^2 - y^2}} \quad \text{if } |y| < A \tag{3.28}$$
$$= 0 \qquad \text{otherwise?}$$

3.5. The image interference distribution. The amplitude a of a signal received at a distance from a transmitter may fluctuate because the signal is both directly received and reflected (reflected either from the ionosphere or the ocean floor, depending on whether it is being transmitted through the air or the ocean). Assume that the amplitude of the direct signal is a constant a_1 and the amplitude of the reflected signal is a constant a_2 but that the phase difference θ between the two signals changes randomly and is uniformly distributed over the interval 0 to π. The amplitude a of the received signal is then given by $a^2 = a_1^2 + a_2^2 + 2a_1a_2\cos\theta$. Assuming these facts, show that the characteristic function of a^2 is given by

$$\phi_{a^2}(u) = e^{iu(a_1^2 + a_2^2)} J_0(2a_1a_2u). \tag{3.29}$$

Use this result and the preceding exercise to deduce the probability density function of a^2.

4. SOLUTION OF THE PROBLEM OF THE ADDITION OF INDEPENDENT RANDOM VARIABLES BY THE METHOD OF CHARACTERISTIC FUNCTIONS

By the use of characteristic functions, we may give a solution to the problem of addition of independent random variables. Let $X_1, X_2, \ldots, X_n$ be n independent random variables, with respective characteristic functions $\phi_{X_1}(\cdot), \ldots, \phi_{X_n}(\cdot)$. Let $S_n = X_1 + X_2 + \ldots + X_n$ be their sum. To know the probability law of S_n, it suffices to know its characteristic function $\phi_{S_n}(\cdot)$. However, it is immediate, from the properties of independent random variables, that for every real number u

$$\phi_{S_n}(u) = \phi_{X_1}(u) \cdots \phi_{X_n}(u) \tag{4.1}$$

or, equivalently, $E[e^{iu(X_1+\cdots+X_n)}] = E[e^{iuX_1}] \cdots E[e^{iuX_n}]$. Thus, in terms of characteristic functions, the problem of addition of independent random variables is given by (4.1) a simple and concise solution, which may also be stated in words: *the probability law of a sum of independent random variables has as its characteristic function the product of the characteristic functions of the individual random variables.*

In this section we consider certain cases in which (4.1) leads to an exact evaluation of the probability law of S_n. In Chapter 10 we show how (4.1) may be used to give a general approximate evaluation of the probability law of S_n.

There are various ways, given the characteristic function $\phi_{S_n}(\cdot)$ of the sum S_n, in which one can deduce from it the probability law of S_n.

It may happen that $\phi_{S_n}(\cdot)$ will coincide with the characteristic function of a known probability law. For example, for each $k = 1, 2, \ldots, n$ suppose that X_k is normally distributed with mean m_k and variance σ_k^2. Then, $\phi_{X_k}(u) = \exp(ium_k - \frac{1}{2}u^2\sigma_k^2)$, and, by (4.1),

$$\phi_{S_n}(u) = \exp[iu(m_1 + \cdots + m_n) - \tfrac{1}{2}u^2(\sigma_1^2 + \cdots + \sigma_n^2)].$$

We recognize $\phi_{S_n}(\cdot)$ as the characteristic function of the normal distribution with mean $m_1 + \ldots + m_n$ and variance $\sigma_1^2 + \ldots + \sigma_n^2$. Therefore, the sum S_n is normally distributed with mean $m_1 + \ldots + m_n$ and variance $\sigma_1^2 + \ldots + \sigma_n^2$. By using arguments of this type, we have the following theorem.

THEOREM 4A. Let $S_n = X_1 + \ldots + X_n$ be the sum of independent random variables.

(i) If, for $k = 1, \ldots, n$, X_k is $N(m_k, \sigma_k^2)$, then S_n is $N(m_1 + \ldots + m_n, \sigma_1^2 + \ldots + \sigma_n^2)$;

(ii) If for $k = 1, \ldots, n$, X_k is binomial distributed with parameters N_k and p, then S_n is binomial distributed with parameters $N_1 + \ldots + N_n$ and p.

(iii) If, for $k = 1, \ldots, n$, X_k is Poisson distributed with parameter λ_k, then S_n is Poisson distributed with parameter $\lambda_1 + \ldots + \lambda_n$.

(iv) If, for $k = 1, \ldots, n$, X_k is χ^2 distributed with N_k degrees of freedom, then S_n is χ^2 distributed with $N_1 + \ldots + N_n$ degrees of freedom.

(v) If, for $k = 1, \ldots, n$, X_k is Cauchy distributed with parameters a_k and b_k, then S_n is Cauchy distributed with parameters $a_1 + \ldots + a_n$ and $b_1 + \ldots + b_n$.

One may be able to invert the characteristic function of S_n to obtain its distribution function or probability density function. In particular, if $\phi_{S_n}(\cdot)$ is absolutely integrable, then S_n has a probability density function for any real number x given by

$$f_{S_n}(x) = \frac{1}{2\pi}\int_{-\infty}^{\infty} e^{-iux}\phi_{S_n}(u)\,du. \tag{4.2}$$

In order to evaluate the infinite integral in (4.2), one will generally have to use the theory of complex integration and the calculus of residues.

Even if one is unable to invert the characteristic function to obtain the probability law of S_n in closed form, the characteristic function can still be used to obtain the moments and cumulants of S_n. Indeed, cumulants assume their real importance from the study of the sums of independent random variables because they are additive over the summands. More precisely, *if $X_1, X_2, \ldots, X_n$ are independent random variables whose* rth *cumulants exist, then the* rth *cumulant of the sum exists and is equal to the sum of the* rth *cumulants of the individual random variables.* In symbols,

$$K_r[X_1 + \cdots + X_n] = K_r[X_1] + \cdots + K_r[X_n]. \tag{4.3}$$

Equation (4.3) follows immediately from the fact that the rth cumulant is (up to a constant) the rth derivative at 0 of the logarithm of the characteristic function and the log-characteristic function is additive over independent summands, since the characteristic function is multiplicative.

The moments and central moments of a random variable may be expressed in terms of its cumulants. In particular, the first cumulant and the mean, the second cumulant and the variance, and the third cumulant and the third central moment, respectively, are equal. Consequently, the means, variances, and third central moments are additive over independent summands; more precisely,

$$\begin{aligned} E[X_1 + \cdots + X_n] &= E[X_1] + \cdots + E[X_n] \\ \operatorname{Var}[X_1 + \cdots + X_n] &= \operatorname{Var}[X_1] + \cdots + \operatorname{Var}[X_n] \\ \mu_3[X_1 + \cdots + X_n] &= \mu_3[X_1] + \cdots + \mu_3[X_n], \end{aligned} \tag{4.4}$$

where, for any random variable X, we define $\mu_3[X] = E[(X - E[X])^3]$; (4.4) may, of course, also be proved directly.

EXERCISES

4.1. Prove theorem 4A.

4.2. Find the probability laws corresponding to the following characteristic functions: (i) e^{-u^2}, (ii) $e^{-|u|}$, (iii) $e^{(e^{iu}-1)}$, (iv) $(1 - 2iu)^{-2}$.

4.3. Let $X_1, X_2, \ldots, X_n$ be a sequence of independent random variables, each uniformly distributed over the interval 0 to 1. Let $S_n = X_1 + X_2 + \ldots + X_n$. Show that for any real number y, such that $0 < y < n + 1$,

$$f_{S_{n+1}}(y) = \int_{y-1}^{y} f_{S_n}(x)\, dx;$$

hence prove by mathematical induction that

$$\begin{aligned} f_{S_n}(x) &= \frac{1}{(n-1)!} \sum_{j=0}^{[x]} \binom{n}{j} (-1)^j (x - j)^{n-1} \qquad \text{if } 0 \le x \le n. \\ &= 0 \qquad \text{if } x < 0 \text{ or } x > n. \end{aligned}$$

4.4. Let $X_1, X_2, \ldots, X_n$ be a sequence of independent random variables, each normally distributed with mean 0 and variance 1. Let $S_n = X_1^2 + X_2^2 + \ldots + X_n^2$. Show that for any real number y and integer $n = 1, 2, \ldots$

$$f_{S_{n+2}}(y) = \int_0^y f_{S_2}(y - x) f_{S_n}(x)\, dx.$$

Prove that $f_{S_2}(y) = \frac{1}{2}e^{-\frac{1}{2}y}$ for $y > 0$; hence deduce that S_n has a χ^2 distribution with n degrees of freedom.

4.5. Let $X_1, X_2, \ldots, X_n$ be independent random variables, each normally distributed with mean m and variance 1. Let $S = \sum_{j=1}^{n} X_j^2$.

(i) Find the cumulants of S.

(ii) Let $T = aY_\nu$ for suitable constants a and ν, in which Y_ν is a random variable obeying a χ^2 distribution with ν degrees of freedom. Determine a and ν so that S and T have the same means and variances. *Hint:* Show that each X_j^2 has the characteristic function

$$\phi_{X_j^2}(u) = \frac{1}{(1 - 2iu)^{\frac{1}{2}}} \exp\left[-\frac{1}{2} m^2\left(1 - \frac{1}{1 - 2iu}\right)\right]$$

5. PROOFS OF THE INVERSION FORMULAS FOR CHARACTERISTIC FUNCTIONS

In order to study the properties of characteristic functions, we require the following basic facts concerning the conditions under which various limiting operations may be interchanged with the expectation operation. These facts are stated here without proof (for proof see any text on measure theory or modern integration theory).

We state first a theorem dealing with the conditions under which, given a convergent sequence of functions $g_n(\cdot)$, the limit of expectations is equal to the expectation of the limit.

THEOREM 5A. Let $g_n(\cdot)$ and $g(\cdot)$ be Borel functions of a real variable x such that at each real number x

$$\lim_{n\to\infty} g_n(x) = g(x). \tag{5.1}$$

If a Borel function $G(\cdot)$ exists such that

$$|g_n(x)| \leq G(x) \qquad \text{for all real } x \text{ and integers } n \tag{5.2}$$

and if $E[G(X)] = \int_{-\infty}^{\infty} G(x)\, dF_X(x)$ is finite, then

$$\lim_{n\to\infty} E[g_n(X)] = E[\lim_{n\to\infty} g_n(X)] = E[g(X)]. \tag{5.3}$$

In particular, it may happen that (5.2) will hold with $G(x)$ equal for all x to a finite constant C. Since $E[C] = C$ is finite, it follows that (5.3) will hold. Since this is a case frequently encountered, we introduce a special terminology for it: *the sequence of functions $g_n(\cdot)$ is said to converge boundedly to $g(\cdot)$ if (5.1) holds and if there exists a finite constant C such that*

$$|g_n(x)| \leq C \qquad \text{for all real } x \text{ and integers } n. \tag{5.4}$$

From theorem 5A it follows that (5.3) will hold for a sequence of functions converging boundedly. This assertion is known as the Lebesgue bounded convergence theorem. Theorem 5A is known as the Lebesgue dominated convergence theorem.

Theorem 5A may be extended to the case in which there is a function of two real variables $g(x, u)$ instead of a sequence of functions $g_n(x)$.

THEOREM 5B. Let $g(x, u)$ be a Borel function of two variables such that at all real numbers x and u

$$\lim_{u' \to u} g(x, u') = g(x, u). \tag{5.5}$$

Note that (5.5) says that $g(x, u)$ is *continuous* as a function of u at each x. If a Borel function $G(x)$ exists such that

$$|g(x, u)| \leq G(x) \qquad \text{for all real } x \text{ and } u \tag{5.6}$$

and if $E[G(X)]$ is finite, then for any real number u

$$\lim_{u' \to u} E[g(X, u')] = E[g(X, u)]. \tag{5.7}$$

Note that (5.7) says that $E[g(X, u)]$ is continuous as a function of u.

We next consider the problem of differentiating and integrating a function of the form of $E[g(X, u)]$.

THEOREM 5C. Let $g(x, u)$ be a Borel function of two variables such that the partial derivative $[\partial g(x, u)]/\partial u$ with respect to u exists at all real numbers x and u. If a Borel function $G(\cdot)$ exists such that

$$\left|\frac{\partial g(x, u)}{\partial u}\right| \leq G(x), \qquad \text{for all } x \text{ and } u \tag{5.8}$$

and if $E[G(X)]$ is finite, then for any real number u

$$\frac{d}{du} E[g(X, u)] = E\left[\frac{\partial}{\partial u} g(X, u)\right]. \tag{5.9}$$

As one consequence of theorem 5C, we may deduce (2.10).

THEOREM 5D. Let $g(x, u)$ be a Borel function of two variables such that (5.5) will hold. If a Borel function $G(\cdot)$ exists such that

$$\int_{-\infty}^{\infty} |g(x, u)|\, du \le G(x) \qquad \text{for all } x \tag{5.10}$$

and if $E[G(X)]$ is finite, then

$$\begin{gathered}\int_{-\infty}^{\infty} E[g(X, u)]\, du = E\left[\int_{-\infty}^{\infty} g(X, u)\, du\right], \\ \int_{-\infty}^{\infty} du \int_{-\infty}^{\infty} dF_X(x) g(x, u) = \int_{-\infty}^{\infty} dF_X(x) \int_{-\infty}^{\infty} du\, g(x, u).\end{gathered} \tag{5.11}$$

It should be noted that the integrals in (5.11) involving integration in the variable u may be interpreted as Riemann integrals if we assume that (5.5) holds. However, the assertion (5.11) is valid even without assuming (5.5) if we interpret the integrals in u as Lebesgue integrals.

Finally, we give a theorem, analogous to theorem 5A, for Lebesgue integrals over the real line.

THEOREM 5E. Let $h_n(\cdot)$ and $h(\cdot)$ be Borel functions of a real variable such that at each real number u

$$\lim_{n \to \infty} h_n(u) = h(u). \tag{5.12}$$

If a function $H(u)$ exists such that

$$|h_n(u)| \le H(u) \qquad \text{for all real } u \text{ and integers } n \tag{5.13}$$

and if $\int_{-\infty}^{\infty} H(u)\, du$ is finite, then

$$\lim_{n \to \infty} \int_{-\infty}^{\infty} h_n(u)\, du = \int_{-\infty}^{\infty} h(u)\, du. \tag{5.14}$$

Theorem 5E, like theorem 5A, is a special case of a general result of the theory of abstract Lebesgue integrals, called the Lebesgue dominated convergence theorem.

We next discuss the proofs of the inversion formulas for characteristic functions. In writing out the proofs, we omit the subscript X on the distribution function $F_X(\cdot)$ and the characteristic function $\phi_X(\cdot)$.

We first prove (3.13). We note that

$$\begin{aligned}\frac{1}{2U}\int_{-U}^{U} e^{-iux}\phi(u)\, du &= \frac{1}{2U}\int_{-U}^{U} du\left[\int_{-\infty}^{\infty} e^{iu(y-x)}\, dF(y)\right] \\ &= \int_{-\infty}^{\infty} dF(y)\, \frac{1}{2U}\int_{-U}^{U} du\, e^{iu(y-x)},\end{aligned}$$

in which the interchange of the order of integration is justified by theorem 5D. Now define the functions

$$g(y, U) = \frac{1}{2U}\int_{-U}^{U} e^{iu(y-x)}\,du = \frac{\sin U(y-x)}{U(y-x)} \qquad \text{if } y \neq x$$
$$= 1 \qquad \text{if } y = x.$$
$$g(y) = 0 \qquad \text{if } y \neq x$$
$$= 1 \qquad \text{if } y = x.$$

Clearly, at each y, $g(y, U)$ converges boundedly to $g(y)$ as U tends to ∞. Therefore, by theorem 5A,

$$\lim_{U\to\infty} \frac{1}{2U}\int_{-U}^{U} e^{-iux}\phi(u)\,du = \lim_{U\to\infty}\int_{-\infty}^{\infty} g(y, U)\,dF(y)$$
$$= \int_{-\infty}^{\infty} g(y)\,dF(y) = F(y+0) - F(y-0).$$

We next prove (3.12). It may be verified that

$$\text{Im}\,[e^{-iux}\phi(u)] = E[\sin u(X-x)]$$

for any real numbers u and x. Consequently, for any $U > 0$

$$(5.15) \qquad \frac{2}{\pi}\int_{1/U}^{U} \frac{\text{Im}\,[e^{-iux}\phi(u)]}{u}\,du = \int_{-\infty}^{\infty} dF(y)\,\frac{2}{\pi}\int_{1/U}^{U} \frac{\sin u(y-x)}{u}\,du,$$

in which the interchange of integrals in (5.15) is justified by theorem 5D. Now it may be proved that

$$(5.16) \qquad \lim_{U\to\infty} \frac{2}{\pi}\int_{1/U}^{U} \frac{\sin ut}{u}\,du = 1 \qquad \text{if } t > 0$$
$$= 0 \qquad \text{if } t = 0$$
$$= -1 \qquad \text{if } t < 0,$$

in which the convergence is bounded for all U and t.

A proof of (5.16) may be sketched as follows. Define

$$G(a) = \int_0^{\infty} e^{-au}\,\frac{\sin ut}{u}\,du.$$

Verify that the improper integral defining $G(a)$ converges uniformly for $a \geq 0$ and that this implies that

$$\int_0^{\infty} \frac{\sin ut}{u}\,du = \lim_{a\to 0+} G(a).$$

Now

$$\int_0^\infty e^{-au} \cos ut \, du = \frac{a}{a^2 + t^2}, \tag{5.17}$$

in which, for each a the integral in (5.17) converges uniformly for all t. Verify that this implies that $G(a) = \tan^{-1}(t/a)$, which, as a tends to 0, tends to $\pi/2$ or to $-\pi/2$, depending on whether $t > 0$ or $t < 0$. The proof of (5.16) is complete.

Now define

$$\begin{aligned} g(y) &= -1 \qquad \text{if } y < x \\ &= 0 \qquad \text{if } y = x \\ &= 1 \qquad \text{if } y > x. \end{aligned}$$

By (5.16), it follows that the integrand of the integral on the right-hand side of (5.15) tends to $g(y)$ boundedly as U tends to ∞. Consequently, we have proved that

$$\frac{2}{\pi} \int_0^\infty \frac{\operatorname{Im}(e^{-iux}\phi(u))}{u} \, du = \int_{-\infty}^\infty g(y) \, dF(y) = 1 - 2F(x).$$

The proof of (3.12) is complete.

We next prove (3.4). We have

$$\begin{aligned} &\int_{-U}^{U} \left(1 - \frac{|u|}{U}\right) \gamma(u)\phi(u) \, du \\ &= \int_{-\infty}^{\infty} dF(x) \int_{-U}^{U} du \, e^{iux} \left(1 - \frac{|u|}{U}\right) \frac{1}{2\pi} \int_{-\infty}^{\infty} e^{-iuy} g(y) \, dy \\ &= \int_{-\infty}^{\infty} dF(x) \int_{-\infty}^{\infty} dy \, g(y) U K[U(x - y)] \end{aligned} \tag{5.18}$$

in which we define the function $K(\cdot)$ for any real number z by

$$K(z) = \frac{1}{2\pi} \left(\frac{\sin(z/2)}{z/2} \right)^2 = \frac{1}{\pi} \int_0^1 dv \, (1 - v) \cos vz; \tag{5.19}$$

(5.18) follows from the fact that

$$\begin{aligned} \frac{1}{2\pi} \int_{-U}^{U} du \, e^{iu(x-y)} \left(1 - \frac{|u|}{U}\right) &= \frac{U}{2\pi} \int_{-1}^{1} dv \, e^{ivU(x-y)} (1 - |v|) \\ &= \frac{U}{\pi} \int_0^1 (1 - v) \cos vU(x - y) \, dv. \end{aligned}$$

To conclude the proof of (3.4), it suffices to show that

$$g_U(x) = \int_{-\infty}^{\infty} dy\, g(y) UK[U(x - y)] \tag{5.20}$$

converges boundedly to $g^*(x)$ as U tends to ∞. We now show that this holds, using the facts that $K(\cdot)$ is even, nonnegative, and integrates to 1; in symbols, for any real number u

$$K(-u) = K(u), \qquad K(u) \geq 0, \qquad \int_{-\infty}^{\infty} K(u)\, du = 1. \tag{5.21}$$

In other words, $K(\cdot)$ is a probability density function symmetric about 0.

In (5.20) make the change of variable $t = y - x$. Since $K(\cdot)$ is even, it follows that

$$g_U(x) = \int_{-\infty}^{\infty} g(x + t) UK(Ut)\, dt. \tag{5.22}$$

By making the change of variable $t' = -t$ in (5.22) and again using the fact that $K(\cdot)$ is even, we determine that

$$g_U(x) = \int_{-\infty}^{\infty} g(x - t) UK(Ut)\, dt. \tag{5.23}$$

Consequently, by adding (5.22) and (5.23) and then dividing by 2, we show that

$$g_U(x) = \int_{-\infty}^{\infty} dt\, UK(Ut) \frac{g(x + t) + g(x - t)}{2}. \tag{5.24}$$

Define $h(t) = [g(x + t) + g(x - t)]/2 - g^*(x)$. From (5.24) it follows that

$$g_U(x) - g^*(x) = \int_{-\infty}^{\infty} dt\, UK(Ut) h(t). \tag{5.25}$$

Now let C be a constant such that $2|g(y)| \leq C$ for any real number y. Then, for any positive number d and for all U and x

$$\begin{aligned} |g_U(x) - g^*(x)| &\leq \sup_{|t| \leq d} |h(t)| \int_{|t| \leq d} UK(Ut)\, dt \\ &\quad + \sup_{|t| > d} |h(t)| \int_{|t| \geq d} UK(Ut)\, dt \leq \sup_{|t| \leq d} |h(t)| + C \int_{|s| \geq Ud} K(s)\, ds. \end{aligned} \tag{5.26}$$

For d fixed $\int_{|s| \geq Ud} K(s)\, ds$ tends to 0 as U tends to ∞. Next, by the definition of $h(t)$ and $g^*(t)$, $\sup_{|t| \leq d} |h(t)|$ tends to 0 as d tends to 0. Consequently, by letting first U tend to infinity and then d tend to 0 in (5.26), it follows that $g_U(x)$ tends boundedly to $g^*(x)$ as U tends to ∞. The proof of (3.4) is complete.

CHAPTER 10

Sequences of Random Variables

The basic concepts of probability theory, such as the probability of a random event (or the mean of a random variable), have been given intuitive meanings as approximately representing certain averages computed from a large sample of independent observed values of the event (or of the random variable). In this chapter we treat the problem of giving an exact mathematical meaning to the word "approximately" as it is employed in the foregoing sentence. At the same time, our discussion leads to an answer to the question of what constitutes an approximate solution to the problem of finding the probability law of the sum of random variables. A basic role in this study is played by the notion of the *convergence of a sequence of random variables*.

1. MODES OF CONVERGENCE OF A SEQUENCE OF RANDOM VARIABLES

Consider a sequence of jointly distributed random variables $Z_1, Z_2, \ldots, Z_n$ defined on the same probability space S on which a probability function $P[\cdot]$ has been defined. Let Z be another random variable defined on the same probability space. The notion of the convergence of the sequence of random variables Z_n to the random variable Z can be defined in several ways.

We consider first the notion of *convergence with probability one*. We say that Z_n converges to Z with probability one if $P[\lim_{n\to\infty} Z_n = Z] = 1$ or, in words, if for almost all members s of the probability space S on which the random variables are defined $\lim_{n\to\infty} Z_n(s) = Z(s)$. To prove that a sequence of random variables Z_n converges with probability one is often technically a difficult problem. Consequently, two other types of convergence of random variables, called, respectively, *convergence in mean square* and *convergence in probability*, have been introduced in probability theory. These modes of convergence are simpler to deal with than convergence with probability one and at the same time are conceptually similar to it.

The sequence $Z_1, Z_2, \ldots, Z_n$ is said to *converge in mean square to the random variable* Z, denoted $\operatorname*{l.i.m.}_{n\to\infty} Z_n = Z$ if $\lim_{n\to\infty} E[(Z_n - Z)^2] = 0$ or, in words, if the mean square difference between Z_n and Z tends to 0.

The sequence $Z_1, Z_2, \ldots, Z_n$ is said to *converge in probability to the random variable* Z, denoted $\operatorname*{plim}_{n\to\infty} Z_n = Z$ if for every positive number ϵ

$$\lim_{n\to\infty} P[|Z_n - Z| > \epsilon] = 0. \tag{1.1}$$

Equation (1.1) may be expressed in words: for any fixed difference ϵ the probability of the event that Z_n and Z differ by more than ϵ becomes arbitrarily close to 0 as n tends to infinity.

Convergence in probability derives its importance from the fact that, like convergence with probability one, no moments need exist before it can be considered, as is the case with convergence in mean square. It is immediate that if convergence in mean square holds then so does convergence in probability; one need only consider the following form of Chebyshev's inequality: for any $\epsilon > 0$

$$P[|Z_n - Z| > \epsilon] \leq \frac{1}{\epsilon^2} E[|Z_n - Z|^2]. \tag{1.2}$$

The relation that exists between convergence with probability one and convergence in probability is best understood by considering the following characterization of convergence with probability one, which we state without proof. Let $Z_1, Z_2, \ldots, Z_n$ be a sequence of jointly distributed random variables; Z_n converges to the random variable Z with probability one if and only if for every $\epsilon > 0$

$$\lim_{N\to\infty} P\left[\left(\sup_{n\geq N} |Z_n - Z|\right) > \epsilon\right] = 0. \tag{1.3}$$

On the other hand, the sequence $\{Z_n\}$ converges to Z in probability if and only if for every $\epsilon > 0$ (1.1) holds. Now, it is clear that if $|Z_N - Z| > \epsilon$, then $\sup_{n \geq N} |Z_n - Z| > \epsilon$. Consequently,

$$P[|Z_N - Z| > \epsilon] \leq P[\sup_{n \geq N} |Z_n - Z| > \epsilon],$$

and (1.3) implies (1.1). Thus, if Z_n converges to Z with probability one, it converges to Z in probability.

Convergence with probability one of the sequence $\{Z_n\}$ to Z implies that one can make a *probability statement simultaneously* about all but a finite number of members of the sequence $\{Z_n\}$: given any positive numbers ϵ and δ, an integer N exists such that

$$(1.4) \quad P[|Z_N - Z| < \epsilon, |Z_{N+1} - Z| < \epsilon, |Z_{N+2} - Z| < \epsilon, \ldots] > 1 - \delta.$$

On the other hand, convergence in probability of the sequence $\{Z_n\}$ to Z implies only that one can make *simultaneous probability statements* about each of all but a finite number of members of the sequence $\{Z_n\}$: given any positive numbers ϵ and δ an integer N exists such that

$$(1.5) \quad P[|Z_N - Z| < \epsilon] > 1 - \delta, \qquad P[|Z_{N+1} - Z| < \epsilon] > 1 - \delta,$$
$$P[|Z_{N+2} - Z| < \epsilon] > 1 - \delta, \cdots.$$

One thus sees that convergence in probability is implied by both convergence with probability one and by convergence in mean square. However, without additional conditions, convergence in probability implies neither convergence in mean square nor convergence with probability one. Further, convergence with probability one neither implies nor is implied by convergence in mean square.

The following theorem gives a condition under which convergence in mean square implies convergence with probability one.

THEOREM 1A. If a sequence Z_n converges in mean square to 0 in such a way that

$$(1.6) \qquad \sum_{n=1}^{\infty} E[Z_n^2] < \infty,$$

then it follows that Z_n converges to 0 with probability one.

Proof: From (1.6) it follows that

$$(1.7) \qquad E\left[\sum_{n=1}^{\infty} Z_n^2\right] = \sum_{n=1}^{\infty} E[Z_n^2] < \infty,$$

since it may be shown that for an infinite series of nonnegative summands the expectation of the sum is equal to the sum of the expectations. Next, from the fact that the infinite series $\sum_{n=1}^{\infty} Z_n^2$ has finite mean it follows that it is finite with probability one; in symbols,

$$P\left[0 \leq \sum_{n=1}^{\infty} Z_n^2 < \infty\right] = 1. \tag{1.8}$$

If an infinite series converges, then its general term tends to 0. Therefore, from (1.8) it follows that

$$P\left[\lim_{n\to\infty} Z_n = 0\right] = 1. \tag{1.9}$$

The proof of theorem 1A is complete. Although the proof of theorem 1A is completely rigorous, it requires for its justification two basic facts of the theory of integration over probability spaces that have not been established in this book.

2. THE LAW OF LARGE NUMBERS

The fundamental empirical fact upon which are based all applications of the theory of probability is expressed in the empirical law of large numbers, first formulated by Poisson (in his book, *Recherches sur le probabilité des jugements*, 1837):

> In many different fields, empirical phenomena appear to obey a certain general law, which can be called the Law of Large Numbers. This law states that the ratios of numbers derived from the observation of a very large number of similar events remain practically constant, provided that these events are governed partly by constant factors and partly by variable factors whose variations are irregular and do not cause a systematic change in a definite direction. Certain values of these relations are characteristic of each given kind of event. With the increase in length of the series of observations the ratios derived from such observations come nearer and nearer to these characteristic constants. They could be expected to reproduce them exactly if it were possible to make series of observations of an infinite length.

In the mathematical theory of probability one may prove a proposition, called the mathematical law of large numbers, that may be used to gain insight into the circumstances under which the empirical law of large numbers is expected to hold. For an interesting philosophical discussion of the relation between the empirical and the mathematical laws of large numbers and for the foregoing quotation from Poisson the reader should consult Richard von Mises, *Probability, Statistics, and Truth*, second revised edition, Macmillan, New York, 1957, pp. 104–134.

A sequence of jointly distributed random variables, $X_1, X_2, \ldots, X_n$, with finite means, is said to obey the (classical) law of large numbers if

$$(2.1) \qquad Z_n = \frac{X_1 + X_2 + \cdots + X_n}{n} - \frac{E(X_1 + \cdots + X_n)}{n} \to 0$$

in some mode of convergence as n tends to ∞. The sequence $\{X_n\}$ is said to obey the strong law of large numbers, the weak law of large numbers, or the quadratic mean law of large numbers, depending on whether the convergence in (2.1) is with probability one, in probability, or in quadratic mean. In this section we give conditions, both for independent and dependent random variables, for the law of large numbers to hold.

We consider first the case of independent random variables with finite means. We prove in section 3 that a sequence of independent identically distributed random variables obeys the weak law of large numbers if the common mean $E[X]$ is finite. It may be proved (see Loève, *Probability Theory*, Van Nostrand, New York, 1955, p. 243) that the finiteness of $E[X]$ also implies that the sequence of independent identically distributed random variables obeys the strong law of large numbers.

In theoretical exercise 4.2 we indicate the proof of the law of large numbers for independent, not necessarily identically distributed, random variables with finite means: if, for some $\delta > 0$

$$(2.2) \qquad \lim_{n\to\infty} \frac{1}{n^{1+\delta}} \sum_{k=1}^{n} E|X_k - E[X_k]|^{1+\delta} = 0$$

then

$$(2.3) \qquad \operatorname*{plim}_{n\to\infty} \frac{1}{n} \sum_{k=1}^{n} (X_k - E[X_k]) = 0.$$

Equation (2.2) is known as *Markov's condition* for the validity of the weak law of large numbers for independent random variables.

In this section we consider the case of dependent random variables X_k, with finite means (which we may take to be 0), and finite variances $\sigma_k^2 = E[X_k^2]$. We state conditions for the validity of the quadratic mean law of large numbers and the strong law of large numbers, which, while not the most general conditions that can be stated, appear to be general enough for most practical applications. Our conditions are stated in terms of the behavior, as n tends to ∞, of the covariance

$$(2.4) \qquad C_n = E[Z_n X_n] = \frac{1}{n} \sum_{k=1}^{n} E[X_k X_n]$$

between the nth summand X_n and the nth sample mean

$$Z_n = (X_1 + X_2 + \cdots + X_n)/n.$$

Let us examine the possible behavior of C_n under various assumptions on the sequence $\{X_n\}$ and under the assumption that the variances Var $[X_n]$ are uniformly bounded; that is, there is a constant M such that

$$\sigma_n^2 = \text{Var}\,[X_n] \leq M \qquad \text{for all } n. \tag{2.5}$$

If the random variables $\{X_n\}$ are independent, then $E[X_kX_n] = 0$ if $k < n$. Consequently, $C_n = \sigma_n^2/n$, which, under condition (2.5), tends to 0 as n tends to ∞. This is also the case if the random variables $\{X_n\}$ are assumed to be orthogonal. *The sequence of random variables $\{X_n\}$ is said to be orthogonal if, for any integer k and integer $m \neq 0$, $E[X_kX_{k+m}] = 0$.* Then, again, $C_n = \sigma_n^2/n$.

More generally, let us consider random variables $\{X_n\}$ that are *stationary* (in the wide sense); this means that *there is a function $R(m)$*, defined for $m = 0, 1, 2, \ldots$, *such that, for any integers k and m,*

$$E[X_kX_{k+m}] = R(m). \tag{2.6}$$

It is clear that an orthogonal sequence of random variables (in which all the random variables have the same variance σ^2) is stationary, with $R(m) = \sigma^2$ or 0, depending on whether $m = 0$ or $m > 0$. For a stationary sequence the value of C_n is given by

$$C_n = \frac{1}{n}\sum_{k=0}^{n-1} R(k). \tag{2.7}$$

We now show that under condition (2.5) a necessary and sufficient condition for the sample mean Z_n to converge in quadratic mean to 0 is that C_n tends to 0. In theorem 2B we state conditions for the sample mean Z_n to converge with probability one to 0.

THEOREM 2A. A sequence of jointly distributed random variables $\{X_n\}$ with zero mean and uniformly bounded variances obeys the quadratic mean law of large numbers (in the sense that $\lim_{n\to\infty} E[Z_n^2] = 0$) if and only if

$$\lim_{n\to\infty} C_n = \lim_{n\to\infty} E[X_nZ_n] = 0. \tag{2.8}$$

Proof: Since $E^2[X_nZ_n] \leq E[X_n^2]E[Z_n^2]$, it is clear that if the quadratic mean law of large numbers holds and if the variances $E[X_n^2]$ are bounded uniformly in n, then (2.8) holds. To prove the converse, we prove first the following useful identity:

$$E[Z_n^2] + \frac{1}{n^2}\sum_{k=1}^{n} E[X_k^2] = \frac{2}{n^2}\sum_{k=1}^{n} kE[X_kZ_k]. \tag{2.9}$$

To prove (2.9), we write the familiar formula

$$E[(X_1 + \cdots + X_n)^2] = \sum_{k=1}^{n} E[X_k^2] + 2\sum_{k=1}^{n} \sum_{j=1}^{k-1} E[X_k X_j] \tag{2.10}$$

$$= 2\sum_{k=1}^{n} \sum_{j=1}^{k} E[X_k X_j] - \sum_{k=1}^{n} E[X_k^2]$$

$$= 2\sum_{k=1}^{n} kE[X_k Z_k] - \sum_{k=1}^{n} E[X_k^2],$$

from which (2.9) follows by dividing through by n^2. In view of (2.9), to complete the proof that (2.8) implies $E[Z_n^2]$ tends to 0, it suffices to show that (2.8) implies

$$\lim_{n\to\infty} \frac{1}{n^2} \sum_{k=1}^{n} kC_k = 0. \tag{2.11}$$

To see (2.11), note that for any $n > N > 0$

$$\frac{1}{n^2} \sum_{k=1}^{n} kC_k \leq \frac{1}{n^2} \sum_{k=1}^{N} |kC_k| + \sup_{N\leq k} |C_k|. \tag{2.12}$$

Letting first n tend to infinity and then N tend to ∞ in (2.12), we see that (2.11) holds. The proof of theorem 2A is complete.

If it is known that C_n tends to 0 as some power of n, then we can conclude that convergence holds with probability one.

THEOREM 2B. *A sequence of jointly distributed random variables $\{X_n\}$ with zero mean and uniformly bounded variances obeys the strong law of large numbers* $\left(\text{in the sense that } P\left[\lim_{n\to\infty} Z_n = 0\right] = 1\right)$ *if positive constants M and q exist such that for all integers n*

$$|E[X_n Z_n]| = |C_n| \leq \frac{M}{n^q}. \tag{2.13}$$

Remark: For a stationary sequence of random variables [in which case C_n is given by (2.7)] (2.13) holds if positive constants M and q exist such that for all integers $m \geq 1$

$$|R(m)| \leq \frac{M}{m^q}. \tag{2.14}$$

Proof: If (2.13) holds, then (assuming, as we may, that $0 < q \leq 1$)

$$\frac{1}{n^2} \sum_{k=1}^{n} kC_k \leq \frac{M}{n^2} \sum_{k=1}^{n} k^{1-q} \leq \frac{M}{n^2} \int_1^{n+1} x^{1-q}\, dx \leq \frac{4M}{2-q} \frac{1}{n^q}. \tag{2.15}$$

By (2.15) and (2.9), it follows that for some constant M' and $q > 0$

$$E[Z_n^2] \leq \frac{M'}{n^q} \qquad \text{for all integers } n. \tag{2.16}$$

Choose now any integer r such that $r > (1/q)$ and define a sequence of random variables $Z_1', Z_2', \ldots, Z_m'$ by taking for Z_m' the m^rth member of the sequence $\{Z_n\}$; in symbols,

$$Z_m' = Z_{m^r} \qquad \text{for } m = 1, 2, \cdots. \tag{2.17}$$

By (2.16), the sequence $\{Z_m'\}$ has a mean square satisfying

$$E[Z_m'^2] \leq \frac{M'}{m^{rq}}. \tag{2.18}$$

If we sum (2.18) over all m, we obtain a convergent series, since $rq > 1$:

$$\sum_{m=1}^{\infty} E[Z_m'^2] \leq M' \sum_{m=1}^{\infty} m^{-rq} < \infty. \tag{2.19}$$

Therefore, by theorem 1A, it follows that

$$P\left[\lim_{m\to\infty} Z_m' = 0\right] = P\left[\lim_{m\to\infty} Z_{m^r} = 0\right] = 1. \tag{2.20}$$

We have thus shown that a properly selected subsequence $\{Z_{m^r}\}$ of the sequence $\{Z_n\}$ converges to 0 with probability one. We complete the proof of theorem 2B by showing that the members of the sequence $\{Z_n\}$, located between successive members of the subsequence $\{Z_{m^r}\}$, do not tend to be too different from the members of the subsequence. More precisely, define

$$\begin{aligned} U_m &= \max_{m^r \leq n < (m+1)^r} \left| Z_{m^r} - \frac{n}{m^r} Z_n \right| \\ V_m &= \max_{m^r \leq n < (m+1)^r} \left| Z_n - \frac{n}{m^r} Z_n \right| \\ W_m &= \max_{m^r \leq n < (m+1)^r} |Z_{m^r} - Z_n|. \end{aligned} \tag{2.21}$$

We claim it is clear, in view of (2.20), that to show that $P\left[\lim_{n\to\infty} Z_n = 0\right] = 1$ it suffices to show that $P\left[\lim_{m\to\infty} W_m = 0\right] = 1$. Consequently, to complete the proof it suffices to show that

$$P\left[\lim_{m\to\infty} U_m = 0\right] = P\left[\lim_{m\to\infty} V_m = 0\right] = 1. \tag{2.22}$$

In view of theorem 1A, to show that (2.22) holds, it suffices to show that

$$\sum_{m=1}^{\infty} E[U_m^2] < \infty, \qquad \sum_{m=1}^{\infty} E[V_m^2] < \infty. \tag{2.23}$$

We prove that (2.23) holds by showing that for some constants M_U and M_V

$$E^{1/2}[U_m^2] \leq \frac{M_U}{m}, \qquad E^{1/2}[V_m^2] \leq \frac{M_V}{m} \qquad \text{for all integers } m. \tag{2.24}$$

To prove (2.24), we note that

$$|U_m| \leq \frac{1}{m^r} \sum_{k=m^r}^{(m+1)^r-1} |X_k|. \tag{2.25}$$

from which it follows that

$$E^{1/2}[U_m^2] \leq \frac{1}{m^r} \sum_{k=m^r}^{(m+1)^r-1} E^{1/2}[X_k^2] \leq \frac{(m+1)^r - m^r}{m^r} M, \tag{2.26}$$

in which we use M as a bound for $E^{1/2}[X_k^2]$. By a calculus argument, using the law of the mean, one may show that for $r \geq 1$ and $m \geq 1$

$$\left(1 + \frac{1}{m}\right)^r - 1 \leq \frac{1}{m} r2^{r-1}. \tag{2.27}$$

Consequently, (2.26) implies the first part of (2.24). Similarly,

$$|V_m| \leq \left| \left(1 + \frac{1}{m}\right)^r - 1 \right| \frac{1}{m^r} \sum_{k=1}^{(m+1)^r-1} |X_k|, \tag{2.28}$$

$$E^{1/2}[V_m^2] \leq \left(\frac{1}{m} r2^{r-1}\right) \frac{1}{m^r} \sum_{k=1}^{(m+1)^r-1} E^{1/2}[X_k^2], \tag{2.29}$$

from which one may infer the second part of (2.24). The proof of theorem 2B is now complete.

EXERCISES

2.1. Random digits. Consider a discrete random variable X uniformly distributed over the numbers 0 to $N-1$ for any integer $N \geq 2$; that is, $P[X = k] = 1/N$ if $k = 0, 1, 2, \ldots, N-1$. Let $\{X_n\}$ be a sequence of independent random variables identically distributed as X. For an integer k from 0 to $N-1$ define $F_n(k)$ as the fraction of the observations $X_1, X_2, \ldots, X_n$ equal to k. Prove that

$$P\left[\lim_{n\to\infty} F_n(k) = \frac{1}{N}\right] = 1.$$

2.2. The distribution of digits in the decimal expansion of a random number. Let Y be a number chosen at random from the unit interval (that is, Y is a random variable uniformly distributed over the interval 0 to 1). Let $X_1, X_2, \ldots$ be the successive digits in the decimal expansion of Y; that is,

$$Y = \frac{X_1}{10} + \frac{X_2}{10^2} + \cdots + \frac{X_n}{10^n} + \cdots.$$

Prove that the random variables $X_1, X_2, \ldots$ are independent discrete random variables uniformly distributed over the integers 0 to 9. Consequently, conclude that for any integer k (say, the integer 7) the relative frequency of occurrence of k in the decimal expansion of any number Y in the unit interval is equal to $\frac{1}{10}$ for all numbers Y, except a set of numbers Y constituting a set of probability zero. Does the fact that only 3's occur in the decimal expansion of $\frac{1}{3}$ contradict the assertion?

2.3. Convergence of the sample distribution function and the sample characteristic function of dependent random variables. Let $X_1, X_2, \ldots, X_n$ be a sequence of random variables identically distributed as a random variable X. The sample distribution function $F_n(y)$ is defined as the fraction of observations among $X_1, X_2, \ldots, X_n$ which are less than or equal to y. The sample characteristic function $\phi_n(u)$ is defined by

$$\phi_n(u) = M_n[e^{iux}] = \frac{1}{n} \sum_{k=1}^{n} e^{iuX_k}.$$

Show that $F_n(y)$ converges in quadratic mean to $F_X(y) = P[X \le y]$, as $n \to \infty$, if and only if

$$\frac{1}{n} \sum_{k=1}^{n} P[X_k \le y, X_n \le y] \to |F_X(y)|^2. \tag{2.30}$$

Show that $\phi_n(u)$ converges in quadratic mean to $\phi_X(u) = E[e^{iuX}]$ if and only if

$$C_n = \frac{1}{n} \sum_{k=1}^{n} E[e^{iu(X_k - X_n)}] \to |\phi_X(u)|^2. \tag{2.31}$$

Prove that (2.30) and (2.31) hold if the random variables $X_1, X_2, \ldots$ are independent.

2.4. The law of large numbers does not hold for Cauchy distributed random variables. Let $X_1, X_2, \ldots, X_n$ be a sequence of independent identically distributed random variables with probability density functions $f_{X_n}(x) = [\pi(1 + x^2)]^{-1}$. Show that no finite constant m exists to which the sample means $(X_1 + \ldots + X_n)/n$ converge in probability.

2.5. Let $\{X_n\}$ be a sequence of independent random variables identically distributed as a random variable X with finite mean. Show that for any bounded continuous function $f(\cdot)$ of a real variable t

$$\lim_{n \to \infty} E\left[f\left(\frac{X_1 + \cdots + X_n}{n}\right)\right] = f(E[X]). \tag{2.32}$$

Consequently, conclude that

$$\lim_{n\to\infty} \int_0^1 \cdots \int_0^1 f\left(\frac{x_1 + \cdots + x_n}{n}\right) dx_1 \cdots dx_n = f(\tfrac{1}{2}) \tag{2.33}$$

$$\lim_{n\to\infty} \sum_{k=0}^{n} f\left(\frac{k}{n}\right) \binom{n}{k} t^k(1-t)^{n-k} = f(t), \qquad 0 \le t \le 1. \tag{2.34}$$

2.6. A probabilistic proof of Weierstrass' theorem: Extend (2.34) to show that to any continuous function $f(\cdot)$ on the interval $0 \le t \le 1$ there exists a sequence of polynomials $P_n(t)$ such that $\lim_{n\to\infty} P_n(t) = f(t)$ uniformly on $0 \le t \le 1$.

3. CONVERGENCE IN DISTRIBUTION OF A SEQUENCE OF RANDOM VARIABLES

In this section we define the notion of *convergence in distribution of a sequence of random variables* $Z_1, Z_2, \ldots, Z_n$ *to a random variable* Z, which is the notion of convergence most used in applications of probability theory. The notion of convergence in distribution of a sequence of random variables can be defined in a large number of equivalent ways, each of which is important for certain purposes. Instead of choosing any one of them as the definition, we prefer to introduce all the equivalent concepts simultaneously.

THEOREM 3A. DEFINITIONS AND THEOREMS CONCERNING CONVERGENCE IN DISTRIBUTION. For $n = 1, 2, \ldots$, let Z_n be a random variable with distribution function $F_{Z_n}(\cdot)$ and characteristic function $\phi_{Z_n}(\cdot)$. Similarly, let Z be a random variable with distribution function $F_Z(\cdot)$ and characteristic function $\phi_Z(\cdot)$. We define the sequence $\{Z_n\}$ as converging in distribution to the random variable Z, denoted by

$$\lim_{n\to\infty} \mathscr{L}(Z_n) = \mathscr{L}(Z), \qquad \text{or} \qquad \mathscr{L}(Z_n) \to \mathscr{L}(Z), \tag{3.1}$$

and read "the law of Z_n converges to the law of Z" if any one (and consequently all) of the following equivalent statements holds:

(i) For every bounded continuous function $g(\cdot)$ of a real variable there is convergence of the expectation $E[g(Z_n)]$ to $E[g(Z)]$; that is, as n tends to ∞,

$$E[g(Z_n)] = \int_{-\infty}^{\infty} g(z)\, dF_{Z_n}(z) \to \int_{-\infty}^{\infty} g(z)\, dF_Z(z) = E[g(Z)]. \tag{3.2}$$

(ii) At every real number u there is convergence of the characteristic functions; that is, as n tends to ∞,

$$E[e^{iuZ_n}] = \phi_{Z_n}(u) \to \phi_Z(u) = E[e^{iuZ}]. \tag{3.3}$$

(iii) At every two points a and b, where $a < b$, at which the distribution function $F_Z(\cdot)$ of the limit random variable Z is continuous, there is convergence of the probability functions over the interval a to b; that is, as n tends to ∞,

$$(3.4) \quad P[a < Z_n \leq b] = F_{Z_n}(b) - F_{Z_n}(a) \to F_Z(b) - F_Z(a) = P[a < Z \leq b].$$

(iv) At every real number a that is a point of continuity of the distribution function $F_Z(\cdot)$ there is convergence of the distribution functions; that is, as n tends to ∞, if a is a continuity point of $F_Z(\cdot)$,

$$P[Z_n \leq a] = F_{Z_n}(a) \to F_Z(a) = P[Z \leq a].$$

(v) For every continuous function $g(\cdot)$, as n tends to ∞,

$$P_{Z_n}[\{z: \quad g(z) \leq y\}] = F_{g(Z_n)}(y) \to F_{g(Z)}(y) = P_Z[\{z: \quad g(z) \leq y\}]$$

at every real number y at which the distribution function $F_{g(Z)}(\cdot)$ is continuous.

Let us indicate briefly the significance of the most important of these statements. The practical meaning of convergence in distribution is expressed by (iii); the reader should compare the statement of the central limit theorem in section 5 of Chapter 8 to see that (iii) constitutes an exact mathematical formulation of the assertion that the probability law of Z "approximates" that of Z_n. From the point of view of establishing in practice that a sequence of random variables converges in distribution, one uses (ii), which constitutes a criterion for convergence in distribution in terms of characteristic functions. Finally, (v) represents a theoretical fact of the greatest usefulness in applications, for it asserts that if Z_n converges in distribution to Z then a sequence of random variables $g(Z_n)$, obtained as functions of the Z_n, converges in distribution to $g(Z)$ if the function $g(\cdot)$ is continuous.

We defer the proof of the equivalence of these statements to section 5.

The Continuity Theorem of Probability Theory. The inversion formulas of section 3 of Chapter 9 prove that there is a *one-to-one correspondence between distribution and characteristic functions;* given a distribution function $F(\cdot)$ and its characteristic function

$$(3.5) \qquad \phi(u) = \int_{-\infty}^{\infty} e^{iux}\, dF(x),$$

there is no other distribution function of which $\phi(\cdot)$ is the characteristic function. The results stated in theorem 3A show that the one-to-one correspondence between distribution and characteristic functions, regarded as a transformation between functions, is *continuous* in the sense that a sequence of distribution functions $F_n(\cdot)$ converges to a distribution function

$F(\cdot)$ at all points of continuity of $F(\cdot)$ if and only if the sequence of characteristic functions

$$\phi_n(u) = \int_{-\infty}^{\infty} e^{iux}\, dF_n(x) \tag{3.6}$$

converges at each real number u to the characteristic function $\phi(\cdot)$ of $F(\cdot)$. Consequently, theorem 3A is often referred to as the *continuity theorem of probability theory.*

Theorem 3A has the following extremely important extension, of which the reader should be aware. *Suppose that the sequence of characteristic functions $\phi_n(\cdot)$, defined by* (3.6), *has the property of converging at all real u to a function $\phi(\cdot)$, which is continuous at $u = 0$. It may be shown that there is then a distribution function $F(\cdot)$, of which $\phi(\cdot)$ is the characteristic function.* In view of this fact, the continuity theorem of probability theory is sometimes formulated in the following way:

Consider a sequence of distribution functions $F_n(x)$, with characteristic functions $\phi_n(u)$, defined by (3.6). *In order that a distribution function $F(\cdot)$ exist such that*

$$\lim_{n\to\infty} F_n(x) = F(x)$$

at all points x, which are continuity points of $F(x)$, it is necessary and sufficient that a function $\phi(u)$, continuous at $u = 0$, exist such that

$$\lim_{n\to\infty} \phi_n(u) = \phi(u) \qquad \textit{at all real } u.$$

Expansions for the Characteristic Function. In the use of characteristic functions to prove theorems concerning convergence in distribution, a major role is played by expansions for the characteristic function, and for the logarithm of the characteristic function, of a random variable such as those given in lemmas 3A and 3B. Throughout this chapter we employ this convention regarding the use of the symbol θ. *The symbol θ is used to describe any real or complex valued quantity satisfying the inequality* $|\theta| \le 1$. It is to be especially noted that the symbol θ does not denote the same number each time it occurs, but only that the number represented by it has modulus less than 1.

LEMMA 3A. Let X be a random variable whose mean $E[X]$ exists and is equal to 0 and whose variance $\sigma^2[X] = E[X^2]$ is finite. Then (i) for any u

$$\phi_X(u) = 1 - \tfrac{1}{2}u^2 E[X^2] - u^2 \int_0^1 dt(1-t)E[X^2(e^{iutX} - 1)]; \tag{3.7}$$

(ii) for any u such that $3u^2E[X^2] \le 1$, $\log \phi_X(u)$ exists and satisfies

$$\log \phi_X(u) = -\tfrac{1}{2}u^2 E[X^2] - u^2 \int_0^1 dt(1-t)E[X^2(e^{iutX} - 1)] + 3\theta u^4 E^2[X^2] \tag{3.8}$$

for some number θ such that $|\theta| \le 1$. Further, if the third absolute moment $E[|X|^3]$ is finite, then for u such that $3u^2E[X^2] \le 1$

$$\log \phi_X(u) = -\frac{1}{2}u^2E[X^2] + \frac{\theta}{6}|u|^3E[|X|^3] + 3\theta|u|^4E^2[X^2]. \tag{3.9}$$

Proof: Equation (3.7) follows immediately by integrating with respect to the distribution function of X the easily verified expansion

$$e^{iux} = 1 + iux - \tfrac{1}{2}u^2x^2 - u^2x^2\int_0^1 dt(1-t)(e^{iutx}-1). \tag{3.10}$$

To show (3.8), we write [by (3.7)] that $\log \phi_X(u) = \log(1-r)$, in which

$$r = \tfrac{1}{2}u^2E[X^2] + u^2\int_0^1 dt(1-t)E[X^2(e^{iutx}-1)]. \tag{3.11}$$

Now $|r| \le 3u^2E[X^2]/2$, so that $|r| \le \frac{1}{2}$ if u is such that $3u^2E[X^2] \le 1$. For any complex number r of modulus $|r| \le \frac{1}{2}$

$$\log(1-r) = -r\int_0^1 \frac{1}{1-rt}\,dt,$$

$$\log(1-r) + r = -r^2\int_0^1 \frac{t}{1-rt}\,dt, \tag{3·12}$$

$$|\log(1-r) - (-r)| \le |r|^2 \le (\tfrac{9}{4})u^4E^2[X^2],$$

since $|1-rt| \ge 1 - |rt| \ge \frac{1}{2}$. The proof of (3.8) is completed.

Finally, (3.9) follows immediately from (3.8), since

$$-u^2\int_0^1 dt(1-t)E[X^2(e^{iutX}-1)] = \frac{(iu)^3}{2}\int_0^1 dt(1-t)^2E[X^3e^{iutX}].$$

LEMMA 3B. In the same way that (3.7) and (3.8) are obtained, one may obtain expansions for the characteristic function of a random variable Y whose mean $E[Y]$ exists:

$$\phi_Y(u) = 1 + iuE[Y] + iu\int_0^1 dtE[Y(e^{iutY}-1)] \tag{3.13}$$

$$\log \phi_Y(u) = iuE[Y] + iu\int_0^1 dtE[Y(e^{iutY}-1)] + 9\theta u^2E^2[|Y|]$$

for u such that $6|u|E[|Y|] \le 1$.

▶ **Example 3A. Asymptotic normality of binomial random variables.** In section 2 of Chapter 6 it is stated that a binomial random variable is approximately normally distributed. This assertion may be given a precise formulation in terms of the notion of convergence in distribution. Let S_n be the number of successes in n independent repeated Bernoulli trials, with probability p of success at each trial, and let

$$Z_n = \frac{S_n - E[S_n]}{\sigma[S_n]} = \frac{S_n - np}{\sqrt{npq}}. \tag{3.14}$$

Let Z be any random variable that is normally distributed with mean 0 and variance 1. We now show that the sequence $\{Z_n\}$ converges in distribution to Z. To prove this assertion, we first write the characteristic function of Z_n in the form

$$\begin{aligned} \phi_{Z_n}(u) &= \exp\,[-iu(np/\sqrt{npq})]\phi_{S_n}\left(\frac{u}{\sqrt{npq}}\right) \\ &= [q \exp\,(-iu\sqrt{p/nq}) + p \exp\,(iu\sqrt{q/np})]^n. \end{aligned} \tag{3.15}$$

Therefore,

$$\log \phi_{Z_n}(u) = n \log \phi_X(u), \tag{3.16}$$

where we define

$$\phi_X(u) = q \exp\,(-iu\sqrt{p/nq}) + p \exp\,(iu\sqrt{q/np}). \tag{3.17}$$

Now $\phi_X(u)$ is the characteristic function of a random variable X with mean, mean square, and absolute third moment given by

$$\begin{aligned} E[X] &= q(-\sqrt{p/nq}) + p\sqrt{q/np} = 0, \\ E[X^2] &= q(-\sqrt{p/nq})^2 + p(\sqrt{q/np})^2 = \frac{p+q}{n} = \frac{1}{n}, \\ E[|X|^3] &= q|-\sqrt{p/nq}|^3 + p|\sqrt{q/np}|^3 = \frac{q^2+p^2}{(n^3pq)^{1/2}}. \end{aligned} \tag{3.18}$$

By (3.9), we have the expansion for $\log \phi_X(u)$, valid for u, such that $3u^2E[X^2] = 3u^2/n \leq 1$:

$$\begin{aligned} \log \phi_X(u) &= -\frac{1}{2}u^2E[X^2] + \frac{\theta}{6}|u|^3E[|X|^3] + 3\theta|u|^4E^2[X^2] \\ &= -\frac{1}{2n}u^2 + \frac{\theta}{6}u^3\frac{q^2+p^2}{(n^3pq)^{1/2}} + 3\theta|u|^4\frac{1}{n^2}, \end{aligned} \tag{3.19}$$

in which θ is some number such that $|\theta| \leq 1$.

In view of (3.16) and (3.19), we see that for fixed $u \neq 0$ and for n so large that $n \geq 3u^2$,

$$\text{(3.20)} \qquad \log \phi_{Z_n}(u) = -\frac{1}{2}u^2 + \frac{\theta}{6}u^3 \frac{q^2 + p^2}{(npq)^{1/2}} + \frac{3\theta|u|^4}{n},$$

which tends to $\log \phi_Z(u) = -\frac{1}{2}u^2$ as n tends to infinity. By statement (ii) of theorem 3A, it follows that the sequence $\{Z_n\}$ converges in distribution to Z. ◀

Characteristic functions may be used to prove theorems concerning convergence in probability to a constant. In particular, the reader may easily verify the following lemma.

LEMMA 3C. A sequence of random variables Z_n converges in probability to 0 if and only if it converges in distribution to 0, which is the case if and only if, for every real number u,

$$\text{(3.21)} \qquad \lim_{n \to \infty} \phi_{Z_n}(u) = 1.$$

THEOREM 3B. *The law of large numbers for a sequence of independent, identically distributed random variables $X_1, X_2, \ldots, X_n$ with common finite mean m.* As n tends to ∞, the sample mean $(1/n)(X_1 + \ldots + X_n)$ converges in probability to the mean $m = E[X]$, in which X is a random variable obeying the common probability law of $X_1, X_2, \ldots, X_n$.

Proof: Define $Y = X - E[X]$ and

$$Z_n = \frac{1}{n}(X_1 + X_2 + \cdots + X_n) - E[X].$$

To prove that the sample mean $(1/n)(X_1 + X_2 + \ldots + X_n)$ converges in probability to the mean $E[X]$, it suffices to show that Z_n converges in distribution to 0. Now, for a given value of u and for n so large that $n > 6|u|E[|Y|]$

$$\text{(3.22)} \qquad \log \phi_{Z_n}(u) = n \log \phi_Y\left(\frac{u}{n}\right)$$

$$= n\left\{i\frac{u}{n}\int_0^1 dt E[Y(e^{iutY/n} - 1)] + 9\theta\frac{u^2}{n^2}E^2[|Y|]\right\},$$

which tends to 0 as n tends to ∞, since, for each fixed t, u, and y, $e^{iuty/n}$ tends to 1 as n tends to ∞. The proof is complete.

EXERCISES

3.1. Prove lemma 3C.

3.2. Let $X_1, X_2, \ldots, X_n$ be independent random variables, each assuming each of the values $+1$ and -1 with probability $\frac{1}{2}$. Let $Y_n = \sum_{j=1}^{n} X_j/2^j$. Find the characteristic function of Y_n and show that, as n tends to ∞, for each u, $\phi_{Y_n}(u)$ tends to the characteristic function of a random variable Y uniformly distributed over the interval -1 to 1. Consequently, evaluate $P[-2 < Y_n \le \frac{1}{2}]$, $P[\frac{1}{4} < Y_n \le \frac{1}{2}]$ approximately.

3.3. Let $X_1, X_2, \ldots, X_n$ be independent random variables, identically distributed as the random variable X. For $n = 1, 2, \ldots$, let

$$Z_n = \frac{S_n - E[S_n]}{\sigma[S_n]}, \qquad S_n = X_1 + X_2 + \cdots X_n.$$

Assuming that X is (i) binomial distributed with parameters $n = 6$ and $p = \frac{1}{3}$, (ii) Poisson distributed with parameter $\lambda = 2$, (iii) χ^2 distributed with $\nu = 2$ degrees of freedom, for each real number u, show that $\lim_{n\to\infty} \log \phi_{Z_n}(u) = -\frac{1}{2}u^2$. Consequently, evaluate $P[18 \le S_{10} \le 20]$ approximately.

3.4. For any integer r and $0 < p < 1$ let $N(r, p)$ denote the minimum number of trials required to obtain r successes in a sequence of independent repeated Bernoulli trials, in which the probability of success at each trial is p. Let Z be a random variable χ^2 distributed with $2r$ degrees of freedom. Show that, at each u, $\lim_{p\to 0} \phi_{2pN(r,p)}(u) = \phi_Z(u)$. State in words the meaning of this result.

3.5. Let Z_n be binomial distributed with parameters n and $p = \lambda/n$, in which $\lambda > 0$ is a fixed constant. Let Z be Poisson distributed with parameter λ. For each u, show that $\lim_{n\to\infty} \phi_{Z_n}(u) = \phi_Z(u)$. State in words the meaning of this result.

3.6. Let Z be a random variable Poisson distributed with parameter λ. By use of characteristic functions, show that as λ tends to ∞

$$\mathscr{L}\left(\frac{Z - \lambda}{\sqrt{\lambda}}\right) \to \mathscr{L}(Y)$$

in which Y is normally distributed with mean 0 and variance 1.

3.7. Show that $\operatorname{plim}_{n\to\infty} X_n = X$ implies that $\lim_{n\to\infty} \mathscr{L}(X_n) = \mathscr{L}(X)$.

4. THE CENTRAL LIMIT THEOREM

A sequence of jointly distributed random variables $X_1, X_2, \ldots, X_n$ with finite means and variances is said to obey the (classical) central limit theorem if the sequence $Z_1, Z_2, \ldots, Z_n$, defined by

$$(4.1) \qquad Z_n = \frac{S_n - E[S_n]}{\sigma[S_n]}, \qquad S_n = X_1 + X_2 + \cdots + X_n,$$

converges in distribution to a random variable that is normally distributed with mean 0 and variance 1. In terms of characteristic functions, the sequence $\{X_n\}$ obeys the central limit theorem if for every real number u

$$\lim_{n\to\infty} \phi_{Z_n}(u) = e^{-\frac{1}{2}u^2}. \tag{4.2}$$

The random variables $Z_1, Z_2, \ldots, Z_n$ are called the sequence of normalized consecutive sums of the sequence $X_1, X_2, \ldots, X_n$.

That the central limit theorem is true under fairly unrestrictive conditions on the random variables $X_1, X_2, \ldots$ was already surmised by Laplace and Gauss in the early 1800's. However, the first satisfactory conditions, backed by a rigorous proof, for the validity of the central limit theorem were given by Lyapunov in 1901. In the 1920's and 1930's the method of characteristic functions was used to extend the theorem in several directions and to obtain fairly unrestrictive necessary and sufficient conditions for its validity in the case in which the random variables $X_1, X_2, \ldots$ are independent. More recent years have seen extensive work on extending the central limit theorem to the case of dependent random variables.

The reader is referred to the treatises of B. V. Gnedenko and A. N. Kolmogorov, *Limit Distributions for Sums of Independent Random Variables*, Addison-Wesley, Cambridge, Mass., 1954, and M. Loève, *Probability Theory*, Van Nostrand, New York, 1955, for a definitive treatment of the central limit theorem and its extensions.

From the point of view of the applications of probability theory, there are two main versions of the central limit theorem that one should have at his command. One should know conditions for the validity of the central limit theorem in the cases in which (i) the random variables $X_1, X_2, \ldots$ are independent and identically distributed and (ii) the random variables $X_1, X_2, \ldots$ are independent but not identically distributed.

THEOREM 4A. THE CENTRAL LIMIT THEOREM FOR INDEPENDENT IDENTICALLY DISTRIBUTED RANDOM VARIABLES WITH FINITE MEANS AND VARIANCES. For $n = 1, 2, \ldots$ let X_n be identically distributed as the random variable X, with finite mean $E[X]$ and standard deviation $\sigma[X]$. Let the sequence $\{X_n\}$ be independent, and let Z_n be defined by (4.1) or, more explicitly,

$$Z_n = \frac{(X_1 + \cdots + X_n) - nE[X]}{\sqrt{n}\sigma[X]}. \tag{4.3}$$

Then (4.2) will hold.

THEOREM 4B. THE CENTRAL LIMIT THEOREM FOR INDEPENDENT RANDOM VARIABLES WITH FINITE MEANS AND $(2 + \delta)$th CENTRAL MOMENT, FOR SOME $\delta > 0$. For $n = 1, 2, \ldots$ let X_n be a random variable with finite mean $E[X_n]$ and finite $(2 + \delta)$th central moment $\mu(2 + \delta; n) = E[|X_n - E[X_n]|^{2+\delta}]$.

Let the sequence $\{X_n\}$ be independent, and let Z_n be defined by (4.1). Then (4.2) will hold if

$$\lim_{n\to\infty} \frac{1}{\sigma^{2+\delta}[S_n]} \sum_{k=1}^{n} \mu(2+\delta; k) = 0, \tag{4.4}$$

in which $\sigma^2[S_n] = \sum_{k=1}^{n} \text{Var}[X_k]$.

Equation (4.4) is called *Lyapunov's condition* for the validity of the central limit theorem for independent random variables $\{X_n\}$.

We turn now to the proofs of theorems 4A and 4B. Consider first independent random variables $X_1, X_2, \ldots, X_n$, identically distributed as the random variable X, with mean 0 and variance σ^2. Let Z_n be their normalized sum, given by (4.1). The characteristic function of Z_n may be written

$$\phi_{Z_n}(u) = \left[\phi_X\left(\frac{u}{\sigma[S_n]}\right)\right]^n. \tag{4.5}$$

Now $\sigma[S_n] = \sqrt{n}\sigma[X]$ tends to ∞ as n tends to ∞. Therefore, for each fixed u, $\log \phi_X(u/\sigma[S_n])$ exists (by lemma 3A) when $n \geq 3u^2$. For n as large as this, using the expansion given by (3.8),

$$\log \phi_{Z_n}(u) = n\left\{-\frac{1}{2}\frac{u^2}{n} - \frac{u^2}{n\sigma^2}\int_0^1 dt(1-t)E[X^2(e^{ituX/\sqrt{n}\sigma} - 1)] + 3\theta\frac{u^4}{n^2\sigma^4}\sigma^4\right\}. \tag{4.6}$$

Theorem 4A will be proved if we prove that

$$\lim_{n\to\infty} \log \phi_{Z_n}(u) = -\tfrac{1}{2}u^2. \tag{4.7}$$

It is clear that to prove (4.7) will hold we need prove only that the integral in (4.6) tends to 0 as n tends to infinity. Define $g(x, t, u) = x^2(e^{itux/\sqrt{n}\sigma} - 1)$. Then, for any $M > 0$

$$E[g(X, t, u)] = \int_{|x|<M} g(x, t, u)\, dF_X(x) + \int_{|x|\geq M} g(x, t, u)\, dF_X(x). \tag{4.8}$$

Now, $|g(x,t,u)| \leq x^2|utx|/\sigma\sqrt{n} \leq x^2|Mut|/\sigma\sqrt{n}$ for $|x| < M$ and $|g(x,t,u)| \leq 2x^2$ for $|x| \geq M$, in view of the inequalities $|e^{iw} - 1| \leq |w|$, $|e^{iw} - 1| \leq 2$. From these facts we may conclude that for any $M > 0$ and real numbers u and t

$$|E[g(X, t, u)]| \leq \sigma\frac{M|ut|}{\sqrt{n}} + 2\int_{|x|\geq M} x^2\, dF_X(x). \tag{4.9}$$

Then

$$\left|\int_0^1 dt(1-t)E[g(X, t, u)]\right| \leq \sigma\frac{M|u|}{\sqrt{n}} + 2\int_{|x|\geq M} x^2\, dF_X(x), \tag{4.10}$$

which tends to 0, as we let first n tend to ∞ and then M tend to ∞. The proof of the central limit theorem for identically distributed independent random variables with finite variances is complete.

We next prove the central limit theorem under Lyapunov's condition. For $k = 1, 2, \ldots$, let X_k be a random variable with mean 0, finite variance σ_k^2, and $(2 + \delta)$th central moment $\mu(2 + \delta; k)$. We have the following expansion of the logarithm of its characteristic function, for u such that $3u^2\sigma_k^2 \leq 1$:

$$(4.11) \qquad \log \phi_{X_k}(u) = -\tfrac{1}{2}u^2\sigma_k^2 + 2\theta|u|^{2+\delta}\mu(2 + \delta; k) + 3\theta u^4\sigma_k^4.$$

To prove (4.11), merely use in (3.8) the inequality $|e^{iw} - 1| \leq 2|w|^{\delta}$, valid for any real number w and $0 \leq \delta \leq 1$.

Now, (4.4) and theoretical exercise 4.3 imply that

$$(4.12) \qquad \left(\max_{1 \leq k \leq n} \frac{\sigma_k^2}{\sigma^2[S_n]}\right)^{(2+\delta)/2} \leq \left(\max_{1 \leq k \leq n} \frac{\mu(2 + \delta; k)}{\sigma^{2+\delta}[S_n]}\right) \to 0.$$

Then, for any fixed u it holds for n sufficiently large that $3u^2\sigma_k^2/\sigma^2[S_n] \leq 1$ for all $k = 1,2, \ldots, n$. Therefore, $\log \phi_{Z_n}(u)$ exists and is given by

$$(4.13) \qquad \log \phi_{Z_n}(u) = \sum_{k=1}^{n} \log \phi_{X_k}\left(\frac{u}{\sigma[S_n]}\right) = -\frac{1}{2}u^2 \sum_{k=1}^{n} \frac{\sigma_k^2}{\sigma^2[S_n]}$$
$$+ 2\theta|u|^{2+\delta} \sum_{k=1}^{n} \frac{\mu(2 + \delta; k)}{\sigma^{2+\delta}[S_n]} + 3\theta u^4 \frac{1}{\sigma^4[S_n]} \sum_{k=1}^{n} \sigma_k^4.$$

The first sum in (4.13) is equal to 1, whereas the second sum tends to 0 by Lyapunov's condition, as does the third sum, since

$$\left(\frac{\sigma_k}{\sigma[S_n]}\right)^4 \leq \left(\frac{\sigma_k}{\sigma[S_n]}\right)^{2+\delta} \leq \frac{\mu(2 + \delta; k)}{\sigma^{2+\delta}[S_n]}.$$

The proof of the central limit theorem under Lyapunov's condition is complete.

THEORETICAL EXERCISES

4.1. Prove that the central limit theorem holds for independent random variables $X_1, X_2, \ldots$ with zero means and finite variances obeying *Lindeberg's* condition: for every $\epsilon > 0$

$$(4.14) \qquad \lim_{n \to \infty} \frac{1}{\sigma^2[S_n]} \sum_{k=1}^{n} \int_{|x| \geq \epsilon\sigma[S_n]} x^2 \, dF_{X_k}(x) = 0.$$

Hint: In (4.8) let $M = \epsilon\sigma[S_n]$, replacing $\sigma\sqrt{n}$ by $\sigma[S_n]$. Obtain thereby an estimate for $E[X_k^2(e^{iutX_k/\sigma[S_n]} - 1)]$. Add these estimates to obtain an estimate for $\log \phi_{Z_n}(u)$, as in (4.13).

4.2. Prove the *law of large numbers under Markov's condition. Hint:* Adapt the proof of the central limit theorem under Lyapunov's condition, using the expansions (3.13).

4.3. Jensen's inequality and its consequences. Let X be a random variable, and let I be a (possibly infinite) interval such that, with probability one, X takes its values in I; that is $P[X \text{ lies in } I] = 1$. Let $g(\cdot)$ be a function of a real variable that is twice differentiable on I and whose second derivative satisfies $g''(x) \geq 0$ for all x in I. The function $g(\cdot)$ is then said to be *convex* on I. Show that the following inequality (Jensen's inequality) holds:

$$g(E[X]) \leq E[g(X)]. \tag{4.15}$$

Hint: Show by Taylor's theorem that $g(x) \geq g(x_0) + g'(x_0)(x - x_0)$. Let $x_0 = E[X]$ and take the expectation of both sides of the inequality. Deduce from (4.15) that for any $r \geq 1$ and $s > 0$

$$|E[X]|^r \leq E^r[|X|] \leq E[|X|^r] \tag{4.16}$$

$$E^r[|X|^s] \leq E[|X|^{rs}]. \tag{4.17}$$

Conclude from (4.17) that if $0 < r_1 < r_2$ then

$$E^{1/r_1}[|X|^{r_1}] \leq E^{1/r_2}[|X|^{r_2}]. \tag{4.18}$$

In particular, conclude that

$$E[|X|] \leq E^{1/2}[|X|^2] \leq E^{1/3}[|X|^3] \leq \cdots. \tag{4.19}$$

4.4. Let $\{U_n\}$ be a sequence of independent random variables, each uniformly distributed on the interval 0 to π. Let $\{A_n\}$ be a sequence of positive constants. State conditions under which the sequence $X_n = A_n \cos U_n$ obeys the central limit theorem.

5. PROOFS OF THEOREMS CONCERNING CONVERGENCE IN DISTRIBUTION

In this section we prove the equivalence of the statements in theorem 3A by showing that each implies its successor. For ease of writing, on occasion we write $F_n(\cdot)$ for $F_{Z_n}(\cdot)$, $\phi_n(\cdot)$ for $\phi_{Z_n}(\cdot)$, $F(\cdot)$ for $F_Z(\cdot)$, and $\phi(\cdot)$ for $\phi_Z(\cdot)$.

It is immediate that (i) implies (ii), since the function $g(z) = e^{iuz}$ is a bounded continuous function of z.

To prove that (ii) implies (iii), we make use of the basic formula (3.6) of Chapter 9. For any $d > 0$ define the function $g_d(\cdot)$ for any real number z by

$$\begin{aligned} g_d(z) &= 1 && \text{if } a \leq z \leq b \\ &= 1 - \left(\frac{a - z}{d}\right) && \text{if } a - d \leq z \leq a \\ &= 1 - \left(\frac{z - b}{d}\right) && \text{if } b \leq z \leq b + d \\ &= 0 && \text{otherwise.} \end{aligned}$$

The function $g_d(\cdot)$ is continuous and integrable. Its Fourier transform $\gamma_d(\cdot)$ is given for any u by

$$(5.1)\qquad \gamma_d(u) = \frac{1}{2\pi}\int_{-\infty}^{\infty} e^{-iuz} g_d(z)\,dz = \frac{1}{2\pi iu}\int_{-\infty}^{\infty} e^{-iuz} g_d'(z)\,dz.$$

Therefore,

$$(5.2)\qquad \gamma_d(u) = \frac{1}{2\pi iud}\left(\int_{a-d}^{a} e^{-iuz} - \int_{b}^{b+d} e^{-iuz}\right)$$

$$= \frac{1}{2\pi u^2 d}(e^{-iua} - e^{-iu(a-d)} - e^{-iu(b+d)} + e^{-iub}).$$

Thus we see that the Fourier transform $\gamma_d(\cdot)$ is integrable. Consequently, from (3.6), of Chapter 9 we have

$$(5.3)\qquad \int_{-\infty}^{\infty} g_d(z)\,dF_{Z_n}(z) - \int_{-\infty}^{\infty} g_d(z)\,dF_Z(z) = \int_{-\infty}^{\infty} du\,\gamma_d(u)[\phi_{Z_n}(u) - \phi_Z(u)].$$

By letting n tend to ∞ in (5.3) and using the hypothesis of statement (ii), we obtain for any $d > 0$, as n tends to ∞,

$$(5.4)\qquad \int_{-\infty}^{\infty} g_d(z)\,dF_{Z_n}(z) \to \int_{-\infty}^{\infty} g_d(z)\,dF_Z(z).$$

Next, define the function $g_d^*(\cdot)$ for any z by

$$\begin{aligned} g_d^*(z) &= 1 && \text{if } a + d \le z \le b - d \\ &= \left(\frac{z-a}{d}\right) && \text{if } a \le z \le a + d \\ &= \left(\frac{b-z}{d}\right) && \text{if } b - d \le z \le b \\ &= 0 && \text{otherwise.} \end{aligned}$$

By the foregoing argument, one may prove that (5.4) holds for $g_d^*(\cdot)$. Now, the expectations of the functions $g_d(\cdot)$ and $g_d^*(\cdot)$ clearly straddle the quantity $F_{Z_n}(b) - F_{Z_n}(a)$:

$$(5.5)\qquad \int_{-\infty}^{\infty} g_d^*(z)\,dF_{Z_n}(z) \le F_{Z_n}(b) - F_{Z_n}(a) \le \int_{-\infty}^{\infty} g_d(z)\,dF_{Z_n}(z).$$

From (5.5), letting n tend to ∞, we obtain

$$(5.6)\qquad \begin{aligned} \int_{-\infty}^{\infty} g_d^*(z)\,dF_Z(z) &\le \liminf_n F_{Z_n}(b) - F_{Z_n}(a) \\ &\le \limsup_n F_{Z_n}(b) - F_{Z_n}(a) \\ &\le \int_{-\infty}^{\infty} g_d(z)\,dF_Z(z). \end{aligned}$$

Now, let d tend to 0 in (5.6); since

$$
\begin{aligned}
0 &\le F_Z(b) - F_Z(a) - \int_{-\infty}^{\infty} g_d^*(z)\, dF_Z(z) \le F_Z(a+d) \\
&\qquad - F_Z(a) + F_Z(b) - F_Z(b-d) \to 0, \\
0 &\le \int_{-\infty}^{\infty} g_d(z)\, dF_Z(z) - [F_Z(b) - F_Z(a)] \le F_Z(a) \\
&\qquad - F_Z(a-d) + F_Z(b+d) - F_Z(b) \to 0.
\end{aligned}
\tag{5.7}
$$

as d tends to 0, it follows that (3.4) holds. Note that (5.7) would not hold if we did not require a and b to be points at which $F_Z(\cdot)$ is continuous.

We next prove that (iii) implies (iv). Let M be a positive number such that $F(\cdot)$ is continuous at M and at $-M$. Then, for any real number a

$$
\begin{aligned}
|F_n(a) - F(a)| &\le |F_n(a) - F_n(-M) - F(a) + F(-M)| \\
&\qquad + F_n(-M) + F(-M).
\end{aligned}
\tag{5.8}
$$

Since statement (iii) holds, it follows that if a is a continuity point of $F(\cdot)$

$$
\limsup_n |F_n(a) - F(a)| \le F(-M) + \limsup_n F_n(-M). \tag{5.9}
$$

Now, also by (iii), since $F_n(M) - F_n(-M)$ tends to $F(M) - F(-M)$,

$$
\begin{aligned}
\limsup_n F_n(-M) &\le \limsup_n (1 - F_n(M) + F_n(-M)) \\
&\le 1 - F(M) + F(-M).
\end{aligned}
\tag{5.10}
$$

Consequently,

$$
\limsup_n |F_n(a) - F(a)| \le 2F(-M) + 1 - F(M), \tag{5.11}
$$

which tends to 0, as one lets M tend to ∞. The proof that (iii) implies (iv) is complete.

We next prove that (iv) implies (i). We first note that a function $g(\cdot)$, continuous on a closed interval, is uniformly continuous there; that is, for every positive number ϵ there is a positive number, denoted by $d(\epsilon)$, such that $|g(z_1) - g(z_2)| \le \epsilon$ for any two points z_1 and z_2 in the interval satisfying $|z_1 - z_2| \le d(\epsilon)$. Choose M so that $F(\cdot)$ is continuous at M and $-M$. On the closed interval $[-M, M]$, $g(\cdot)$ is continuous. Fix $\epsilon > 0$, and let $d(\epsilon)$ be defined as in the foregoing sentence. We may then choose $(K+1)$ real numbers $a_0, a_1, \ldots, a_K$ having these properties: (i) $-M = a_0 < a_1 < \ldots < a_K = M$, (ii) $a_k - a_{k-1} \le d(\epsilon)$ for $k = 1, 2, \ldots, K$, (iii) for $k = 1, 2, \ldots, K$, $F(\cdot)$ is continuous at a_k. Then define a function $g(\cdot; \epsilon, M)$:

$$
\begin{aligned}
g(x; \epsilon, M) &= 0 && \text{if } |x| > M, \\
&= g(a_k) && \text{if } a_{k-1} < x \le a_k, \quad \text{for some } k = 1, 2, \cdots, K, \\
&= g(-M) && \text{if } x = -M.
\end{aligned}
\tag{5.12}
$$

It is clear that for $|x| \leq M$

$$|g(x) - g(x; \epsilon, M)| \leq \epsilon. \tag{5.13}$$

Now

$$\left| \int_{-\infty}^{\infty} g(x)\, dF_n(x) - \int_{-\infty}^{\infty} g(x)\, dF(x) \right| \leq |I_n| + |J_n| + |I|, \tag{5.14}$$

where
$$I_n = \int_{-\infty}^{\infty} [g(x) - g(x; \epsilon, M)]\, dF_n(x)$$
$$I = \int_{-\infty}^{\infty} [g(x) - g(x; \epsilon, M)]\, dF(x)$$
$$J_n = \int_{-\infty}^{\infty} g(x; \epsilon, M)\, dF_n(x) - \int_{-\infty}^{\infty} g(x; \epsilon, M)\, dF(x)$$

Let C be an upper bound for $g(\cdot)$; that is, $|g(x)| \leq C$ for all x. Then

$$|J_n| \leq C \sum_{k=1}^{K} [|F_n(a_k) - F(a_k)| + |F_n(a_{k-1}) - F(a_{k-1})|]. \tag{5.15}$$

Next, we may write I_n as a sum of two integrals, one over the range $|x| \leq M$ and the other over the range $|x| > M$. In view of (5.13), we then have

$$|I_n| \leq \epsilon + C[1 - F_n(M) + F_n(-M)].$$

Similarly

$$|I| \leq \epsilon + C[1 - F(M) + F(-M)].$$

In view of (5.14), (5.15), and the two preceding inequalities, it follows that

$$\limsup_n \left| \int_{-\infty}^{\infty} g(x)\, dF_n(x) - \int_{-\infty}^{\infty} g(x)\, dF(x) \right| \leq 2\epsilon + 2C[1 - F(M) + F(-M)]. \tag{5.16}$$

Letting first ϵ tend to 0 and then M tend to ∞, it follows that (5.13) will hold. The proof that (iv) implies (i) is complete.

The reader may easily verify that (v) is equivalent to the preceding statements.

THEORETICAL EXERCISES

5.1. Convergence of the means of random variables convergent in distribution. If Z_n converges in distribution to Z, show that for $M > 0$ such that $F_Z(\cdot)$ is continuous at $\pm M$, as n tends to ∞.

$$\int_{-M}^{M} z\, dF_{Z_n}(z) \to \int_{-M}^{M} z\, dF_Z(z).$$

From this it does not follow that $E[Z_n]$ converges to $E[Z]$. *Hint:* Let $F_{Z_n}(z) = 0, 1 - (1/n), 1$, depending on whether $z < 0$, $0 \leq z < n$, $n \leq z$; then $E[Z_n] = 1$ does not tend to $E[Z] = 0$. But if Z_n converges in distribution to Z and, in addition, $E[Z_n]$ exists for all n and

$$\lim_{M\to\infty} \limsup_{n\to\infty} \int_{|z|\geq M} |z| \, dF_{Z_n}(z) = 0, \tag{5.17}$$

then $E[Z_n]$ converges to $E[Z]$.

5.2. On uniform convergence of distribution functions. Let $\{Z_n\}$ be a sequence of random variables converging in distribution to the random variable Z, so that, for each real number z, $\lim_n F_{Z_n}(z) = F_Z(z)$. Show that if Z is a continuous random variable, so that $F_Z(\cdot)$ has no points of discontinuity, then the distribution functions converge uniformly: more precisely

$$\lim_{n\to\infty} \underset{-\infty<z<\infty}{\text{supremum}} |F_{Z_n}(z) - F_Z(z)| = 0.$$

Hint: To any $\epsilon > 0$, choose points $-\infty = z_0 < z_1 < \ldots < z_K = \infty$, so that $F_Z(z_j) - F_Z(z_{j-1}) < \epsilon$ for $j = 1, 2, \ldots, K$. Verify that

$$\underset{-\infty<z<\infty}{\text{supremum}} |F_{Z_n}(z) - F_Z(z)| \leq \max_{j=0,1,\cdots,K} |F_{Z_n}(z_j) - F_Z(z_j)| + \epsilon.$$

Tables

TABLE I

Area under the Normal Density Function

A table of $\Phi(x) = \frac{1}{\sqrt{2\pi}} \int_{-\infty}^{x} e^{-\frac{1}{2}y^2}\, dy$

x	0.00	0.01	0.02	0.03	0.04	0.05	0.06	0.07	0.08	0.09
0.0	.5000	.5040	.5080	.5120	.5160	.5199	.5239	.5279	.5319	.5359
0.1	.5398	.5438	.5478	.5517	.5557	.5596	.5636	.5675	.5714	.5753
0.2	.5793	.5832	.5871	.5910	.5948	.5987	.6026	.6064	.6103	.6141
0.3	.6179	.6217	.6255	.6293	.6331	.6368	.6406	.6443	.6480	.6517
0.4	.6554	.6591	.6628	.6664	.6700	.6736	.6772	.6808	.6844	.6879
0.5	.6915	.6950	.6985	.7019	.7054	.7088	.7123	.7157	.7190	.7224
0.6	.7257	.7291	.7324	.7357	.7389	.7422	.7454	.7486	.7517	.7549
0.7	.7580	.7611	.7642	.7673	.7704	.7734	.7764	.7794	.7823	.7852
0.8	.7881	.7910	.7939	.7967	.7995	.8023	.8051	.8078	.8106	.8133
0.9	.8159	.8186	.8212	.8238	.8264	.8289	.8315	.8340	.8365	.8389
1.0	.8413	.8438	.8461	.8485	.8508	.8531	.8554	.8577	.8599	.8621
1.1	.8643	.8665	.8686	.8708	.8729	.8749	.8770	.8790	.8810	.8830
1.2	.8849	.8869	.8888	.8907	.8925	.8944	.8962	.8980	.8997	.9015
1.3	.9032	.9049	.9066	.9082	.9099	.9115	.9131	.9147	.9162	.9177
1.4	.9192	.9207	.9222	.9236	.9251	.9265	.9279	.9292	.9306	.9319
1.5	.9332	.9345	.9357	.9370	.9382	.9394	.9406	.9418	.9429	.9441
1.6	.9452	.9463	.9474	.9484	.9495	.9505	.9515	.9525	.9535	.9545
1.7	.9554	.9564	.9573	.9582	.9591	.9599	.9608	.9616	.9625	.9633
1.8	.9641	.9649	.9656	.9664	.9671	.9678	.9686	.9693	.9699	.9706
1.9	.9713	.9719	.9726	.9732	.9738	.9744	.9750	.9756	.9761	.9767
2.0	.9772	.9778	.9783	.9788	.9793	.9798	.9803	.9808	.9812	.9817
2.1	.9821	.9826	.9830	.9834	.9838	.9842	.9846	.9850	.9854	.9857
2.2	.9861	.9864	.9868	.9871	.9875	.9878	.9881	.9884	.9887	.9890
2.3	.9893	.9896	.9898	.9901	.9904	.9906	.9909	.9911	.9913	.9916
2.4	.9918	.9920	.9922	.9925	.9927	.9929	.9931	.9932	.9934	.9936
2.5	.9938	.9940	.9941	.9943	.9945	.9946	.9948	.9949	.9951	.9952
2.6	.9953	.9955	.9956	.9957	.9959	.9960	.9961	.9962	.9963	.9964
2.7	.9965	.9966	.9967	.9968	.9969	.9970	.9971	.9972	.9973	.9974
2.8	.9974	.9975	.9976	.9977	.9977	.9978	.9979	.9979	.9980	.9981
2.9	.9981	.9982	.9982	.9983	.9984	.9984	.9985	.9985	.9986	.9986
3.0	.9987	.9987	.9987	.9988	.9988	.9989	.9989	.9989	.9990	.9990
3.1	.9990	.9991	.9991	.9991	.9992	.9992	.9992	.9992	.9993	.9993
3.2	.9993	.9993	.9994	.9994	.9994	.9994	.9994	.9995	.9995	.9995
3.3	.9995	.9995	.9995	.9996	.9996	.9996	.9996	.9996	.9996	.9997
3.4	.9997	.9997	.9997	.9997	.9997	.9997	.9997	.9997	.9997	.9998
3.6	.9998	.9998	.9999	.9999	.9999	.9999	.9999	.9999	.9999	.9999

TABLE II

Binomial Probabilities

A table of $\binom{n}{x} p^x(1-p)^{n-x}$ for $n = 1, 2, \ldots, 10$ and

$p = 0.01, 0.05(0.05)0.30, \frac{1}{3}, 0.35(0.05)0.50$, and $p = 0.49$

n	x \ p	.01	.05	.10	.15	.20	.25	.30	$\frac{1}{3}$	.35	.40	.45	.49	.50
2	0	.9801	.9025	.8100	.7225	.6400	.5625	.4900	.4444	.4225	.3600	.3025	.2601	.2500
	1	.0198	.0550	.1800	.2550	.3200	.3750	.4200	.4444	.4550	.4800	.4950	.4998	.5000
	2	.0001	.0025	.0100	.0225	.0400	.0625	.0900	.1111	.1225	.1600	.2025	.2401	.2500
3	0	.9703	.8574	.7290	.6141	.5120	.4219	.3430	.2963	.2746	.2160	.1664	.1327	.1250
	1	.0294	.1354	.2430	.3251	.3840	.4219	.4410	.4444	.4436	.4320	.4084	.3823	.3750
	2	.0003	.0071	.0270	.0574	.0960	.1406	.1890	.2222	.2389	.2880	.3341	.3674	.3750
	3	.0000	.0001	.0010	.0034	.0080	.0156	.0270	.0370	.0429	.0640	.0911	.1176	.1250
4	0	.9606	.8145	.6561	.5220	.4096	.3164	.2401	.1975	.1785	.1296	.0915	.0677	.0625
	1	.0388	.1715	.2916	.3685	.4096	.4219	.4116	.3951	.3845	.3456	.2995	.2600	.2500
	2	.0006	.0135	.0486	.0975	.1536	.2109	.2646	.2963	.3105	.3456	.3675	.3747	.3750
	3	.0000	.0005	.0036	.0115	.0256	.0469	.0756	.0988	.1115	.1536	.2005	.2400	.2500
	4	.0000	.0000	.0001	.0005	.0016	.0039	.0081	.0123	.0150	.0256	.0410	.0576	.0625
5	0	.9510	.7738	.5905	.4437	.3277	.2373	.1681	.1317	.1160	.0778	.0503	.0345	.0312
	1	.0480	.2036	.3280	.3915	.4096	.3955	.3602	.3292	.3124	.2592	.2059	.1657	.1562
	2	.0010	.0214	.0729	.1382	.2048	.2637	.3087	.3292	.3364	.3456	.3369	.3185	.3125
	3	.0000	.0011	.0081	.0244	.0512	.0879	.1323	.1646	.1811	.2304	.2757	.3060	.3125
	4	.0000	.0000	.0004	.0022	.0064	.0146	.0284	.0412	.0488	.0768	.1128	.1470	.1562
	5	.0000	.0000	.0000	.0001	.0003	.0010	.0024	.0041	.0053	.0102	.0185	.0283	.0312
6	0	.9415	.7351	.5314	.3771	.2621	.1780	.1176	.0878	.0754	.0467	.0277	.0176	.0156
	1	.0571	.2321	.3543	.3993	.3932	.3560	.3025	.2634	.2437	.1866	.1359	.1014	.0938
	2	.0014	.0305	.0984	.1762	.2458	.2966	.3241	.3292	.3280	.3110	.2780	.2437	.2344
	3	.0000	.0021	.0146	.0415	.0819	.1318	.1852	.2195	.2355	.2765	.3032	.3121	.3125
	4	.0000	.0001	.0012	.0055	.0154	.0330	.0595	.0823	.0951	.1382	.1861	.2249	.2344
	5	.0000	.0000	.0001	.0004	.0015	.0044	.0102	.0165	.0205	.0369	.0609	.0864	.0938
	6	.0000	.0000	.0000	.0000	.0001	.0002	.0007	.0014	.0018	.0041	.0083	.0139	.0156

TABLE II *(Continued)*

n	x													
7	0	.9321	.6983	.4783	.3206	.2097	.1335	.0824	.0585	.0490	.0280	.0152	.0090	.0078
	1	.0559	.2573	.3720	.3960	.3570	.3115	.2471	.2048	.1848	.1306	.0872	.0603	.0547
	2	.0020	.0406	.1240	.2097	.2753	.3115	.3177	.3073	.2985	.2613	.2140	.1740	.1641
	3	.0000	.0036	.0230	.0617	.1147	.1730	.2269	.2561	.2679	.2903	.2918	.2786	.2734
	4	.0000	.0002	.0026	.0109	.0287	.0577	.0972	.1280	.1442	.1935	.2388	.2676	.2734
	5	.0000	.0000	.0002	.0012	.0043	.0115	.0250	.0384	.0455	.0774	.1172	.1543	.1641
	6	.0000	.0000	.0000	.0001	.0004	.0013	.0036	.0064	.0084	.0172	.0320	.0494	.0547
	7	.0000	.0000	.0000	.0000	.0000	.0001	.0002	.0005	.0006	.0015	.0037	.0068	.0078
8	0	.9227	.6634	.4305	.2725	.1678	.1001	.0576	.0390	.0319	.0168	.0084	.0046	.0039
	1	.0746	.2793	.3826	.3847	.3355	.2670	.1977	.1561	.1373	.0896	.0548	.0352	.0312
	2	.0026	.0515	.1488	.2376	.2936	.3115	.2965	.2731	.2587	.2090	.1569	.1183	.1094
	3	.0001	.0054	.0331	.0839	.1468	.2076	.2541	.2731	.2786	.2787	.2568	.2273	.2188
	4	.0000	.0004	.0046	.0185	.0459	.0865	.1361	.1707	.1875	.2322	.2627	.2730	.2734
	5	.0000	.0000	.0004	.0026	.0092	.0231	.0467	.0683	.0808	.1239	.1719	.2098	.2188
	6	.0000	.0000	.0000	.0002	.0011	.0038	.0100	.0171	.0217	.0413	.0703	.1008	.1094
	7	.0000	.0000	.0000	.0000	.0001	.0004	.0012	.0024	.0033	.0079	.0164	.0277	.0312
	8	.0000	.0000	.0000	.0000	.0000	.0000	.0001	.0002	.0002	.0007	.0017	.0033	.0039
9	0	.9135	.6302	.3874	.2316	.1342	.0751	.0404	.0260	.0207	.0101	.0046	.0023	.0020
	1	.0830	.2985	.3874	.3679	.3020	.2253	.1556	.1171	.1004	.0605	.0339	.0202	.0176
	2	.0034	.0629	.1722	.2597	.3020	.3003	.2668	.2341	.2162	.1612	.1110	.0776	.0703
	3	.0001	.0077	.0446	.1069	.1762	.2336	.2668	.2731	.2716	.2508	.2119	.1739	.1641
	4	.0000	.0006	.0074	.0283	.0661	.1168	.1715	.2048	.2194	.2508	.2600	.2506	.2461
	5	.0000	.0000	.0008	.0050	.0165	.0389	.0735	.1024	.1181	.1672	.2128	.2408	.2461
	6	.0000	.0000	.0001	.0006	.0028	.0087	.0210	.0341	.0424	.0743	.1160	.1542	.1641
	7	.0000	.0000	.0000	.0000	.0003	.0012	.0039	.0073	.0098	.0212	.0407	.0635	.0703
	8	.0000	.0000	.0000	.0000	.0000	.0001	.0004	.0009	.0013	.0035	.0083	.0153	.0176
	9	.0000	.0000	.0000	.0000	.0000	.0000	.0000	.0001	.0001	.0003	.0008	.0016	.0020
10	0	.9044	.5987	.3487	.1969	.1074	.0563	.0282	.0173	.0135	.0060	.0025	.0012	.0010
	1	.0914	.3151	.3874	.3474	.2684	.1877	.1211	.0867	.0725	.0403	.0207	.0114	.0098
	2	.0042	.0746	.1937	.2759	.3020	.2816	.2335	.1951	.1757	.1209	.0763	.0495	.0439
	3	.0001	.0105	.0574	.1298	.2013	.2503	.2668	.2601	.2522	.2150	.1665	.1267	.1172
	4	.0000	.0010	.0112	.0401	.0881	.1460	.2001	.2276	.2377	.2508	.2384	.2130	.2051
	5	.0000	.0001	.0015	.0085	.0264	.0584	.1029	.1366	.1536	.2007	.2340	.2456	.2461
	6	.0000	.0000	.0001	.0012	.0055	.0162	.0368	.0569	.0689	.1115	.1596	.1966	.2051
	7	.0000	.0000	.0000	.0001	.0008	.0031	.0090	.0163	.0212	.0425	.0746	.1080	.1172
	8	.0000	.0000	.0000	.0000	.0001	.0004	.0014	.0030	.0043	.0106	.0229	.0389	.0439
	9	.0000	.0000	.0000	.0000	.0000	.0000	.0001	.0003	.0005	.0016	.0042	.0083	.0098
	10	.0000	.0000	.0000	.0000	.0000	.0000	.0000	.0000	.0000	.0001	.0003	.0008	.0010

TABLE III

Poisson Probabilities

A table of $e^{-\lambda}\lambda^x/x!$ for $\lambda = 0.1(0.1)2(0.2)4(1)10$

λ \ x	0	1	2	3	4	5	6	7	8	9	10	11	12
.1	.9048	.0905	.0045	.0002	.0000								
.2	.8187	.1637	.0164	.0011	.0001	.0000							
.3	.7408	.2222	.0333	.0033	.0002	.0000							
.4	.6703	.2681	.0536	.0072	.0007	.0001	.0000						
.5	.6065	.3033	.0758	.0126	.0016	.0002	.0000						
.6	.5488	.3293	.0988	.0198	.0030	.0004	.0000						
.7	.4966	.3476	.1217	.0284	.0050	.0007	.0001	.0000					
.8	.4493	.3595	.1438	.0383	.0077	.0012	.0002	.0000					
.9	.4066	.3659	.1647	.0494	.0111	.0020	.0003	.0000					
1.0	.3679	.3679	.1839	.0613	.0153	.0031	.0005	.0001	.0000				
1.1	.3329	.3662	.2014	.0738	.0203	.0045	.0008	.0001	.0000				
1.2	.3012	.3614	.2169	.0867	.0260	.0062	.0012	.0002	.0000				
1.3	.2725	.3543	.2303	.0998	.0324	.0084	.0018	.0003	.0001	.0000			
1.4	.2466	.3452	.2417	.1128	.0395	.0111	.0026	.0005	.0001	.0000			
1.5	.2231	.3347	.2510	.1255	.0471	.0141	.0035	.0008	.0001	.0000			
1.6	.2019	.3230	.2584	.1378	.0551	.0176	.0047	.0011	.0002	.0000			
1.7	.1827	.3106	.2640	.1496	.0636	.0216	.0061	.0015	.0003	.0001	.0000		
1.8	.1653	.2975	.2678	.1607	.0723	.0260	.0078	.0020	.0005	.0001	.0000		
1.9	.1496	.2842	.2700	.1710	.0812	.0309	.0098	.0027	.0006	.0001	.0000		
2.0	.1353	.2707	.2707	.1804	.0902	.0361	.0120	.0034	.0009	.0002	.0000		

TABLE III (*Continued*)

λ													
2.2	.1108	.2438	.2681	.1966	.1082	.0476	.0174	.0055	.0015	.0004	.0001	.0000	
2.4	.0907	.2177	.2613	.2090	.1254	.0602	.0241	.0083	.0025	.0007	.0002	.0000	
2.6	.0743	.1931	.2510	.2176	.1414	.0735	.0319	.0118	.0038	.0011	.0003	.0001	.0000
2.8	.0608	.1703	.2384	.2225	.1557	.0872	.0407	.0163	.0057	.0018	.0005	.0001	.0000
3.0	.0498	.1494	.2240	.2240	.1680	.1008	.0504	.0216	.0081	.0027	.0008	.0002	.0001
3.2	.0408	.1304	.2087	.2226	.1781	.1140	.0608	.0278	.0111	.0040	.0013	.0004	.0001
3.4	.0334	.1135	.1929	.2186	.1858	.1264	.0716	.0348	.0148	.0056	.0019	.0006	.0002
3.6	.0273	.0984	.1771	.2125	.1912	.1377	.0826	.0425	.0191	.0076	.0028	.0009	.0003
3.8	.0224	.0850	.1615	.2046	.1944	.1477	.0936	.0508	.0241	.0102	.0039	.0013	.0004
4.0	.0183	.0733	.1465	.1954	.1954	.1563	.1042	.0595	.0298	.0132	.0053	.0019	.0006
5.0	.0067	.0337	.0842	.1404	.1755	.1755	.1462	.1044	.0653	.0363	.0181	.0082	.0034
6.0	.0025	.0149	.0446	.0892	.1339	.1606	.1606	.1377	.1033	.0688	.0413	.0225	.0113
7.0	.0009	.0064	.0223	.0521	.0912	.1277	.1490	.1490	.1304	.1014	.0710	.0452	.0264
8.0	.0003	.0027	.0107	.0286	.0573	.0916	.1221	.1396	.1396	.1241	.0993	.0722	.0481
9.0	.0001	.0011	.0050	.0150	.0337	.0607	.0911	.1171	.1318	.1318	.1186	.0970	.0728
10.0	.0000	.0005	.0023	.0076	.0189	.0378	.0631	.0901	.1126	.1251	.1251	.1137	.0948

λ \ x	13	14	15	16	17	18	19	20	21	22	23	24
5.0	.0013	.0005	.0002									
6.0	.0052	.0022	.0009	.0003	.0001							
7.0	.0142	.0071	.0033	.0014	.0006	.0002	.0001					
8.0	.0296	.0169	.0090	.0045	.0021	.0009	.0004	.0002	.0001			
9.0	.0504	.0324	.0194	.0109	.0058	.0029	.0014	.0006	.0003	.0001		
10.0	.0729	.0521	.0347	.0217	.0128	.0071	.0037	.0019	.0009	.0004	.0002	.0001

Answers to Odd-numbered Exercises

CHAPTER 1

4.1. $S = \{(D, D, D), (D, D, G), (D, G, D), (D, G, G), (G, D, D), (G, D, G), (G, G, D), (G, G, G)\}$, $A_1 = \{(D, D, D), (D, D, G), (D, G, D), (D, G, G)\}$, $A_1A_2 = \{(D, D, D), (D, D, G)\}$, $A_1 \cup A_2 = \{(D, D, D), (D, D, G), (D, G, D), (D, G, G), (G, D, D), (G, D, G)\}$.

4.3. (i), (xvi) $\{1, 2, 3\}$; (ii), (viii) $\{1, 2, 3, 7, 8, 9\}$; (iii), (iv), (vii), (xiii) $\{10, 11, 12\}$; (v), (vi), (xiv) $\{1, 2, 3, 7, 8, 9, 10, 11, 12\}$; (ix), (xii), $\{4, 5, 6\}$; (xi) $\{1, 2, 3, 4, 5, 6, 7, 8, 9\}$; (x), (xv) S.

4.5. (i) $\{10, 11, 12\}$; (ii) $\{1, 2, 3)$; (iii) $\{4, 5, 6, 7, 8, 9\}$; (iv) ϕ; (v) S; (vi) $\{1, 2, 3, 4, 5, 6, 7, 8, 9\}$; (vii) $\{4, 5, 6, 7, 8, 9\}$; (viii) ϕ; (ix) $\{10, 11, 12\}$; (x) $\{1, 2, 3, 10, 11, 12\}$; (xi) S; (xii) S.

5.5. P[exactly 0] $= 1 + P[AB] - P[A] - P[B]$. P[exactly 1] $= P[A] + P[B] - 2P[AB]$. P[exactly 2] $= P[AB]$. P[at least 0] $= 1$. P[at least 1] $= P[A] + P[B] - P[AB]$. P[at least 2] $= P[AB]$. P[at most 0] $= 1 + P[AB] - P[A] - P[B]$. P[at most 1] $= 1 - P[AB]$. P[at most 2] $= 1$.

5.7.

(i)	(ii)	(iii)
$\frac{1}{2}$	$\frac{4}{9}$	$\frac{1}{3}$
$\frac{1}{3}$	$\frac{4}{9}$	$\frac{2}{3}$
$\frac{1}{6}$	$\frac{1}{9}$	0
1	1	1
$\frac{1}{2}$	$\frac{5}{9}$	$\frac{2}{3}$
$\frac{1}{6}$	$\frac{1}{9}$	0
$\frac{1}{2}$	$\frac{4}{9}$	$\frac{1}{3}$
$\frac{5}{6}$	$\frac{8}{9}$	1
1	1	1

5.9. N[exactly 0] $= 400$. N[exactly 1] $= 400$. N[exactly 2] $= 100$. N[at least 0] $= 900$. N[at least 1] $= 500$. N[at least 2] $= 100$. N[at most 0] $= 400$. N[at most 1] $= 800$. N[at most 2] $= 900$.

5.11. Let M, W, and C denote, respectively, a set of college graduates, males and married persons. Show $N[M \cup W \cup C] = 1057 > 1000$.

7.1. 12/21.

7.3. (i) 0.14, (ii) 0.07.

7.5. $\frac{1}{3}$.

CHAPTER 2

1.1. 450.

1.3. 10, 32.

1.5. 10.

1.7. (i) 70; (ii) 2.

1.9. $n = 18, r = 10$.

1.11. 204, 54, 108, 98.

1.13. 2205.

2.1. Without replacement, (i) $\frac{5}{14}$, (ii) $\frac{13}{28}$, (iii) $\frac{25}{28}$; with replacement, (i) $\frac{25}{64}$, (ii) $\frac{34}{64}$, (iii) $\frac{55}{64}$.

2.3.

k	2, 12	3, 11	4, 10	5, 9	6, 8	7
with replacement	$\frac{1}{36}$	$\frac{2}{36}$	$\frac{3}{36}$	$\frac{4}{36}$	$\frac{5}{36}$	$\frac{6}{36}$
without replacement	0	$\frac{2}{30}$	$\frac{2}{30}$	$\frac{4}{30}$	$\frac{4}{30}$	$\frac{6}{30}$

2.5. 0.026, $(\frac{7}{240})$

2.7. 0.753.

2.9. $\frac{385}{1728} \doteq 0.223$.

2.11. (i) $\frac{1}{6}$; (ii) $\frac{1}{30}$; (iii) $\frac{3}{10}$.

2.13. (i) 0; (ii) $\frac{18}{35}$.

3.1. With replacement (i) $\frac{1}{16}$, (ii) $\frac{9}{16}$, (iii) $\frac{7}{16}$, (iv) $\frac{3}{8}$; without replacement (i) $\frac{1}{17}$, (ii) $\frac{19}{34}$, (iii) $\frac{15}{34}$, (iv) $\frac{13}{34}$.

3.3. (i), (ii) 2^{-10}; (iii) $\frac{126}{512}$; (iv), (v) $\frac{319}{512}$.

3.5. $\frac{11}{36}$.

3.7. $(45)_5/(50)_5$.

3.9. 0.1.

3.11. Manufacturer would prefer plan (*a*), consumer would prefer plan (*b*).

3.13. $(900)_5/(1000)_5 \doteq (0.9)^5 = 0.59$.

3.15. $\frac{21}{32}$.

3.17. (i) $\frac{4}{9}$, (ii) $\frac{1}{27}$.

4.1. $\frac{1}{3}$.

4.3. (i), (ii) $\frac{66}{149}$; (iii) $\frac{99}{149}$

4.5. (i) False, since $P[AB] = \frac{1}{8}$; (ii) false; (iii) true; (iv) false.

4.9. $\frac{10}{11}$.

4.11. (i) $\frac{1}{3}$; (ii) $\frac{2}{7}$; (iii), (iv) $\frac{1}{6}$; (v) $\frac{2}{9}$; (vi) 0; (vii) $\frac{1}{8}$; (viii) undefined.

4.13. $\frac{2}{3}$, $\frac{1}{2}$.

5.3. $(6)_2$, 6^2, $\binom{6}{2}$, $\binom{7}{2}$.

5.5. (i) $\dfrac{\binom{52}{13} - \binom{48}{13} - 4\binom{48}{12}}{\binom{52}{13} - \binom{48}{13}}$; (ii) $\dfrac{\binom{51}{12} - \binom{48}{12}}{\binom{51}{12}}$.

6.1. (i) $\binom{12}{7}\Big/\binom{16}{10}$; (ii) $\binom{9}{3}\Big/\binom{16}{10}$; (iii) $9\binom{7}{5}\Big/\binom{16}{10}$.

6.3. (i) $\binom{10}{6}\Big/4^{10}$; (ii) $\binom{10}{4\ 3\ 2\ 1}\Big/4^{10}$; (iii) $\sum_{k=0}^{4}(-1)^k\binom{4}{k}\left(1 - \frac{k}{4}\right)^{10}$.

6.5. (i) $P[B_0] = 1 - S_1 + S_2$, $P[B_1] = S_1 - 2S_2$, $P[B_2] = S_2$.
(ii) $P[B_0] = 1 - S_1 + S_2 - S_3$, $P[B_1] = S_1 - 2S_2 + 3S_3$, $P[B_2] = S_2 - 3S_3$, $P[B_3] = S_3$. (iii) $P[B_0] = 1 - S_1 + S_2 - S_3 + S_4$, $P[B_1] = S_1 - 2S_2 + 3S_3 - 4S_4$, $P[B_2] = S_2 - 3S_3 + 6S_4$, $P[B_3] = S_3 - 4S_4$, $P[B_4] = S_4$. P [at least 1] $=$ $S_1 - S_2 + S_3 - \cdots \pm S_M$. P [at least 2] $= S_2 - 2S_3 + 3S_4 - \cdots \pm MS_M$. P [at least 3] $= S_3 - 3S_4 + \cdots \pm \frac{1}{2}M(M-1)S_M$. P [at least M] $= S_M$.

CHAPTER 3

1.1. Yes, since $P[AB] = (\frac{4}{6})^2$ and $P[A] = P[B] = \frac{4}{6}$
(No, since $P[AB] = \frac{2}{5}$ and $P[A] = P[B] = \frac{4}{6}$).

1.3. No.

1.5. (i) 0.729; (ii) 0.271; (iii) 0.028; (iv) 0.001.

1.9. Possible values for $(P[A], P[B])$ are $(\frac{1}{2}, \frac{1}{3})$ and $(\frac{1}{3}, \frac{1}{2})$.

2.1. (i) T; (2) F; (3) F; (4) T; (5) F; (6) T.

2.3. (i) $\frac{32}{75}$; (ii), (iii) $\frac{44}{75}$.

3.1. (i) 0.240; (ii) 0.260; (iii) 0.942; (iv) 0.932.

3.3. (i) 0.328; (ii) 0.410; (iii) 0.262.

3.5. (i) 0.133; (ii) 0.072.

3.7. (i) 0.197; (ii) 0.803; (iii) 0.544.

3.9. Choose n such that $(0.90)^n < 0.01$; therefore, choose $n = 44$.

3.11. (i) $(1 - q_{60})^5 = 0.881$; (ii) $(1 - q_{60})^5 + 5q_{60}(1 - q_{60})^4 = 0.994$;
(iii) $(1 - q_{60})^5(1 - q_{65}) = 0.846$; (iv) $(1 - q_{60})^5 q_{65} = 0.035$.

3.13. (i), (ii) 0.2456; (iii) 0.4096.

3.15. $5^{n-1}/6^n$.

3.17. $\frac{3}{8}$.

3.19. 0.010.

3.21. (i) 0.1755; (ii) 0.5595.

3.23. (i) 0.0256; (ii) 0.0081; (iii) 0.1008.

3.25. (i) 0.3770; (ii) 9.

3.27. 0.379.

3.29. (i) 1; (ii) 4 or 5.

4.1. (i) $\frac{13}{24}$; (ii) $\frac{1}{2}$.

4.3. (i) T; (ii) F; (iii) T; (iv) T; (v) F; (vi) F.

4.5. $\frac{2}{3}$.

4.7. $\frac{1}{9}$.

4.9. Let the event that a student wears a tie, comes from the East, comes from the Midwest, or comes from the Far West be denoted, respectively by A, B, C, D. Then $P[B \mid A] = \frac{20}{33}$, $P[C \mid A] = \frac{9}{33}$, $P[D \mid A] = \frac{4}{33}$.

4.11. (i) $\frac{1}{3}$; (ii) box A, $\frac{2}{3}$; box B, 0; box C, $\frac{1}{3}$.

4.13. $\frac{8}{9}$.

4.15. (i) $\frac{3}{13}$; (ii) $\frac{5}{13}$; (iii) $\frac{5}{52}$.

4.17. (i) $1/k$; (ii) $\frac{25}{48}$; (iii) 0.24.

5.1. $P_3(s,f) = \frac{31}{54}$, $P_3(f,s) = \frac{31}{72}$.

5.3. $\frac{22}{43}$.

5.5. 0.35.

5.7. (i) $\frac{1}{2} + \frac{1}{2}(1 - 2p)^3$; (ii) $\frac{1}{2} + \frac{1}{2}(1 - 2p)^4$; (iii) $\frac{1}{2}$ if $p < 1$.

5.9. $\frac{2}{5}$.

6.1. (i), (iii) $P_2 = P_3 = P$;

(ii)
$$P_2 = \begin{bmatrix} \frac{1}{2} & \frac{1}{2} & 0 \\ \frac{1}{2} & \frac{1}{2} & 0 \\ \frac{1}{4} & \frac{1}{2} & \frac{1}{4} \end{bmatrix}, \qquad P_3 = \begin{bmatrix} \frac{1}{2} & \frac{1}{2} & 0 \\ \frac{1}{2} & \frac{1}{2} & 0 \\ \frac{3}{8} & \frac{1}{2} & \frac{1}{8} \end{bmatrix};$$

(iv)
$$P_2 = P_3 = \begin{bmatrix} \frac{1}{4} & \frac{1}{2} & \frac{1}{4} \\ \frac{1}{4} & \frac{1}{2} & \frac{1}{4} \\ \frac{1}{4} & \frac{1}{2} & \frac{1}{4} \end{bmatrix}.$$

6.3. (i), (iii) $\pi_1 = \pi_2 = \frac{1}{2}$; (ii) $\pi_1 = \frac{2}{3}$, $\pi_2 = \frac{1}{3}$.

6.5. P has rows $(p, q, 0, 0)$, $(0, 0, p, q)$, $(p, q, 0, 0)$, $(0, 0, p, q)$. For $n > 1$ the rows are (p^2, pq, pq, q^2).

6.7. $\frac{4}{7}$ if $p = q = \frac{1}{2}$, $(q^7 - p^4q^3)/(q^7 - p^7)$ if $p \neq q$.

6.9. 1.

CHAPTER 4

1.1. P [exactly 0] $= \frac{1}{3}$. P [exactly 1] $= \frac{1}{3}$. P [exactly 2] $= \frac{1}{3}$. P [at least 0] $= 1$. P [at least 1] $= \frac{2}{3}$. P [at least 2] $= \frac{1}{3}$. P [at most 0] $= \frac{1}{3}$. P [at most 1] $= \frac{2}{3}$. P [at most 2] $= 1$.

2.9. (i) $A = \frac{1}{5}$; (iii) (*a*) 0.1353, (*b*) 0.6321, (*c*) 0.2326; (iv) $P[A(b)] = e^{-b/5}$.

2.11. (i) $A = \frac{2}{3}$; (iii) (*a*) $(\frac{1}{3})^6$, (*b*) $\frac{3}{4}$, (*c*) $\frac{1}{4}$; (iv) $P[A(b)] = (\frac{1}{3})^b$.

3.9. (ii) $f(x) = \dfrac{|x|}{2500} e^{-(x/50)^2}$; (iii) (*a*), (*b*) 0.184, (*c*) 0.632, (*d*) 0; (iv) (*a*) $1 - e^{-1}$, (*b*) $(e^{-1} - e^{-4})/(2 - e^{-4})$.

3.11. (ii) $f(x) = \frac{1}{2}$ for $0 < x < 1$; $= \frac{1}{4}$ for $2 < x < 4$; $= 0$ otherwise; (iii) (*a*) $\frac{1}{4}$, (*b*) $\frac{3}{4}$, (*c*) $\frac{1}{4}$; (iv) (*a*) $\frac{1}{2}$, (*b*) $\frac{1}{2}$.

4.1. (i) Hypergeometric with parameters $N = 200$, $n = 20$, $p = 0.05$; (ii) binomial with parameters $n = 30$, $p = 0.51$; (iii) geometric with parameter $p = 0.51$; (iv) binomial with parameters $n = 35$, $p = 0.75$.

4.3. $p(x) = \binom{6}{x}(\frac{2}{3})^x(\frac{1}{3})^{6-x}$ for $x = 0, 1, \cdots, 6$; 0 otherwise.

4.5. $p(x) = (x - 1)\binom{12 - x}{4} \Big/ \binom{12}{6}$ for $x = 2, \cdots, 12$; 0 otherwise.

4.7. $p(x) = (\frac{2}{3})(\frac{1}{3})^{x-1}$ for $x = 1, 2, \cdots$; 0 otherwise.

4.9. $p(x) = (x - 1)(\frac{2}{3})^2(\frac{1}{3})^{x-2}$ for $x = 2, 3, \cdots$; 0 otherwise.

5.1. $\frac{3}{5}$.

5.3. (i) $\frac{10}{37}$; (ii) $\frac{18}{37}$; (iii) $\frac{1}{37}$.

5.5. $P[x < z] = P[\tan(-\theta) < z] = \frac{1}{2} + \dfrac{1}{\pi}\tan^{-1} z$.

5.7. (i) $P[0.3 < \sqrt{x} < 0.4] = 0.07$; (ii) $P[-\ln x < 3] = 1 - e^{-3}$.

6.1.

	α	0.05	0.10	0.50	0.90	0.95	0.99
(i)	$J(\alpha)$	1.645	1.282	0.000	−1.282	−1.645	−2.326
	$K(\alpha)$	0.063	0.126	0.675	1.645	1.960	2.576
(ii)	$J(\alpha)$	3.290	2.564	0.000	−2.564	−3.290	−4.652
	$K(\alpha)$	0.126	0.252	1.350	3.290	3.920	5.152

6.3. 0.512.

6.5. (i), (ii) 0.2866; (iii) 0.0456.

6.7. $H(x) = \Phi\left(\frac{x-1}{2}\right) - \Phi\left(\frac{-x-1}{2}\right)$.

7.3. (i) $\pi/16$; (ii) $\pi/64$.

CHAPTER 5

1.1. (i) 13; (ii) 24.4; (iii) 63.6; (iv) 0; (v) 4.4.

1.3. (i) 1010; (ii) 9100; (iii) 63,600; (iv) 0; (v) 840.

2.1. Mean (i) $\frac{2}{3}$, (ii) 0, (iii) $\frac{4}{15}$; variance (i) $\frac{1}{18}$, (ii) $\frac{1}{2}$, (iii) $\frac{52}{225}$.

2.3. Mean (i) does not exist, (ii) 0, (iii) 0; variance (i) does not exist, (ii) 3, (iii) 1.

2.5. Mean (i) $\frac{2}{3}$, (ii) 4, (iii) 4; variance (i) $\frac{2}{9}$, (ii) $\frac{4}{3}$, (iii) $\frac{8}{11}$.

2.7. Mean (i) $\frac{2}{3}$, (ii) $\frac{1}{3}$; variance (i) $\frac{1}{18}$, (ii) $\frac{4}{45}$.

2.9. (i) $r > 2$; (ii) $r > 3$.

3.1. (i) $1/(1-t)$; (ii) $e^{5t}/(1-t)$.

3.3. (i) $2e^t/(3-e^t)$; (ii) $e^{2(e^t-1)}$.

3.5. (i) 1, 1, 1, 4; (ii) 1, 1, 1, 3.

4.1. 250.

4.3. (i) $1 - (\frac{1}{2})^4 = 0.9375$; (ii) $1 - (\frac{15}{16})^{47} \doteq 1$, Chebyshev bound 0.75.

5.1. Chebyshev bound, (i): (*a*) 50,000, (*b*) 500; (ii) (*a*) 250,000, (*b*) 2500. Normal approximation, (i): (*a*) 9600, (*b*) 96; (ii) (*a*) 16,600, (*b*) 166.

5.3. Chebyshev bound, (i) 8000; (ii) 12,500. Normal approximation, (i) 1537; (ii) 2400.

6.1. (i) m.g.f., $\frac{1}{3}\frac{1}{1-3t} + \frac{2}{3}\frac{1-e^{-1}}{1-e^{3t-1}}e^{3t}$; (ii) m.g.f., $\frac{1}{2}\left(1-\frac{t}{4}\right)^{-2} + \frac{1}{2}e^{2(e^t-1)}$; (iii) mean $\frac{5}{2}$, variance ∞, m.g.f. does not exist. (iv) mean ∞; variance, m.g.f. does not exist.

CHAPTER 6

2.1. (i) (*a*) 0.003; (*b*) 0.007; (ii) (*a*) 0.068; (*b*) 0.695.

2.5. (i) 0.506; (ii) 0.532.

2.7. (i) 423; (ii) 289.

2.9. Choose n so that (i), (ii) $\Phi\left(\frac{\sqrt{n}+9.8}{\sqrt{24}}\right) - \Phi\left(\frac{\sqrt{n}-9.8}{\sqrt{24}}\right) \le 0.05$;

(iii) $\Phi\left(\frac{2\sqrt{n}+9.8}{\sqrt{21}}\right) - \Phi\left(\frac{2\sqrt{n}-9.8}{\sqrt{21}}\right) \le 0.05$. One may obtain an upper bound for n: (i), (ii) $(1.645\sqrt{24} + 9.8)^2 \doteq 319$; (iii) $\frac{1}{4}(1.645\sqrt{21} + 9.8)^2 \doteq 75$.

2.11. (i) 0.983; (ii) 0.979.

3.1. 0.0671, 0.000.

3.3. 0.8008.

3.5. (i) 0.111; (ii) 0.968.

3.7. (i) 0.632; (ii) not surprising, since the number of 2 minute intervals in an hour in which either no one enters or 2 or more enter obeys a binomial probability law with mean 19.0 and variance 6.975.

3.9. (i) 0.1353; (ii) 0.3233.

3.11. 15.

4.1. $T = 10$ hours.

4.3. $N - r$ obeys a negative binomial probability law with parameters $p = \frac{1}{2}$ and (i) $r = 1$, (ii) $r = 2$, (iii) $r = 3$.

4.5. (i) 0.0067; (ii) 0.0404.

4.7. $1 - (1 - p)^n$; $n = \log(0.9)/\log(0.9999) = 1054$.

4.9. (i) 0.368; (ii) 0.865; (iii) 0.383.

CHAPTER 7

2.1. $p_X(x) = \binom{4}{x}\binom{48}{13-x}\Big/\binom{52}{13}$ for $x = 0, 1, \cdots, 4$; $= 0$ otherwise.

2.3. Without replacement $p_X(x) = \frac{2}{30}(x - 1)$ for $x = 1, 2, \cdots, 6$; $= 0$ otherwise;

with replacement $p_X(x) = \dfrac{2x - 1}{36}$ for $x = 1, 2, \cdots, 6$; $= 0$ otherwise.

2.5. $p_X(x) = \frac{1}{10}$ for $x = 0, 1, \cdots, 9$; $= 0$ otherwise.

2.7. (i) $13\binom{39}{x-1}\Big/(53 - x)\binom{52}{x-1}$ for $x = 1, 2, \cdots, 40$; $= 0$ otherwise.

(ii) $4\binom{48}{x-1}\Big/(53 - x)\binom{52}{x-1}$ for $x = 1, 2, \cdots, 49$; $= 0$ otherwise.

2.9. 0.3413.

2.11. 0.5811.

2.13. $\frac{5}{9}$.

2.15. $\frac{1}{2}$.

3.1. $\frac{1}{3}$.

4.1. $f_Y(y) = (y-5)^2/125$ if $0 \le y \le 5$
$= (25 - y^2)/125$ if $-5 \le y \le 0$
$= 0$ otherwise.

5.1.

(i), (ii)	(x_1, x_2, x_3)	$p_{X_1,X_2,X_3}(x_1, x_2, x_3)$.
with	(0, 0, 0)	$(\frac{2}{3})^3$
	(1, 0, 0), (0, 1, 0), (0, 0, 1)	$\frac{1}{3}(\frac{2}{3})^2$
	(1, 1, 0), (1, 0, 1), (0, 1, 1)	$(\frac{1}{3})^2\,\frac{2}{3}$
	(1, 1, 1)	$(\frac{1}{3})^3$
	otherwise	0;
without,	(1, 0, 0), (0, 1, 0), (0, 0, 1)	$\frac{1}{3}$
	otherwise	0.

5.3. With, $p_{Y_1}(y) = \binom{3}{y}\left(\frac{1}{3}\right)^y\left(\frac{2}{3}\right)^{3-y}$ if $y = 0, 1, 2, 3$; $= 0$ otherwise;
$p_{Y_2}(y) = (\frac{2}{3})^3$ if $y = 0$; $= 1 - (\frac{2}{3})^3$ if $y = 1$; $= 0$ otherwise;
$p_{Y_3}(y) = 1 - (\frac{1}{3})^3$ if $y = 0$; $= (\frac{1}{3})^3$ if $y = 1$; $= 0$ otherwise;
without, $p_{Y_1}(1) = p_{Y_2}(1) = p_{Y_3}(0) = 1$.

5.5. (*a*) (i) $\frac{3}{4}$, (ii) $\frac{1}{2}$, (iii) 0; (*b*) (i) $\frac{2}{3}$, (ii) e^{-1}, (iii) 0.

5.7. Yes.

5.9. (*a*) (i) $\frac{4}{5}$, (ii) $\frac{1}{2}$, (iii) $\frac{7}{30}$; (*b*) (i) $\frac{17}{30}$, (ii) $\frac{1}{2}$, (iii) $\frac{1}{6}$.

6.1. $1 - (1 - e^{-2})^5$.

6.3. (i) Yes; (ii) yes; (iii) yes; (iv) $1 - e^{-2}$; (v) yes; (vi) 0.8426; (vii) $f_{X^2}(y) = \frac{1}{\sqrt{2\pi y}} e^{-y/2}$ for $y > 0$; $= 0$ otherwise; (viii) $f_{X^2,Y^2}(u, v) = \frac{1}{2\pi\sqrt{uv}} e^{-(u+v)/2}$ for $u, v > 0$; $= 0$ otherwise; (ix) yes; (x) no.

6.5. (i) True; (ii) false; (iii) true; (iv) false; (v) false.

6.7. (i) 0.125; (ii) 0.875.

6.9. (i) 0.393; (ii) $1 - \ln 2 \doteq 0.307$; (iii) $\frac{2}{3}$.

7.1. $\frac{11}{36}$.

7.3. (i) 0; (ii) $(\frac{1}{6})^6$; (iii) $(\frac{3}{8})^6$.

7.5. (i) $\frac{1}{4}(1 + \ln 4)$; (ii) 0.

7.7. $\frac{1}{4}$.

7.9. (i) $1 - (0.6)^n$; (ii) $1 - (0.4)^n$; (iii) $(0.4)^n + n(0.6)(0.4)^{n-1}$.

8.3. $f_E(x) = \frac{2}{\sqrt{\pi}} \frac{\sqrt{x}}{(kT)^{3/2}} e^{-x/kT}$ for $x > 0$; $= 0$ otherwise.

χ^2 distribution with parameters $n = 3$ and $\sigma = (\frac{1}{2}kT)^{1/2}$

8.5. $\frac{1}{\pi}(1 - x^2)^{-1/2}$ for $|x| < 1$; $= 0$ otherwise.

8.7. $(y\sigma\sqrt{2\pi})^{-1}\exp\left[-\frac{1}{2\sigma^2}(\log y - m)^2\right]$ for $y > 0$; $= 0$ otherwise.

8.9. (i): (a) $\frac{1}{y}$ for $1 < y < e$; $= 0$ otherwise; (b) $\frac{1}{2y}$ for $e^{-1} < y < e$; $= 0$ otherwise; (ii) e^{-y} for $y > 0$; $= 0$ otherwise.

8.11. (a): (i) $\frac{1}{2}$ for $1 < y < 3$; $= 0$ otherwise; (ii) $\frac{1}{4}$ for $-1 < y < 3$; $= 0$ otherwise; (b) $\frac{1}{4}\left(\frac{y-1}{2}\right)^{-\frac{1}{2}}$ for $1 < y < 3$; $= 0$ otherwise.

8.13. (i) $\frac{4y}{\sqrt{2\pi}}e^{-\frac{1}{2}y^4}$ for $y > 0$, 0 otherwise; (ii) $\frac{6y^2}{\sqrt{2\pi}}e^{-\frac{1}{2}y^6}$ for $y > 0$, 0 otherwise.

8.15. (i) $[2\pi^3(1 - y^2)]^{-\frac{1}{2}} \sum_{k=-\infty}^{\infty} e^{-\frac{1}{2}x_k^2}$ where $y = \sin \pi x_k$ for $|y| \leq 1$; $= 0$ otherwise;

(ii) $\frac{1}{\sqrt{2\pi}}\sec^2 y e^{-\frac{1}{2}\tan^2 y}$ for $|y| \leq \frac{\pi}{2}$; $= 0$ otherwise.

8.17. (a) $\frac{1}{2\sqrt{y}}$ for $0 < y < 1$; 0 otherwise; (b) $\frac{1}{\sigma\sqrt{2\pi y}}e^{-y/2\sigma^2}$ for $y > 0$; 0 otherwise;

(c) $\frac{1}{2\sigma^2}e^{-y/2\sigma^2}$ for $y > 0$; 0 otherwise.

8.19. Distribution function $F_X(x)$:

(a) 0 for $x < 0$; $\frac{1}{2}$ for $x = 0$; $\frac{x+1}{2}$ for $0 < x < 1$; 1 for $x > 1$; (b) 0 for $x < 0$;

$\frac{1}{2}$ for $x = 0$; $\Phi\left(\frac{x}{\sigma}\right)$ for $x > 0$; (c) 0 for $x < 0$; $1 - e^{-x^2/2\sigma^2}$ for $x > 0$.

9.1. 0.9772.

9.3. (i) $2y$, $0 < y < 1$; 0 otherwise. (ii) $2(1 - y)$, $0 < y < 1$; 0 otherwise;

9.5. (i), (ii) Normal with mean 0, variance $2\sigma^2$; (iii) $\frac{1}{\sigma\sqrt{\pi}}e^{-y^2/4\sigma^2}$ for $y > 0$;

0 otherwise; (iv), (v) normal with mean 0, variance $\frac{1}{2}\sigma^2$.

9.7. $\{\pi(y^2 + 1)\}^{-1}$.

9.9. (i) Gamma with parameters $r = 3$ and $\lambda = \frac{1}{2}$; (ii) exponential with $\lambda = \frac{3}{2}$; (iii) $\frac{3}{2}e^{-y/2}(1 - e^{-y/2})^2$ for $y > 0$; 0 otherwise; (iv) $(1 + y)^{-2}$ for $y > 0$; 0 otherwise.

9.11. $n\frac{y}{\sigma^2}e^{-y^2/2\sigma^2}(1 - e^{-y^2/2\sigma^2})^{n-1}$ for $y > 0$.

9.17. $1 - n(0.8)^{n-1} + (n - 1)(0.8)^n$.

9.19. See the answer to exercise 10.3.

9.21. $3u^2/(1 + u)^4$ for $u > 0$; 0 otherwise.

10.1. (i) $\frac{1}{8}e^{-\frac{1}{2}y_1}$ if $y_1 > 0$, $|y_2| \leq y_1$; 0 otherwise; (ii) $\frac{1}{2}e^{-\frac{1}{2}(y_1+y_2)}$ if $0 \leq y_2 < y_1$ and $y_1 \geq 0$; 0 otherwise.

10.3. $f_{R,\Theta}(r, \alpha) = r$ if $0 < r\cos\alpha, r\sin\alpha < 1$; 0 otherwise; $f_R(r) = \dfrac{\pi}{2} r$ for $0 < r \leq 1$, $\left(2\csc^{-1} r - \dfrac{\pi}{2}\right)$ for $1 \leq r \leq \sqrt{2}$; 0 otherwise; $f_\Theta(\theta) = \frac{1}{2}\sec^2\theta$ for $0 \leq \theta \leq \dfrac{\pi}{4}$; $\frac{1}{2}\csc^2\theta$ for $\dfrac{\pi}{4} \leq \theta \leq \dfrac{\pi}{2}$; 0 otherwise.

11.1. (i) 1; (ii), (iii), (iv) $\frac{1}{2}$.

11.3. (i) 0.865; (ii) 0.632; (iii) 0.368; (iv) 0.5.

11.5. (i) 0.276; (ii) 0.5; (iii) 0.2; (iv) 0.5, (v) $\frac{1}{2}\phi(v/2)$.

11.7. (i) 0.28; (ii) 0.61.

CHAPTER 8

1.1. Mean, $\frac{1}{6}$; variance, $\frac{5}{36}$.

1.3. Mean winnings, 20 dollars.

1.5. Mean, 1 dollar 60 cents; variance, 1800 cents^2.

1.7. Mean, 58.75 cents; variance, 26 cents^2.

1.9. (i) Mean, 5.81, variance, 1.03; (ii) mean, 5.50, variance, 1.11.

1.11. With replacement, mean, 4.19, variance, 0.92; without replacement, mean, 4.5, variance 0.45.

1.13. Mean, $\sqrt{\pi/2}$; variance, $2 - (\pi/2)$.

1.15. $E[v^n] = \beta^{-n/2} \dfrac{2}{\sqrt{\pi}} \Gamma\left(\dfrac{n+3}{2}\right)$.

2.1. Mean, $\frac{2}{3}$; variance $\frac{2}{9}$, covariances $-\frac{2}{99}$.

2.3. $f_Y(y) = 2(1 - y)$ for $0 < y < 1$; $E[Y] = \frac{1}{3}$, $\text{Var}[Y] = \frac{1}{18}$, $E[Y^3] = \frac{1}{10}$, $E[Y^4] = \frac{1}{15}$.

2.5. $((1 - \rho)/\pi)^{1/2}$.

2.7. Means, 1; variances, 0.5; covariance, $2a - 0.5$.

2.9. Means, 4; variances, 6; covariance, $6e^{-(a_2 - a_1)}$.

3.1. $E[X] = \frac{1}{4}$, $E[Y] = \frac{1}{2}$, $\text{Var}[X] = \frac{3}{16}$, $\text{Var}[Y] = \frac{1}{4}$, $\rho[X, Y] = 0$; X and Y are independent.

3.3. $\sqrt{2/3}$.

3.5. $(\sigma_1^2 - \sigma_2^2)/(\sigma_1^2 + \sigma_2^2)$.

3.7. $4a - 1$.

3.9. $e^{-(a_2-a_1)}$.

4.1. 0.8413.

4.5. $E[L] = 150$. (i) Var $[L] = 16$; (ii) Var $[L] = 25.6$.

5.1. (i) throws more doubt than (ii).

5.3. 62.

5.5. 25 or more.

5.7. 0.70; 7.4.

5.9. 38.

5.11. $\eta = 0.10$.

6.1. (i) $n \geq 1537$; (ii) $n \geq 385$; (iii) $n \geq 16$.

6.3. $E[v]/\sigma[v] \doteq 10^5$

7.1. (i) $0.8x_2$; (ii) $-0.6x_2$; (iii) $\frac{1}{6}$; (iv) $\frac{7}{8}x_2 + \frac{1}{8}x_3$; (v) 0.35; (vi) 0.36.

7.3. (i) Var $[Y] = 0.5$; (ii) 0.

CHAPTER 9

2.1. (i) $(\frac{3}{4} + \frac{1}{4}e^{iu})^{12}$; (ii) $e^{3(e^{iu}-1)}$; (iii) $e^{iu}/4(1 - \frac{3}{4}e^{iu})$; (iv) $e^{3iu-(9/8)u^2}$; (v) $(1 - \frac{1}{3}iu)^{-2}$.

3.3. $\frac{2}{3} - y^2 + \frac{1}{2}|y|^3$ for $|y| \leq 1$; $\frac{4}{3} - 2|y| + y^2 - \frac{1}{6}|y|^3$ for $1 \leq |y| \leq 2$; 0 otherwise.

3.5. $(\pi^2[4a_1^2a_2^2 - x^2])^{-1/2}$ for $|x - a_1^2 - a_2^2| < 2a_1a_2$; 0 otherwise.

4.5. (i) kth cumulant of S is $n2^{k-1}(k-1)!(1 + km^2)$; (ii) $\nu = (1 + m^2)^2/1 + 2m^2$, $a = (1 + 2m^2)/(1 + m^2)$.

Index